# VIBRATION AND ACOUSTIC
# MEASUREMENT HANDBOOK

# VIBRATION AND ACOUSTIC MEASUREMENT HANDBOOK

MICHAEL P. BLAKE, EDITOR
*Director, Research and Development*
*Lovejoy, Inc.*

WILLIAM S. MITCHELL, CO–EDITOR
*Assoc. Prof., Tennessee Technological University*
*President, Physical Consultants Co.*

SPARTAN BOOKS

New York–Washington

Library of Congress Catalog Card Number 71-150339
International Standard Book Number 0-87671-561-7

Printed in the United States of America.
Sole distributor in Great Britain, the British Common-
wealth, and the Continent of Europe:

The Macmillan Press Ltd.
4 Little Essex Street
London WC2R 3LF

# CONTENTS

# PREFACE

Sooner or later, top management in industry becomes involved with the well-being, safety, economy, and general control of the cost of performance of existing machinery. Lovejoy, Inc., because of its long association in the power transmission field, discerned a lack of available and useable data on vibration.

After developing an internal staff of engineers with actual field experience in vibration and vibration measuring, we discovered others in industry were faced with the same lack of information on this subject. To encourage wider dissemination of the information we were gathering, Lovejoy, Inc., established, as part of its continuing program in industrial vibration technology, the Lovejoy Vibration Foundation, to cover manuscript costs and related field tests and experiments utilized in the preparation of this handbook. The manuscript was prepared through a special grant from the Lovejoy Foundation. This important handbook reflects the need for more industry-related information.

It is our hope that industry, through this handbook, will become more familiar with the philosophy, hazards, and methods of measurement associated with problems in owning, using, or maintaining machinery.

C. P. Hennessy, President
Lovejoy, Inc.

# INTRODUCTION

Many books are devoted to vibration and acoustics. But if we arrange them on a shelf in our mind we find that they lie mostly towards the theoretical end of the shelf and that the practical end, the applied end, the measuring end, is almost empty. The shelf that is here imagined is a continuum or spectrum, and it should not exhibit significant gaps. There is no necessary antipathy or difference in interest between theory and practice. All that is idealized in theory was firstly noticed in the phenomenal world. It is noticed there in an approximate form and tidied up and gathered together into models that are called theory. The concept of impedance is for example an idealization of the excitation-response relations that are manifest every second and minute in the daily world of dynamics.

This present handbook is an effort to start squarely at the measuring end of the spectrum. It tries to start with the industrial or technological problem and throw light on it by discussion of known practical methods and of such theory as may be unavoidable. Such is the aim of the book; but in places the reader may opine that the aim has not been fulfilled. For example, much of the discussion of rubber is plainly theoretical. Why is this? Because rubber is a material of paramount importance, quite unique in the control of vibration; and very little substantial data about rubber is to be found in available literature. Thus, the theoretical aspects of rubber or of acoustics or of other topics, contained in this book, are best viewed as reading to be undertaken after an understanding of related problems and opportunities are gleaned from life experience or from other parts of the handbook.

During perhaps the first four decades of Radio Telephony there was a discernible body of knowledge, often called 'Audio' for want of a better word. It was a hodgepodge of theory and practice. Often no one knew why this or that worked; but, that it did work was well known. Practice was often ahead of theory. There were few satisfactory books about 'Audio'; and the newcomer was forced to learn the art by a sort of rub-off apprenticeship. Vibration measuring is now at the stage whereat Audio found itself about forty years ago. A vast amount of measuring has been done in the past decade or two, and a vast body of theory has been brought to bear in some areas. Nevertheless, the theoretical and measuring activities are still, as it were, two separate worlds, particularly as regards the vibratory and acoustic control of existing machines, as distinguished from design work. This present handbook tries to join the two worlds. It tries to form one bridge among the many that are still needed. The field is vast, and so expanding that its progress outpaces the capacity of writers to document it. For this reason, and as will be noticed, the handbook is simply a selection. It lacks the extension and variety that any editor would like to achieve. But this lack is the natural result of starting to document a field like vibration measuring which has grown from almost nothing in a decade or two to a complete world that now includes maintenance, medicine, design, ecology, transportation, space travel, and so on almost without end. Out of all the possible areas of specialization, the present book tends to focus on maintenance or on the well being and control of existing machines more than on anything else.

Incredible as it may seem, vibration measuring is only in its infancy. Such giants as Rayleigh and Helmholtz, who gave to the world in the last century the respective classics *The Theory of Sound* and *The Sensations of Tone*, had in effect no measuring equipment at all at their disposal, except perhaps a siren, a few resonators or sensitive flames, or a mica disc and silk filament. The microphone, the transducer in general, and the electronic amplifier together with other marvels of measurement such as the oscillograph and wave analyzer were unknown to them. The field of modern measurement dates from the first reliable useable electronic instrumentation. In fact, the field is only two decades old. In the two decades, the gap between theory and practice has, as it seems, widened instead of closed. An attempt is made in these pages to close the gap.

Just as an effort is made to consider always the practical problem and the industrial need, every effort is also made to address the maintenance engineer, the informed technician, or any one who is charged with the task of using vibration and acoustic measurements in the day-to-day world of practice. Indeed it is hoped that the industrial operator will find that the authors meet him as far as possible on his own ground and in his own terms. In its constant effort to accentuate the need and character of the actual field situation, the present handbook is unique.

The first eleven chapters include some of the fundamentals of acoustics and vibration together with what seemed to be some of the more useful highlights of analytical methods and of vibration isolation. The next eight chapters are devoted entirely to examples drawn from the field of machine maintenance. In this field, vibration measuring covers the entire gamut from safety to economy of operation. Throughout, the topics of velocity measuring and of severity levels have been given the prominence that their importance deserves. Seven more chapters are then devoted to miscellaneous measuring examples and measuring instrumentation. Finally, a chapter is devoted to such standards and general information as seemed likely to prove most valuable in the industrial situation.

The handbook does not replace nor attempt to replace existing handbooks or other books. It will be found to be a useful supplement for reference in a variety of spheres of work, such as in understanding some of the implication of codes relating to noise and vibration, in following courses of instruction devoted to vibration and acoustics, or for example in design and test work.

The editors are deeply conscious of all the help that has been given by unmentioned industries and persons. Wherever possible, acknowledgement has been made. We extend our thanks particularly to Dolores Klein, Harriet Hammonds and Helen Mitchell for their extensive help in the preparation of a large manuscript. All concerned have taken pains to avoid errors. But some will be found, no doubt; and the editors would appreciate hearing about these, and indeed hearing comments in general from the reader. Thank you.

Michael P. Blake, South Haven, Mich.,
W. S. Mitchell, Cookeville, Tennessee,
September 1971

# ABBREVIATIONS

| | | | |
|---|---|---|---|
| **AC** | alternating current | **mil** | thousandth of an inch |
| **ASME** | American Society of Mechanical Engineers | **mm** | millimeter |
| | | **oz** | ounce |
| **cm** | centimeter | **P** | peak |
| **DC** | direct current | **PP** | peak-to-peak |
| **FM** | frequency modulation | **psi** | pounds per square inch |
| **ft** | foot | **rms** | root-mean-square |
| **g** | gravity 32.2 ft/sec$^2$ | **rpm** | revolutions per minute |
| **hp** | horsepower | **TIR** | total indicator run |
| **Hz** | Hertz (cycles per second) | **PWL** | sound-power level |
| **dB** | decibel, 1/10 bel unit | **SPL** | sound-pressure level |
| **in** | inch | **USASI,** | United States of America |
| **IPSP** | inches per second peak | **ANSI** | Standards Institute (now American National Standards Institute) |
| **kcps** | kilo (thousand) cycles per second | | |
| **khz** | kilo (thousand) cycles per second | **$\mu$B** | microbar, $10^{-6}$ atmospheres |
| **lbs** | pounds | | |

# TERMINOLOGY

**absorption**—The change of sound energy into some other form, usually heat, in passing through a medium or on striking a surface.

**acceleration**—A vector quantity that specifies the time rate of change of velocity.

**accidental (nuisance) impact**—Impacts other than functional are so described. They are unwanted results of a functional process, usually resulting from looseness and controllable by machine adjustment or correction.

**active isolation**—Attenuation wherein a correction is made at the transmitter location.

**all-pass**—In a vibration meter or sound meter, the reading that is obtained is called all-pass if no filter is interposed between the signal and that part of the meter comprising the read-out and its associated circuitry.

**angular frequency**—In radians per unit time, the frequency multiplied by $2\pi$. The usual symbol is $\omega$.

**antinode**—A point, line or surface in a standing wave where some characteristic of the wave field has maximum amplitude.

**attenuation**—Synonymous with damping, but is a more accurate description.

**audio frequency (range)**—The frequency range from about 20 to 20,000 cps.

**background noise**—The total of all sources of interference in a system used for the production, detection, measurement or recording of a signal, independent of the presence of the signal.

**bandpass filter**—A wave filter that has a single transmission band extending from a lower cutoff frequency greater than zero to a finite upper cutoff frequency.

**beats**—Periodic variations that result from the superposition of two simple harmonic quantities of different frequencies $f_1$ and $f_2$. They involve the periodic increase and decrease of amplitude at the beat frequency $(f_1 - f_2)$.

**beneficial (functional) impact**—Impacts that are purposeful, so as to fulfill a useful function, are so described.

**blip**—An oscillograph-trace excursion, relatively instantaneous as compared with adjacent excursions, or a knock, rattle or impact of negligible duration compared with the cyclic duration of vibration from the same source.

**broadband**—Containing many frequencies. Often used as a synonym for "all-pass."

**centimil**—A unit used to avoid decimals: one centimil equals 0.00001 of an inch.

**complex excitation**—An excitation is complex if to every real excitation that is a simple harmonic function of the time there is combined an imaginary excitation having the same amplitude and frequency but differing in phase by one-quarter of a cycle.

**complex quantity**—A mathematical quantity containing imaginary elements such as j which indicates the square root of minus one.

**complex tone**—A sound wave containing simple sinusoidal components of different frequencies. A complex tone is a sound sensation characterized by more than one pitch.

**compliance**—The reciprocal of stiffness, the softness of a system.

**compound**—A substance formed by the union of two or more elements.

**continuous spectrum**—The spectrum of a wave, the components of which are continuously distributed over a frequency region.

**coordinates**—The vertical coordinate of amplitude; the horizontal coordinate of amplitude, time.

**critical damping**—The minimum viscous damping that will allow a displaced system to return to its initial position without oscillation.

**critical frequency (panels)**—The lowest frequency at which wave coincidence occurs; that is, the frequency at which the wavelength of incident sound equals the wavelength of the flexural response of a surface, such as a solid wall.

**cycle**—The completed sequence of a periodic motion that occurs during a period; a minimum oscillation.

**damping**—The dissipation of energy from a dynamic system due to internal (material friction) and external (air or liquid resistance) effects.

**damping factor**—A ratio of actual system damping to critical damping for the system (twice system mass times radian natural frequency). A frequency-dependent quantity inversely proportional to the Q factor of the system, thus indicating sharpness of system response.

**decibel**—One-tenth of a bel. The decibel is a unit of level when the base of the logarithm is ten and the quantities concerned are proportional to power. If the level of a number Z is expressed in decibels by saying that the level is X decibels with reference to a number Y, that is itself of reference level zero dB, this usually connotes for nonpower-like quantities: $X = 20 \log_{10} (Z/Y)$. For power-like quantities, the connotation is: $X = 10 \log_{10} (Z/Y)$.

**degree-of-freedom**—The number of degrees of freedom of a mechanical system is equal to the minimum number of independent generalized coordinates required to define completely the positions of all parts of the system at any instant of time. In general, it is equal to the number of independent generalized displacements that are possible.

**differentiation**—A mathematical procedure which determines, for example, the ratio of a small change in a function to the small change in the independent variable that causes the change in the function. Denoting a function of x as f(x) and denoting a small change in x by delta x, then f(x + delta x) − f(x) all divided by delta x is the ratio of interest. The limit of this ratio as delta x becomes small without limit is called the first derived function or derivative.

**diffuse sound field**—One in which the time average of the mean-square sound pressure is everywhere the same and the flow of energy in all directions is equally probable.

**diffracted wave**—One whose front has been changed in direction by an obstacle or other nonhomogeneity in a medium, otherwise than by reflection or refraction.

**directional microphone**—A microphone whose response varies significantly with the direction of sound incidence.

**directional response pattern**—The directional response pattern of a transducer used for sound emission or reception is a description, often presented graphically, of the response of the transducer as a function of the direction of the transmitted or incident sound waves in a specified plane and at a specified frequency.

**directivity factor**—The directivity factor of a transducer used for sound emission is the ratio of the sound pressure squared, at some fixed distance and specified direction, to the mean-square sound pressure at the same distance averaged over all directions from the transducer. The distance must be great enough so that the sound appears to diverge spherically from the effective acoustic center of the sources. Unless otherwise specified, the reference direction is understood to be that of maximum response.

**directivity index**—A quantity which is used for specifying the directionality of a sound source expressed in dB; it is defined as ten times the logarithm (base 10) of the ratio of the axial sound intensity to some reference intensity.

**discrete spectrum**—A graphical presentation of the distribution of energy in a complex wave. The energy is allocated to those specific frequencies which comprise the wave.

**displacement**—A vector quantity that specifies the change of position of a body or particle. It is usually measured from the mean position or position of rest.

**distortion**—An undesired change in waveform. Noise and certain desired changes in waveform, such as those resulting from modulation or detection, are not usually classed as distortion.

**draft gear**—A device used to reduce longitudinal shock in rail cars. It is installed between the coupling and the car body.

**dynamic modulus**—The ratio of stress to strain under vibratory conditions.

**ergodic vibration**—A stationary random vibration whose statistical averages can be obtained from a single time record and still reflect an ensemble of vibration time-histories.

**excitation**—An external force (or other input) applied to a system that causes the system to respond in some way.

**far field**—That part of a sound field for which spherical divergence occurs; that is, SPL decreases by $-6$ dB for each doubling of distance. As a general rule, it is also considered as that part of a sound field which is beyond a distance of 3 to 4 times the largest dimension of the source or greater than the maximum wavelength of sound for the lowest frequency of interest.

**flanking (acoustical)**—Transmission of acoustic sound energy around or past a barrier such as a wall by way of openings in the structure face or joints; also as a structural transmission of vibrations to elements which reradiate the sound in the acoustic range. Flanking is the bypassing of an acoustic or vibration treatment such as an isolator through contacting members (structure-borne vibration) or enclosure/surface openings (acoustic radiation).

**flutter**—Any deviation in frequency in reproduced sound from the original frequency; it usually results from nonuniform motion of the recording medium during recording, duplication or reproduction.

**foundation**—A structure that supports the gravity load of a mechanical system.

It may be fixed in space or it may undergo a motion that provides excitation for the supported system.

**free sound field**—A field in which the effects of the boundaries are negligible over the frequency region of interest.

**frequency spectrum**—The frequency spectrum of a measured vibration is a description of its resolution into frequency components with associated amplitudes.

**functional (beneficial) impact**—Impacts that are purposeful, so as to fulfill a useful function, are so described.

**functional vibration**—As for example in a sifter or vibrator conveyor, wherein the vibration is specifically intended.

**fundamental frequency**—The fundamental frequency of an oscillating system is the lowest operative frequency, a periodic quantity.

**furring strip**—A relatively thin strip of wood used for leveling or attaching finishing material such as plasterboard or acoustical tiles.

**geometrical damping**—Vibratory attenuation by virtue of mere distance.

**geophone**—Usually, a small inexpensive seismic-type velocity transducer.

**gradient**—A rate of increase or decrease of a variable magnitude or the curve that represents it.

**harmonic**—If a series of periodic time functions (usually sinusoidal) are related in frequency by ratios that are all integers, then the series is harmonically related. The lowest frequency is called the fundamental, and frequencies 3 or 5 times the fundamental are called the third and fifth harmonics or orders, and so on.

**histogram**—A graphical record of acoustic energy with respect to frequency for an operating machine. These data are recorded at various periods in the life of a machine, and they indicate its history as well as its current operating condition.

**hysteresis**—A retardation of response which accompanies a change of forces. A lag in the previously established stress-strain relationship of a deformed system when left unstressed.

**hysteretic damping**—This term is often used to signify material damping; however, true hysteretic damping relates to the internal properties of homogeneous materials such as steel or rubber, rather than to the intergranular movements of soil.

**ideal spectrum**—Such that a pure tone appears as a vertical line because of ideally sharp filtering. Actual filters (wave analyzers) give curves as if the sloping sides of an isoceles triangle had sagged.

**impact (shock)**—Relatively rapid transient transmissions of mechanical energy into or out of a mechanical system.

**impedance**—The ratio of two complex quantities whose arguments increase linearly with time and whose real (or imaginary) parts represent a force-like and velocity-like quantity, respectively.

**impulse**—The time integral of force (the area under the force-time curve).

**incidental vibration**—As for example in a reciprocating compressor or turbine, wherein the observed vibration is not intended.

**initial**—Herein used interchangeably with transient; the term applies to a response state that is caused by an impact and is not typical of the system and quickly comes to an end.

**integer**—A whole number such as 1, 9, 72, and not such as 7.6.

**intensity level**—The intensity level, in decibels, of a sound is ten times the logarithm to the base 10 of the ratio of the intensity of this sound to a reference intensity. The reference intensity is stated explicitly.

**integration**—A mathematical procedure that is more or less the inverse of differentiation. For example, velocity may be obtained by differentiating displacement because it is the instantaneous rate of change of displacement with respect to time. Conversely, integration of velocity may be used to compute displacement.

**isolation**—A reduction, in the capacity of a system to respond to an excitation, attained by the use of a resilient support. The force or motional separation of two inertia systems.

**jerk**—The time rate of change of the acceleration; the third derivative of the displacement with respect to time. Sometimes used to assess system performance, e.g. an automobile ride.

**level**—In acoustics, the level of a quantity is the logarithm of the ratio of that quantity to a reference quantity of the same kind. The base of the logarithm, the reference quantity, and the kind of level must be specified.

**line spectrum**—A spectrum whose components occur at a number of discrete frequencies (see discrete spectrum).

**lineal**—An old word, used to distinguish running feet from square or cubic feet. It is used here to distinguish to-and-fro or Lissajou-type vibration from torsional vibration. The words lateral and linear are invariably used instead of lineal.

**linear**—In a mathematical sense this usually connotes a straight line functional relationship such as y = Cx, where C is constant. In a sense of vibration classification, it connotes lineal.

**linear system**—A system is linear if for every element in the system the response is proportional to the excitation.

**Lissajou figure**—The figure obtained, for example, in an oscillograph from two mutually perpendicular signals acting on the beam to form a combined trace. Usually the signals are periodic and the harmonics in one have a whole-number relation with those of the other.

**longitudinal wave**—A wave in which the direction of displacement of the medium is in the same direction as the propagation.

**loudness**—The intensive attribute of an auditory sensation, in terms of which sounds may be ordered on a scale extending from soft to loud.

**loudness level**—The loudness level of a sound, in phons, is numerically equal to the median sound-pressure level, in decibels, relative to 0.0002 microbars, of a free progressive wave of frequency 1000 cycles per second presented to listeners facing the source, which in a number of trials is judged by the listeners to be equally loud.

**masking**—The process by which the threshold of audibility for one sound is raised by the presence of another (masking) sound.

**material damping**—That due to energy loss in the material.

**mechanical impedance**—The impedance obtained from the ratio of force to velocity during simple harmonic motion. If the force and velocity are measured at the same point, the ratio is designated driving point impedance; if they are measured at different points, the ratio is designated transfer impedance.

**microphone**—An electroacoustic transducer that responds to sound waves and delivers essentially equivalent electric waves.

**mil**—One thousandth of an inch.

**millisecond**—One thousandth of a second.

**millivolt**—One thousandth of a volt.

**modulation**—The variation in value of some parameter characterizing a periodic oscillation. Thus, amplitude modulation of a sinusoidal oscillation is a variation in the amplitude of the sinusoidal oscillation.

**molecule**—The smallest particle of an element or compound which preserves the properties of the original.

**momentum**—The product MV, where M is weight W divided by g; V is velocity.

**natural frequency**—The frequency of free oscillation of a system. For a multiple-degree-of-freedom system, the natural frequencies are the frequencies of the normal modes of vibration.

**node**—A point, line or surface in a standing wave where some characteristic of the wave field has essentially zero amplitude.

**noise**—Any undesired sound. By extension, noise is any unwanted disturbance within a useful frequency band.

**nominal bandwidth**—The nominal bandwidth of a filter is the difference between the nominal upper and lower cutoff frequencies. The difference may be expressed (1) in cycles per second (cps); (2) as a percentage of the passband center frequency; or (3) as the interval between the upper and lower nominal cut-offs in octaves.

**nominal passband center frequency**—The geometric mean of the nominal cutoff frequencies.

**nonlinear damping**—Damping due to a damping force that is not proportional to velocity.

**normal mode**—A normal mode of vibration is a mode of free vibration of an undamped system. In general, any composite motion of an undamped system is analyzable into a summation of its normal modes.

**nuisance (accidental) impact**—Impacts other than functional are so described. They are unwanted results of a functional process, usually resulting from looseness and controllable by machine adjustment or correction.

**octave**—An octave considered as a ratio is 2. Considered as a frequency, for example, 100 cps is the octave of 50 cps; and 25 cps is its suboctave.

**oscillograph**—The read-out of an oscilloscope or other graphic recorder of the "DC" or analogous type, for example, the pen oscillograph. The read-out of a level (AC) recorder is not called an oscillograph.

**oscilloscope**—An electronic instrument in which a spot on a fluorescent screen responds more or less linearly to the instantaneous voltage of one or more input signals.

**parameter**—A word that is often used when a simpler one would do. It is about synonomous with the word measure. It may connote value. As for example, acceleration and frequency are the parameters that define the value of a vibration. Or it may connote form or function, for example, the amplitude-frequency spectrum is a useful parameter for the evaluation of the condition of a machine.

**particle velocity**—In a sound field, the velocity of a very small part of the medium, with reference to the medium as a whole, due to the sound wave.

**passive isolation**—Attenuation wherein a correction is made at the receiver location.

**peak-to-peak value**—The peak-to-peak value of an oscillating quantity is the algebraic difference between the extremes of the value of the quantity.

**periodic quantity**—An oscillating quantity whose values recur for certain increments of the time variable.

**periodic signal, simple**—A voltage or other alternating event that repeats at time intervals called the period. The usual connotation is a sinusoidal signal without harmonics.

**periodic signal, complex**—A signal with more than one sinusoidal component, or a nonsinusoidal signal such as a square wave. In one sense there is no such thing as a nonsinusoidal periodic signal because any such signal may be imagined to be composed of a suitable group of sine signals having the appropriate frequencies, amplitudes and phase relations so that the combined group exactly reproduces the nonsinusoidal signal.

**phase**—The difference in time or position between forces and/or motions as implied by the angular difference as noted in mathematical expressions or phasor diagrams. If motion $x = \sin(\omega t + \theta)$ is the result of an excitation $F = \sin \omega t$, then $\theta$ is the angular difference or phase between excitation force and response motion.

**pitch**—That attribute of auditory sensation in terms of which sounds may be ordered on a scale extending from low to high. Pitch depends primarily upon the frequency of the sound stimulus, but it also depends upon the sound pressure and waveform of the stimulus.

**plane wave**—A wave in which the wavefronts are everywhere parallel planes normal to the direction of propagation.

**plate**—The lower, horizontal member of a wood-framed wall partition which is attached to the floor and on which the vertical wall studs rest.

**Poisson's ratio**—Ratio of lateral unit strain to longitudinal unit strain for uniform uniaxial longitudinal stress in the elastic range. Unit strain means change in length caused by stress, divided by working length.

**polymer**—High molecular weight compounds formed by the combination of two or more molecules.

**power density**—Power density $S(f)$ is the mean-square value of vibration magnitude per unit bandwidth of the output of an ideal filter with unit gain.

**power level**—Power level, in decibels, is ten times the logarithm to the base 10 of the ratio of a given power to a reference power. The reference power must be indicated.

**power-density spectrum**—A graphical presentation of power density versus frequency. It indicates the distribution of energy in a vibration.

**probabilistic process**—A probabilistic process is a mathematically described vibration phenomenon for which the instantaneous values of amplitude (displacement, velocity, acceleration) cannot be specified uniquely at any given instant of time.

**proximity measurement**—Some probes have the capability to detect the distance between the probe face and a surface such as that of a shaft. This measuring is called proximity measuring.

**pulse**—Not exactly defined. A term more generic than impact and including impact and impulse. A term usually used in a more generic sense than pulse is excitation, which includes pulse and vibration.

**pure tone**—A simple or pure tone is a sound wave, the instantaneous sound pressure of which is a simple sinusoidal function of the time. A simple tone is a sound sensation characterized by its singleness of pitch.

**random noise**—An oscillation whose instantaneous magnitude is not specified for any given instant of time. The instantaneous magnitudes of a random noise are specified only by probability distribution functions giving the fraction of the total time that the magnitude, or some sequence of magnitudes, lies within a specified range.

**random signal**—A signal that does not repeat periodically, and is usually persistent rather than transient.

**Rayleigh wave**—A vibratory wave, propagated in a cylindrical front on the earth's surface.

**read-out**—The meter reading or oscillographic display or other final expression of information within the capability of the eye. Sometimes the term call-out is used in relation to the ear.

**receiver**—The affected person or equipment.

**residual**—Applies to the state that follows the initial state.

**resilience**—In a rubber or rubber-like body, the ratio of energy given up on recovery from deformation to the energy required to produce the deformation.

**resonance**—Resonance of a system in forced oscillation exists when any change, however small, in the frequency of excitation causes a decrease in the response of the system.

**response**—The response of a device or system is the motion (or other output) resulting from an excitation (stimulus) under specified conditions.

**response flatness**—An accelerometer transducer or measuring system is said to have a flat response if the signal or read-out is constant for the same acceleration as the frequency varies.

**reverberation**—The persistence of sound in an enclosed space, as a result of multiple reflections after the sound source has been quieted.

**ripple tank**—A shallow pool of water, usually with a transparent bottom and underside lighting, used in studying the phenomena of wave motion and optical effects.

**seismic**—The word is used in two ways: to describe matters related to vibration and shock in the earth, and to describe particular transducers wherein the inertia of a mass is used as a base of reference, instead of using a fixed external structure as is done in the case of the rarer "driven or relative" transducers.

**shaft-stick measuring**—Shaft movement is measured rather than the usual bearing-housing movement. Shaft movement is usually somewhat greater and sometimes, in the case of plain bearings, may be three times the bearing-housing movement.

**shock (impact)**—Relatively rapid transient transmissions of mechanical energy into or out of a mechanical system.

**shock spectrum**—The plot as a function of $T_p/T$ of the maximum response due to an impact, $X_i$. The maximum transient displacement may occur at the instant corresponding to $T_p/T$ or before that instant. $X_r$ may occur after, or at that instant. Instead of plotting the maximum response, the ratio of maximum response to statically applied response, a dimensionless number, ranging from about 1 to 2, is often plotted.

**signal form**—The graphic functional relation of the instantaneous signal values and time, where time is the independent variable.

**simple harmonic quantity**—A periodic quantity that is a sinusoidal function of the time variable.

**simple sound source**—A source that radiates sound uniformly in all directions under free-field conditions.

**snubber**—A device used to increase the stiffness of an elastic system (usually by a large factor) whenever the displacement becomes larger than a specified amount.

**sound**—An oscillation in pressure, stress, particle displacement, particle velocity, etc., in a medium with internal forces (e.g. elastic, viscous), or the superposition of such propagated oscillations. Sound is an auditory sensation evoked by the oscillation described above.

**sound energy**—The sound energy of a given part of a medium is the total energy in this part of the medium minus the energy which would exist in the same part of the medium with no sound waves present.

**sound field**—A region containing sound waves.

**sound level**—Sound level is a weighted sound-pressure level, obtained by the use of metering characteristics and the weightings A, B or C as specified in USASI Standard, Sound Level Meters for Measurement of Noise and Other Sounds, Z24.3–1944. The weighting employed must always be stated. The reference pressure is 0.0002 microbar.

**sound-pressure level**—The sound-pressure level, in decibels, of a sound is twenty times the logarithm to the base 10 of the ratio of the pressure of the sound to a reference pressure. The reference pressure is explicitly stated.

**sound recording system**—A combination of transducing devices and associated equipment suitable for storing sound in a form capable of subsequent reproduction.

**spectrum**—The spectrum of a function of time is a description of its resolution into components, each of different frequency and (usually) different amplitude and phase. "Spectrum" is also used to signify a continuous range of components, usually wide in extent, within which waves have some specified common characteristic; e.g. "audio-frequency spectrum."

**spectrum, complex**—Such as provides phase and amplitude information.

**spectrum, simple**—Usually a graphic presentation of the amplitudes of the various signal components, plotted as a function of the frequency of these components.

**spectrum, frequency-time**—Such as presents the amplitudes of the assembly of component frequencies as a function of time.

**spectrum density**—The spectrum density of an oscillation is the mean-square amplitude of the output of an ideal filter with unity gain responding to the oscillation, per unit bandwidth; i.e. the limit for vanishingly small bandwidth of the quotient of the mean-square amplitude divided by the bandwidth.

**spectrum level**—The spectrum level of a specified signal at a particular frequency is the level of that part of the signal contained within a band 1 cycle wide, centered at the particular frequency.

**spherical wave**—A wave within a system of waves in which the wavefronts are concentric spheres.

**standing wave**—A periodic wave having a fixed distribution in space which is the result of interference of progressive waves of the same frequency and kind. Such waves are characterized by the existence of nodes or partial nodes and antinodes that are fixed in space.

**stationary random vibration**—A type of random vibration for which such properties as mean value, mean square and rms values, spectral density and probability distribution are independent of time.

**stiffness**—The ratio of the change of force (or torque) to the corresponding change in translational (or rotational) displacement of an elastic element.

**stud**—One of the vertical wood framing members used to form the structure for gypsum dry-wall construction; usually made from 2 x 4-in. lumber.

**symptomatic vibration**—Wherein the vibration is mostly of interest as an indication of a faulty or nonfaulty condition of a machine, rather than being of interest itself.

**tachometer**—An instrument for measuring rotative frequency.

**timbre**—That attribute of auditory sensation in terms of which a listener can judge that two sounds similarly presented and having the same loudness and pitch are dissimilar.

**tone**—A sound wave capable of exciting an auditory sensation having pitch.

**transmission loss**—The degree of attenuation of a sound that passes through a wall or other part of a structure; usually expressed in dB and defined as ten times the logarithm (base 10) of the ratio of incident intensity to transmitted intensity.

**transmissibility**—The nondimensional ratio of the response amplitude of a system in steady-state forced vibration to the excitation amplitude. The ratio may be one of forces, displacements, velocities or accelerations.

**transmitter**—The source of the vibration.

**undulation**—An oscillation, fluctuation, vibration, pulsation or wave-like motion.

**velocity**—The time rate of change of displacement.

**vibration**—An oscillation or alternating mechanical motion of an elastic system.

**vibration isolator**—A resilient support that tends to separate a mechanical system from its support structure in terms of mechanical vibration.

**vortex**—A mass of fluid such as air which moves about an axis, where the axis itself can be moving in free space. If each particle in the vortex moves in a circular path with a speed varying inversely as the distance from the axis, it is a free vortex.

**wave**—A disturbance which is propagated in a medium in such a manner that at any point in the medium the quantity serving as a measure of disturbance is a function of the time, while at any instant the displacement at a point is a function of the position of the point.

**waveform**—The instantaneous value of a signal, as a function of time.

**wavefront**—The wavefront of a progressive wave in space is a continuous surface which is a locus of points having the same phase at a given instant.

**wavelength**—The wavelength of a periodic wave is the perpendicular distance between two wavefronts at points of comparable amplitude and a phase difference of one complete period.

**white noise**—A random noise whose spectrum level is independent of frequency over a specified range.

**wideband random vibration**—An oscillation whose instantaneous magnitude is not specified for any given instant of time, and whose frequency spectrum contains many frequencies as opposed to a few such as in narrowband noise. The instantaneous magnitudes of a random noise are specified only by probability distribution functions giving the fraction of the total time that the magnitude, or some sequence of magnitudes, lies within a specified range.

---

* Some terminology has been abstracted from American Standard S1.1–1960, Acoustical Terminology (including Mechanical Shock and Vibration) ANSI, New York, 1960. Other has been contrived by the authors to suit new concepts or be in agreement with typical industrial usage.

# SECTION I

# VIBRATION AND ACOUSTIC FUNDAMENTALS

# SECTION I CONTENTS

# SYMBOLS

| | | |
|---|---|---|
| **A** | area | ft$^2$ |
| **c** | velocity of sound | ft/sec |
| **c$_p$, c$_v$** | specific heat | BTU/lbm°R |
| **dB** | decibel | —— |
| **f** | frequency | cycles/sec |
| **g$_c$** | gravitational constant | lbm·ft/lbf·sec$^2$ |
| **Hz** | frequency (acoustic terminology) | cycles/sec |
| **I** | intensity | watts/m$^2$ |
| **IL** | intensity level | watts |
| **k** | wave number $2\pi/\lambda$ | ft$^{-1}$ |
| **L** | length | ft |
| **m** | meter | —— |
| **n** | integer 1, 2, 3, ... | —— |
| **P** | instantaneous acoustic pressure | —— |
| **p** | acoustic pressure, phon | dyne/cm$^2$ or lbf/in$^2$ |
| **p$_{rms}$** | effective acoustic pressure | dyne/cm$^2$ or lbf/in$^2$ |
| **R** | gas constant | lbf·ft/lbm·°R |
| **r** | radius distance | ft |
| **s** | sone unit | —— |
| **s** | condensation | —— |
| **S** | sensation | —— |
| **SPL** | sound-pressure level | dB |
| **T** | absolute temperature | °Rankine |
| **t** | time | seconds |
| **u** | particle velocity | cm/sec or ft/sec |
| **W** | acoustic sound power | watts |
| **$\beta$** | elastic modulus | lbf/in$^2$ |
| **$\gamma$** | specific heat ratio $c_p/c_v$ | —— |
| **$\lambda$** | wavelength | ft |
| **$\rho$** | density | lbm/ft$^3$ |

| $\rho_0$ | ambient, undisturbed density | lbm/ft³ |
| $\phi$ | phase angle | radians |
| $\tau$ | one-half wave period | seconds |
| $\omega$ | angular frequency $2\pi f$ | radian/sec |

# 1

# SOUND AND NOISE IN AIR

## W. S. Mitchell

## 1. SOURCES OF SOUND

The objects that constitute our living environment have one thing in common—they vibrate. In some cases, such as the ground, vibration is of a low frequency seldom exceeding 100 Hz. On the other hand, operating machinery can vibrate in excess of 20 kHz. When we walk on the floor, it vibrates; when we place a machine tool or turbine into operation, it vibrates. Motions of vibrating objects may be too small for the eye to recognize but, nevertheless, they exist. Consider the air in which we live. Whether in an office or out-of-doors, to get from one location to another, we displace a quantity of that air. Moreover, if we cause an object such as a book, telephone, or table to move, the air in contact with the object is set in motion. The motion of the air propagates throughout the surrounding volume until the energy of the motion is dissipated into heat.

But man is only one of many living creatures that moves in air. Insects produce varying motions of air with such diverse mechanisms as fluttering wings, vibrating membranes and the rubbing of rough-surfaced legs against serrated body panels. In each case a surface is caused to move in some form of mechanical motion which is imparted to the surrounding air. Likewise, machine tools and process equipment are capable of producing localized air motion by means of their moving surfaces. Air motions will result from rotation of wheels, gears and shafts, running drive belts, reciprocating slides and vibrating body panels. Where the motion is periodic and of a moderate frequency, we experience a sustained acoustic response in the area of the vibrating surface. This response we recognize as sound.

Sound is also produced by motions other than the sustained periodic vibrations as described above. There are, for instance, sounds caused by a fallen wrench, a broken valve lifter or a worn-out crankshaft bearing. These are transient sources. They produce sound in air which occurs over a short, finite period of time (although they may be considered periodic for an extended interval of time). Such sources are immediately recognized due to the impulsive nature of their sound. This type of sound decays rapidly to a level below perception, by processes similar to those of the transient vibration described in a later section of this handbook.

Shock waves are the result of yet another type source of sound and noise in air. Although they are not related to the sounds and noise usually associated with machine and equipment radiation, they should be considered for a complete understanding of airborne noise phenomena. Shock waves are produced by the extremely rapid motion of a mechanical surface. Often, they originate in explosions, impacts and supersonic motions. They are caused by overflights of high-speed aircraft and projectiles. In all cases, the result is a sharp impulse of sound, that is, a spasmodic release of energy.

Not all mechanical motion results in the perception of sound. For example, if we wave our arms back and forth as rapidly as we can, our effort results in nothing more than a circulation of air. So first we will define what is meant by sound. Audible sound is a physical event that is caused by the vibration of a mechanical body resulting in an organized to-and-fro movement of air. Physiologically, sound has meaning when there has been perception by the human ear. When the frequency of the air motion is between 20 Hz and 20 kHz the ear is consequently stimulated and we are aware of the sensation called sound.

## 2. CHARACTERISTICS OF HEARING

The ear is an extremely sensitive instrument which detects sounds produced by the smallest of insects and the largest of man-made machines. Below a measured sound-pressure level of 85 dBA* the ear functions without risk of permanent damage; physical feeling occurs at 120 dB; above a level of 140 dB, pain and hearing impairment occur. For exposure between the levels, 85 to 120 dB, one should consult published data which relates exposure time and sound-pressure level because disagreement exists among today's authorities concerning noise within this region. The threshold of hearing occurs around an rms pressure level of $2.9 \times 10^{-9}$ psi. The pressure associated with the upper limit of hearing tolerance is approximately $2.9 \times 10^{-3}$ psi. Consequently, the normal ear is sensitive to a fluctuating pressure range of about one million to one. Thus, we note the remarkable range of the human ear with respect to pressure. Typical sound-pressure levels have been listed in Fig. 1.1. Although the ear can detect a wide variation of pressure, it is more restricted in its frequency response. At very low frequencies, that is below 5 Hz, one does not perceive any audible sound. This is the acoustic region of infrasonics. Approaching 20 Hz, acoustic pulses can be recognized. Above 20 Hz, which is the audio acoustic range, we hear continuous acoustic sound. Beyond 15kHz our ability to recognize sound becomes restricted depending on age. At 20 kHz we surpass the hearing capability of the human ear. This latter frequency is considered by some to be the beginning of the ultrasonic range of acoustics. For the average listener, if a surface has periodic, vibrational motion between 20 Hz and 15 kHz, then that surface acts as a source of sound and noise in air since its motion produces sound in the audible range.

Loudness is a subjective quality of the ear. It is an auditory sensation produced by sound that is found to depend on intensity and frequency. Ear sensitivity

---

*All decibel values relative to the threshold of hearing, .0002 microbar.

| SPL  dB | SOURCE | SPL  dB | SOURCE |
|---|---|---|---|
| 150 | Pneumatic Siren at 1' | 70 | Conversation |
| 140 | Threshold of Pain | 60 | Department Store—Large |
| 130 | Pneumatic Chipper | 50 | Private Business Office |
| 120 | Threshold of Sensation | 40 | Quiet Street—Night |
| 110 | Riveter | 30 | Whisper |
| 100 | Subway | 20 | Broadcasting Studio |
| 90 | Boiler Room | 10 | Rustle of Leaves |
| 80 | Inside City Bus | 0 | Threshold of Hearing |

Fig. 1.1   Sound-pressure levels and corresponding acoustical sources of
common sounds (re .0002 microbar)

varies as shown in Fig. 1.2, which relates contours of constant loudness to intensity level and frequency. For a given loudness, the intensity level (and consequently sound-pressure level) of a perceived sound must be increased in both
low- and high-frequency regions because of the mid-frequency sensitivity of the
ear. A listener is more aware of frequencies centered around 3 kHz than he is of
frequencies at 100 Hz or 10 kHz. An approximate law of psychology states that

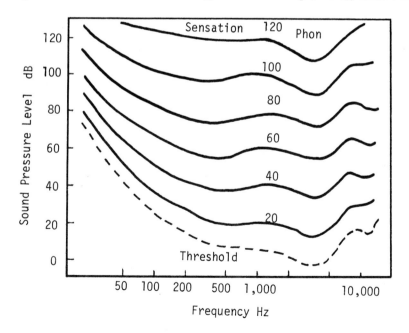

Fig. 1.2   Hearing response curves for a pure tone in a free-field (re 0.0002
μBar). (after Robinson and Dadson, NPL, England)

the magnitude of a sensation, $S$, is proportional to the logarithm of the intensity (relative to a reference intensity $I_o$). Mathematically, this is expressed as

$$S = \text{Log}_{10}\left(\frac{I}{I_o}\right)$$

The subjective characteristic of *loudness* is defined in terms of a "sone" unit. One sone is defined as the level of loudness heard by a typical listener confronted with a 1 kHz pure tone at a sound-pressure level of 40 dB. The *loudness level* of a sound is defined in terms of a "phon," the value of which is numerically equal to the sound-pressure level of an equally loud 1 kHz pure tone. For example, if equipment noise is judged to be as loud as a 1 kHz pure tone at a sound-pressure level of 60 dB, then the loudness level is defined as 60 phons. A corresponding loudness of 4 sones is found from the relation

$$S = 2^{(p-40)/10} \text{ sones}$$

where $p$ is the number of phons.

Since one sone is associated with a sound-pressure level of 40 dB, it follows that 1 sone equals 40 phons. Loudness level in phons is therefore an arbitrary yardstick related nonlinearly to loudness. In other words, phons are related to hearing stimulus; sones are related to loudness.

As noted previously, the ear is sensitive to a very wide range of sound. If we use the logarithm of a ratio such as that given above to indicate this physiological sensation, then one unit of an intensity level (called one BEL after Alexander G. Bell) would correspond to an increase in intensity of an order of ten. This is noted to be an inconveniently large increment of measure since the ear is sensitive to sounds in the range of intensity: $10^{-12}$ to 10 watts/m². Thus, 1/10 of a unit is used, which is called a "decibel." The decibel is not a unit of measure; it is not like horsepower, feet, or pounds. Rather, it is a dimensionless number for indicating a relative value (relative to a reference) of one power-like quantity to another. Hence, intensity levels but not intensities are rated in terms of decibels. For example, the difference between a level of 1 watt and 2 watts (3dB) is the same as the difference of 50 watts and 100 watts (3dB) since by definition

$$\text{Intensity Level (IL)} = 10 \text{ Log}_{10}\frac{I}{I_o} \text{ dB } (I_o = 10^{-12} \text{ watts/m}^2)$$

An intensity level difference of 1 decibel (1dB) is about the smallest change in energy level that the ear can ordinarily detect. It is about half again as good as values obtained in typical field measurements. Under closely controlled laboratory conditions a 0.3 dB change can be detected. Although the decibel is a convenience number, it possesses one major disadvantage: its numbers cannot be algebraically added or subtracted. For example, 50 dB plus 50 dB equals 53 dB rather than 100 dB. Such mathematical operations are best handled by consulting tabulated decibel values.

## 3. WAVE THEORY

Noise is considered to be an unwanted sound as contrasted to interesting conversation, music and pleasing tones. However, sound is the result of physiological stimulation; and to some persons it is noise, while to others it is sound. Hence, we use these terms interchangeably in this section of the handbook. When a mechanical surface vibrates, the surrounding air is set into motion in a manner characteristic of the surface vibration. If the surface moves slowly, so does the air. When surface motion is rapid, the motion of the air is likewise. But, the pressure wave that is produced in air by the to-and-fro motion of the surface propagates from the vibrating source at a velocity of sound which depends only on the temperature of the air. At 68°F, that velocity is 1125 fps, approximately.

The simplest type of surface motion is plane sinusoidal. This is a smoothly varying periodic motion resulting from a limited number of pure sinusoidal components which produces sinusoidal-pressure pulsations in air. Such pulsations have instantaneous longitudinal displacement amplitudes that can be described by an equation of the form

$$x(t) = X \cos (\omega t + \phi)$$

where $X$ is the maximum displacement amplitude, $\omega$ is the angular frequency $2\pi f$, and $\phi$ is a phase angle of the motion. If one listens carefully to a sound of this type it will be found to have a smooth, clear and pleasing timbre* which is a subjective characteristic of the wave that depends on the harmonic content of the waveform. A pure sine wave has no harmonics. It is indicative of a simple, single source of mechanical vibration. However, such sounds are rarely experienced with operating machinery. Because of the interactions of various structural components of a machine, complex periodic waves and random waves are normally radiated as sound.

According to Fourier analysis, a complex periodic wave can be reduced to a fundamental wave and a system of progressive harmonics. Thus, a wave having the form $x(t) = X(t) \cos (\omega t + \phi)$ where $x(t)$ is a time-varying amplitude of the wave can be analyzed into a series of sinusoids

$$x(t) = \frac{a_o}{2} + \sum_{n=1}^{\infty} a_n \cos \frac{n\pi t}{\tau} + \sum_{n=1}^{\infty} b_n \sin \frac{n\pi t}{\tau}$$

where the amplitude coefficients are

$$a_n = \frac{1}{\tau} \int_{-\tau}^{\tau} x(t) \cos \frac{n\pi t}{\tau} \, dt \qquad (n = 1, 2, 3, \ldots)$$

$$b_n = \frac{1}{\tau} \int_{-\tau}^{\tau} x(t) \sin \frac{n\pi t}{\tau} \, dt \qquad (n = 1, 2, 3, \ldots)$$

---

* Timbre is currently used in place of "quality" and has been adopted as standard acoustic terminology (USASI S1.1–1960).

and $\tau$ is one-half the period of the wave. This type wave is, therefore, a distorted sine wave, distorted to the degree of its harmonic content. Sounds containing many harmonics are not pleasing to hear, as they exhibit annoying characteristics. For example, a triangular waveform is composed entirely of odd harmonics. When such a sound wave is associated with operating equipment, it is sometimes indicative of an internal degradation of a mechanism.

When two pieces of machinery operate in close proximity, there is often an interaction between the mechanical vibrations of the machines and a phasing of the propagating sound. This results in a slowly varying sound level at some frequency that depends on the operating speeds of the machines. The process, known as signal modulation, produces a system of unrelated harmonics. Because the system is not coordinated harmonically, the timbre of the sound is found to be irritating to the listener. If the modulation of the signal is slow, there is a distinctive frequency beat. Jet airplane travelers are usually spared the annoyance of these beats which are always experienced in multi-engine propeller-driven aircraft. Such frequency beats are the direct result of the sum of two independently vibrating devices described by

$$x(t) = X \cos \omega_1 t \quad \text{and} \quad y(t) = Y \cos (\omega_2 t + \phi)$$

where $X$ and $Y$ are the displacement amplitudes and $\phi$ is a phase angle between the vibrations. The sum of these motions becomes $z(t) = x(t) + y(t)$

$$z(t) = Z \cos \frac{(\omega_1 - \omega_2)t}{2} \cos \frac{(\omega_1 + \omega_2)t}{2}$$

Considering $Z \cos (\omega_1 - \omega_2) t / 2$ as a time-varying amplitude of the surface motion, the above equation can be expressed as

$$z(t) = Z_o(t) \cos \frac{2\pi(f_1 + f_2)t}{2}$$

where $Z_o(t)$ is a function of $(f_1 - f_2)/2$. Whenever the cosine in $Z_o(t)$ equals $(0, \pi)$ a maximum amplitude (beat) occurs. Since this happens twice in every cycle, the beat frequency (number of beats per second) is $f_B = (f_1 - f_2)$. When $f_1$ is close to $f_2$, the perceived sound has an average frequency $(f_1 + f_2)/2$ and a beat frequency of the difference $(f_1 - f_2)$. But for $f_1$ far removed from $f_2$, the sound is a complex tone with an apparent frequency equal to the frequency difference. As the modulating frequency increases the sound becomes decidedly harsh. Beating is then very rapid, and the unpleasant whine often associated with high-pitch gear noise is produced. In some cases, this beating and resulting sound has been traced to the eccentricities of mating gears.

The characteristics of sounds as described above are the result of surface motions that have a fundamental frequency component. Thus, they are referred to as *pitched* sounds. Sounds without a fundamental frequency component are called unpitched or *random* sounds. Random sound is nonperiodic sound having acoustic energy distributed in numerous frequency bands, and its time-displacement amplitude is completely unpredictable. Hence, no two random noise sources are identical in time or displacement. In addition each noise has its own frequency spectrum, and the perceived sound of the noise changes with spectral distribution. Random noise is recognized by its *hissing* sound. Examples of such noise are the sound of a television set when a station ends its broadcasting day (without displaying a test pattern) and the noise from an opened steam pressure release valve.

Shock waves are neither pitched nor random; they are impulse noise. This phenomenon involves a transmission of energy in an elastic medium which occurs in a very short period of time. In some cases, this is followed by a natural decay. Generally, the pressure wave is characterized by an "N" shape. Analysis of such a wave by Fourier techniques shows that energy is contained in all frequency bands which is similar to the case of a transient pulse. The characteristics of the above waveforms are shown in Figs. 1.3, 1.4 and 1.5. Because of their wide variation in form, it is helpful to associate each one with some type of vibratory motion or machine. Acoustical-vibration specialists and equipment operators can develop an awareness of the relationship between acoustical sounds and vibrating mechanical systems. This knowledge is an invaluable asset for identifying probable mechanical condition from a perceived sound.

## 4. PROPAGATION THEORY

The pressure waves that propagate through air due to the motion of a vibrating surface are called acoustic waves. These waves have a specific form depending on the time-displacement relationship of the source and the amplitude of the displacement. Acoustic radiation is a three-dimensional-type phenomenon and it is generally complicated in its behavior. There are two basic types of wave motion —transverse and longitudinal. Transverse waves are commonly produced, for example, on the surface of a pool of water or by flexing a slender beam. In such cases the motion of the particles of the vibrating medium is up and down, that is, perpendicular to the direction of propagation of the waves. On the other hand, longitudinal waves are propagated by the back-and-forth motion of the molecular particles of the medium. Such motions produce alternate regions of compression (high density) and rarefaction (low density) parallel to the direction of wave motion rather than crests and troughs as experienced in transverse waves.

Acoustic waves are of the longitudinal type. There are two types of acoustic waves: plane waves and spherical waves. The former type wave occurs in directed systems, a good example being sound propagating in a ventilation duct or in a gas-filled rigid pipe. The characteristic property of a plane wave is that the acoustic pressure, particle displacements and density changes have common phases and amplitudes at all points in any given plane of the medium normal to the direction of propagation. Thus the designation, plane waves.

When a propagating wave compresses a local region in an elastic medium, such as air, due to the excess (acoustic) pressure $P = P_{wave} - P_o$, where $P_o$ is the ambient pressure, the resulting condensation $s$ (defined as the ratio of the excess

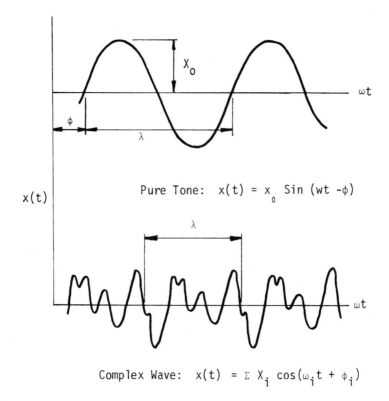

Fig. 1.3   Periodic waves, propagating characteristics

density and the undisturbed density, $\delta\rho/\rho_o$) is related to the instantaneous acoustic pressure in the form

$$P = \rho_o c^2 s$$

where $c$ is the velocity of sound in the medium. This expression is valid for both plane and spherical waves. If the wave is a simple harmonic motion, then the instantaneous pressure can be described by a simple time-varying expression

$$P = \rho_o c \omega A \sin (\omega t + kx) - \rho_o c \omega B \sin (\omega t - kx)$$

where $k$ is the wave number $2\pi/\lambda$, $\lambda$ being the length of the wave and $A$, $B$ are amplitude constants. The above expression accounts for pressure waves traveling in opposite directions due to reflections or independent sources. The particle velocity of the moving medium is given as $\mu = P/\rho_o c$. Therefore

$$-\mu = \omega A \sin (\omega t + kx) + \omega B \sin (\omega t - kx)$$

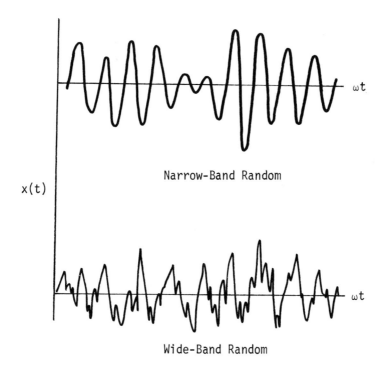

Narrow-Band Random

x(t)

Wide-Band Random

Fig. 1.4 Random waves, propagating characteristics

Since acoustic pressure propagates as a wave, its macroscopic properties obey the general wave equation $c = \lambda f$, where the velocity of sound is defined as

$$c = \sqrt{\frac{\text{Elastic Modulus}}{\text{density}}} \quad \text{ft/sec}$$

However, the elastic modulus of air for small changes in pressure and density $(\Delta p, \Delta \rho)$ is expressed as $\beta = \rho_0(\Delta p/\Delta \rho)$. Considering air as an ideal gas in which there is negligible heat loss in the propagation, it follows from the ideal gas law that $c = (\gamma RT)^{1/2}$ where $\gamma$ is the ratio of specific heats of the medium $c_p/c_v$, $R$ is the gas constant, and $T$ is the absolute temperature in degrees Rankine. The ratio of pressure and density in air is a function of temperature only. Thus, the velocity of sound is considered to be independent of pressure as well as frequency. For an atmospheric pressure of 14.7 psia and an ambient temperature of 68°F, the velocity of sound is 1125 fps. At any other temperature $T = (T_{\text{Fahr}} + 460)$ the velocity of sound is

$$c = 49 \sqrt{T} \quad \text{ft/sec}$$

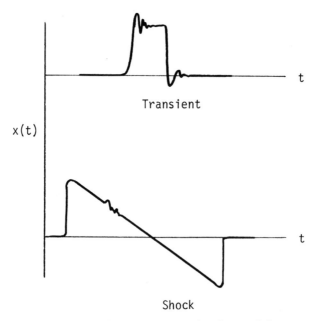

x(t)

Transient

Shock

Fig. 1.5　Impulse waves, propagating characteristics

The intensity of a physical quantity is defined as the average rate of energy flow through a unit area normal to the direction of the flow. Hence, for a propagating acoustic wave at a given distance from a vibrating source, intensity is given as

$$I = \frac{p^2}{\rho_o c} \ \text{W/m}^2$$

where $p$ is the effective rms pressure of the wave. Like many quantities associated with physical events, we are not usually interested in knowing or measuring absolute pressure, condensation or intensity as they relate to the investigation of equipment vibration through the measurement of acoustical effects. Rather, we are interested in knowing intensity level (IL), sound-power level (PWL) and sound-pressure level (SPL). The ear judges the relative loudness of sounds as the ratio of their intensities in a logarithmic behavior. Intensity level becomes

$$\text{IL} = 10 \ \text{Log}_{10} \frac{I}{I_o} = 10 \ \text{Log}_{10} I + 120 \ \text{dB}$$

relative to an intensity $I_o = 10^{-12} \ \text{W/m}^2$. Since intensity is proportional to the square of the acoustic pressure, sound-pressure level is expressed as

$$\text{SPL} = 20 \ \text{Log}_{10} \frac{p}{p_o} = 20 \ \text{Log}_{10} p + 74 \ \text{dB}$$

relative to a pressure $p_o = 0.0002$ microbars where $p$ is an rms pressure in microbars. Sound-power is defined as the product

$$W = IA \text{ watts}$$

where $A$ is the area through which the sound is passing. This could be the cross-sectional area of a duct for plane waves or a hemispherical area for diverging spherical waves. Sound-power level is the logarithmic ratio

$$\text{PWL} = 10 \text{ Log}_{10} \frac{W}{W_o} = 10 \text{ Log}_{10} W + 120 \text{ dB}$$

relative to a power $W_o = 10^{-12}$ watts. Typical power levels are shown in Fig. 1.6. From the above noted expressions of levels, the following relations exist for a free progressive wave:

$$\text{SPL} = \text{IL} + 0.2 \text{ dB} \qquad \text{(re 0.0002 microbar)}$$

$$\text{SPL} = \text{PWL} - 10 \text{ Log}_{10} A + 10.5 \text{ dB} \qquad \text{(re 0.0002 microbar)}$$

| WATTS | PWL dB | SOURCE | WATTS | PWL dB | SOURCE |
|---|---|---|---|---|---|
| $10^7$ | 190 | Large Rocket | $10^{-2}$ | 100 | |
| $10^6$ | 180 | | $10^{-3}$ | 90 | |
| $10^5$ | 170 | | $10^{-4}$ | 80 | Vacuum Cleaner |
| $10^4$ | 160 | Turbo-Jet | $10^{-5}$ | 70 | Conversation |
| $10^3$ | 150 | | $10^{-6}$ | 60 | |
| $10^2$ | 140 | | $10^{-7}$ | 50 | |
| $10^1$ | 130 | Pipe Organ | $10^{-8}$ | 40 | |
| 1 | 120 | | $10^{-9}$ | 30 | Whisper |
| $10^{-1}$ | 110 | Blaring Radio | | | |

Fig. 1.6 Acoustic-power levels for various sources (re $10^{-12}$ watts)

where $A$, the area through which sound is propagated uniformly, is expressed in square feet. When a system is known to possess simple sinusoidal plane waves, numerical values of intensity level and sound-pressure level are the same. Unfortunately, plane waves apply to a limited number of situations. Complex, diverging sound fields generally exist in industrial environments such that intensity level and sound-pressure level are not so simply related as above.

Spherical waves are more complex than plane waves and commonly experienced in the locality of small noise-producing machines. Characteristic of this type wave is the phenomenon of divergence whereby amplitude properties decrease with increasing distance from the vibrating source. However, the physical situation for the propagating wave is still similar to that of a plane wave except for this effect of divergence. This is seen in the expression for acoustic pressure

$$p = \frac{P_o}{r} \cos (\omega t - \phi)$$

in which the wave amplitude $P_o/r$ becomes small as the propagation distance $r$ becomes large. Here $P_o$ is the peak value of the acoustic pressure. Moreover, as $r$ becomes large the curvature of the spherical wave decreases so that at large distances from the source, spherical waves can be analyzed as plane waves. As in the case of plane waves, the intensity of a spherical wave is $I = p^2/\rho_o c$, where the effective acoustic pressure must account for spherical divergence. This leads to values for sound-power level and sound-pressure level as described above. A free field is a sound field characterized by acoustic waves traveling outward from the source, free from reflecting boundaries or surfaces, which implies spherical spreading of the acoustic waves. The sound pressure is reduced by a factor of two for each doubling of distance from the source. This is equivalent to a SPL change of $-6$ dB. For a uniformly diverging spherical wave (no local reflecting surfaces or sources), the intensity is inversely proportional to the square of the distance from the source. As an example, at 10 ft from a small source of sound the intensity will be one-one-hundredth of its value at a distance of only one foot. If the source cannot be considered small, the intensity will decrease less rapidly than predicted by the inverse square law. Reflected sound sometimes contributes to overall intensity, especially in the area of other equipment, walls and floors. Acoustic power in a spherical wave is the product of intensity and total surface area. It is noted from the expression for the average power of a diverging wave,

$$W = \frac{2\pi p^2}{\rho_o c}$$

that it is independent of distance from the sound source. Close to a small source of spherical waves the velocity of air particles, even for low pressures, is excessively large since the velocity is inversely proportional to $r^2$. Consequently, small sources cannot generate spherical waves of large intensity. We also note that because pressure is inversely proportional to the product of wavelength and source size, it is impossible to have a moderate size source produce large amounts of power at low frequencies. From this it follows that high-intensity, low-frequency sounds are produced only by large acoustic radiators such as surface panels or massive operating machinery.

Sound composed predominately of low frequencies has a propagation pattern similar to that observed when an object is thrown into a quiet pool of water. Ripples spread outward in circular arcs from the site of the initial disturbance, and those closest to the splash have larger amplitudes than those located some distance away. Of course, if reflecting surfaces are present, then the waves can be reflected back toward the splash, redirected off into some new orientation, or reflected back across the pattern of outwardly traveling waves in which case the pattern becomes quite confused or in acoustical terms, *diffuse*. This is an analogous picture of the spreading of low-frequency sound in air. As frequency increases, the propagating sound becomes more directed; for large ratios of the size of the source to the wavelength of the sound, acoustic distribution of the sound assumes a central beam-like configuration as shown in Fig. 1.7. For the intermediate-frequency range, various side lobes occur depending on the size and shape of the source. Additional information on sound field patterns and source directivity can be found in most acoustical texts.[1] Figure 1.7 also indicates that discrepancies can occur when collecting acoustical data if sound measurement equipment is located haphazardly in a high-frequency sound field. It should be recognized that high-frequency components of a sound source are more likely to be located along the axis of the source. In addition, this will be

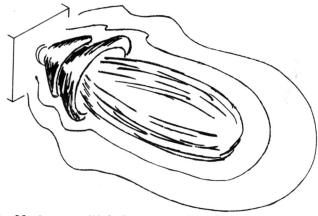

Small Source, High-Frequency Sound

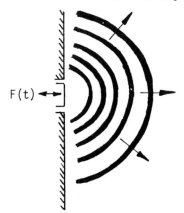

$F(t)$

Small Source, Low-Frequency Sound

Fig. 1.7    Free-field sound propagation patterns

a region of higher intensity. Phenomena such as these are not encountered when measuring vibrating mechanical systems directly.

When an acoustic wave impinges normally on a hard-surfaced area, the ratio of the pressure amplitude of the reflected wave to that of the incident wave is approximately

$$\frac{P_r}{P_i} = \frac{\rho c - \rho_o c_o}{\rho c + \rho_o c_o}$$

where the subscript ($o$) refers to air and $\rho$, $c$ refers to properties of the reflecting surface. Since $\rho c$ is greater than $\rho_o c_o$, the reflected wave will have an amplitude almost equaling that of the incident wave. Thus, the wave is reflected with a slight attenuation of its amplitude, and there is no phase change because a condensation is reflected by a hard surface as a condensation. For continuous sound this results in a pattern of standing waves which has a displacement node and a pressure antinode at the reflecting surface, since the surface can support the excess acoustic pressure. The magnitude of this pressure antinode is almost twice that of the pressure in the incident wave. This phenomenon produces a confusing situation when a listener attempts to pinpoint a source of equipment noise while standing close to a large, hard-surfaced wall.

A listener's discriminatory capability is also seriously impaired by audio masking. This occurs when background sounds have a frequency content similar to that of the sound being investigated. For example, if the background sound is wideband random noise, then it will mask wideband random sounds. This can be quite disturbing when recording low-frequency data in a low-frequency background sound field. However, at times this phenomenon can be useful. For example, it is currently being used in some office surroundings to alleviate the psychological effects of a "too-quiet" environment. This is accomplished by introducing low-intensity wideband random noise into the local environment through an electronic intercom system.

When acoustic waves travel outward from a source they are usually reflected in part by large surfaces. Suppose a diverging wave is reflected by independent surfaces into two paths and later recombined so that a total path difference equal to one wavelength occurs. The sound intensity at the point where they come together is reinforced by the action of the waves and the loudness of the sound is enhanced. However, if the path difference equals one-half of a wavelength, then interference occurs and partial cancellation of the sound is experienced. This is what sometimes causes the loud and quiet zones that exist in large offices and factory spaces that are bounded by hard-surfaced walls.

Consider the propagation of sound in a pipe of length $L$ as shown in Fig. 1.8. When the end of the pipe is open, there exists at the end an antinode of displacement and, consequently, a pressure node. When the end is closed, there then exists a displacement node and an antinode of pressure. In both cases the driven end of the pipe is considered open. Since a closed pipe has a displacement node at one end and an antinode at the other, the associated fundamental resonant wavelength is four times the length of the pipe. The open pipe has a displacement antinode at each end; thus, its fundamental resonant wavelength

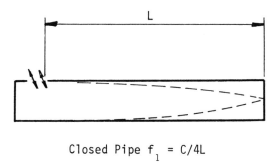

Closed Pipe $f_1$ = C/4L

Open Pipe $f_1$ = C/2L

Fig. 1.8   Displacement modes of open and closed resonating pipes

must be twice the length of the pipe. Substituting these relationships into the expression, $c = f\lambda$, the fundamental or lowest frequency and harmonics of sound-producing pipes are

Open pipe     $f_n = \dfrac{nc}{2L}$   $(n = 1, 2, 3, \ldots)$       Hz

Closed pipe   $f_n = \dfrac{nc}{4L}$   $(n = 1, 3, 5, \ldots)$       Hz

Because the closed pipe has only odd harmonics, its pitch or frequency response is one octave lower than that of an open pipe. Without going into the more technical aspects of resonators, it can be noted that open pipes, closed pipes and tubes act as sources of equipment noise. In particular long intake ducts on reciprocating compressors can radiate excessive acoustic sound in addition to exciting resonances in neighboring equipment and structures. This is also true of exhaust tubes.

## 5. LIMITATIONS OF THEORY

The theory of airborne noise as described above is not without limitation.

Phenomena have been described for which pressure fluctuations of the acoustic wave are small. Under such conditions, acoustical response is linear and concepts of superpositioning apply in full, that is, waves can pass one over the other without either one being affected as observed with ripples on water. High-intensity noise, explosions and impulse noise do not obey these theories exactly. For these cases frequency content, thermal effects and nonlinear interaction, among other things, have a decided effect on the propagation of the wave. However, one can still apply the basic physical concepts of airborne sound and noise to such cases for an understanding of the mechanics of a propagation problem.

## 6. MEASUREMENT

As with any technology, theory is of limited use without the complementing effect of measurement. To hear a noise and know from theory that it is composed of pressure fluctuations that propagate throughout the environment, reflecting off hard surfaces while being absorbed by soft, porous surfaces, and is decreased by proper damping and isolation techniques, is of little practical value. We are interested generally in knowing (and this can be done only through procedures of measurement):

(1) the origin of the noise,
(2) sound-pressure level,
(3) the total radiated acoustic sound power,
(4) frequency spectrum of the noise and
(5) radiation pattern or directivity.

These quantities are subsequently found with the measurement of one specific quantity: acoustic-sound pressure. This is the function of the microphone. Sound-pressure level is measured directly with a sound-pressure-level meter. Precision meters that meet requirements of USASI S1.4-1961 are commercially available from such companies as the General Radio Company and Bruel and Kjaer. Knowing the SPL at a point in a free field at a distance $r$(ft) from a small source (the characteristic or average overall dimension of the source $D \ll \lambda_{min}$), sound-power level can be calculated from the expression

$$PWL = SPL + 20 \, Log_{10} \, r + 0.5 \, dB \quad (re \, 10^{-12} \, watts)$$

For the more common sound source where directivity must be accounted for

$$PWL = SPL - 10 \, Log_{10} \, D_\theta + 20 \, Log_{10} \, r + 0.5 \, dB$$

where $D_\theta$ is the directivity factor and SPL is the sound-pressure level associated with $D_\theta$ and $r$. A comparison of the expression for sound-pressure level and that of power level shows that a doubling of sound pressure results in an increased sound-pressure level of $+6$ dB, but only a 3 dB increase in the level of power.

The directivity patterns of a sound source can cause numerous measurement problems. For instance, in the near field (close to the source) the distribution of acoustic energy is not the same for both large and small sources. Therefore, measurements should be performed in what is termed the "far field." This is the region remote from the source at a distance $r$ greater than $\lambda_{max}$, where $\lambda$ is the wavelength of the lowest frequency of interest and $r$ is at least three times the largest dimension of the source. For example, a 100 Hz sound has a wavelength of $\lambda = 11.25$ feet. Thus, for a source whose characteristic dimension is less than 3-1/2 ft, measurements should be made at a minimum distance of 11-1/4 ft from the source.

As mentioned above, radiation patterns will affect the collection of acoustical data. Single-point measurements of SPL in the vicinity of a source are useful in establishing the topology of the radiation field. Moreover, this data can be used for the calculation of directivity which is a function of distance and angular position from the source.

Unless measurement is made in a spherical, nondirective free field, a one-point acoustical measurement is insufficient for calculating acoustic sound-power level. This task requires quite specialized techniques. Imagine a large hemisphere surrounding the sound source with a number of microphones located uniformly in the surface of the imaginary hemisphere. It is recommended that sets of 2, 4, 8, 12, 16 or 20 microphones be used in obtaining this measurement. The indicated sound-pressure levels are averaged over the surface of the hemisphere and a value of PWL calculated from

$$PWL = \overline{SPL} + 20 \, Log_{10} \, r - 2.5 \, dB$$

for a source on a hard-surfaced plane or foundation and

$$PWL = \overline{SPL} + 20 \, Log_{10} \, r + 0.5 \, dB$$

for a source suspended far above a surface plane, where $\overline{SPL}$ is the mean sound-pressure level calculated from the mean-square sound pressures when SPL variations exceed 6 dB. This measurement becomes tedious and expensive to obtain because of the number of microphones and their necessary location. An alternate method uses a single microphone in the far field with rotation of the noise source when the source is movable. Another method suggests rotating a set of microphones about a stationary source. However the measurement is made, the PWL is reduced by $-3$ dB to account for the fact that a free-field hemisphere is being measured rather than a sphere.

When it is necessary to analyze the condition of a piece of operating equipment by way of acoustical measurement, data obtained at a single location is generally sufficient. These data contain frequency response information that can be correlated to operating machine speeds, thus indicating possible sources of mechanical malfunction. It is possible also to measure the relative energy levels within the sound spectrum, but only for qualitative analysis, since this information will be biased by directivity. These measurements can be obtained with a handheld portable sound-level meter or a remote microphone and cable system. When a meter is handheld in a sound field, the presence of the investigator is sufficient to alter the local character of the field. For this type measurement the meter should be held away from the investigator's body and properly directed relative to the source of sound.

## 7. MICROPHONES, CABLES

Where low-level sound fields (below 40 dB) are to be measured, a microphone should be used which has a low self-noise. Microphones containing lead-zirconate-titanate (PZT) elements are well suited for this type measurement. Conversely, high-level fields, for example, from 100 dB to 155 dB can be appropriately measured with a high-quality condenser microphone. For noise levels in excess of 155 dB, special type microphones such as a blast microphone or high-intensity quartz microphone should be used. Because the intermediate sound levels present no particular problem, they can be measured accurately with either PZT or condenser microphones. If a choice has to be made, it should be noted that condenser microphones, although very accurate, are affected by long exposure to high levels of local humidity and they are not as rugged for field use as the PZT devices.

Low-frequency measurements down to 20 Hz are performed routinely with PZT microphones. Many condenser microphones are unsuited for this application because of electrical leakage in the dielectric. Conversely, high-frequency measurements require use of a condenser microphone. When the dominant wavelength of the sound being measured is small, for example, a frequency of 10 kHz has a wavelength of only 0.112 inches, inaccuracies will appear in the data unless the size of the microphone is of the order of the wavelength of the sound or less. For these higher frequencies, Bruel and Kjaer manufactures a high-quality 1/4-in diameter condenser microphone. As to relative size, the author has constructed 1/16-in diameter microphones from hollow PZT tubes, which had good response characteristics over the audible frequency range.

Outdoor measurements are often taken during periods of mild wind. To suppress the effects of wind noise, commercially available windscreens can be placed over the sensitive element of the microphone. Shields are also available for protecting the microphone from rain. Whenever a microphone system is used out-of-doors, it is important that it be recalibrated at its working position, since the sensitivity of a microphone is affected by changes in temperature. Microphones are also direction sensitive. For this reason the local sound field should be surveyed first before deciding on a location and orientation at which to obtain acoustical data. There are now available small, handheld sound-pressure-level meters which are becoming popular for survey work. These units are especially desirable for obtaining a reasonable indication of sound-pressure levels and acoustic-radiation patterns at a minimum of equipment cost. Microphone systems are available that adapt easily to most sound-level meters. However, use of this measuring apparatus introduces a few additional problems, the most serious being that of cable noise. Long cables are sensitive to stray electric and magnetic fields. Thus, they act themselves as sources of signal noise. A common solution for this problem is to install a signal preamplifier in the cable as near to the microphone as possible. This results in the transmission of a strong signal in the cable so that the signal-to-noise ratio is high. Long cables also generate thermal noise which may approach the level of the acoustic signal when measuring very low-level acoustic fields. But, this problem is not normally encountered in field measurements unless cable lengths are extremely long and sound levels are very low. A long cable with its associated capacitance has an opposing effect on frequency measurements of the acoustic field. The addition of cable capacitance to the measuring system limits its high-frequency response. But on the other hand, this same capacitance enhances measurement in the lower-frequency range. The latter effect can be used to good advantage when measuring infrasonic frequencies. Dynamic microphones were formerly preferred for measurements requiring long cables since their signal response is good. Now, because of their relative size and weight, and with the availability of special microphone preamplifiers, dynamic microphones no longer enjoy their former popularity. Besides, these microphones are very sensitive to hum pickup. This is signal interference caused by the induction of stray signals from the electric-magnetic fields of local electrical equipment such as electric motors and transformers. The PZT and condenser microphones are not affected by this phenomenon. Fig. 1.9 shows typical frequency response curves for condenser and PZT microphones.

Accurate data can be obtained from a sound field if the measuring microphone has a uniform frequency response over the frequency range of the measured sound. When measuring overall sound levels it should be remembered that sounds composed mainly of the mid-frequencies have the greatest effect on sound measurement. High frequencies will be more directional, and they are attenuated more rapidly than other frequencies.

## 8. INSTRUMENTATION

Sound-level meters contain various weighting networks: A, B, C and more recently D. These weighting networks are used to reduce the sensitivity of the meter in the low- and mid-frequency ranges. For example, A-weighting reduces

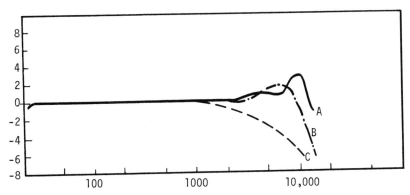

Fig. 1.9  Frequency response of various microphones with random incidence: (a) ceramic, (b) 1 in. condenser critically-damped, (c) 1 in. condenser over-damped

meter sensitivity by −50 dB at 20 Hz; B- and C-weighting are not so restrictive at the lower audio frequencies. Also, a linear network is usually included that has an all-pass or flat response from 2 Hz to 20 kHz. This response is shown in Fig. 1.10 which reflects USASI STD S1.4–1961. The result of the various weightings is that the meter can indicate either sound-pressure levels in a manner indicative of the human ear or measure the perceived noise directly in a linear manner. Generally, A-weighting is used in obtaining simple ratings for comparison of similar devices. To insure that such ratings are valid, care should be taken to maintain similar measurement distances and angular orientations from the sources. This weighting is also used for measuring nondirectional sources out-of-doors and for the preliminary rating of ambient noise to which a human receiver will be subjected. Thus, the instrument tends to simulate human response. If a system is used for measuring low noise levels, circuit noise can be troublesome below 40 dB. In such instances best results are obtained with either A- or B-weighting. A major problem associated with acoustical measurements is that of measuring narrow-band high-level noise and comparing it with noise that is wideband random; or comparing pure tone sounds to random noise. This always results in poor correlation of the sound fields. It is far better to measure and compare only those noises of similar character.

Following the microphone and sound-level meter, the next most important piece of measuring apparatus is a frequency analyzer. This piece of equipment is required to analyze the complex spectrum of a sound field and identify the included frequencies. Dominant frequencies can then often be correlated to the operating vibrational frequencies of a machine element, thus pinpointing major

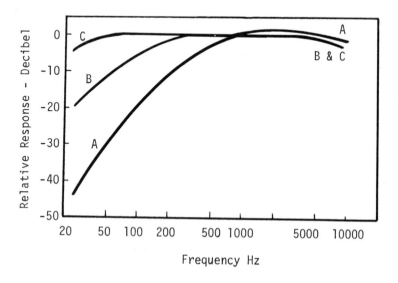

Fig. 1.10 Frequency response characteristics, sound level meters (USASI S1.4, 1961)

sources of vibration and noise. Analyzers can be purchased with differing measurement capacities. The most common contain 1/1-octave band, 1/3-octave band or 1/10-octave band filtering. These have filtering widths of 70, 23 and 7 percent, respectively. Nonstandard analyzers containing 1 percent filters or 10 Hz, 50 Hz or 200 Hz bandwidths are also available. However, their use is more common in laboratories than in typical field-measurement arrangements. It is common practice to analyze the acoustic radiation of vibrating mechanical sources by measuring the signal in preferred discrete frequency bands or octaves. An octave is a doubling of frequency. The preferred center frequencies for full octave-band analyses are 31.5, 63, 125, 250, . . . 16,000 Hz. Sound-survey measurements are usually performed with 1/1-octave band analyses in order to secure frequency related information with a minimum of effort, time and investment. When a more comprehensive investigation of a sound source is required, analyses can be made with a 1/3-octave analyzer. However, even 1/3-octave analysis is not always sufficient for obtaining the discrimination required for special spectra. It is then recommended to use either a 1/10-octave or 10 Hz bandwidth analyzer. But in this respect, it should be noted that too narrow an analyzer can be self-defeating in the time required for analysis and the overly abundant data thus obtained. Some measurement arrangements make use of low-pass and high-pass filters. For example, sound-level measurements in a range of interest around 1000 Hz in an area of concentrated electrical equipment and low-frequency noise can be enhanced with a high-pass filter set at 500 Hz, placed in the line between microphone and analyzer. The filter will block 60 Hz and 120 Hz electrical noise in the signal. Consequently, the measured spectrum of the sound reflects more accurately the investigated noise.

In many instances of data acquisition it becomes necessary to use a tape-recording device capable of fidelity reproduction in the audio range 20 Hz to 20 kHz. Few recorders are sufficiently rugged to withstand continued field service and yet possess the required acoustic reproducibility. There are some relatively fine, small portable tape recorders that can be purchased for about $100. These are made by such firms as the Sony Company and the Craig Company, Inc. Certain of their models have a usable frequency range that approaches 9 kHz whereas most commercial portable tape recorders seldom exceed a range of 4 kHz. In addition to a wide frequency response, these noted recorders have an acceptable noise level for recording signals above 50 dB and good control of *wow* and *flutter*, which adversely affect low-speed recording quality. With knowledge of a recorder's characteristics, it is possible to record useful data with an inexpensive, better-quality recorder. However, absolute accuracy requires a much higher-quality recorder. The General Radio Company manufactures an excellent portable data recorder which is priced under $2800. The recorder is basically a tape, two-channel, direct-recording instrument built by the Ampex Corporation for General Radio Company and fitted with a General Radio sound-pressure-level measurement system. Moreover, it is well suited for rugged field service. When a recorder is not subjected to excessive transport for securing data or infrasonics comprise a major part of the spectrum to be recorded, a frequency-modulated tape recorder manufactured by Bruel and Kjaer is an excellent choice to make. This instrument contains two-channel tape capacity with a commentary channel separating the two data channels. Thus, two channels are available for acquiring data and cross-talk between channels is minimized by the location of the commentary strip. High-capacity recorders for permanent installation that provide multiple-channel capacity are supplied by Honeywell, Ampex, Hewlitt-Packard and others.

Tape recorders are an invaluable aid to data processing. By taking advantage of the variable speed of the recorder, the sounds of mechanical sources can be expanded or compressed. For example, ultrasonic signals which are inaudible to the human ear can be recorded at very fast speeds and then slowed down for audio playback. In this way it is possible to *listen* to ultrasonic sounds. Conversely, infrasonic vibrations are recorded at slow tape speeds and then played back at higher speeds, in effect, compressing the time scale. With appropriate equipment, scaling ratios of one million to one can be achieved. At times, it may be necessary to record signals with a small degree of time scaling. If the time scaling is equivalent to a one-octave change in frequency then the signals possess remarkable similarity—a phenomenon known as octave confusion. This can be avoided by time scaling recorded data with a minimum change of two octaves. For example, a ten-cycle source can be scaled by five octaves to sound like a source of 320 Hz which is well within the audible range.

One of the least considered yet most important pieces of acoustical apparatus is a good set of earphones with high-quality, complete fitting sound cushions. The phone elements should have a high impedance similar to that of crystal phones. The basic reason for having earphones is to monitor the sound signal being recorded by the instrumentation. This is to insure that proper recording is occurring, and in some instances it may save the embarrassment (and expense) of recording for an extended period of time during which part of the instrumentation system is inoperative. This could be particularly crucial if the sound source requires special excitation. Earphones are also helpful in picking up microphonics and hum during initial setup.

Hum pickup has directional characteristics. If it is found occurring in the system, it must be identified in either the microphone or the instrumentation. In either case reorientation or a relocation of equipment will usually alleviate the problem. A problem occurring with the measurement of high-intensity fields is that of microphonics. Even the best vacuum tubes and transistors are sensitive to the pressure vibrations of a sound field, and they can be induced to resonate when the sound-pressure level is in excess of 100 dB. This results in an apparent increase in the level of the output signal.

Various other devices can be used for making qualitative acoustical measurements. One such device, a physician's stethoscope, is an excellent detector of low-frequency sound or vibration. Another useful tool is a portable resonant tube. When the effective length of a hollow tube is changed so that it resonates at the dominant frequency of a source then the frequency of the mechanical vibration can be identified from the previously stated relationship between frequency and length for a closed tube. Basic measuring arrangements are shown in Figs. 1.11, 1.12 and 1.13. These arrangements increase in complexity with the measurement problem.

Great care must be exercised in the handling and maintenance of acoustical

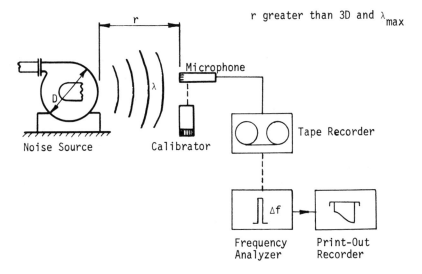

Fig. 1.11   Field measurement system for equipment noise spectrum
analysis

and vibration measuring equipment. The corrosive atmosphere of today's
industrial plants is sufficient to severely damage these instruments. If they are
not properly stored, contacts corrode, circuit resistance changes and accuracy
is no longer assured. For this reason all acoustical measurement systems must
be calibrated at the microphone or pickup device before attempting to secure
data from sound or noise. It should be kept in mind that acoustical field measurements are not absolute. Reliability of sound-pressure-level measurements is
often questionable; for instance, there is error associated with the accuracy of

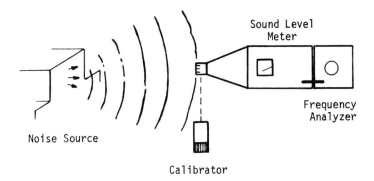

Fig. 1.12   Field measurement system for equipment noise survey or
on-location spectrum analysis

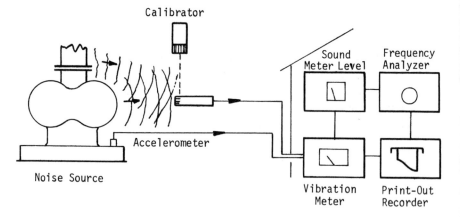

Fig. 1.13   Measurement system for narrowband random sound and
vibration analysis

the calibration of microphones, meters and amplifiers. These units have individual errors alone of approximately one-half decibel. Moreover, when an averaging meter is used instead of a true rms meter to measure some sound levels this causes an additional error of one decibel, since the average level of a random-type noise is one decibel less than its rms value. But aside from all this, if care is taken during acoustical measurement, the resulting data can be invaluable for system analysis.

# REFERENCES

1. Kinsler, L. E. and Frey, A. R. *Fundamentals of Acoustics.* New York: John Wiley and Sons, 1962.

# BIBLIOGRAPHY

American Standard S1.1–1960. "Acoustical Terminology (including Mechanical Shock and Vibration)." New York: The United States of America Standards Institute, 1960.

Beranek, L. L. *Acoustic Measurements*. New York: John Wiley and Sons, 1949.
Beranek, L. L. *Noise Reduction*. New York: McGraw-Hill Book Co., 1960.
Brock, J. T. "FM Magnetic Tape Recording." *Bruel and Kjaer Technical Review* 1 (1967): 3–15.
Buscarello, R. T. "Practical Solutions for Vibration Problems." *Chemical Engineering* (August 12, 1968): 147–166.
Keast, D. N. *Measurements in Mechanical Dynamics*. New York: John Wiley and Sons, 1967.
Nittinger, R. H. "Vibration Analysis Can Keep Your Plant Humming." *Chemical Engineering*, (April 17, 1964): 152–158.
Peterson, A. and Gross, E. Jr. *Handbook of Noise Measurement*, Massachusetts: General Radio Company, 1967.
Ruffini, A. J. "Bearing Noise, Part 1: Analysis of Rolling-Element Bearings." *Machine Design* 35, no. 11 (1963): 232–235.
Sears, F. and Zemansky, M. *University Physics*. Addison-Wesley Publishing Co., Inc., 1964.
Skode, F. "Windscreening of Outdoor Microphones." *Bruel and Kjaer Technical Review*, 1 (1966): 3–9.
Stevens, S. S. and Warshofsky, F. *Sound and Hearing*. New York: Time Incorporated, 1965.
Wirt, L. S. "Acoustics as an Aid to Measurement." In *Measurement Engineering*, 3rd ed., edited by P. K. Stein, Chap. 10. Phoenix: Stein Engineering Services, Inc., 1964.
Wood, A. B. *A Textbook of Sound*. New York: Dover Publications, 1955.

# SYMBOLS

| | | |
|---|---|---|
| A | area | ft$^2$ |
| a | total absorptive power | ft$^2$ |
| c | velocity of sound | ft/sec |
| d | ball diameter | inch |
| D | ball (bearing) pitch diameter | inch |
| D | decay rate | dB/sec |
| E | Young's modulus | lbf/in$^2$ |
| f | frequency (Hz, cps) | sec$^{-1}$ |
| $g_c$ | gravitational constant | lbm·ft/lbf·sec$^2$ |
| hp | horsepower | —— |
| H | height | ft |
| I | sound intensity | watts/m$^2$ |
| L | length, wall thickness | ft |
| lbm | pound mass unit | —— |
| lbf | pound force unit | —— |
| m | integer | —— |
| n | speed (shaft, bearing) | rpm |
| N | number of teeth, balls, blades | —— |
| p | rms acoustic pressure | lbf/in$^2$, dynes/cm$^2$ |
| PWL | sound-power level | dB |
| Q | fluid flow rate | ft$^3$/min |
| r | radius | ft |
| R | absorptive power $A_T\bar{\alpha}/(1 - \bar{\alpha})$ | ft$^2$ |
| S | Strouhal number | —— |
| S | number of slots | —— |
| STC | standard transmission class | —— |
| SPL | sound-pressure level | dB |
| TL | transmission loss | —— |
| v | velocity | ft/sec |
| V | volume | ft$^3$ |

| | | |
|---|---|---|
| W | width | ft |
| W | acoustic power | watts |
| $\theta$ | ball-raceway angle of contact | degrees |
| $\alpha$ | absorption coefficient | —— |
| $\rho_0$ | air density | $lbm/ft^3$ |
| $\eta$ | coefficient of viscosity | $lbf \cdot sec/ft^2$ |
| $\omega$ | angular frequency | $sec^{-1}$ |
| $\psi$ | attenuation constant | $ft^{-1}$ |
| $\lambda$ | wavelength | ft |
| $\alpha$ | sound-power reflection coefficient | —— |
| $\alpha$ | transmission coefficient | —— |

# 2

# SOUND AND NOISE
# IN STRUCTURES

## W. S. Mitchell

## 1. MACHINE SOURCES

In the discussion of sound and noise in structures, we divide our attention between two specific types of structures: machines and buildings, both of which
are extremely important to today's society. We must be aware of the various
causes of sound and noise in structures since these causes must be controlled if
meaningful noise reduction is to be attained. In some cases only machine structures are of interest. In others interest is centered on building response. There
has been concern over the past several years about the effect of machine structures on buildings, which is in reality a situation of *structure-inside-of-structure*.

Consider first machine structures. Sound and noise is generated by various
mechanical components of a machine structure. Likewise, physical phenomena
associated with fluid (compressible gas) motion contribute to the generation of
audible sound. Some of the more common sources of sound and noise generation are

(1) machine unbalance,
(2) gear transmission systems,
(3) lubrication,
(4) bearing elements; ball, roller and journal and
(5) air turbulence.

The first noted source, machine unbalance, is produced in both rotary and
reciprocating motion. For instance, a machine such as a fan, blower or pump
has a primarily rotary motion. If n is the speed of the machine in rpm, then the
frequency of the radiated noise is given by

$$f = \frac{n}{60} \text{ cps}$$

This noise will be of high or low frequency depending on the speed of the individual machine. Reciprocating air compressors are a good example of a source
of reciprocating unbalance. In these machines the time-dependent motion of the
piston is controlled by the length of the connecting rod so that the harmonic
motion of the driven end of the connecting rod becomes nonsinusoidal motion
at the piston. Fourier analysis of the piston motion shows that the motion has

a fundamental vibration with higher harmonics, the second being the most prominent of the harmonics.

Sound is generated in machine structures due to the dynamic motion of gear meshes. For example, some gearing arrangements are quite efficient for transmitting large amounts of power; however, they are also extremely noisy. This is true of spur gear meshes which are quite noisy when compared to helical gears. The meshing action of the gears is influenced by both the tooth spacing and the transmitted load. When the load is passed from one tooth to the next, the pitch-line clearance of the teeth produces shock loading at the tooth-mesh frequency. As this shock energy is transmitted to a resonant part of the structure, the radiated noise becomes very pronounced. For a single countershaft, the tooth-mesh frequency is expressed as

$$f = \frac{N_g n_g}{60} \text{ cps}$$

where $N_g$ is the number of teeth on a gear and $n_g$ is the gear shaft speed in revolutions per minute (rpm). For a planetary gear arrangement the tooth-mesh frequency is given by

$$f = \frac{N_r}{60} (n_r \pm n_o) \text{ cps}$$

where $N_r$ is the number of teeth in the reference gear, $n_r$ is its speed in rpm, $n_o$ is the rpm speed of the cage and the ring gear is considered fixed. When there is a defective tooth on both pinion and gear in a running set, then the frequency of the maximum radiated sound due to the imperfections contacting is

$$f = \frac{n_g}{60 N_p} \text{ cps}$$

where $N_p$ is the number of teeth on the pinion.

Another source of sound in gearing comes from the error in concentricity of the mounting. This causes a periodic change in the depth of tooth engagement. In certain conditions a low-frequency beating is produced by the noise from the gears. However, at high speeds this same mechanical error produces the familiar sound of gear whine. Other sources of gear noise result from the basic shape of the gear. Large, relatively thin gears resonate as thin discs or plates, and rather than radiate sound directly in this mode, they flex and create inaccuracies in the mesh geometry. Consequently, this generates tooth noise and wear. As a gear tooth moves in and out of mesh with its mate, air is alternately compressed and then given a negative pressure in the tooth cavity. This is especially true at high speeds in gearboxes containing closely spaced gears. The result is loading pulsations or even directly radiated sound. If the gears are operating in oil and the mesh has a minimum bottom clearance, then hydraulic shock loading can occur due to the incompressibility of the oil trapped in the mesh. Such loading is ultimately transmitted to the structure where it is radiated as noise by a surface vibration. The phenomena mentioned above do not usually occur in low-speed, lightly loaded gearboxes.

Many types and classes of gears are available for transmitting mechanical motion and power. The proper selection of a gear for a specific function thus becomes an important factor in the success of a usable product. Improperly applied gears can produce:

(1) component failure,
(2) environmental noise,
(3) structural vibration and
(4) operator fatigue.

For example, in the latter case a carelessly matched set of spur gears can operate with a countershaft arrangement without seriously affecting shaft-mode response. But when the inaccuracies of the mesh are transmitted in the form of a mechanical vibration to the machine structure, they excite various panel resonances. The responding panels generate an irritating audible hum that increases operator fatigue.

Ball-bearings represent another type of noise in machines; it is usually analyzed through the use of accelerometers and filters. A search of the literature shows that writers tend to emphasize the frequencies that are related to the rotative frequency ball-defect. Thus, much attention has been given to the frequency of impact of a ball-defect with the race, for example. But recent research by the present author indicates quite clearly that the rotation-related frequencies are not significantly important, except in a minority of instances. Nevertheless, they are considered hereunder as an aid to general understanding. The resonant frequencies of the bearing rings appear to be more important in noise analysis than the rotation-related frequencies.

With a measure of eccentricity in an operating bearing, the frequency of the radiated sound is like that of a rotating unbalance, in that there is in the sound a frequency component corresponding with rotative speed:

$$f = \frac{n}{60} \text{ cps}$$

where n is the shaft speed in rpm. For a stationary outer bearing race, the frequency of the bearing noise due to the traveling of the ball elements is

$$f_0 = \frac{n}{120}\left(1 - \frac{d}{D}\cos\theta\right) \text{ cps}$$

where d is the ball diameter, D is the pitch diameter of the ball assembly and $\theta$ is the angle of contact of a ball in the bearing raceway (see Fig. 2.1). When the inner race is stationary, the radiated noise has a frequency given by

$$f_i = \frac{n}{120}\left(1 + \frac{d}{D}\cos\theta\right) \text{ cps}$$

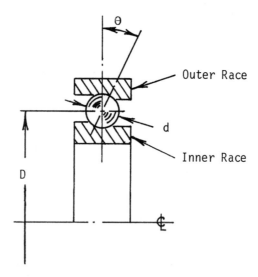

Fig. 2.1   Ball bearing element

The rotational or spin frequency of a rolling ball is

$$f_s \ = \ \frac{n}{60} \frac{D}{d} \left( 1 \ - \ \frac{d^2}{D^2} \cos^2 \theta \right) \ cps$$

If a ball has m defects on its surface, then the frequency of the resulting sound is

$$f_d \ = \ \frac{mn}{60} \frac{D}{d} \left( 1 \ - \ \frac{d^2}{D^2} \cos^2 \theta \right) cps$$

The radiated sound caused by the relative motion between rolling balls and a rotating ring has a frequency

$$f_{ri} \ = \ \frac{n}{120} \left( 1 \ - \ \frac{d}{D} \cos \theta \right) cps$$

when the inner race is stationary and

$$f_{ro} \ = \ \frac{n}{120} \left( 1 \ + \ \frac{d}{D} \cos \theta \right) cps$$

when the outer race is stationary. If there is a defect on the rotating bearing raceway, the frequency of the radiated sound is

$$f'_{ri} = \frac{nN}{120}\left(1 - \frac{d}{D}\cos\theta\right) \text{cps}$$

or

$$f'_{r0} = \frac{nN}{120}\left(1 + \frac{d}{D}\cos\theta\right) \text{cps}$$

depending on whether the inner or outer ring is held stationary, respectively, where N is the number of balls in the bearing. On the other hand, when the defect is on the stationary raceway, the frequency of the resulting sound is expressed as

$$f_{sri} = \frac{nN}{120}\left(1 + \frac{d}{D}\cos\theta\right) \text{cps}$$

or

$$f_{sr0} = \frac{nN}{120}\left(1 - \frac{d}{D}\cos\theta\right) \text{cps}$$

where inner race and outer race are stationary, respectively.

Electric motors contain various sources of sound. Aside from the usual bearing and unbalance noise, they also generate electrical noise, a soft humming sound that has a frequency spectrum composed of harmonics of the line frequency. In induction motors the dominant frequency is twice that of the line frequency. The existence of rotor slots in induction motors creates instantaneous drop-offs in the motor torque at frequencies of

$$f_s = \frac{nS}{60} \pm 2f_L \text{ cps}$$

where S is the number of slots on the rotor and $f_L$ is the line supply frequency. A consequence of rotor slots in forced-air cooled motors is a modulation of the flow of cooling air through the motor. This creates air flow pulsations similar to those in a siren having a wide rotor clearance. The predominant frequencies of the modulated sound are given by

$$f = \frac{nS}{60} \text{ cps}$$

Air turbulence is a major source of noise in machine structures, particularly in those pieces of equipment associated with air-moving systems. In air-moving equipment, such as fans and blowers, noise is generated at both inlet and outlet ducts. However, the latter is the more dominant source since the greatest relative motion occurs at this location as the blade delivers the compressed air. The next applies to those cases where air flows smoothly over an elastic system, such as a stretched wire, a high-voltage line or a long suspension bridge.

$$f = S\frac{v}{D} \text{ cps}$$

where $0.14 \leq S \leq 0.21$ (Strouhal number), v is the air velocity and D is the characteristic dimension of the body, e.g. a fan blade thickness. As air vortices (swirls) separate from the system, alternating first from one side then the other, the frequency of the separation induces mechanical vibration of the body as shown in Fig. 2.2. This vibration becomes a source of sound when the frequency response is in the audible range 20 Hz to 20kHz. The frequency of the radiated sound of a turbine is dictated by the number of blades on the turbine wheel. For instance, if there are N turbine blades, then the primary frequency of the resulting sound is

$$f = \frac{nN}{60}$$

Due to the turbulence of the flowing medium each blade will itself vibrate to produce sound. When these sounds are combined with the aerodynamic sounds of the moving air plus the siren-like sounds from the turbine blades passing the stator openings, the result is a multi-frequency random noise.

## 2. SOUND RADIATION

The aforementioned sources are the cause of many sounds and noises in air which reflect on the mechanical condition of a machine. In order to make use of these sounds for analysis, they are normally monitored with a high-quality microphone. When the measured signal is passed through a frequency analyzer, valuable data are obtained for locating a specific mechanical source of excitation. These same data are also useful in constructing an audio frequency histogram that can be used as a reference or base for a preventative maintenance program similar to those described by Baxter,[1] Blake,[2] Bowen and Graham[3] and others.

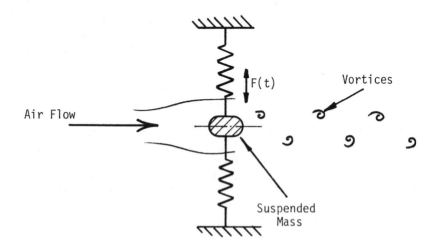

Fig. 2.2   Blade vibration

Structure-borne noise is encountered in all types of machinery housings, in weldments as well as castings where contacting members of the structure afford excellent acoustical transmission paths. It is not uncommon for noise to travel from one part of a structure to another where it is radiated as audible sound, although it may be hardly noticeable in the area of the excitation force. This occurs when a compliant portion of a structure has a natural frequency that coincides with the frequency of the disturbing vibration. The result is a noticeable amplification of the induced vibration. Building structures can be made to resonate in this same way, that is, by a mechanically vibrating source activating some portion of the building into a state of resonance. This causes mechanical vibration as well as audible sound. Consequently, mechanical isolation of offending machinery and equipment often provides a first solution of structure-borne noise. Isolation is effectively obtained by properly inserting compliant springs and damping materials between the excitation source and the responding structure. Since structure-borne noise in machines results from sound passing from one machine element to another, discontinuities which interrupt a path of sound propagation often result in an improved condition. This is one reason for utilizing such items as specially placed gaskets, nonmetallic gears and plastic bushings. This reasoning is also applied in building construction for floating machine bases, floors and general supports.

## 3. SOUND IN ENCLOSURES

Sound and noise in buildings and enclosures depend on such items as structural transmissibility, frequency, sound sources, air and structural damping, isolation and geometric shape. Consider first a long hallway or tube of length L. Steady-state sound propagation in such a duct occurs in the form of plane waves and the normal mode frequencies of the space are expressed as

$$f = \frac{nc}{2L} \text{ Hz} \qquad (n = 1, 2, 3, \ldots)$$

where c is the velocity of sound. This is true when the wavelength of the sound is greater than 1.7 times the effective diameter of the given duct. Furthermore, the hallway or tube acts as a waveguide or transmission duct for sound between the spaces that it connects. At low frequencies and in short ducts, $L_c$ must be substituted for L in the above equation. When the duct has a flanged outlet, such as a hallway opening into a wall

$$L_c = L + 0.85 \, r \quad \text{ft}$$

where r is the radius of the duct. For an unflanged or plain-ended duct,

$$L_c = L + 0.60 \, r \quad \text{ft}$$

Factory office spaces which contain pairs of parallel walls are typical of small and medium-size (those having volumes of less than 50,000 ft³) rectangular enclosures. If noise sources in such a volume are sufficient to excite a mode of vibration of the enclosed air, then a system of acoustical standing waves (plane waves) will exist between the parallel walls. These standing waves will have frequencies given by the same expression as for the tube resonances except that L is considered as a distance between a given pair of walls. For spaces with flat, unpitched ceilings, it is possible to excite the frequencies of three distinct sets of standing waves. Each set includes an infinite number of integral multiples of the fundamental wave where n equals 1, and they are the harmonics of n greater than 1.

Suppose one end of an enclosed space is a full vertical wall that contains only one open area such as a window. When low-frequency plane waves (standing waves) impinge on the wall and opening, that part of the wave in contact with the wall is partially reflected and partially transmitted, but the portion of the wave entering the opening acts like a small mechanical source of sound to the adjoining space. If the open area is small compared to the area of the wall, then the plane waves entering the adjoining space propagate as spherical waves. Such transformation of plane waves to spherical waves is called *diffraction*, and it is commonly observed in optical research and ripple tank studies. This phenomenon is shown in Fig. 2.3.

## 4. FREQUENCY MODES OF ROOMS

An enclosed volume responds in other ways. For instance, if a sound source is located in a small, hard-wall room, then as its sound waves travel throughout the enclosure and are reflected at each containing surface, there are periods of time during which the waves repeat on themselves. This creates a resonant system, a system of standing waves. These normal modes of vibration are classified as:

1. Axial modes—the waves propagate parallel to two pairs of surfaces such as previously described in terms of a tube.

2. Tangential modes—the wave system moves parallel to one pair of surfaces and oblique to all others.

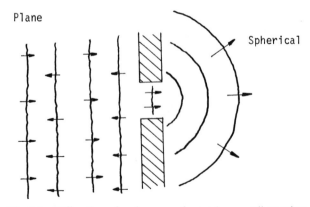

Fig. 2.3   Diffraction of a plane sound wave by a small opening

3. Oblique modes—the wave motion is oblique to all surfaces.
For clarification of the directions of propagation, see Fig. 2.4.

These mode frequencies are expressed as

$$f(m_x, m_y, m_z) = \frac{c}{2}\left[\left(\frac{m_x}{L_x}\right)^2 + \left(\frac{m_y}{L_y}\right)^2 + \left(\frac{m_z}{L_z}\right)^2\right]^{1/2} \qquad (m = 1, 2, 3, \ldots)$$

where the m are independent integer numbers, and $L_x$, $L_y$, $L_z$ are the appropriate dimensions of the enclosure in feet. The angles $\theta_x$, $\theta_y$, $\theta_z$ are the angles
formed by the axis along which a wave propagates and the x, y and z axes, respectively. These angles are given by the following relations:

$$\text{Tan } \theta_x = \frac{[(m_y/L_y)^2 + (m_z/L_z)^2]^{1/2}}{m_x/L_x}$$

$$\text{Tan } \theta_y = \frac{[(m_x/L_x)^2 + (m_z/L_z)^2]^{1/2}}{m_y/L_y}$$

$$\text{Tan } \theta_z = \frac{[(m_x/L_x)^2 + (m_y/L_y)^2]^{1/2}}{m_z/L_z}$$

Sound distribution within the volume will be such that for any mode of vibration, maximum sound pressures will occur in the corners of the enclosure. In
addition, whenever any independent value of $m_x$, $m_y$, or $m_z$ is odd, the center

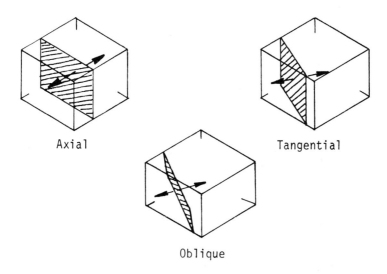

Axial                                    Tangential

Oblique

Fig. 2.4   Normal modes of room vibration

of the enclosure will be a point of minimum sound-pressure level. If any two of the m's are odd, then a minimum sound-pressure level occurs at the center of any one wall. Thus, location of sources or measuring systems in a small or medium size room will determine the extent of acoustical excitation. For example, a microphone or sound source in the corner of a rectangular room will couple to the maximum number of acoustical responses in the volume. But when these devices are placed in the center of the room, they react with only 12 percent, approximately, of the possible modes of the room. This failure of the equipment to couple to all room modes should be considered. In large enclosures there are many complex variations of the sound field, and these produce situations which preclude any detailed mathematical analysis of the enclosure. High and low frequencies resonate different modes of vibration in the enclosure. Moreover, the effect of air absorption becomes important because of the lengths of the propagating paths involved between reflecting surfaces. There are also paths of constructive and destructive interference in the volume along which the intensity of the sound is either reduced or reinforced. Frequency and the type of excitation, such as wideband random noise or pure tones, determine whether the resulting sound field is diffuse or nondiffuse. Hence, sound-pressure levels can vary markedly throughout the space.

## 5. ABSORPTION

Much of the foundation of acoustics is credited to the work of Sabine, who conducted primary investigations into the effects of materials on the absorption of sound in air. His results showed that as propagating sound waves impinge on soft, porous surfaces the energy of the wave is absorbed. Further, the degree to which it is absorbed depends on the material and the frequency of the sound. In situations requiring only a first approximation of a solution, or in which sounds are of a low-frequency character, or in small enclosures it is permissible to neglect the effect of sound-wave attenuation by the air. But for high frequencies or large enclosures the attenuation of the air can be quite significant. For example, in large enclosures such as auditoria, the effects of air attenuation may equal or exceed that of surface absorption.

There are three basic mechanisms responsible for the observed attenuation of sound in air. These are viscous losses, heat conduction losses and losses due to molecular exchanges of energy. Viscous losses occur when microscopic particles of the elastic medium (air) are displaced in to-and-fro motion as they react in transmitting the longitudinal acoustic wave from one set of particles to another. The resulting relative motion between particles manifests the viscous loss. Accompanying each localized compression and rarefaction within the propagating wave is an increase and decrease of temperature, respectively. These localized temperature differentials promote a transfer of heat between alternating pressure regions by the process of heat conduction which contributes to a loss of acoustic energy. Molecular energy exchange, which considers the energies of vibration and rotation of molecules, is best explained in terms of kinetic theory which is not in the scope of this handbook.

Consider an enclosure containing a sound source and acoustically absorbing surfaces. The absorption coefficients $\alpha_i$ of the absorbing surfaces, which have values between zero and one, denote the fraction of randomly incident energy that is absorbed by each surface (wall or other) of area $A_i$ in the enclosure. When the products of $\alpha_i \cdot A_i$ are summed, the result is a quantity called the absorptive power of the room

$$\mathbf{a} = \sum \alpha_i \cdot A_i \qquad (i = 1, 2, 3, \ldots) \text{ ft}^2$$

Enclosures having small values of absorptive power are referred to as reverberant rooms. Suppose a sound source is excited in a hard-wall room where there are partial reflections of the sound at each wall. After a sufficient period of time, the room will support a level of sound intensity I. Two factors control the speed with which the room reaches its final acoustic level: absorption of the acoustic energy by the medium and absorption by the boundary surfaces. As previously noted, only high frequencies or large volumes contribute to appreciable attenuation of the sound by the medium. Therefore, sound in small and medium size enclosures is most noticeably affected by the absorptive power of the room. A consequence of this is that rooms with high absorption quickly reach a maximum level of sound intensity for a given source, whereas enclosures with negligible absorption require comparatively longer periods of time to attain maximum sound levels. Sound reflections at the walls and interior surfaces of the room produce an internal distribution of sound energy that is sometimes assumed to be diffuse. In other words, the energy distribution is uniform throughout the volume and essentially random in its direction of flow. This is approximately true for wideband noise excitation or one-half octave wide warbled tones. If the excitation is a pure tone or of a single frequency, then the room will contain a system of standing waves and large variations of sound-pressure level will exist from point to point in the room. For example, when there is a wideband noise source in an enclosure, the resulting acoustic pressure at a point will depend on the pressure due to the directly radiated waves as well as the reflected waves. This total pressure is expressed by the sum

$$P^2 = \rho_0 cW \left( \frac{1}{4\pi r^2} + \frac{4}{\mathbf{a}} \right)$$

where W is the acoustic power of the source, c is the velocity of sound, $\rho_0$ is the density of the air (the absorption is considered here to be small) and r is the distance from the source of sound to the location where pressure is measured. In real rooms the absorption of sound is not always small and $\mathbf{a}$ is usually replaced by a room constant $R_T$ defined as

$$R_T = \frac{\mathbf{a}}{1 - \bar{\alpha}} = \frac{A_T \bar{\alpha}}{1 - \bar{\alpha}}$$

where $A_T$ is the total room surface area and $\bar{\alpha}$ is the average absorption coefficient for the room. Solving the above expression for acoustic power and noting that in a hard-wall or reverberant room the absorptive power is small, the acoustic power is approximately

$$W = \frac{a\ P^2}{4\rho_0 c}$$

The manner in which a reverberant or lightly damped room reaches a given level of sound intensity, I, is seen in the expression

$$I = \frac{W}{a}(1 - e^{act/4v})$$

where W is the rate at which sound energy is generated in the room and V is the volume of the enclosure. Thus, intensity growth is exponential with a time constant of 4V/ac. This shows the marked effect of surface absorption on the time required to produce a given level of sound. When the sound source is quieted acoustic energy traverses the enclosure until it is finally absorbed by either the air or boundary surfaces. The rate of absorption or the decay rate is given as

$$D = 1.087\frac{ac}{V}$$

which is proportional to the absorptive power of the room.

## 6. REVERBERATION TIME

Reverberation time is defined as the time in seconds that is required for an established sound level in an enclosure to decay by 60 dB when all sound excitation ceases. During this period of time the interior surfaces and the air enclosed in the volume absorb the acoustic energy of the sound. Consider a simplified situation in which there is no sound attenuation due to the presence of the air. This is approximately true in small volumes with highly absorptive surfaces. In this case the volume influences the time required for wave propagation between absorptive surfaces and the reverberation time is thus given as

$$T = 0.049\frac{V}{a}\ \text{sec}$$

Unless stated otherwise reverberation time relates to sound fields whose center frequency is 500 Hz. The above equations are subjected to the following conditions: first, it is assumed that the established sound field is diffuse; second, the sound field is produced in a regular-shaped, small to medium size enclosure where the absorption coefficients all have approximately the same value. A regular enclosure is one that does not contain deep recesses, hallways, L-shapes or other such geometries. If these conditions do not exist, as would be the case in oddly shaped rooms such as curved or circular rooms, then the acoustic intensity fluctuates about the room and theoretical expressions become meaningless. When interior surfaces are highly absorptive, sufficient reflections do not occur to establish a diffuse-type sound field. It is then necessary to calculate a reverberation time based on a corrected absorptive coefficient which accounts for the excess absorption of the room. If the coefficients of all the interior surfaces are approximately the same, then the absorptive power is expressed as

$$\mathbf{a} = -A_T \, \text{Log}_e \, (1 - \bar{\alpha}) \quad \text{ft}^2$$

When the coefficients vary widely from surface to surface, the absorptive power of the room must be taken for the m absorptive surfaces as

$$\mathbf{a} = \Sigma_m \, -A_m \, \text{Log}_e \, (1 - \alpha_m) \quad \text{ft}^2$$

The previous value of reverberation time considered sound attenuation due to surface reflections only. If a room is highly reverberant, then it is most probable that air absorption will be the more dominant part of the attenuation. The intensity of a sound wave traveling through a viscous medium such as air is attenuated exponentially by a factor $e^{-\psi x}$ where x is the distance traveled by the wave. For air the attenuation constant $\psi$ is given as

$$\psi = 0.5 \times 10^{-6} \, \frac{\omega^2}{\rho_0 c^3} \quad \text{ft}^{-1}$$

Consideration of viscous attenuation results in a corrected reverberation time for a large room or auditorium

$$T = \frac{0.049 \, V}{\mathbf{a} + 4\psi V} \quad \text{sec}$$

Because of the dependence of T on the attenuation constant $\psi$, it becomes obvious that reverberation time is shorter for the higher-frequency sounds.

The reverberation time of an enclosure indicates its response to such sources of sound as speech and music. For example, if reverberation time is large (indicative of minimal absorptive surfaces), then a spoken word is heard first as direct sound, then as a series of reflections. The reflections persist until the acoustic energy is absorbed by the room. This causes audio confusion when internal reflections of the first word are heard at the same time as new words are heard from direct propagation. On the other hand, if reverberation time is very short due to excessive interior absorption, then the first word decays before it has had a chance to propagate throughout the volume. Consequently, it is not heard distinctly everywhere in the room. With some reflection at the interior walls it is possible to receive sound remote from a source almost as clearly as it is nearby because of the enhancement of the direct sound by sound via surface reflection. Therefore, it can be said that although some interior sound reflection is good too much is undesirable.

Experiments have shown that for small and medium size rooms used as offices or similar spaces, reverberation time should be about 0.5 sec; large conference rooms and small auditoria should have reverberation times of 0.8 sec; and very large volumes require reverberation times of 1.5 sec or greater. For example, the acoustical redesign of Carnegie Hall in New York City (1969–70) will lead to a reverberation time of 2.15 sec. The latter enclosures are large enough that they often require electronic sound distribution systems. It is suggested that a quiet, industrial, working environment should have a reverberation time that is relatively low, that is, close to one-half second. However, two facts must be kept in mind. Reverberation time is dependent on the frequency of the sound. In some large enclosures, boundary surface treatment produces negligible reduction of localized sound, although there is some reduction of the overall sound level. Regarding the size of a room, a room is considered large when $\lambda$, the dominant wavelength of the included sound, is much smaller than $4V/A_T$. Note that a room can be considered small at the lower frequencies and large at the higher frequencies. Rooms that have an average absorption coefficient of $\alpha$ greater than 0.2 are considered to be highly damped or "dead." When $\alpha$ is less than 0.2, the rooms are considered "live" with an associated long reverberation time. Moreover, when the average coefficient $\alpha$ is greater than 0.3, it is reasonable to assume that propagating waves are attenuated at the surface boundaries and a diffuse sound field will not exist.

Table 1 lists effective sound-absorption coefficients for numerous construction materials. The values can only be considered approximate, since the many factors affecting the measurements of such quantities are not always properly con-

TABLE 1
ABSORPTION COEFFICIENTS FOR ARCHITECTURAL USE

| MATERIAL | FREQUENCY Hz | | |
| --- | --- | --- | --- |
| | 125 | 500 | 2000 |
| Brick wall, unpainted | 0.02 | 0.03 | 0.05 |
| Brick wall, painted | 0.01 | 0.02 | 0.02 |
| Poured concrete, unpainted | 0.01 | 0.02 | 0.02 |
| Plaster, gypsum | 0.02 | 0.02 | 0.04 |
| Marble | 0.01 | 0.01 | 0.02 |
| Wood paneling | 0.10 | 0.10 | 0.08 |
| Draperies, light | 0.04 | 0.11 | 0.30 |
| Draperies, heavy | 0.10 | 0.50 | 0.82 |
| Carpet (wool) | 0.09 | 0.21 | 0.27 |
| Carpet and pad | 0.20 | 0.35 | 0.50 |
| Chair, upholstered | 3.50 | 3.50 | 3.50 |
| People, standing | 2.00 | 4.70 | 5.00 |
| People, seated | 0.70 | 0.50 | 1.60 |

trolled. It is worthwhile to comment here on the effect of open doors and windows. Since these spaces are open to the sound wave, it may seem a priori that there is no reflection of incident sound; that is, the absorption coefficient has a value of one. This is true only when the dimensions of an opening are several times greater than the wavelength of the sound. At other times, especially at low frequencies, there is reflection of the incident sound which implies an absorption coefficient of less than one.

## 7. TRANSMISSION THROUGH WALLS

The sound and noise that is experienced in an enclosure must come from one of three sources: an internal sound source, structurally transmitted vibrations which resonate a surface of the enclosure, or transmission through the walls. Consider the latter case. When sound of intensity $I_i$ is incident on a partially reflecting surface such as a wall, sound of intensity $I_r$ is reflected. The ratio of these two intensities is the sound-power reflection coefficient

$$\alpha_r = \frac{I_r}{I_i}$$

This coefficient can also be written in terms of the amplitudes of the incident and reflected acoustic pressures $P_i$ and $P_r$, respectively, or $\alpha_r = P_r^2/P_i^2$. The transmission coefficient $\alpha_T$ is expressed from the reflection coefficient as

$$\alpha_T = (1 - \alpha_r)$$

Thus, it can be said that the sum of the transmitted energy and the reflected energy is equal to one. If normal incidence is assumed where the sound waves impinge on the wall from a direction perpendicular to the surface of the wall, then for three media such as earth-wall-air indicated by the subscripts 1, 2 and 0 respectively, the transmission coefficient is expressed as

$$\alpha_T = \frac{4(\rho c)_0 \, (\rho c)_1}{[(\rho c)_0 + (\rho c)_1]^2 \cos^2 k_2 L + [(\rho c)_2 + (\rho c)_0 \, (\rho c)_1/(\rho c)_2]^2 \sin^2 k_2 L}$$

where $k = 2\pi/\lambda$ is the wave number of the sound and L is the thickness of the wall. A very common case in structures is that in which the medium on each side of the wall is air. Then $(\rho c)_1 = (\rho c)_0$. However, the wall is still much more dense than the surrounding air so that $(\rho c)_2$ is much greater than $(\rho c)_0$. With air on either side of an average wall, such as shown in Fig. 2.5 (approximately 4-in thick), and conversation as the incident sound

$$\alpha_T \doteq \frac{4(\rho c)_0^2}{(\rho c)_2^2 \sin^2 k_2 L}$$

For low-frequency sounds and thin walls, the product $k_2 L$ is small so that $\sin^2 k_2 L$

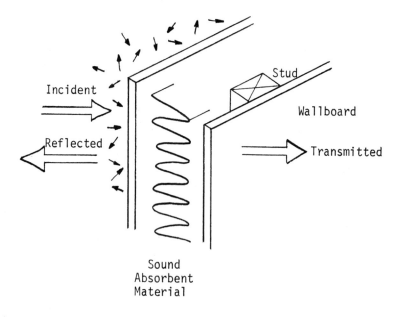

Fig. 2.5   Typical drywall construction

equals $(kL)^2$, approximately. Considering the product $\rho_2 L$ as an area density $\sigma$ that has units of lbm/ft², the transmission coefficient for a typical wall can be expressed in terms of an area density and frequency as

$$\alpha_T \doteq \left[ \frac{(\rho c)_0}{\pi \sigma f} \right]^2$$

Table 2 shows some of the approximate surface densities of common building materials.

**TABLE 2**
**SURFACE DENSITY OF COMMON BUILDING MATERIALS**

| MATERIAL | SURFACE AREA DENSITY $\sigma(\text{lb}_m/\text{ft}^2)$* | MATERIAL | SURFACE AREA DENSITY $\sigma(\text{lb}_m/\text{ft}^2)$* |
|---|---|---|---|
| Gypsum wallboard | 5 | Brick | 10 |
| Concrete | 12 | Aluminum | 14 |
| Wood | 2–4 | Lead | 65 |
| Glass | 15 | Steel | 40 |

*Density per inch of thickness

One of the requirements of an enclosure is that it have a high transmission loss to insure that intense noise sources placed within such an enclosure will be acoustically isolated from the local environment. Conversely, exterior sounds can be effectively attenuated before passing into the interior of an acoustically quiet room. Sound and noise in a structure can be satisfactorily controlled.

Airborne sound is produced by sources which radiate directly into the air, penetrating into or emanating from an enclosure by way of small openings in the walls, floor and ceiling. Airborne sound causes structural panels to vibrate so that noise is ultimately radiated from the opposite side. Noise, if it is to be properly controlled, requires such treatments as the sealing of paths of direct transmission (acoustical flanking) and a reducing of the degree of sound transmission via mechanical vibrations. A very effective way of reducing acoustical transmission of the latter type is to use appreciable mass. For instance, a single, heavy wall of cinder block or poured concrete effectively reduces sound transmission. Each doubling of the weight of a wall increases attenuation by approximately 5 dB. However, there are obviously economical and structural limitations associated with this treatment. Another method of attenuating sound is to use construction practices which physically separate interior and exterior surfaces. The resulting discontinuity of a structure with no connection between wall faces significantly reduces the transmission of sound. The discontinuity can be formed from resilient mountings, staggered wall studs or separated and independent walls. These constructions are made more effective by adding acoustically absorbent material in the void of the discontinuity. Material such as $3/4$ lbf/ft$^3$ Fiberglas wool can increase the resulting transmission loss by 5 dB or more.

## 8. TRANSMISSION THROUGH FLOORS

Impact noise is produced by footsteps, dropped objects and the impulse vibrations of mechanical equipment. This noise is generally observed in the form of sound transmission between the floor of one room and the ceiling of the room below it. The transmission of impact noise is reduced by severing the direct structural connections which carry vibration and transmit noise between the rooms. It is also reduced by absorbing the energy of the impact motion with carpeting or other resilient material. One method for control of this noise in industrial construction is to use floating ceiling construction. Isolation of a ceiling from its structure by use of resilient metal channels breaks the vibration path and reduces sound transmission 6 to 8 dB. This is true for impact noise as well as for airborne sound. If acoustically-absorbent material is placed in the voids above the floating ceiling, then transmission loss between rooms can be increased by an additional 6 to 8 dB. Carpet and padding placed on the upper floor are also effective and will produce a 10-to 15-dB reduction of impact noise. However, there is negligible reduction due to carpet and padding of the airborne sound. Impact noise can be isolated from the ceiling of a lower room when the basic structure is formed with solid, reinforced concrete. Good sound isolation results when the concrete slab is covered with a resilient underlayment of Fiberglas wool over which a very lightweight floor is placed. In this latter type of construction many variations exist. For instance, one method uses a resilient quilt to completely cover the slab. Then a raft of 2 x 2-in battens (furring strips) is placed on the quilt and the flooring is nailed to the battens. A baseboard molding is used to close the space between the walls and the floor. The resulting floor is completely isolated from the building structure and the impact noise is reduced because of the discontinuity (see Fig. 2.6). Another construction method recommends the pouring of a thin concrete slab over a resilient quilt. If the edges of the quilt are turned up at the walls, then the slab will be totally isolated from the structure of the room. Again a baseboard molding is required to close off the space around the edges of the floating floor.

Suspended ceilings are used in a great many modern office buildings. They are composed of a grid of tee-shaped metal strips which form rectangular openings of 2 x 2 ft, 2 x 4 ft, or 2 x 8 ft. Fiberglas acoustical panels, available with a variety of facing patterns and lighting panels, are then inserted into the grid for an attractive and low-cost sound-absorbent ceiling. The grid is suspended with a series of thin wires attached to the underside of the structural ceiling. Suspended ceilings are of little value in attenuating impact noise from the rooms above or in affecting an increase in the transmission loss of a structure. They are important only in absorbing sound in a given enclosure. Consider a large enclosure that has been partitioned into individual office spaces with a common suspended ceiling used to form the upper barrier. This condition creates a large plenum chamber above all the office spaces. When one of the rooms is used for noisy office equipment, or if a noise source such as a blower is coupled to the plenum chamber, then it is possible for sound to propagate through the upper space and result in an increased sound level in one or all of the offices. For this reason, it is suggested that sound-absorbing baffles be placed in these plenum-like chambers to reduce levels of transmitted sound.

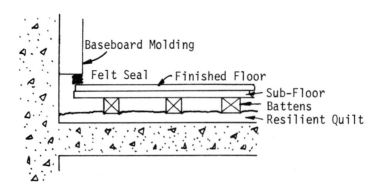

Fig. 2.6   Floating floor construction on a concrete slab structure

## 9. STRUCTURE-BORNE NOISE

Numerous theoretical and empirical expressions have been developed relating power levels and sound-pressure levels in a room. Some of the more important ones are noted here. For a small, simple, sound source that radiates acoustic energy uniformly in free space the power level of the source is

$$PWL = SPL + 10 \, Log_{10} \, A - 10.5 \, dB \quad (re \; 10^{-12} \, watts)$$

where A is the total surface area in square feet through which the sound energy propagates. When the acoustic radiation is directional and the source is located in a not too reverberant room,

$$PWL = SPL_r - 10 \, Log_{10} \left( \frac{D_\theta}{4\pi r^2} + \frac{4}{R_T} \right) - 10.5 \, dB$$

where r is the distance to the point of measurement, $D_\theta$ is the directivity factor along the axis and $R_T$ is the total room absorptivity or room constant $A_T \bar{\alpha}/(1 - \bar{\alpha})$. If a room is a very large enclosure with highly absorptive walls where wall effects are negligible on the interior sound field in the room and the floor surface is hard and reflective, then for measurements averaged on an idealized hemisphere of radius r about the source.

$$PWL = \overline{SPL_H} + 20 \, Log_{10} \, r - 2.5 \, dB$$

For a properly suspended source with measurements averaged on an idealized sphere of radius r (free-field propagation)

$$PWL = \overline{SPL_S} + 20 \, Log_{10} \, r + 0.5 \, dB$$

In a reverberant room, the power level of a source is calculated from

$$PWL = \overline{SPL} + 10 \, Log_{10} \, V - 10 \, Log_{10} \, T - 29 \, dB$$

where $\overline{SPL}$ is the space-averaged sound-pressure level, V is the volume of the room in ft$^3$ and T is the reverberation time in seconds.

The success of wall construction for sound attenuation often depends on the background noise of the receiving side, since transmitted noise is normally kept below background levels. Therefore, higher performance is required where background noise is minimal. For example, suppose an office must be maintained at a low existing background level when a piece of machinery is installed in an adjoining room. Since it is unlikely that an existing wall would be much more than 6-in single cinder block or 1/2-in plasterboard on 2 x 4-in studs which have transmission losses of approximately 40 dB, it would not be unusual in the case of a noisy machine to incorporate local sound isolation of the machinery at the time of installation.

The performance of a sound barrier is denoted in terms of STC (sound transmission class). This is a single number rating in dB for determining sound transmission loss performance of a wall or floor as related to airborne sound; the higher the rating number, the better the implied performance. The transmission loss is defined by the expression

$$TL = 10 \, Log_{10} \left( \frac{I_i}{I_T} \right) dB$$

where $I_i$ and $I_T$ refer to the incident and transmitted intensities of a sound wave, respectively. The ratio of the sound intensities is the transmission coefficient. Hence, transmission loss becomes

$$TL = 20 \, Log_{10} \frac{\pi}{\rho_0 c} + 20 \, Log_{10} \sigma f \, dB$$

Values of sound-transmission class have been defined in ASTM E90–61T and are incorporated in such standards as the Federal Housing Administration's Minimum Property Standards No. 2600 which establishes criteria for various walls as related to background noise.

Sound-transmission class values with respect to frequency are determined in the following manner. The basic contour of STC-20 starts with 0 dB at 125 Hz, increasing by 9 dB/octave to 14 dB at 350 Hz; then it increases by 3 dB/octave to 20 dB at 1400 Hz after which it remains constant to 4 kHz. All other STC curves are parallel to this initial contour, but at different levels of sound transmission loss as shown. (Fig. 2.7). Walls are classified so that in the mid-frequency range, 350 Hz to 4 kHz, the TL of the wall falls on or above a given STC contour. In the lower- and higher-frequency regions. TL values may be less than the STC contour by not more than a total of 3 dB.

The transmission loss through homogeneous solid walls constructed with common building materials depends on the product of $\sigma$ f. With each doubling of the value of this product, due to either frequency or surface density, the transmission loss increases by approximately 5 dB. Hence, the sound-attenuating property of a wall is most pronounced with high-frequency sound and when using dense materials for construction. Multi-wall construction is effectively used for enclosures when high values of transmission loss (for TL greater than 40 dB) must be obtained. Typical construction details of different STC-type walls are shown in Fig. 2.8. Experience has demonstrated that for comparable weights of construction, multi-wall construction has a transmission loss of about 8 dB greater than that which can be obtained with a single wall. The transmission loss through a wall or area that separates a room or space (subscript 1) containing a sound source from one that receives the sound (subscript 2) is expressed as

$$TL = \overline{SPL_1} - SPL_2 + 10 \, Log_{10} \left( \frac{1}{4} + \frac{A_W}{R_2} \right) dB$$

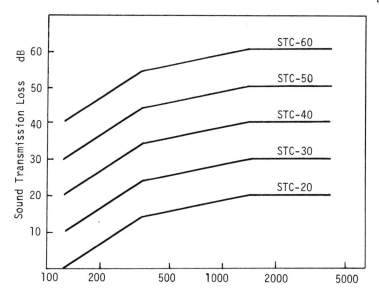

Fig. 2.7   Sound transmission contours

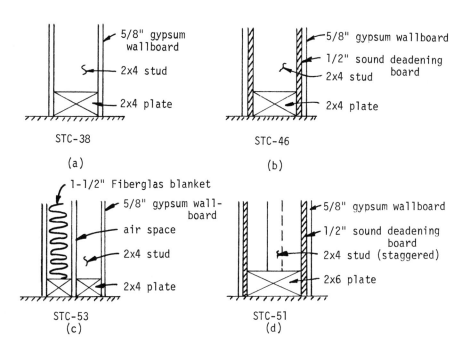

Fig. 2.8   Various STC rate wall constructions

where $A_W$ is the area of the wall or opening through which the sound propagates and $R_2$ is the room absorptivity. The sound-pressure level $SPL_2$ is measured and averaged close to the transmitting wall and $SPL_1$ is the rms average of the sound level in the space containing the source. If the sound-receiving room is reverberant ($\bar{\alpha}_2$ less than 0.2), then $A_W/R_2$ is greater than $1/4$ and transmission loss becomes

$$TL = \overline{SPL_1} - SPL_2 + 10 \, Log_{10} \frac{A_W}{A_2 \bar{\alpha}_2} \quad dB$$

where $A_2$ is the total surface area of the room receiving the sound.

For the case where the receiving enclosure is highly damped ($\bar{\alpha}_2$ greater than 0.3), or the receiving space is the out-of-doors,

$$TL = \overline{SPL_1} - SPL_2 - 6 \, dB.$$

The previous discussion implies that if a structural wall is placed between rooms, one of which contains a source of sound, then effective transmission loss or sound attenuation will occur. But Beranek[4,5] points out that wall panels resonate at all frequencies above some critical frequency for oblique incidence. With low excitation frequencies the panel is stiffness controlled. As the frequency increases, the panel passes through various structural resonances after which response becomes mass controlled with a transmission loss of approximately 6 dB per octave. At a frequency of two to three times the fundamental resonant frequency of the wall, the critical frequency occurs at which the flexural wavelengths of the panel equal the wavelengths of the transmitted sound. This critical frequency is given by the expression

$$f_c = \frac{13.9 \, c^2}{L} \sqrt{\frac{\rho}{E}} \quad Hz$$

where L is the wall thickness in inches, $\rho$ is the mass density of the wall (lbm/ft³) and E is Young's Modulus (lbf/in²). The above equation is exact only when the wavelength of the flexural wave of the wall is about six times the thickness L. In the frequency range where a damped panel or wall is mass controlled, transmission loss for normally incident sound is expressed as

$$TL_N = 10 \, Log_{10} \left[ 1 + \left( \frac{\pi \sigma f}{\rho c} \right)^2 \right] \quad dB$$

If the sound impinging on a wall is randomly incident as it would be in a live or reverberant room, then the latter expression becomes

$$TL = TL_N - 10 \, Log_{10} \, (0.23 \, TL_N) \quad dB$$

Inspection of this latter equation reveals that transmission loss from a diffuse sound field is less than that occurring from a field with normal incidence. This is because a larger amount of reflection is associated with the obliquely incident waves.

Because of their influence on the sound and noise in a structure, it is worthwhile to mention here air conditioning and ventilation systems and their contributions to the general problems of sound. A typical system contains some means for moving air, such as a fan or blower, and a network of air-distributing ducts. The air is drawn through a main return duct to the fan inlet and distributed beyond the outlet through the system of ducts to various working spaces and offices. As air travels through the duct work, mechanical vibrations are produced which result in radiated sound. In addition, sounds from any one air-conditioned space can propagate back through the duct work into other office environments.

Sound is also produced in the motion of the fan or blower wheel, and this is distributed throughout the system from both the fan inlet and outlet duct work. The sound-power level generated in either duct given in terms of the fan horse-power and static pressure is

$$PWL \doteq 90 + 10 \, Log \, hp + 10 \, Log \, p \, dB \quad (re \, 10^{-12} \, watts)$$

where p is the static pressure at the fan outlet in inches of water. This level can also be expressed in terms of pressure and flow rate as

$$PWL \doteq 55 + 10 \, Log \, Q + 20 \, Log \, p \quad dB$$

where Q is the volume flow rate in cfm or

$$PWL = 125 + 20 \, Log \, hp - 10 \, Log \, Q \quad dB$$

These equations are for approximation only and apply to conventional air-moving fans. Axial fans have frequency spectra that are nearly flat. On the other hand, centrifugal fans with forward curved blades have spectra that drop off at a rate of approximately 5 dB per octave with increasing frequency. Centrifugal fans with backward curved blades radiate more noise in the higher frequencies than those with forward curved blades. Therefore, they are more amenable to systems with long duct work in which acoustical treatment can be used to absorb the higher-frequency sound.

Noise is created by different components of an air-distribution system. For example, air passing through the inlet grill in a room generates noise because of air turbulence. Such noise contains frequencies that coincide with those of normal speech. Hence, there is a potential of speech interference due to the phenomena of masking. Noise in duct work can be attenuated by various methods. Fan inlet and outlet duct work, when fitted with *package attenuators*, can yield noise-reduction levels of from 2 to 16 dB in the 20 to 75 Hz frequency band. Approximately 12 to 68 dB of attenuation can be obtained in the 2400 to 4800 Hz frequency band. This varies with each manufacturer where the larger, more complex and expensive units produce the higher values of noise reduction.

A plenum chamber is a type of package attenuator used to reduce turbulence in the flow of air. It consists of a volume expansion in the duct system that is heavily lined with sound-absorbing material. Typically, these chambers are placed downstream of heating and cooling equipment where air turbulence is more pronounced. Airborne sound and noise is reduced as the velocity of the airflow decreases in the expansion and the sound waves reflect about the space.

For plane waves in a duct, the power level of the sound is approximately

$$PWL \doteq \overline{SPL} + 10 \text{ Log}_{10} A_c - 10 \text{ dB} \quad (\text{re } 10^{-12} \text{ watts})$$

where $\overline{SPL}$ is the value of the sound-pressure level averaged across the duct and $A_c$ is the cross-sectional area in square feet. Further, noise radiated through the duct wall will have a power level on the outside of

$$PWL_0 = PWL_i - TL + 10 \text{ Log}_{10} \frac{A_W}{A_c} \quad \text{dB(re } 10^{-12} \text{ watts)}$$

where $A_W$ is the area of the wall through which the sound is transmitted. System noise is reduced through the use of acoustical lining in the duct work. A common practice is to install a layer of 1-in thick Fiberglas wool on each surface in the duct for a distance of several feet from each outlet and inlet of the system. This provides reasonable sound reduction at low cost. For a more detailed discussion of noise control in ventilation systems the reader is referred to the listed references.

# REFERENCES

1. Baxter, R. L. and Bernard, D. L. "Vibration, an Indicating Tool." *Mechanical Engineering* (March 1968): 36–41.
2. Blake, M. P. "New Vibration Standards for Maintenance." *Hydrocarbon Processing and Petroleum Refiner* 43, no. 1 (January 1964): 111–114.
3. Bowen, K. A. and Graham, T. S. "Noise Analysis: A Maintenance Indicator." *Mechanical Engineering* (October 1967): 31–33.
4. Beranek, L. L., Reynolds, J. L. and Wilson, K. E. *J. Acoust. Soc. Am.* 25 (1953): 313.
5. Beranek, L. L. *Noise Reduction.* New York: McGraw-Hill Book Co., 1960.

# BIBLIOGRAPHY

American Standard S1.1–1960. Acoustical Terminology (including Mechanical Shock and Vibration). New York: The United States of America Standards Institute, 1960.

Crede, C. Shock and Vibration Concepts in Engineering Design. Englewood Cliffs: Prentice-Hall, Inc., 1965.

Gregory, R. W., et al. "Dynamic Behavior of Spur Gears," Proc. Instn. Mech. Engrs. 178, no. 8 (1963–64): 207–226.

Kinsler, L. E. and Frey, A. R. Fundamentals of Acoustics. New York: John Wiley and Sons, Inc., 1962.

Kenny, R. J. "Noise in Air-Moving Systems." Machine Design, September 26, 1968.

Jacobsen, L. S. and Ayre, R. S. Engineering Vibrations. New York: McGraw-Hill Book Co., 1958.

Johnson, D. C. and Bishop, R. E. D. "A Note on the Excitation of Vibrating Systems by Gearing Errors." Jour. of the Roy. Aero. Soc. 59 (1955): 434–435.

London, A. "Methods for Determining Sound Transmission Loss in the Field." J. Res. Nat. Bur. Stds. 26 (1941): 419–453.

Miller, Thomas D. "Machine Noise Analysis and Reduction." Sound and Vibration (March, 1967): 8–14.

Moeller, K. "Bearing Noise." In Handbook of Noise Control, edited by C. Harris, Chap. 10. New York: McGraw-Hill Book Co., 1957.

Rieger, N. F. "Vibration in Geared Systems." Machine Design, September 16, 1965.

Wells, R. J. "Apparatus Noise Measurement." AIEE, Power Apparatus and Systems, December 1955.

Quiet Comfort for Multi-Family Housing Wood Frame Construction. Owens-Corning Fiberglas Corp. Pub. No. I–BL–3863A.

Noise Control With Insulation Board, 4th. ed. Insulation Board Institute, A.I.A. File No. 39–B.

Palmgren, A. Ball and Roller Bearing Engineering. Philadelphia: SKF Industries Inc., 1945.

# SYMBOLS

| | | |
|---|---|---|
| **D** | deviation | —— |
| **f** | frequency | cycles/sec |
| **m** | meter | —— |
| **n** | integer, total possible events | —— |
| **N** | actual number of events, Newtons | —— |
| **p** | probability, cumulative probability | —— |
| **P** | probability density | —— |
| **S** | spectral density | —— |
| **T** | period | second |
| **t** | time, total time | second |
| **V** | velocity | ft/sec |
| **x** | instantaneous magnitude | —— |
| **X** | peak amplitude | —— |
| $\omega$ | angular frequency $2\pi f$ | radians/sec |
| $\phi$ | phase angle | radian |
| $\sigma$ | standard deviation | —— |
| $\lambda$ | positive number | —— |
| $\mu$ | micro- | —— |

# 3

# RANDOM VIBRATION

*William S. Mitchell*

## 1. INTRODUCTION

The time-history characteristics of a random vibration are such that magnitudes are never repeated periodically in time due to the random nature of the vibration. Random vibration is evidenced by a nonperiodic waveform having a continually changing magnitude that varies in an unpredictable way. This is quite unlike a harmonic or complex periodic vibration. Whereas periodic vibration is completely described by its amplitude, frequency and phase characteristics, random vibration can be described only by such quantities as root-mean-square (rms) magnitude, power spectral density and probability distribution of magnitudes. If we possessed a tape recording of the sounds of a city street corner over a period of time, say one day, we would have a record of a random event. This record would contain the sounds of ringing bells, passing cars, footsteps, conversation and all other associated sounds. It would be a random record because there would be no predictability of future events. Indeed, the events are neither constant nor periodic; therefore, they must be random.

Two types of random vibration are distinguished: wideband and narrowband vibration. Wideband random vibration is composed of a continuous spectrum of frequencies rather than the harmonics of any one particular frequency. Because of its unpredictable response, there is no analytical expression for wideband random noise. All frequencies are equally possible; the function is not associated with any particular frequency band. Narrowband random vibration is essentially a continuous, nonperiodic single frequency wave that is sometimes referred to as random sine vibration. This is because the basic signal appears to be sinusoidal in form, yet its amplitude variation is quite random. An instantaneous narrowband random vibration is expressed by a relation of the form

$$x(t) = X(t)[\cos \omega t + \phi(t)]$$

where $X(t)$ is an unpredictable time-dependent amplitude and $\phi(t)$ is a randomly varying phase angle which has negligible meaning for a random signal. These latter variations can be assumed small with respect to the angular frequency $\omega$.

This implies that such a function is a slowly varying, almost single-frequency wave which resembles in form an amplitude-modulated carrier wave. Narrowband and wideband random waveforms can be seen in Fig. 3.1.

## 2. STATISTICAL PROCESSES

The instantaneous magnitude of a random vibration cannot be described uniquely at any given instant of time by ordinary mathematics or analysis. The more specialized theories of statistics and probability must be used to analyze the vibratory response and predict its effective values. With respect to the fundamentals of statistical analysis, we first define some basic concepts.

Consider n events (an event could be, for instance, a value of displacement of which there are many), each event being described by a value $x_i$. The average or arithmetic mean value $\bar{x}$ of many similar events (n events) is expressed* as

$$\bar{x} = \frac{1}{n}\sum_{i=1}^{n} x_i = \frac{1}{n}(x_1 + x_2 + x_3 + \ldots x_n)$$

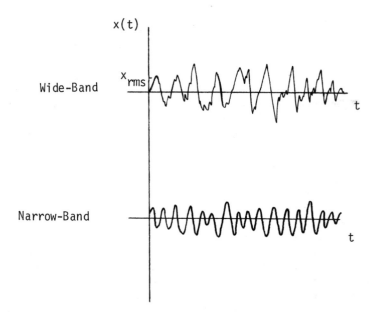

Fig. 3.1    Amplitude-time relationship for random signals

---

* Where the symbol $\sum_{i=1}^{n}$ implies an addition of the n values, $x_1 + x_2 + \ldots x_n$, and n is the number of individual events.

Now, the difference between the average value and any specific value of an event is called the deviation

$$D_i = x_i - \bar{x}$$

Some of the specific events $x_i$, such as velocity or acceleration amplitudes, will be greater than the average value and some will, obviously, be less. The consequence of this relation is that the average deviation

$$\bar{D} = \frac{1}{n} \sum_{i=1}^{n} D_i = \frac{1}{n} [(x_1 - \bar{x}) + (x_2 - \bar{x}) + \ldots (x_n + \bar{x})]$$

is identically zero.

When the squares of the deviations are summed for the n events and divided by the quantity n, the result is the variance or mean square of the data

$$\bar{x^2} = \frac{1}{n} \sum_{i=1}^{n} (x_i - \bar{x})^2$$

The square root of the variance is the standard deviation or rms

$$x_{rms} = \sqrt{\frac{1}{n} \sum_{i=1}^{n} (x_i - \bar{x})^2}$$

What has been described above from statistical theory is related to simple, specific, individual events. Random vibration, however, is generally a continuous series of events showing that when the theory is extended to a continuous process the complexity of the mathematics is increased.

In its simplest form probability theory provides a way of designating a single number to indicate the chances of one single event or amplitude value occurring out of many possibilities. For example, suppose it is desired to withdraw a red ace from a deck of cards. There are fifty-two cards, two of which are red aces. The probability of such a success on the first draw is 1/26, a probable chance of one in twenty-six. Suppose three withdrawals are allowed. On the first draw the probability is 1/26; the second is 2/51; and the third is 1/25, since each withdrawal reduces the number of possible selections by one. In this case probability increases with each selection. However, if a total of three selections are allowed to pick the red ace, then the probability is increased to 1/9. Thus, if events can occur in various ways, it is possible to ascribe probability numbers between zero and one to the occurrence of the specific event.

Consider now ten amplitude events, random and independent, occurring simultaneously. At a given instant of time, each of the ten events has a particular value; hence, each will have an associated probability of occurrence. If these probability values are combined so as to present a probability of their occurrence within a specified amplitude range, then this is called a probability distribution, usually presented in graphical form. Probability distribution says that at a given instant of time there is a probability (or chance) that one of the amplitudes in question will be within a specified amplitude range. At some later period of time, a different set of random amplitudes exists. When this latter set is combined to yield a probability distribution as before, we should expect to obtain a new set of values different from those of the first set. In other words, it can be expected that the statistical results will vary with time. It is possible that in certain random systems, statistical values will be the same for every given period of time. Suppose that for an extended period of time the probability parameters remain constant. Such random processes whose statistical properties are independent of time are called "stationary processes." For example, the graphic response of a wideband random vibration appears to be distributed about a zero mean value. If there exists a statistical average such as the RMS level of magnitude that is independent of time, then the process is said to be stationary. Conversely, those processes that are not identified as stationary are called nonstationary processes. For instance, the sound level at a street corner in a country town is statistically stationary from midnight to five o'clock in the morning. No one passes by; only background noise is present and this is generally constant. However, during the daytime hours the noise level is constantly changing due to the traffic and passing pedestrians. This is a random, nonstationary condition. Because of the randomness of wideband and narrowband response, it is important that statistical analysis be performed with a large number of data points. This requires an ample supply of recorded data to insure dependable results. Random response does not need to be stationary for a total recorded signal. It may happen that statistical averages change slowly with time. In these particular instances, it is permissible to subdivide the recorded data into shorter intervals of time and process each interval that satisfies, in itself, the condition of stationarity.

In those instances where numerous data are not available, it is still possible to obtain useful information about a random process from only a few individual records. Rather than considering probability distributions in a group of individual records at specific times $t_1$, $t_2$, $t_3$, ... $t_n$, it suffices to consider a time average of the amplitude values of a single record, assuming that the record reflects the group and the process is stationary. From such a process, all desired statistical properties of a system can be obtained. In order that reliable data will evolve from the random analyses, certain conditions must be satisfied. First, the process being described has to be continuous and finite over a given interval of time. Second, time-averaged parameters of a single record must be the same as the average of a set of records. When these conditions are satisfied, the process is called an "ergodic" random process. Under these conditions only one data record from a set of records need be examined to obtain useful statistical data. This data will consist of such statistical properties of the random process as the rms magnitude, power density spectrum and the magnitude probability distribution. It is seldom possible to acquire multiple sets of vibrational data because of the time and expense involved. Thus, it is generally assumed that a process is ergodic; that is, for various time-histories the statistical parameters will be constant unless the contrary can be shown. Note should be taken of the fact that, while an ergodic process is stationary, the converse is not true.

## 3. STATISTICAL SIGNALS

According to Fourier analysis, the waveform of a periodic vibration is composed of discrete sinusoidal components. Hence, while the amplitude average or mean value of the vibration is zero, that is

$$\bar{x} = \frac{1}{n} \sum_{i=1}^{n} x_i = 0$$

the mean-square value of the amplitude of a harmonic component is given as

$$\overline{x_i^2} = \frac{1}{2} x_i^2$$

The mean-square value of the periodic wave is

$$\overline{x^2} = \frac{1}{n} \sum_{1}^{n} x_i^2$$

Further, the rms value of the periodic wave can be expressed as

$$x_{rms} = \sqrt{\frac{1}{n} \sum_{i=1}^{n} x_i^2}$$

For a random signal processed over a sufficiently long period of time, T, the mathematics of simple statistics now becomes, in terms of time-averaged events:

ARITHMETIC MEAN          $\bar{x} = \frac{1}{T} \int_0^T x(t)\, dt$          *Average*

MEAN-SQUARE              $\overline{x^2} = \frac{1}{T} \int_0^T x(t)^2\, dt$          *Variance*

ROOT-MEAN-SQUARE     $x_{rms} = \sqrt{\underset{T \to \infty}{\text{Limit}} \frac{1}{T} \int_0^T x(t)^2\, dt}$     *Standard Deviation*

The rms value is often referred to in statistical theory as the standard deviation $\sigma$. It is an effective value of the signal analogous to the $AC_{rms}$ value of an alternating electric current. For example, the mean value of a sinusoidal function $x(t) = X_0 \sin \omega t$ is identically zero. In addition, it has an rms value $x_{rms} = X_0/\sqrt{2}$. Thus, its mean-square value is $\overline{x^2} = X_0^2/2$. In a manner similar to that which was used to calculate rms values in terms of time for a given signal, it is likewise possible to describe such values in terms of frequency. For a periodic vibration, the rms value of the signal displacement amplitude is expressed as

$$x_{rms} = \sqrt{\underset{f \to \infty}{\text{Limit}} \frac{1}{f} \int_0^f x(f)^2\, df}$$

where displacement must be known as a function of frequency. This latter calculation is not possible for random signals since displacement is neither a function
of time nor frequency. However, it can be found with respect to frequency when
averaged over a long period of time.

Consider a wideband random vibration that is a function of time, x(t). If transducer data from this motion is passed through a bandpass filter set at a center
frequency $f_0$ with a bandwidth of $\Delta f$, then the resulting instantaneous signal
becomes a function of frequency and time, x(f,t). In addition, the signal is converted at the filter to a narrowband random response. When the filtered signal
is measured for its rms content, it becomes a function of frequency alone,
$x(f)_{rms}$. This, in fact, is a basis for the experimental determination of both the
rms value of a wideband random signal with respect to frequency and the value
of the spectral density S(f) (some authors refer to this quantity as power spectral
density). Block diagrams of measurement systems are shown in Fig. 3.2. The
spectral density of a wideband random signal is defined as the mean-square value
of the signal per frequency bandwidth expressed as

$$S(f) = \lim_{\Delta f \to 1} \frac{\overline{x(f)^2}}{\Delta f}$$

where $\overline{x(f)^2}$ is the mean-square filtered amplitude of the random signal (wideband
signal filtered to narrowband) and f is the bandwidth at the center frequency of
the filter. It becomes obvious that the rms value of a random wave (and, consequently, spectral density) depends on the bandwidth of the filtering process.
A wide bandwidth will include data far removed from the center frequency of
the filter. On the other hand, a very narrow bandwidth restricts signal analysis
to information located in the region immediate to the center frequency. In terms

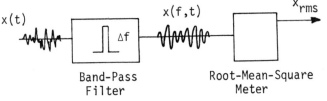

Band-Pass
Filter

Root-Mean-Square
Meter

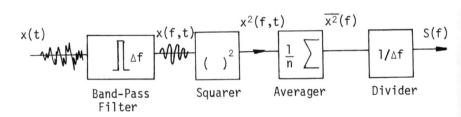

Band-Pass
Filter

Squarer          Averager          Divider

Fig. 3.2   Measurement of root-mean-square and power spectral density

of frequency and spectral density the rms value of the random vibration is now given as

$$x_{rms} = \sqrt{\int_f S(f)df}$$

where integration is taken over the frequency range of the random signal. This is equivalent to a summing process $\Sigma \bar{S}(f) \cdot \Delta f$ and as such is equal to an area. The rms value of a random vibration is therefore equivalent to the square root of the value of the area bounded by the spectral density curve.

Once a large number of spectral density values have been obtained, they can be plotted as a function of frequency. The result is a graphical display of the energy distribution in the wideband random vibration. This is appropriately called a power density spectrum. The power density spectrum depends on the specific distribution of energy. For example, if a complex periodic vibration is analyzed and displayed as noted above, the resulting power density spectrum will be a discrete spectrum rather than a continuous one. In this case all the energy of the signal is allocated to the specific frequencies which comprise the input. This is in accordance with Fourier's theorem which states that for a periodic, continuous wave the signal can be described by a summation of different frequency sinusoids. A wideband random vibration is composed of a sum of many frequencies; hence, the designation wideband. In some cases the band may be biased toward the lower frequencies, in others it may be in the higher range, or it may even contain all frequencies, high and low. Regardless of the distribution, the power density spectrum of a wideband random vibration is a continuous spectrum whose shape depends on the distribution of the frequency components. When a periodic information signal is masked by, or contained in, random noise, a power density analysis leads to yet another form of spectrum. This spectrum is a superpositioning of the periodic or discrete components on the continuous background spectrum of the masking noise. These various spectra are shown in Fig. 3.3.

The manner in which energy is distributed in a wideband random vibration is used to identify various types of signal noise. Pink noise is defined as a wideband random signal in which the spectrum level decreases with increasing frequency so as to produce a constant energy level per octave bandwidth. This amounts to an energy change of $-3$ dB per octave. USASI noise (USASI Standard S1.4–1961) is filtered random noise having an increasing energy level which peaks around 200 cps, followed by attenuation of the signal at the higher frequencies. The level change is $\pm 6$ dB per octave. A third type of noise, shown in Fig. 3.4, is white noise. This is defined as wideband random noise whose spectral density level is constant over the entire frequency spectrum. It has a power density spectrum that is a single, continuous, horizontal line at some specific value of spectral density $S(f)$. If the frequency spectrum is limited, then the power density spectrum is limited also. In this case it becomes band-limited white noise, and its spectrum is a given portion of the white-noise power density spectrum.

Root-mean-square values can be obtained for various alternating quantities. In the measurement of equipment vibration, the important quantities are velocity,

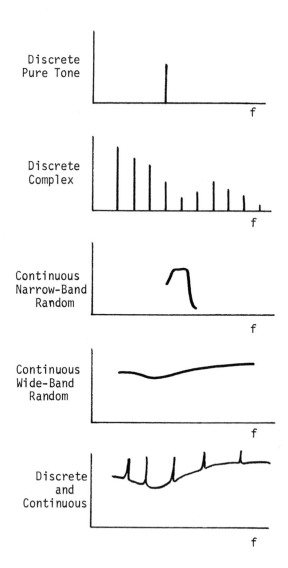

Fig. 3.3   Power density spectra—relative amplitude as a function of
frequency

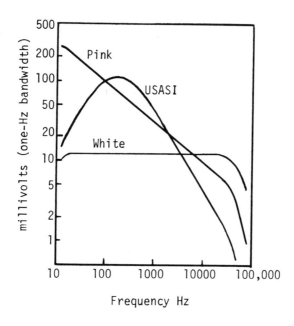

Fig. 3.4   Random noise spectra

displacement and acceleration. Velocity measurements are important when the primary concern is wear or the prevention of internal machine failure. Acceleration measurements yield useful data for the design of structural integrity, and displacement measurements provide data on unbalance and structural deflection. Frequency measurement and power spectra analyses help to locate damaging sources of vibration. Thus, random displacement, velocity and acceleration are all important properties. They can be described in terms of power density spectra since these spectra are a display of the mean-square values of the various amplitudes as a function of vibrational frequency. Random vibration analysis involves such terms as mean value, mean square and rms. It also involves peak values of the random response. The reader is cautioned to understand that peak amplitudes of a random vibration are instantaneous in character and infinite in number and they cannot be treated as a repetitive maximum value like the peak value of a sinusoidal waveform.

The construction of a power density spectrum enables an investigator to visualize the distribution of frequency-related vibrational energy in either an input signal or a vibrational response. Such spectra also make it possible to identify or correct for such special noise spectra as white or pink noise. The above noted information concerns data averaged over an extended period of time, and reveals nothing of the distribution of instantaneous amplitudes in the vibration, whether they be displacement, velocity or acceleration. For this latter type of information, it is necessary to apply the theories of probability.

## 4. PROBABILITY THEORY

It becomes obvious when looking at the display of a random signal $x(t)$ that for some given instant of time the function will have a value between some value of x and $x + \Delta x$. The probability $p(x)$ of the function $x(t)$ occurring in this given amplitude interval, which is defined as the ratio of total time $x(t)$ remains in the interval to the total time of the record, is a characteristic value of the signal required for an analysis of the random vibration. Consider $\Delta N$ individual amplitudes having the same value from a record of events or values which have a total number n where $\Delta N$ is the sum of a limited number of values $N_i$ and $\sum_{i=1}^{n} N_i = n$. Probability of an occurrence for these $N_i$ is defined as the ratio

$$p(x) = \frac{\Delta N}{n}$$

or the number of special amplitudes counted divided by the total number possible or existing. It should be noted that probability depends on the extensiveness of the data since this controls the frequency for which the function $x(t)$ remains in a given interval $\Delta x$, or as it is commonly called, the frequency of occurrence. Thus, probability phenomena have meaning only for very large numbers of events or values. Random data satisfies this condition.

Consider the signal shown in Fig. 3.5. Since probability ranges from zero to one, values of $\Delta x_i = x_i - x(t)_{min}$ in a range $0 \leq \Delta x_i \leq [x(t)_{max} - x(t)_{min}]^*$ yield

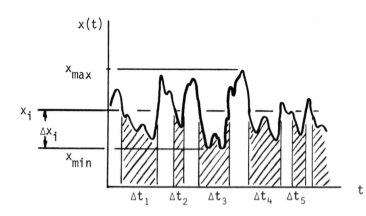

Fig. 3.5   Random signal for cumulative probability calculation

---

\* The notation $\leq$ is read "less than or equal to"; $\geq$ is read "greater than or equal to."

for all $x_i$ a cumulative probability of the total time $t_{xi} = \Sigma \Delta t_i$ that the function $x(t)$ is within $\Delta x_i$

$$p_{xi} = \frac{t_{xi}}{t}$$

where t is the total time of the analysis. It can be seen from the illustration that each $\Delta t_i$ will change in value with a change in $x_i$. A plot of cumulative probability $p_{xi}$ versus $\Delta x_i$ results in a cumulative probability curve such as is shown in Fig. 3.6. Applying these concepts to a random vibration containing amplitude values $x_i$ at times $t_i$ in a range of amplitude x to $x + \Delta x$, where occurrence time for a given $x_i$ is expressed as $\Delta t_i = \Sigma t_i$, the probability of this $x_i$ occurring in time t is given by

$$p(x) = \frac{\Delta t_i}{t}$$

And when $p(x)$ is plotted with respect to $x_i$, the result is a curve of probability versus amplitude as shown by Fig. 3.7. Probability density is defined by the ratio of the probability to the amplitude interval over which the probability is defined. For example, in the limiting process at $x_i$, the probability density approaches the tangent value of the probability curve. Hence, the probability density curve is a plot of the loci of tangents of the probability curve with respect to the amplitudes $x_i$; and it is expressed as

$$P = \frac{p(x)}{x_i} = \frac{\Delta t_i}{tx_i}$$

When probability density is plotted with respect to the $x_i$, where $i \to \infty$ *, the result is the probability density curve of Fig. 3.8, assuming statistical regularity. This is the condition for which the probability approaches a limiting value for an

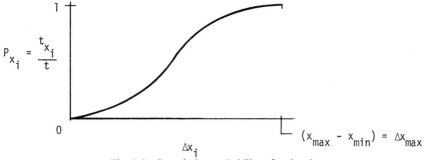

Fig. 3.6 Cumulative probability of a signal

---

* The symbol $\infty$ denotes infinity, a number which is too large to count, unlimited in extent. The symbol $\to$ is read as "approaches or tends to become."

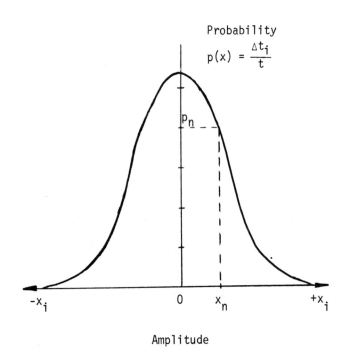

Fig. 3.7   Probability versus amplitude curve

$x_i$ as $i \to \infty$. It should be noted that the probability density curve is centered about the region of $x_i = 0$. Moreover, since the area under the curve is $\Sigma p \cdot \Delta x = \Sigma p(x)$, it follows that this area is always equal to unity.

**Probability Distributions.** Various probability distributions exist in random vibrations because of the many possible combinations of frequency that enter into the makeup of a given response. One particular probability distribution is the binomial distribution. If there are $\Delta N$ existing amplitudes recognized out of n possible independent amplitudes where each amplitude is associated with a probability of its own occurrence $p = p(x)$, the distribution is expressed from the probability function

BINOMIAL $\qquad p(\Delta N) = \dfrac{n!}{(n - \Delta N)!\, \Delta N!}\, p^{\Delta N}(1 - p)^{n-\Delta N}$

where N! is the mathematical expression N-factorial (N) (N − 1) (N − 2) ... (N − N + 1). If n is very large and p, the probability of occurrence is small, the mathematical computation $n!(1 - p)^{n-\Delta N}/(n - \Delta N)!$ becomes extremely labor-

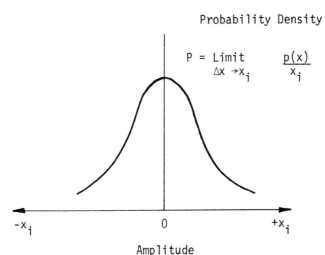

Fig. 3.8    Probability density curve

ious. However, the limit of the binomial distribution for $n \rightarrow \infty$ and $p \rightarrow 0$, where the product $np$ remains constant, that is $np = C$, is expressed by the probability expression

POISSON $$p(x) = \frac{C^{\Delta N} e^{-C}}{\Delta N!}$$

This is the Poisson probability distribution. Wideband random vibrations which normally contain all frequencies have probability distributions that are termed Gaussian or normal distributions. This type of distribution is well-established mathematically and has been found to be applicable to many practicable cases. For a given amplitude $x_i$, the Gaussian normal distribution function gives the probability that $x_i$ will occur in an interval of the response, and this is expressed by

GAUSSIAN $$p(x) = \frac{1}{\sqrt{2\pi} \; x_{rms}} \exp \left[ -(x_i - \bar{x})^2 / 2 \, x^2_{rms} \right]$$

where $\bar{x}$ is the arithmetic mean value. It should be noted that various probability distributions result from special probability functions. Consequently, these expressions of probability are commonly referred to as probability distributions.

Values for the Gaussian error distribution $(1/\sqrt{2\pi})\exp[-(x_i - \bar{x})^2/2x^2_{rms}]$ are readily found in tabulated form. If $x_i$ equals the most probable value of $\bar{x}$ then

$$p(\bar{x}) = p(x) = \frac{1}{\sqrt{2\pi}\ x_{rms}}$$

A plot of $p(x)\sqrt{2\pi}$ versus $\bar{x}$ results in the Gaussian distribution shown in Fig. 3.9, where two characteristics can be noted. First, the rms value is a measure of the width of the distribution curve where large $x_{rms}$ implies a broadness of the curve. Second, the rms value affects the probability value $p(x)$; that is, a small $x_{rms}$ results in a large $p(x)$. However, it should be remembered that the area under the curve is always equal to unity. Given any positive number $\lambda$, the probability of a value of the function $x_i$ being found in the interval $\pm \lambda x_{rms}$ is given by

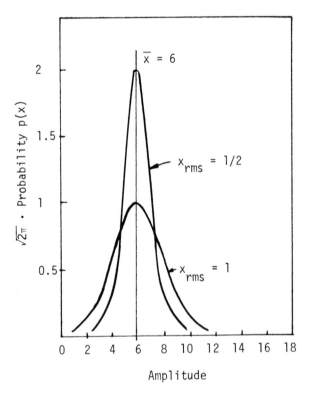

Fig. 3.9   Normal (Gaussian) distribution of a response for a mean value
$\bar{x} = 6$

$$p(x) = \frac{1}{\sqrt{2\pi} \, x_{rms}} \int_{-\lambda x_{rms}}^{\lambda x_{rms}} e^{-(x_i^2/2x_{rms}^2)} \, dx$$

For the values of $\lambda = 1$, 2 and 3, the probabilities that $-\lambda x_{rms} \leq x_i \leq x_{rms}$ are 68.3, 95.4 and 99.7 percent, respectively.

In the analysis of random vibrations, Rice[1] has shown that when the ratio of the number of zero crossings of a random signal to the absolute number of peaks is approximately zero, as it would be for wideband random vibrations, the probability density distribution of the amplitude peaks is Gaussian. However, for narrowband vibrations this same ratio is equal to one. The probability distribution of the amplitude peaks assumes a Rayleigh distribution which, for a mean value of zero, is expressed in standardized form as

RAYLEIGH $$p\left(\frac{x_i}{x_{rms}}\right) = \frac{x_i}{x_{rms}} \cdot e^{-x_i^2/2x_{rms}^2}$$

where $x_i$ greater than zero is the amplitude of a vibration peak.

Figure 3.10 exhibits the obvious fact that probability is highest when a peak value $x_i$ equals the rms value $x_{rms}$. But when Gaussian probability is plotted versus $x_i/x_{rms}$, it is found that probability is highest and centered about the mean value of $\bar{x} = 0$. Note that the Rayleigh distribution is centered about one. The probability of a peak value $x_i$ exceeding a value $\lambda x_{rms}$ in a Rayleigh distribution is

$$p(x) = \int_{\lambda x_{rms}}^{\infty} \frac{x_i}{x_{rms}^2} \cdot e^{-x_i^2/2x_{rms}^2} \, dx$$

For values of $\lambda = 1$, 2 and 3 the probabilities that $x_i$ is greater than $\lambda x_{rms}$ are 60.7, 13.5 and 1.2 percent, respectively.

Probability distribution is related to mean values for stationary processes as follows.

MEAN VALUE $$\bar{x} = \int_{-\infty}^{\infty} x \, p(x) dx = \underset{T \to \infty}{\text{Limit}} \frac{1}{2T} \int_{-T}^{T} x(t) dt$$

MEAN SQUARE $$\overline{x^2} = \int_{-\infty}^{\infty} x^2 \, p(x) dx = \underset{T \to \infty}{\text{Limit}} \frac{1}{2T} \int_{-T}^{T} x(t)^2 dt$$

ROOT-MEAN-SQUARE $$x_{rms} = \sqrt{\int_{-\infty}^{\infty} (x - \bar{x})^2 \, p(x) dx} = \sqrt{\overline{x^2} - \bar{x}^2}$$

For the Rayleigh distribution these values are

MEAN VALUE $$\bar{x} = \sqrt{\frac{\pi}{2}} \, x_{rms}$$

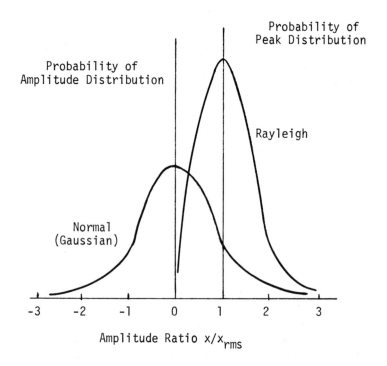

Fig. 3.10  Graphical probability distributions for normal and Rayleigh functions

MEAN SQUARE                       $\overline{x^2} = 2x^2_{rms}$

ROOT-MEAN-SQUARE                  $x_{i_{rms}} = 0.65\, x_{rms}$

Probability distributions for various vibrational responses are depicted in Fig. 3.11.

An example serves to demonstrate the necessity of probabilistic theory. Consider a lightly damped, complex mechanical system excited by a wideband random vibration. It is required to find the number of displacement amplitude peaks in the response exceeding a given value $d_i$. If the response of the system is a narrowband random vibration due to filtering characteristics of the structure, then the probable peak displacement distribution as described by the Rayleigh distribution function is

$$p(d) = \frac{d}{d^2_{rms}} \cdot e^{-d^2/2d^2_{rms}}$$

where d is the amplitude of a given displacement peak and $d_{rms}$ is the rms value

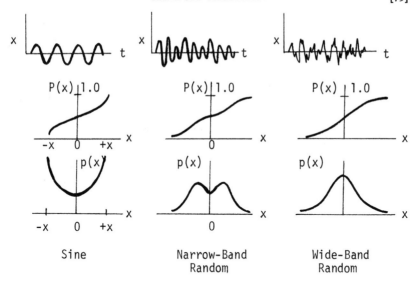

Fig. 3.11   Cumulative probability and probability density curves for sine, narrowband random and wideband random signals

of the narrowband response. Integrating this expression from $d_i$ to infinity results in the probability of a displacement peak exceeding $d_i$

$$p(d_i) = e^{-d_i^2/2d_{i\,rms}^2}$$

Since the maximum number of displacement peaks is the product of excitation time t and response frequency $f_0$, the predicted number of peaks $N_{max}$ exceeding $d_i$ is

$$N_{max} = t \cdot f_0 \cdot p(d_i)$$

This type of information is necessary to determine whether or not a predicted number of displacement peaks is excessive. If it is, then changes such as revising the mechanical design, altering the test environment, or providing for changes in the operational environment have to be made.

## 5. VIBRATION TESTING METHODS

Vibration testing can be divided into two areas: operational and design. Operational testing is performed when a system is found to vibrate in excess of acceptable limits. This involves the measurement and analysis of the vibrational output of the machine with no attention given to the excitation forces. Design vibrational testing, on the other hand, is performed to insure that a system will operate properly in a given environment. Such testing allows design configurations to have minimum weight as well as optimal shape. This is especially important in the aerospace and related industries. For these cases it is necessary that input test excitation simulate as closely as possible actual operating excitation. Then, the measured vibrational response of the device subjected to a specific input will be indicative of its operating character.

When design testing was first performed, it involved only pure sinusoidal excitation at a fixed frequency. But over the years this has been complemented with such tests as sweeping sine wave, fixed-random, sweep-random, random-sweep-random and other more complicated forms of complex-random testing in an effort to duplicate more closely the response from anticipated environments. Wideband random vibration testing simulates the characteristics of vibrational environments encountered in land, sea and air motions much better than that of pure sinusoidal testing. However, there does exist doubt concerning the statistical characteristics of these environments with regard to spectrum shape. Therefore, each excitation spectrum should be considered carefully for its relevance to an operational environment.

The sweeping sine wave test is still quite popular today. It is a highly efficient tool when searching for system resonances, and it can be performed in a reasonable amount of time. In addition the investment in instrumentation for this testing procedure is relatively small, unless very complex systems are involved. In the arrangement shown in Fig. 3.12, the sweeping frequency output of the oscillator is passed through a power amplifier to an electrodynamic shaker where a vibration pickup on the shaker controls the output level of the oscillator through a feedback loop. This maintains a constant level of excitation at the shaker. The vibrational response of the device on the shaker is either recorded from the output of an appropriately placed accelerometer or visually monitored for the onset

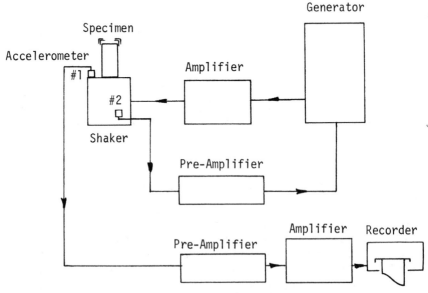

Fig. 3.12   Sweep-sine test apparatus

of fracture or failure. Complexity of this test is increased by the addition of such instrumentation as vibration programmers, vibration meters and additional accelerometers. At times, large complex test specimens require numerous accelerometers to detect resonances in different parts of the structure. Otherwise, some elements in the test specimen can be overtested while others are not tested at all. The sweeping sine wave test is a valuable tool for detecting dangerous resonances during the development stage of a device. However, since it can only test a device sequentially with respect to frequency, it has been judged too time-consuming for most other tests and remains more or less a laboratory tool.

Random-signal vibration testing has become the more accepted method of testing. If a mass is excited by a wideband random excitation, the acceleration response of the mass will have a continuous character. Moreover, it will be completely irregular and not manifest any form of periodicity. Because of the continuous and irregular character of the response, only such concepts as probability density and power spectral density can be applied to give it satisfactory description.

As stated above, if one period of a harmonic motion is known, then the complete future history of the waveform can be predicted. But a random waveform cannot be predicted even with a knowledge of its total history. Probability allows us to find amplitude values within a given range of amplitudes $x + \Delta x$, and with this knowledge, probability density and probability distribution curves can be drawn. When these curves are found to be Gaussian in form and the vibrating system has a linear character, then computation of the power spectral density in terms of frequency is all that is required to form a complete probabilistic description of the vibration.

Wideband random vibration testing, which is very costly, often involves the use of a specific test spectrum such as that specified by military standard MIL-STD-810, USAF and shown in Fig. 3.13. This specification requires white noise excitation between 100 to 1000 cps with an attenuation of $-12dB/octave$ at both ends of the frequency spectrum. Such special excitation of a dynamic shaker system would normally result in a completely different excitation of the test specimen due to structural filtering and mismatching of mechanical impedance. This has the adverse effect of subjecting the specimen to an excitation quite unlike that required or found in normal operation. In addition, it often requires unnecessarily large dynamic shakers and power amplifiers. The type of excitation can be varied by using some form of equalization or a combination excitation of wideband background noise and a superimposed sweeping sine wave or narrowband random signal.

When a wideband random signal excites a shaker system supporting a resonant specimen, the vibrational response of the shaker is found to be far different from that of the excitation. The response contains numerous valleys and peaks spread-out over the frequency spectrum. These maxima and minima of the system response should be equalized with such devices as peak-notch equalizers or multi-band equalization systems to insure that the tested device is, in fact, subjected to random excitation. The latter systems are available with either manual or automatic adjustment. Manual systems have the disadvantage of requiring a large number of filtering elements which leads to a rather considerable amount of

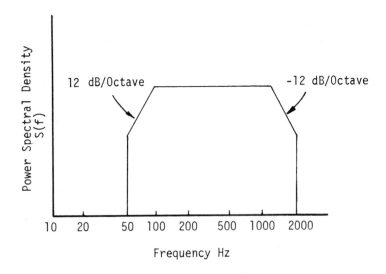

Fig. 3.13   Spectrum shape for wideband random vibration testing in frequency bands: 50–100 Hz, 100–1000 Hz and 1000–2000 Hz

setup time. Setup time can be reduced substantially through the use of an automatic multi-band system. It should be noted that these systems are complex and very expensive. When compared to the time required for a sweeping sine wave test, this latter method of testing is often preferred, especially for qualification. There are other reasons for recommending the wideband random test. It has already been mentioned that the vibrational character of naturally found environments is random. But this in itself does not satisfy the need of simulating frequency spectra. The other advantage of a random excitation is that all specimen resonances are excited at the same time, thus producing interaction effects.

Sweeping narrowband random or, as it is more commonly known, sweep-random testing was originally proposed by M. W. Oleson[2] as an economic substitute for the more costly method of wideband random testing. Sweeping sine wave testing was in general use. It was economical but slow and it failed to produce equivalent distribution of acceleration and stress amplitudes in the testing procedure. The sweep-random test was a compromise between wideband random and sweeping sine wave testing. In principle it replaces low acceleration density wideband random excitation with an intense, narrowband random excitation that sweeps slowly over the frequency range of the test. As an alternate form of testing, it contains a routine to provide an excitation equivalent to that of a wideband random vibration. It also possesses some characteristics of the sweeping sine wave test. Most important, resonances can only be checked sequentially and the test does not allow for observation of interacting resonance effects. In addition, it is more time-consuming than wideband random testing. However, the advantages of sweep-random testing justify its use. For instance, specific test levels can be obtained with much lighter equipment than is required for wideband random testing. Further, setup and control is simplified and the statistical character of the excitation is retained.

The philosophy of sweep-random testing is based on the requirement of producing the same number of stresses and acceleration peaks at each level of response in sweep-random testing as is produced in wideband random testing. In particular the resonant response of the tested specimen should be the same for either method at any given frequency. Using a constant level narrowband excitation, a logarithmic sweep rate insures the same number of stress reversals inside a resonance peak as would occur under wideband excitation. Also, sweep-random testing is designed to obtain an equivalent number of stress reversals at any given level of stress. This is accomplished by generating a gain of 3 dB/octave with increasing frequency. Further details on the mechanics of sweep-random testing can be found in the Bibliography at the end of this chapter. Sweep-random testing manifests certain characteristics that are unique. It has sweep-test simplicity and a statistically defined amplitude distribution. Therefore, it has been recommended that it be used as a separate testing method as opposed to establishing an equivalence to wideband random excitation. A block diagram of a sweep-random measurement system is shown in Fig. 3.14.

Another method of random vibration testing is called multiple sweep-random testing. In this method a sweeping excitation is superimposed on a low-level wideband random background. The excitation can be single or multiple bands of narrowband random noise that sweep or dwell at specific frequencies depending on the requirements of the testing program. This allows excitation of specific resonances while the total test specimen is subjected to an overall excitation of wideband random noise. Such a testing arrangement is shown in Fig. 3.15.

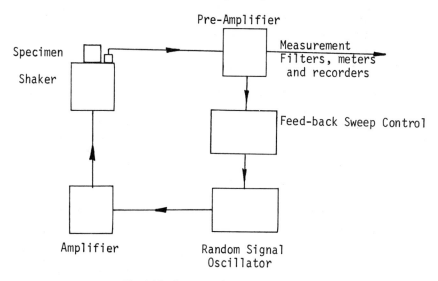

Fig. 3.14   Sweep-random test arrangement

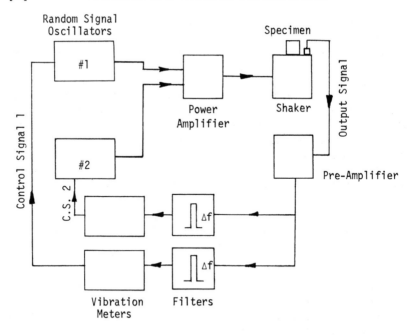

Fig. 3.15   Multi-sweep random test arrangement

## 6. APPLICATION TO MEASUREMENTS

Some ideas should be kept in mind when performing vibration testing. Although the excitation of an electrodynamic shaker may possess a flat input power spectrum, there is nothing to insure that the vibrational spectrum at the point of support of the specimen will be similar. To the contrary, it will most likely contain marked changes in magnitude unless corrected by some of the aforementioned techniques; that is, it will appear to the tested specimen as a narrowband random excitation. It is equally important after completing the arrangement of a vibration test setup to check each component of the system individually and then the system as a whole. The latter can be done with a very low level of excitation which will insure the integrity of the total system. A fact not to be overlooked in design testing is the need for testing large numbers of specimens to insure a reliability of testing results. As mentioned previously, random analyses, statistics and probability are meaningful only when supported by sufficient data which can be obtained through proper testing programs that include a multiplicity of test objects.

Because of the popularity of electrodynamic shakers, we have neglected to dwell on a relatively new method of excitation for test specimens, that is, with acoustically generated noise. This method of inducing vibration acoustically requires sound sources that can produce sound-pressure levels in excess of 140 dB (re .0002 $\mu$B). High-level sound sources are commercially available from such firms as Wyle Laboratories and the Ling-Temco-Vought Company. These firms can manufacture electropneumatic power systems with power levels in excess of 10,000 acoustic watts. Acoustical excitation is said to produce a more realistic input to a vibrational system than that obtained with dynamic shakers, and thereby avoids the problem of mechanical impedance between specimen and source.

Whereas design testing is concerned primarily with ascertaining a specific input excitation spectrum, operational vibration testing is most concerned with the measurement of transducer response, whether it be a velocity pickup or accelerometer. For proper analysis, basic rules of measurement should be adhered to. Velocity pickups are useful for measuring frequencies below 1000 cps where there are no strong electric or magnetic fields present; the higher frequencies are best measured with the aid of an accelerometer. It is also necessary to insure that ground loops do not exist and proper shielding is present on all low-signal conductors. Vibration measurement can range from simple to extremely complex situations. However, most meaningful measurements can be obtained with the use of a preamplifier, filter, frequency analyzer and a true rms meter as shown in Fig. 3.16. When system response is sinusoidal, all necessary information is contained within a single period of the wave. A photograph of the waveform as displayed on a cathode ray oscilloscope is sufficient for subsequent analysis. But, when the response is complex and contains multiple overtones or is random, it becomes necessary to pass the response signal through a filtering and measuring system to identify the frequency components of the response and measure their resulting amplitudes. This data is then presented in either an amplitude versus frequency diagram or a power spectral density diagram, both of which describe the system in the frequency domain rather than in time. In addition, values of mean amplitude and rms enable the investigator to determine on a statistical basis whether or not testing programs are satisfying pretest specifications. They also reveal when and to what extent velocity amplitudes exceed recommended vibrational standards. This leads to the question, "If on statistical analysis ($v_{rms}$ being measured) peak velocities are exceeding the rms value in large number, is the vibrating system still conforming to vibrational standards?" Questions such as this can only be argued from experience.

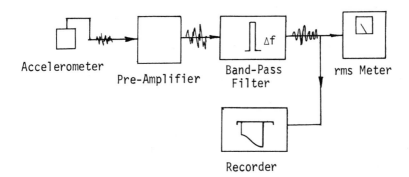

Fig. 3.16   Measurement arrangement for operational testing

# REFERENCES

1. Rice, S. O. *Mathematical Analysis of Random Noise*. New York: Dover Publications, 1954.
2. Oleson, M. W. "A Narrow-Band Random Vibration Test." *Shock and Vibration Bulletin*, vol. 25, no. 1 (1957).

# BIBLIOGRAPHY

Bendat, J. S. *Principles and Applications of Random Noise Theory*. New York: John Wiley and Sons, 1958.

Bendat, J. S. and Piersol, A. G. *Measurement and Analysis of Random Data*. New York: John Wiley and Sons, 1966.

Booth, G. B. "Sweep Random Vibration." *Proc. Inst. Env. Sci.*, April 1960.

Broch, J. T. "An Introduction to Sweep Random Vibration." *Bruel & Kjaer, Technical Review* 2 (1964).

Broch, J. T. "Vibration Testing—The Reasons and the Means." *Bruel & Kjaer, Technical Review* 3 (1967).

Cooper, B. E. *Statistics for Experimentalists*. New York: Pergamon Press, 1969.

Crandall, S. H. *Random Vibration*, vol. 2. Cambridge, Massachusetts: The Technology Press of M.I.T., 1963.

Davenport, W. B., Jr. and Root, W. L. *Random Signals and Noise*. New York: McGraw-Hill Book Co., 1958.

Holman, J. P. *Experimental Methods for Engineers*. New York: McGraw-Hill Book Co., 1966.

Maten, S. "New Vibration Velocity Standards." *Hydrocarbon Processing*, vol. 46, no. 1 (1967): 137–141.

Miles, J. W. and Thomson, W. T. "Statistical Concepts in Vibration." In *Shock and Vibration Handbook*, edited by C. M. Harris and C. E. Crede, chapter 11. New York: McGraw-Hill Book Co., 1961.

Mood, A. M. and Graybill, F. A. *Introduction to the Theory of Statistics*, 2d ed. New York: McGraw-Hill Book Co., 1962.

Ruzicka, J. E. "Characteristics of Mechanical Vibration and Shock," *Sound and Vibration*, April 1967: 14–30.

Thomson, W. T. *Vibration Theory and Applications*. Englewood Cliffs: Prentice-Hall Inc., 1965.

# SYMBOLS

| | | | |
|---|---|---|---|
| x | instantaneous displacement of mass | P, PP, IPSP | peak, peak-to-peak, and inches per second peak |
| u | instantaneous displacement of the base | gP | 32.2 feet per second per second, peak |
| q | instantaneous charge or quantity of electricity | $\omega$ | frequency, radians per second |
| X | peak displacement of mass | X | peak displacement |
| U | peak displacement of the base | \|X\| | the real value of the complex X, absolute |
| Q | peak charge or quantity of electricity | $M_d$ | dynamic magnification factor |
| $\dot{x}, \ddot{u}, \ddot{q}$ | first and second derivatives, usually with respect to time, are indicated by placing one or two dots above the variable | $\epsilon$ | eccentricity of shaft unbalance |

# 4

# LINEAR VIBRATION

*Michael P. Blake*

## 1. SCOPE

This paper has three objectives. The first is to introduce the reader to some basic concepts and systems. The second is to focus attention on the *lightly damped forced system with one-degree-of-freedom* and show the relation of this system to the vast array of systems and vibrations that may be encountered in nature. The third is to provide an analytical understanding of the simplest type of forced system, while providing graphical summaries of system response which may serve as useful reference material for persons concerned mainly with the performance, economics and safety of existing machinery. An effort is first made to classify vibration according to whether the system is continuous or discrete, and then to classify according to frequency and motion. The character of a mathematically linear system is compared to that of a nonlinear system as an introduction to the unique features of a periodic, sinusoidally forced vibration. This type of vibration is found, in practice, to lie at the root of most machinery problems. The typical magnitude and form of periodic and beat vibrations are discussed as an introduction to simple free vibrations, which in turn serves as a basis for understanding the simplest type of forced system. This understanding is the final objective because it enables the person who maintains machinery to control the majority of situations which concern vibration that is considered either a fault symptom or a required function of the machine. The concept of impedance is introduced because it seems the most appropriate concept for the solution of the equations of motion and for manipulating the solutions into convenient forms. It is shown that impedance, in general, connotes the ratio of excitation to response where both are sinusoidal, and that the ratio may be a simple number or a simple number combined with a complex operator that can transform the response vector into the excitation vector both as regards magnitude and phase. Because adequate standard nomenclature does not exist some difficulty is experienced here. A few nonstandard terms are used and explained in the *Terminology* section. The title, Linear Vibration, is intended to describe what is often called lateral vibration. Because the word linear has a mathematical meaning, an old word, lineal, is sometimes used in the text to describe the more or less to-and-fro motions or other vibrations that are often called linear or lateral. Lineal, therefore, connotes to-and-fro, circular, elliptic or other vibratory motion in a plane or curved surface of a machine component, considered as a discrete (lumped) member. This paper concludes with two practical examples.

## 2. CONCEPTS

The concept of vibration arises in connection with a large variety of wave phenomena, and so do the concepts of heat, sound, light and impact. All of these are what might be called spectral concepts. Because the limits of these spectra are "gray" areas, wherein vibration ceases to be practical, we refrain at this time from defining the concept of vibration which is time related.

In general, the more rapid vibrations have greater meaning for us since they occur in a time span of our attention. For example, the jarring of a vehicle or the undulation of a light wave command our attention. On the other hand, a tidal vibration requires a day per cycle, and the heaving of the crust of the earth may occupy an entire geological age. These latter vibrations are outside our immediate perception, just as we could not see a light undulation if it required one lifetime for completion. The number of vibrations per second is called the frequency f. The radian or circular frequency is $2\pi$ times the frequency and is denoted here by $\omega$. A vibration may be classified according to its frequency, such as infra-sonic, sonic (20 to 20,000 cps) and ultra-sonic.

Vibration is also classified according to whether it is discrete or mainly continuous. There is a difference between a ripple on a pond and the bobbing of a leaf that floats on it, and between a thunderclap and the window that it rattles. The pond water and the thunder-filled air exemplify continuous systems and the leaf and window exemplify discrete systems. In one event are seen the internal vibrations of a composite medium or component. In the latter the mass, the elasticity and, perhaps, the damping elements are more or less separated.

Vibrating systems invariably comprise mass and elasticity, together with some damping. The interest of this paper is focused on systems that are linear in the sense that restoring force is directly proportional to displacement and on systems that comprise damping effects where the force is proportional to instantaneous velocity (viscous damping). The interest is further focused on single-degree-of-freedom discrete systems, particularly those that are forced to vibrate because of some external excitation such as unbalance. The vibrations of practical systems that approach the foregoing ideals are more or less to and fro in a plane and are generally accompanied by externally noticeable forces and displacements. On the other hand, it is characteristic of torsional vibration that it may be highly destructive and cause high stresses while giving no readily evident external signal of its presence.

A further concept is that which may be called idealization. This concept and the related lumped parameter concept make possible the building of mathematical models without which analysis is not possible. Suppose a child jumps on a plank bridge. The restoring force may be idealized by defining it as being directly proportional to displacement. The damping may be idealized by defining it as being a force directly proportional to velocity.

The descriptive term, *lumping of parameters*, connotes that all of the mass system is in the child and all of the elasticity and damping is in the plank. Actual systems found in both electrical and mechanical engineering are never either completely ideal or lumped. However, by carefully building a mathematical model that approaches the facts of nature, as regards idealization and lumping, no essential violence is done to truth, with respect to the analytical results that are sought. The systems considered in this paper are essentially all idealized and lumped. Table 1 makes an effort to classify some vibrations by drawing sharp fictitious lines regarding both type and numerical values. Sketches of continuous and discrete systems are shown in Fig. 4.1.

The condition of a machine, as regards its safety or rate of degradation, is often estimated by measuring its vibration. It is an empirical fact that mechanical condition is about proportional to machine velocity, regardless of frequency. This fact appears true irrespective of whether the system is continuous or discrete. For example, a ball bearing showing a frequency in the range of 4000 cps is likely to be in danger at a peak velocity of one inch per second, and so also is an electric motor at a frequency of 60 cps and the same velocity. In Table 3 the velocity ranges in the motion classes are about the same for the continuous and the discrete systems, whereas the respective displacement and acceleration ranges differ considerably, depending on whether the system is continuous or discrete. The frequency range is higher for the continuous systems than it is for the discrete.

Again, it is not possible to draw sharp distinctions. If we like, we may think of continuity when the linear dimension of the vibrating body is long in comparison with the wave-length of the internal vibration. Thus, high frequency is a feature of continuous vibration. Any physical system behaves continuously for sufficiently high frequencies, and discretely for sufficiently low frequencies.

## 3. PRINCIPLES

A simple mass vibrates if excited by a periodic force. When the force is removed the vibration ceases. On the other hand, if the mass is associated with an elastic element, again, it vibrates at the frequency of the exciting force. When the excitation is removed, it vibrates at a specific frequency, called the natural frequency $f_n$, determined by the combination of mass and elasticity. Such a system exhibits a will of its own, a compromise between the generated forces of the elastic and mass members. The elastic member always trys to eliminate its deformation. Stress and strain are inseparable. The elastic member gives up its potential energy to the mass member which resists accelerations and finds itself free of force at the position of zero displacement. It traverses through this position until brought to rest at its extreme position by the stress in the elastic member. This stress returns the mass once more to its position of zero displacement. Such a free vibration goes on for an extended time. In all practical systems some damping is present so that the vibration eventually dies out.

Because mass is of one kind only; because elasticity may be divided into two important types and because some damping is always present, we consider, under the heading of principles, the difference between linear and nonlinear elastic elements and the nature of damping.

From zero to relatively high frequencies, common metallic spring elements

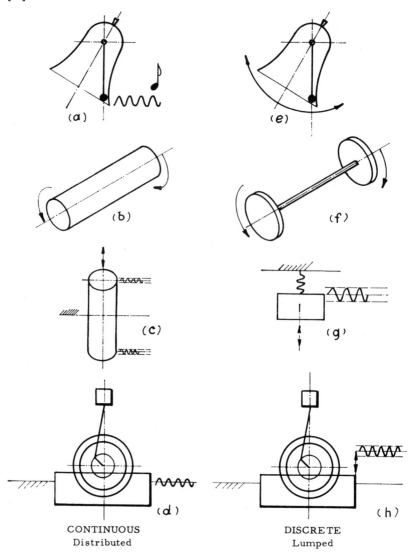

Fig. 4.1   Systems that are mainly continuous or discrete

are linear, by which is meant that $F = Kx$, where $F$ is the force on the spring and $x$ is the displacement. $K$ is a spring constant called the stiffness. In its working range of frequency, the spring element may be considered ideal, meaning that it will show no tendency to vibrate in its own right, which indeed it can do by virtue of its own distributed mass.

When the spring element is linear and when damping is small or negligible, the following important results are observed.

## TABLE 3
## SOME CLASS CONCEPTS

| CONTINUOUS | DISCRETE |
|---|---|

### SYSTEM CLASSES

| CONTINUOUS | DISCRETE |
|---|---|
| Pealing of church bells | Swinging of a church bell |
| Sound and noise in solids and fluids | A child on a swing |
| A sounding organ pipe | A shimmering aspen leaf |
| A bowed violin string | A velocity transducer |

### FREQUENCY: CYCLES PER SECOND

| CONTINUOUS | | DISCRETE | |
|---|---|---|---|
| Light waves | 1,000,000,000,000,000 | Very high speed machine | 1500 |
| Sound waves | 20 to 20,000 | High-speed machine | 150 |
| Musical instruments | 20 to 5,000 | Most machines | 30 to 60 |
| Machine components | 0 to 30,000 | Slow-speed machines | 1 to 10 |
| Seismic waves in earth | 0 to 500 | Clock pendulum | 0.25 to 2 |
| Sea tides | 0.00001 | | |

### MOTION

| CONTINUOUS | DISCRETE |
|---|---|
| Machine component such as a rolling bearing considered continuous. | Machine component, such as a plain bearing housing, considered discrete. |

#### Displacement, inches PP

| | | |
|---|---|---|
| 0.0000005 to 0.00005 | 0.0001 to 0.075 | Incidental |
| | 0.001 to 36 | Functional |

#### Velocity, inches/sec. P

| | | |
|---|---|---|
| 0.01 to 1.5 | 0.002 to 2.0 | Incidental |
| | 0.2 to 10 | Functional |

#### Acceleration, inches/sec$^2$ P

| | | |
|---|---|---|
| 30 to 30,000 | 10 to 1,000 | Incidental |
| | 20 to 2,000 | Functional |

Values given above are indicative of actual systems, but it is not possible to suggest anything more than the order of magnitude, in general, because of the endless variety of systems.

1. The vibration of the free system is sinusoidal so that $x = X \sin \omega_n t$, where x is the instantaneous value and X the peak value of displacement, $\omega_n$ is the radian frequency of free vibration (or natural frequency) and t is time in seconds. The radian frequency is $2\pi$ times the frequency in cps.

2. It follows that the peak velocity $\dot{X}$ is equal to $\omega_n$ times X and that the phase of velocity is ninety degrees in advance of displacement. The peak acceleration $\ddot{X}$ is equal to $\omega_n^2$ times X and acceleration leads velocity by ninety degrees.

3. The natural frequency of free vibration is independent of amplitude for small excursions so that all free vibrations of small displacement take the same time. Such a system is called isochronous, and is exemplified by the pendulum or the balance wheel in a watch. The circular pendulum is isochronous for small arcs only. The cycloidal pendulum is truly isochronous.

Because the system has a natural frequency at which it tends to vibrate, it exhibits a limited response to vibrations coming from the outside, except for those that tend to coincide with the natural frequency. Then the system executes large amplitudes and is said to resonate. The critical frequency of excitation is defined as being that at which, if the frequency is increased or decreased, the amplitude of response will decrease. The critical frequency and the natural frequency are the same when there is no damping. But with damping, the damped natural frequency becomes less than the undamped natural frequency. The critical frequency, in turn, becomes less than the damped natural frequency when the excitation takes the form of force $= F_0 \sin \omega t$, where $F_0$ is constant. Whether the linear spring is part of a damped or an undamped system, the analytical expressions that describe the motion are relatively simple. With a few notable exceptions, all of the day-to-day machine systems may be considered linear and lightly damped. On the other hand, the analysis of nonlinear systems is difficult, but they are so important that they must be considered, however briefly, even in a condensed treatment such as this.

The nonlinear spring elements of machines are very often rubber mounts or cushions of various kinds. A body of gas in an enclosed cylinder being acted on by a piston is an example of a hardening or nonlinear spring. Rubber in compression exhibits hardening. Special arrangements of metal and other types of springs may exhibit hardening or softening. Some arrangements of rubber in shear may exhibit hardening or softening. Another kind of nonlinearity arises from looseness. A mass sliding freely on a rod and bouncing on linear compression springs at each end is an example of such a nonlinearity. This situation arises when a bearing pedestal is loose on its foundation bolts or baseplate.

Even when the value of K is expressed as a simple function of x, the solution of the differential equation of nonlinear motion becomes most difficult. Some such systems are amenable to relatively simple graphical solution. Systems having nonlinear springs are sometimes called asynchronous, aperiodic or nonlinear. Brief graphical hints relating to their behavior are given in Fig. 4.2. Materials of the rubber family exhibit endless variation as regards composition, stiffness and internal damping. The natural frequency of a system moving with large displacement is likely to be higher than one having small amplitude. When the natural frequency depends on amplitude, it may be imagined that the critical

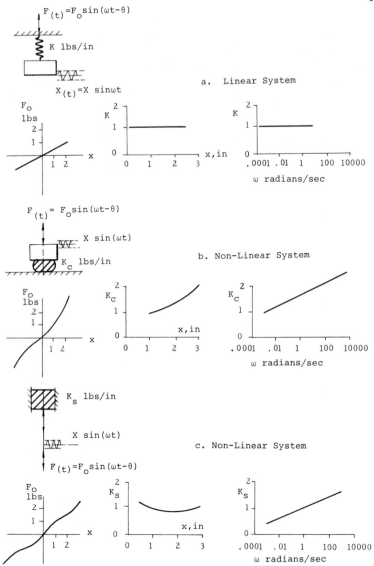

Fig. 4.2  Linear and nonlinear springs, static and dynamic behavior

frequency also depends on amplitude. It follows that if a system is excited at its critical frequency so that the amplitude tends to increase, then the critical frequency itself changes. This leads to a moderation of the vibration response and resonance in the ordinary sense does not endure. For this reason such springs have an attraction as virtual dampers of a very simple kind for shaft couplings or for vehicle suspensions. Such nonlinear systems do exhibit such peculiarities as an instability called *jump*, requiring a study in each case to determine if a proposed system is likely to perform as expected.

Another kind of nonlinearity, typical of rubber, which further complicates analysis, is that stiffness is frequency dependent. The result of this is that slow or so-called static measurements are not the key to dynamic performance. Often the dynamic stiffness is not known, and it is then assumed as a first estimate that the dynamic stiffness is twice the static stiffness.

Damping may be envisaged as the action that dissipates the energy in a system. Sometimes it is desirable, as in the galvanometer or the engine crank system. At other times it is undesirable, as in a clock pendulum or watch balance system. It is present in the systems of nature and makes life possible, as, for example, ordinary friction which makes motion possible.

Damping is measured in terms of force. When analyzing systems, the damping is almost always idealized as being either wet (viscous) or dry (coulomb) friction, most usually wet. Wet friction, meaning friction that increases with velocity, is further idealized as being a simple linear function of velocity, such as c times velocity where c = constant. Dry friction, ideally, is such that the force is independent of the velocity of rubbing. In practice, the assumption $F = c\dot{x}$ leads, generally, to accurate results. Here, F is the instantaneous damping force in pounds and c is the damping constant in pounds per foot per second, and $\dot{x}$ is the instantaneous velocity in feet per second.

For viscous or wet damping, c may be very large, as in a dashpot (shock absorber) filled with highly viscous fluid, or small, as in an air vane. In an analogous electrical system, current corresponds to velocity in the mathematical model so that electrical resistance corresponds to the damping coefficient c. Furthermore, mechanical damping may be induced by forces that are electrically generated as, for example, in the moving coil of a velocity-type pickup which may be partially short-circuited (shunted) by a resistance. For large values of c, a disturbed system can only execute a quarter cycle of a vibration before coming to rest at the mean position. For small values of c, the system may execute hundreds of cycles of vibration. There is an intermediate value of c for which the system may overshoot the mean just once or not at all before coming to rest. For values of c greater than this, overshoot is absent for free vibration. This intermediate value of c is called the critical value of damping. The amount of damping in a system is often expressed in terms of critical damping by stating, for example, that damping is 60 percent of critical. The symbol, $c_c$ is used in this Chapter to denote critical damping.

In the vast majority of instances damping occurs in machine components simply as a result of either their molecular friction or other incidental factors, such as rubbing. However, in many important applications damping is intentionally provided. The objective is usually to limit vibration displacement amplitude and, therefore, limit stress. Diesel engines generally contain one or more viscous torsional dampers or, perhaps, elastic cushions that are made of high-damping rubber in shaft couplings. The galvanometer coils of meters, such as sound meters or wave analyzers, are often damped in order to avoid mechanical damage or to steady the needle reading. Velocity pickups comprising a seismically suspended coil, and various pallographs, are usually damped, both to limit vibration and to shape the response according to the most desirable curve. Intentional damping is found in the steering and spring suspension systems of automobiles and motorcycles as well as in electrical power lines and endless other applications. There are so-called dampers, such as the pendulum damper used in some engines, that are really absorbers for vibration and energy. The mentioned pendulums are pivoted at some radius, comparable to the crank radius, and tend to mitigate unwanted torsional vibration. Some of the cast irons exhibit useful internal damping and for this reason are sometimes used for making crankshafts.

## 4. MAGNITUDE AND FORM

Magnitude connotes here displacement amplitude and form connotes waveform, such as oscillographic, wherein amplitude is exhibited as a function of time. For most purposes this is the usual form to be considered. However, it is to be borne in mind that waveform is generally a mental abstraction, while the form of the vibration in nature is the spatial path of a component or particle. This form might be obtained in an oscillograph by using two pickups with axes mutually perpendicular. One axis is connected to the vertical and the other to the horizontal beam-deflector systems of the cathode ray oscilloscope. This is sometimes done when analyzing shaft motions. The resulting oscillograph is called a Lissajou figure. The figure may be a true ellipse, a circle or, perhaps, a figure eight. As the speed of compressors and other machines increases (15,000 rpm is now not at all uncommon), the use of Lissajou measurements in problem cases is bound to increase, particularly as it is easier to grasp the physical implication of the Lissajou figure than to grasp the significance of the two signal forms that combine to give the Lissajou form. This writer has not measured Lissajou forms, but C. Jackson[1] has communicated some results of his work connected with alignment and vibration control for turbines and compressors.

Practical magnitudes and forms are a matter of experimental observation. Without a knowledge of them we are at a severe disadvantage. Conversely, knowing them, we are at a disadvantage if we do not know the idealized forms used in mathematical analysis. Both must be understood in order to profitably grasp the implications of innovation or correction. In Table 4 consider the typical practical magnitudes, giving two extreme figures and perhaps a third to indicate the most usual magnitudes for machinery.

**TABLE 4**
**TYPICAL MAGNITUDES OF INCIDENTAL VIBRATION**

|  | CONTINUOUS SYSTEMS | DISCRETE SYSTEMS |
|---|---|---|
| Frequency, cps | 0/30; 60/30,000 | 0/30; 60/1000 |
| Displacement, mil PP | 0.0005/0.05 | 0.1/0.5/50 |
| Velocity, IPSP | 0.01/1.5 | 0.002/0.1/2.0 |
| Acceleration, gP | 0.1/100 | 0.02/0.1/5.0 |

It is seen, for example, that a typical displacement amplitude in common machines is about half a mil or about one-eighth of the diameter of a human hair. Despite the smallness of the amplitudes, they are a key to machine condition, particularly when the vibration is considered as a symptom of malfunction. Persons maintaining machinery often find that magnitude is the key to the question of classification. For example, is the machine malfunctioning or not? On the other hand, form is often the key to the question of diagnosis or fault identification. For example, a signal form exhibiting the same frequency as that of rotation may indicate mass unbalance, whereas a form containing twice that frequency might indicate misalignment. Now consider signal form from the points of view of the sinusoidal model, of the effect of read-out in terms of displacement, velocity or acceleration and of the practical results.

Many practical signal forms approach the sinusoidal model, wherein amplitude is a sine function of time. Without an appreciation of the sinusoidal form, practical systems cannot be appreciated. It is a feature of the sinusoidal form that it may be described in terms of four parameters: the frequency, the peak values of displacement, velocity and acceleration. Knowing any two of these, the others may be found as follows.

$$\text{Instantaneous values:} \quad x = X \sin(\omega t)$$
$$\dot{x} = \omega X \cos(\omega t) = \omega X \sin(\omega t + \pi/2)$$
$$\ddot{x} = -\omega^2 X \sin(\omega t) = \omega^2 X \sin(\omega t + \pi)$$
$$\text{Peak values:} \quad X = X, \text{ Displacement, Peak}$$
$$\dot{X} = \omega X, \text{ Velocity, Peak}$$
$$\ddot{X} = \omega^2 X, \text{ Acceleration, Peak}$$

It is seen that the common element between the peak values is the radian frequency $\omega$ in radians per second. In the same way $\omega$ is the multiplying element between the rms, the peak-to-peak and the average values. But these facts are only so when the form is sinusoidal. For other forms $\omega$ may be entertained as a concept, but it is not of constant value. It is to be noted that $\omega$ is usually a concept even for sinusoidal vibration since it generally has no counterpart in the physical machine.

The importance of the mode of detection and read-out will now be considered. A signal may be picked up by a transducer, giving a voltage proportional to either displacement, velocity or acceleration. Conversely, a signal which is acceleration proportional may be conditioned (integrated) to read out in terms of either velocity or displacement. Electronic integration is practical and useful for signals approaching the sinusoidal form. But even for these, differentiations, to obtain, for example, the third derivative of displacement x (called jerk) is often not practical. When a nonsinusoidal signal is electronically integrated, it must be borne in mind that the result may not be at all what is assumed by the observer. A single integration of a sinusoidal-like signal may be expected to decrease the meter reading by a factor $\omega$. It is a useful check to integrate an actual acceleration signal twice by turning the meter knob. The meter reading should decrease each time by the same factor. If the signal is sinusoidal that decrease factor should correspond to the radian frequency observed on the oscillograph or wave analyzer.

In the measuring of machinery it is generally desirable to detect and read out in terms of velocity. For special reasons it may be desirable to detect in terms of displacement, as for example, in mass balancing to avoid phase distortion and the use of filters. Or it may be desirable to read out in terms of acceleration, perhaps, because the velocity pickup is limited to a maximum frequency of about 2000 cps. Suppose that the form of the read-out exhibits the true magnitudes of a signal in a given situation. Consider further the relative significance of the displacement, velocity and acceleration and how these modes of read-out affect the waveform that is observed.

If the fundamental amplitude is X and a harmonic of the $n^{th}$ order has an amplitude rX, then we have the results shown in Table 5.

It is evident that if the tenth harmonic has 10 percent of the amplitude of the fundamental in the displacement mode, it will have ten times the amplitude of the fundamental in the acceleration mode. Thus, the reading of a meter and the form of the oscillographic display are both affected considerably by the choice of mode, whether it is displacement, velocity or acceleration when harmonics are present. This is illustrated in Fig. 4.3. The acceleration mode exaggerates the higher frequencies and acts rather like a high-pass filter, whereas the displacement mode acts like a low-pass filter. In making phase and amplitude measurements to balance a machine, it would be almost impossible to work in terms of acceleration because it is the fundamental that must be suppressed in the balancing procedure, and it is the phase and value of this that is sought. This phase and value could not be found amid the din of the exaggerated higher frequencies. An understanding of the foregoing is particularly necessary as a prelude to planning a monitoring system for a machine exhibiting several frequencies of vibration.

Figure 4.4 illustrates the oscillographic and wave analyzer read-out for the waveform of Fig. 4.3. The wave analyzer reads on a meter the amplitude of any harmonic to which it is tuned by turning a tuning knob. Its read-out here is plotted in the form of a discrete spectrum.

Most of the signal forms for machinery are sinusoidal-like periodic, although they may exhibit spikes or various departures from the sinusoidal form. A periodic form is one that repeats exactly as time passes. In general, a sinusoidal

**TABLE 5**
**FUNDAMENTAL-HARMONIC RELATION**

| | FUNDAMENTAL F | HARMONIC H | RATIO H/F |
|---|---|---|---|
| Displacement, peak | X | rX | r |
| Velocity, peak | $\omega$X | $n\omega$rX | nr |
| Acceleration, peak | $\omega^2$X | $(n\omega)^2$rX | $n^2$r |

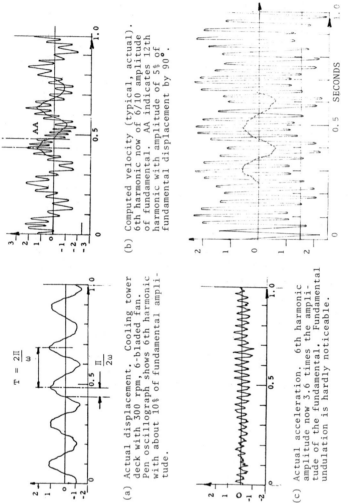

(a) Actual displacement. Cooling tower deck with 300 rpm, 6-bladed fan. Pen oscillograph shows 6th harmonic with about 10% of fundamental amplitude.

(b) Computed velocity (typical, actual). 6th harmonic now of 6/10 amplitude of fundamental. AA indicates 12th harmonic with amplitude of 5% of fundamental displacement by 90°.

(c) Actual acceleration. 6th harmonic amplitude now 3.6 times the amplitude of the fundamental. Fundamental undulation is hardly noticeable.

(d) Typical velocity. 6th harmonic of displacement has amplitude of 50% of that of the fundamental, and 6th harmonic of velocity has amplitude 3 times that of fundamental.

Fig. 4.3   Oscillographs

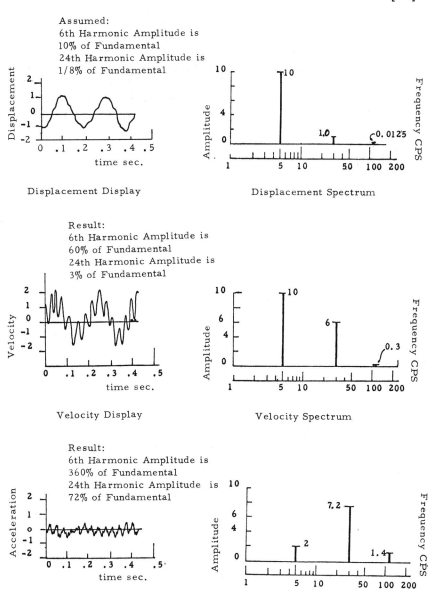

Fig. 4.4   Three actual oscillographs of the same vibration

*Note:* Data illustrates marked suppression and amplification of high frequencies in displacement and acceleration, respectively. Negligible 0.125% of fundamental in 24th harmonic of displacement in acceleration waveform. Assumption: Fundamental and harmonics are pure sine forms; filter: ideal. Result: pure tone (vertical line).

form with or without sinusoidal harmonics tends to suggest normality in the machine and departures from this form tend to suggest the existence of a fault.

The signal form connected with rolling bearings, rubbing and "knocking" processes is often what is called random, meaning that there is no repetition of the same pattern as time proceeds. In the mathematical way of thinking, all transducer-measured periodic forms may be called random. Only mathematical definition can guarantee their periodicity and the machine gives no such guarantee. However, if a brief oscillograph is periodic the vibration is called periodic. Periodic and random forms are illustrated in Fig. 4.5.

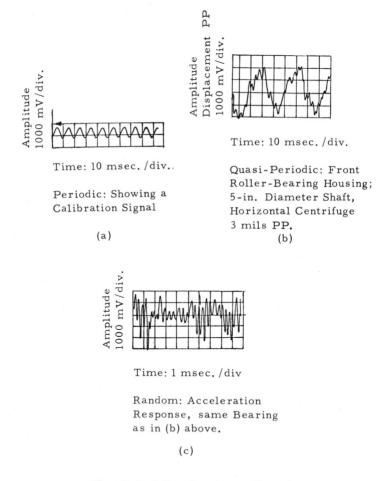

Time: 10 msec./div.

Periodic: Showing a
Calibration Signal

(a)

Time: 10 msec./div.

Quasi-Periodic: Front
Roller-Bearing Housing;
5-in. Diameter Shaft,
Horizontal Centrifuge
3 mils PP.

(b)

Time: 1 msec./div

Random: Acceleration
Response, same Bearing
as in (b) above.

(c)

Fig. 4.5   Periodic and random oscillographs

*Note:* The displacement in 5b has a basic frequency from the oscillograph of about 30 cps and a radian frequency of about 190 radians/sec, which gives by computation a peak acceleration of about 1/8th of one-gP. The frequency in 5c is of the order 3000 cps, and the measured acceleration is of the order 3-gP.

A vibration that exhibits a beat (heterodyne) is often generated in machinery and is illustrated in Fig. 4.6. Two periodic forms of different frequencies combine to generate a frequency which is the arithmetic average of the two and which

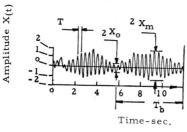

Time-sec.

Oscillograph

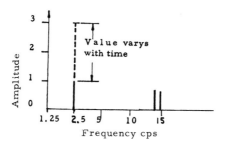

Frequency cps

Spectrum

Motion      $X(t) = A \sin(\omega_a t - \theta_a) + B \sin(\omega_b t - \theta_b)$

Beat Frequency      $f_b = \dfrac{1}{T_b} = \dfrac{|\omega_a - \omega_b|}{2\pi}$

Combined Frequency      $f = \dfrac{1}{T} = \dfrac{1}{2\pi}\dfrac{(\omega_a + \omega_b)}{2}$

Maximum Motion      $X_m = |A| + |B|$

Minimum Motion      $X_0 = ||A| - |B||$

$\omega_a, \omega_b$      radian frequencies of the component vibrations that generate the beat vibration

A, B      semi-amplitudes of the two component vibrations

| |      symbol denoting absolute (positive) value

Fig. 4.6   Beat vibration

*Note:* The pen oscillograph shows a soil vibration displacement caused by two reciprocating compressors operating at different speeds of the order of 150 rpm. On the wave analyzer, which measures the spectrum, the beat frequency appears as a wavering of the read-out needle on the meter. The needle wavers at frequency $f_b$. There is one line in the spectrum, as shown at the lower frequencies identifying the beat frequency and two lines at higher frequency which are the individual component vibration frequencies.

is associated with an amplitude that beats between the limits of the sum and difference of the ingredient amplitudes. The beat phenomenon is something different from the modulation of a fundamental. It poses a severe difficulty, for example, when it is desired to achieve a refined balance in a rotor attached to the shaft of an induction motor. Sometimes in these circumstances, a magnetically induced mechanical vibration leads to a synchronous 60 cps signal, while the unbalance vibration leads to a slip signal of say 59.8 cps. A beat occurs perhaps every five seconds. The two frequencies are so close that the ordinary wave analyzer cannot suppress the synchronous signal in favor of the slip signal. It is the objective of the balancing procedure to evaluate the slip signal as regards both amplitude and phase. This is easily done when the balance is rough. As it becomes better, the slip signal no longer predominates with the result that the phase mark, which should remain steady, rotates steadily and the meter reading goes up and down in time with the rotation. Thus, the ordinary balancing instrumentation becomes incapable of identifying either the phase or the amplitude of the mass-unbalance signal.

## 5. FREE SYSTEMS

A free system behaves as it likes and an understanding of this behavior is a necessary prelude to the understanding of forced vibrations. For the world of machinery, we define a free vibration as that which occurs when the system is not subject to force or motion from an external source.

The preceding definition of free vibration is not to be confused, for example, with that which has been traditional in horological science. In this science a free system connotes one that is free from both internal damping and external force. Damping arises from windage and from viscous friction at pivots or from molecular friction in suspensions and hairsprings. It also arises from the necessity of unlocking the mechanical element that provides the external impulse. On the other hand, a free vibration for a machine is defined as only occurring in the absence of external influence.

The simplest free system is illustrated in Fig. 4.7 and the important engineering aspects are as follows.

1. The system, if left alone, vibrates at a natural frequency that is independent of amplitude of motion (as long as the elastic element operates within its linear range) and depends only on the ratio of stiffness to mass, where K is the stiffness and M is the mass (weight lbs/32.2). Since the ratio M/K determines the static deflection, the natural frequency may be considered to depend only on the static deflection of the suspension.

2. The form of the vibration is sinusoidal.

The next simplest system is illustrated in Fig. 4.8. It has a single degree of freedom, indicated by the guide rods. It differs from Fig. 4.7 only in that it is now damped. The damping is assumed viscous and directly proportional to

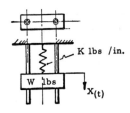

a. The System

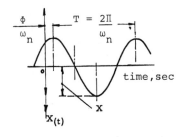

b. Oscillogram, for $X_{(t)}$

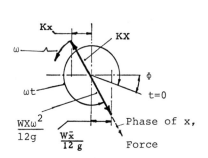

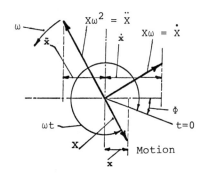

c. Phasor Diagrams

| EQUATION OF MOTION | SOLUTION |
|---|---|

$$\frac{W\ddot{x}}{12g} + Kx = 0$$

$$x = A\sin\omega_n t + B\cos\omega_n t$$

$$\omega_n = \sqrt{\frac{12Kg}{W}} = \sqrt{\frac{g}{X_s}} \quad \begin{array}{l}\text{Radian}\\\text{Frequency}\\\text{(rad/sec)}\end{array}$$

displacement $\quad x = X\cos(\omega_n t - \phi)$
velocity $\quad\quad\;\; \dot{x} = -\omega_n X\sin(\omega_n t - \phi)$
acceleration $\quad \ddot{x} = -\omega_n^2 X\cos(\omega_n t - \phi)$

$$f_n = \frac{1}{2\pi}\sqrt{\frac{12Kg}{W}} \quad \begin{array}{l}\text{Natural Frequency}\\\text{(cps)}\end{array}$$

Peak values

$$T_n = 2\pi\sqrt{\frac{W}{12Kg}} = 2\pi\sqrt{\frac{X_s}{g}} \quad \begin{array}{l}\text{Natural}\\\text{Period}\\\text{(sec.)}\end{array}$$

displacement $\quad x = X$
velocity $\quad\quad\;\; \dot{x} = \omega_n X$
acceleration $\quad \ddot{x} = \omega_n^2 X$

$X^2 = A^2 + B^2 \quad$ Amplitude, Peak

$\phi = \text{Arctan}\left(\dfrac{A}{B}\right) \quad$ Phase angle

$x = X(t)$ instantaneous value of displacement

$X_s = $ Static Deflection (feet)

$g = 32.2$ Ft–sec$^2$ gravity constant

Initial conditions $t = 0$:

$$x = B, \; \dot{x} = A\omega_n$$

Fig. 4.7   The simplest free system—discrete, linear, undamped, periodic

velocity at any instant. This assumption fortunately gives results that correspond well with many real systems, because the differential equation of motion as based on this assumption is easy to manage. Dry friction, or Coulomb friction as it is called, which is assumed constant, leads to equations that are not

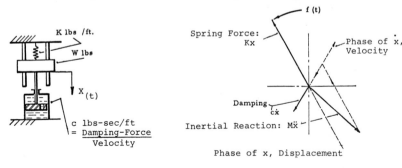

a. The System

c. Phasor Diagram

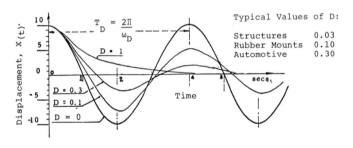

b. Displacement versus Time

Equation of Motion    $M\ddot{x} + c\dot{x} + Kx = 0$

assuming    $x = Ce^{bt}$    then    $b = \dfrac{1}{2M}\left(-c \pm \sqrt{c^2 - 4MK}\right)$

Case 1. $c^2 > 4MK$, overdamped ($D > 1$): b always real and negative; x decreases with time, no change in sign.

$$x = Ae^{(D + \sqrt{D^2-1})\omega_n t} + Be^{D - 1\sqrt{D^2-1})\omega_n t}$$

Case 2. $c^2 = 4MK$, critically damped ($D = 1$): x decreases, no sign change.

$$x = e^{-D\omega_n t}(At + B)$$

Case 3. $c^2 < 4MK$, Lightly damped $D < 1$: x decreases with sign change, oscillation.

$$x = e^{-D\omega_n t}(A \cos \sqrt{1 - D^2}\, \omega_n t + B \sin \sqrt{1 - D^2}\, \omega_n t)$$

Damped Frequency    $\omega_D = \omega_n \sqrt{1 - D^2}$    (Radian/sec)

Natural Frequency    $f = \dfrac{\omega_n}{2\pi}$    (cps)

Damped Natural Period    $T_D = \dfrac{2\pi}{\omega_D} = \dfrac{2\pi}{\omega_n \sqrt{1 - D^2}}$    (sec)

Fig. 4.8   The simplest free damped system—discrete, linear, damped and periodic where $\omega_n = \sqrt{K/M}$, $M = W/32.2$, e is the base of natural logarithms, $c_c = 2\sqrt{MK}$ and $D = c/c_c$

easy to manage. The more important engineering aspects of Fig. 4.8 are as follows.

1. The natural frequency is independent of amplitude (elastic member is linear) and depends only on the ratio K/M and on D, the damping ratio. For values of D found in typical machinery, the damped natural frequency is indistinguishable from the undamped natural frequency.

2. The motion amplitude decreases as time increases.

3. The behavior of the systems in Figs. 4.8 and 4.9 is typical of the vibration decay in a machine or structure, when the external force that caused the vibration is removed.

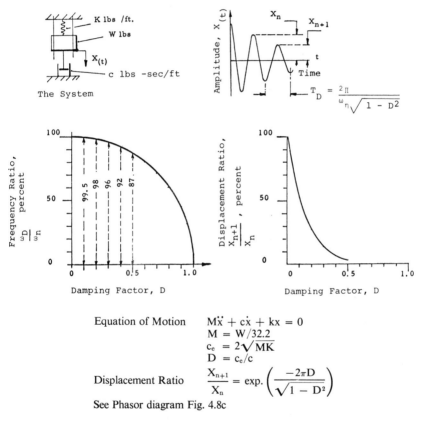

Equation of Motion    $M\ddot{x} + c\dot{x} + kx = 0$
$M = W/32.2$
$c_c = 2\sqrt{MK}$
$D = c_c/c$

Displacement Ratio    $\dfrac{X_{n+1}}{X_n} = \exp\left(\dfrac{-2\pi D}{\sqrt{1 - D^2}}\right)$

See Phasor diagram Fig. 4.8c

Fig. 4.9  How damping affects the natural frequency and the log decrement

*Note:* When D is 0.1, the displacement ratio is 53 percent. Thus, when D exceeds 0.1 the free vibration dies almost immediately. For D less than 0.1, the difference of the damped and undamped natural frequencies is about 1/2 percent. The logarithmic decrement is defined as: natural log $(X_n/X_{n+1})$ which is $2\pi D/\sqrt{1 - D^2}$. When D is 0.2, the error in assuming Log decrement $= 2\pi D$ is about 2 percent and the error in assuming displacement ratio $= \exp.[-2\pi D]$ is about 2 percent. When D is smaller, then error becomes negligible.

The usual vector diagram cannot be drawn for Fig. 4.8 because the vector (phasor) lengths shrink exponentially in time in accordance with the factor: exponential $-D\omega_n$. The radian frequency $\omega_n$ invariably arises as a mathematical convenience and is readily converted to frequency in cps by dividing by $2\pi$. Because the instantaneous forces must add to zero, the vector triangle of Fig. 4.8c, from which the forces are derived by projection, must close. The spring force is in phase opposition to displacement. The inertial reaction of the mass exceeds the spring force by an amount equal to the friction when the displacement is increasing in either direction, and it is less than the spring force by an amount equal to the friction when the displacement is decreasing in either the positive or negative direction. Because the friction reverses its direction when displacement is maximum in either direction, the damping force may be said to lag the displacement by 90 degrees at these instants of maximum displacement. As long as the spring force exceeds the damping force instantaneous velocity increases as displacement in either direction decreases. Before the spring force reaches zero, that is to say, before the displacement reaches zero the friction force exceeds the spring force so that velocity is maximum somewhat before displacement becomes zero. At the instants of maximum velocity, the damping force which is always in phase opposition to velocity lags the displacement by less than 90 degrees, whereas it was seen to lag by 90 degrees at instants of maximum displacement. However, this is only in a manner of speaking and to indicate the particular behavior of the system because phase relations have an indefinite meaning when we speak of phasors having length that diminishes with time.

It will be noticed that the vibration as mathematically described never achieves zero amplitude but merely approaches it. In practical systems the presence of some dry friction tends to eliminate the vibration more quickly than suggested by the equation of Fig. 4.8, and the system comes to rest at zero displacement or at a position of some small displacement.

## 6. FORCED SYSTEMS

The story of the simplest forced systems is told here almost totally in terms of graphic illustration. This is done to take advantage of both the shorthand presentation and, more important, to make the retrieval of the story as easy and as accessible as possible for day-to-day use wherein only the results are required. These must be self-evident as regards meaning and application.

The discussion is almost totally confined to the system that is most typical of machinery, namely the lightly damped system which is forced to vibrate because it is connected with a self-alternating force, such as rotative mass-unbalance. First, though, we will make clearer the nature of the subject system by dismissing two other families of forced vibration.

One family arises from impact. The transient motion that precedes the steady (residual, free) state is a forced vibration, although it is not usually so called. The second family arises because some systems have the property of being able to transform a steady source of energy into an alternating source. Examples of self-induced vibration abound in civil, mechanical and electrical engineering. A suspension bridge may gallop in a steady wind; a turbine blade may hum in a streamline environment; or an oscillator may operate using direct current, so to speak. The systems modulate the energy source in order to draw kinetic (alternating for electricity) energy from it.

Consider the rubbing group of vibration sources such as the violin string, the shrieking chalk or fingernail on the blackboard or the axle that needs grease. The rubbing group is characterized by virtue of a type of dry friction that is maximum when relative rubbing velocity is zero. The violin string is borne along with the bowstring at zero relative velocity. Then the friction force is exceeded by string force, leading to an increasing relative velocity and a collapse of frictional force. If the bowstring is touched with a greasy finger, the friction is no longer dry but viscous. This friction is greatest when relative velocity is greatest, which is exactly what is not wanted. The violinist cannot play. It is evident by this that the failure of lubrication in machinery can be detected by the onset of high-frequency vibration at the rubbing surfaces of bearings. The detection may be unprofitable because the time lapse between detection and failure may be less than a second. So much for the solid-to-solid process that is called rubbing.

Another group arises from the interaction of fluid on fluid and fluid on solid. These characterize what may be called the vortex group. For example, the wind induces waves in a lake and in a field of wheat (fluid to fluid) and in the aspen leaf (fluid to solid). Sometimes the solid responds but little, while the fluid oscillates greatly as in the flue and reed pipes of the musical organ. Sometimes the solid vibrates greatly, such as in the case of the suspension bridge destroyed by wind.

The self-induced vibrations that we witness all around us in tangible nature may have counterparts in their innermost recesses. Science has inferred the eternal vibrations of the molecular, atomic and subatomic worlds. In our present inquiry it seems easier to imagine all these being self-induced in relation to an ultimate absolute steady source rather than being forced by an infinite group of absolute sources, each alternating in its own right. This takes us back to the main topic of the discussion, namely the system that is affected by a self-alternating source.

For our present discussion, we define a forced vibration as that resulting when a system is connected with a source that is self-alternating and periodic, or potentially self-alternating. For example, if the source is an unbalanced rotor, there is no alternation in the spinning rotor until it is connected with the system, namely the bearings and supports. The free-spinning rotor cannot exhibit alternation, either in motion or force.

**The Undamped System.** When any system is forced, the frequency of the response is always that of the exciting source. The matter of most interest is the discovery of what is called the response. For example, in electrical engineering if a current of $i$ amperes is the response of a circuit to an excitation of e volts, then $e$ and $i$ are related so that $e = ri$ when the circuit contains resistive elements only and r may be called the impedance. In this simple example the response is known if the excitation is specified and the impedance is known. Very similar relationships are found in forced mechanical vibrations. An expression is almost invariably sought that has the character of an impedance or, perhaps, the character of the reciprocal of impedance, which is sometimes referred to as admittance or mobility. Reference to Fig. 4.10 in which the system is excited through the mass makes this clear for the undamped system.

Excitation is sinusoidal and response is sinusoidal. The displacement response is described by saying that the instantaneous displacement is x. The response frequency is the same as the exciting frequency $\omega$ (radians per second) so that $x = X \sin \omega t$, where X is the peak value of x. Thus, the response is fully known when X is known and X is, in fact, $M_d(F_0/K)$. Notice that $F_0/K$ is the static response to F. The response is now known in terms of the excitation and $K/M_d$ is a factor analogous to impedance.

It is seen that X depends on both $F_0$ and on the ratio of $\omega$ to $\omega_n$, where $\omega_n$ is the natural frequency. When the system is steady it absorbs no energy. If $F_0$ is then doubled so is X, and some energy is absorbed until the system becomes steady again. As $\omega$ approaches $\omega_n$, then X increases without limit. When $\omega$ exceeds $\omega_n$ and becomes great without limit, then X becomes small and approaches zero. Therefore we can say that very large exciting forces at relatively high frequencies induce almost no response.

When $\omega$ has the value $\omega_n$, the system is critically excited and any change in the value of $\omega$ decreases the response. Because K is constant, we may think of the system as having a mass greater than M when $\omega$ is less than critical and less than M when $\omega$ is greater than critical and vibrating freely (i.e. the excitation is now envisaged as being included in the mass) so that $\omega$ is equal to the square root of $K/M_p$, where $M_p$ is a fictitious equivalent mass. Both $F_0$ and the inertial force are in phase opposition to the restoring force when $\omega$ is less than $\omega_n$. Thus, the total force seen by the spring is

$$F_0 + X \omega^2 M$$

and this, envisaged as an equivalent mass, is given by

$$M_p = \frac{F_0}{X\omega^2} + M$$

Replacing $F_0/X$ with $K(1 - \omega^2/\omega_n^2)$ and K with $M\omega^2$, it is found that

$$M_p = M \frac{\omega_n^2}{\omega^2}$$

Thus, by expressing the force seen by the spring as a resulting mass, this mass corresponds to a free vibration at frequency $\omega$. This system always behaves according to the curve of $M_d$ shown in Fig. 4.10, as long as the spring behaves without fluttering and in a massless way. The motion response and the force transmissibility are both numerically the same. They are defined by the graphs in Figs. 4.10 and 4.11.

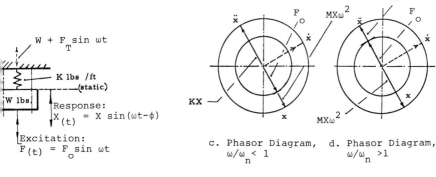

a. The System

c. Phasor Diagram,   d. Phasor Diagram,
   $\omega/\omega_n < 1$              $\omega/\omega_n > 1$

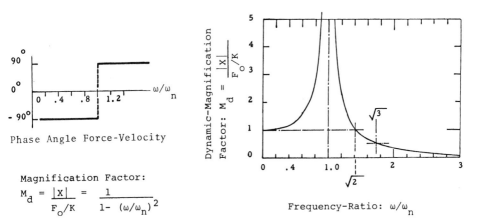

Phase Angle Force-Velocity

Magnification Factor:

$$M_d = \frac{|X|}{F_o/K} = \frac{1}{1- (\omega/\omega_n)^2}$$

Frequency-Ratio: $\omega/\omega_n$

Fig. 4.10   Undamped-mass response to a sinusoidal constant-peak force
applied to the mass

*Note:* $M_d$ is the ratio of dynamic to static deflection response (peak) as
well as the ratio of dynamic to static force response in the base (peak).
Fig. 4.10b gives the phase of force in relation to velocity. In Fig. 4.10c
$\omega/\omega_n$ is less than one, and in Fig. 4.10d it is more than one. For motion
response and force transmissibility see Fig. 4.11.

Excitation of the system through the base instead of through the mass is
illustrated in Fig. 4.11. Here the task is to discover the ratio of the response displacement of the mass to the excitation displacement applied to the base. The
ratio is called motion transmissibility. This, as a function of frequency of excitation, turns out to be the same as the motion response and force transmissibility
functions.

**Damped Systems.** The system shown in Fig. 4.12 is viscously damped and,
therefore, corresponds closely with many practical machines. The damping is
c times the relative velocity in the dashpot. The spring element is linear and the

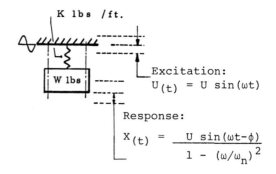

a. The System

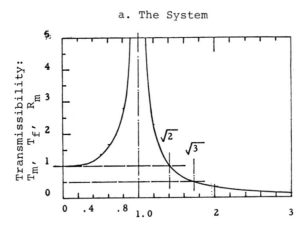

Frequency Ratio, $\omega/\omega_n$

b. Motion Transmissibility

Motion Response    $R_m = \dfrac{X}{F_0/K}$

Force Transmissibility    $T_f = \dfrac{F_T}{F_0}$

Motion Transmissibility    $T_m = \dfrac{X}{U}$

$R_m = T_f = T_m = \dfrac{1}{1 - \left(\dfrac{\omega}{\omega_n}\right)^2}$

U    peak displacement   ft
X    peak displacement   ft
M = W/32.2
$\omega_n = \sqrt{K/M}$
$\omega$    radian frequency of excitation.   rad/sec
(See phasor diagrams Figs. 4.10c and 4.10d.)

Fig. 4.11   Response of an undamped system to a sinusoidal constant-peak amplitude displacement, applied through the base

excitation is sinusoidal through the mass. Again, this situation is typical of many actual machines. An understanding of the illustrated system is the key to understanding and controlling a vast array of actual machines and to understanding somewhat more complicated systems when they arise. Before considering the mechanical model, we must first consider an idealized electrical model.

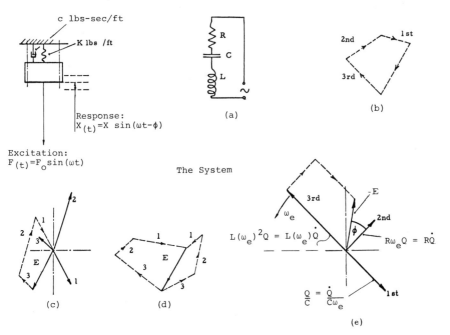

Let $-E \cos \omega_e t$ be the exciting voltage in Fig. 4.12a. Then from Fig. 4.12e

$$L\ddot{q} + R\dot{q} + \frac{1}{C}q + E \cos \omega_e t = 0$$

$$(L\dot{Q}\omega_e - \dot{Q}/\omega_e C)^2 + (R\dot{Q})^2 = E^2$$

$$Z = \sqrt{(L\omega_e - 1/C\omega_e)^2 + R^2} = E/I \qquad \text{Impedance}$$

where Z is defined as $E/\dot{Q}$. Hence [Z exp. (j$\phi$)] (vector $\dot{Q}$) = vector $(-E)$.

$$q = Q \cos(\omega_e t - \phi)$$

$$\phi = \text{Arctan} \frac{(L\omega_e - 1/C\omega_e)}{R}$$

If $L\omega_e > 1/C\omega_e$, Tan $\phi$ is positive and current lags voltage.

Power dissipation (watts) $= EI \cos \phi$

R    resistance  ohms
C    capacitance  farads
L    inductance  henries
Q    peak charge  coulombs
q    instantaneous charge  coulombs
I  = dq/dt current amperes

Fig. 4.12   A force-excited mechanical system and its electrical analog

In Fig. 4.12a an alternating voltage excites a circuit having three discrete elements across each of which a voltage drop takes place so that the vector sum of the drops is equal to the exciting voltage, prefixed by a minus sign. A sinusoidal excitation may be seen as a cosine projection of a phasor of length E which is constant and which rotates at the radian frequency $\omega_e$. At each instant the sum of the voltage drops and E must be zero, which is to say the four-sided polygon must close because its four cosine projections add algebraically to zero. If the angular positions of the voltage drops 1, 2 and 3 are not fixed, there is an endless number of solutions for the polygon. But, if the angular dispositions (phase relations) are known, then there is a unique solution. In fact, the resistive drop $(R\dot{q})$ leads the capacitative drop $(q/C)$ by 90 degrees and lags the inductive drop $(L\ddot{q})$ by the same angle.

The circuit equation is as follows; the vector solution is given in Fig. 4.12e.

$$L\ddot{q} + R\dot{q} + \frac{q}{C} + E\cos(\omega_e t) = 0$$

Each term in the equation is a voltage. The symbol q indicates the instantaneous value of the quantity of charge expressed in coulombs, and $\omega_e$ is the radian frequency of excitation. The vector diagrams make it clear that the three voltage drops must each be a cosine function of time so that the solution of the equation can be in the form $q = Q\cos(\omega_e t - \phi)$, where $\phi$ is a phase angle. The first three terms of the equation become

$$L\ddot{q} = -LQ\omega_e{}^2\cos(\omega_e t - \phi) = LQ\omega_e{}^2\cos(\omega_e t - \phi + \pi)$$

$$R\dot{q} = -RQ\omega_e\sin(\omega_e t - \phi) = RQ\omega_e\cos(\omega_e t - \phi + \pi/2)$$

$$\frac{q}{C} = \frac{Q}{C}\cos(\omega_e t - \phi)$$

From Fig. 4.12e it is now possible to write expressions for peak excitation divided by peak response in terms of the responses $\ddot{Q}$, $\dot{Q}$ or Q. But $\dot{Q}$ is the most meaningful response as it connotes amperes of current. Likewise in a mechanical system, response may be considered in terms of displacement, velocity or acceleration. The expression for $E/\dot{Q} = Z$ is given in Fig. 4.12. The absolute value of the peak current is transformed into the absolute value of the peak voltage when multiplied by this Z; the peak excitation divided by Z gives peak response. Considering $\dot{Q}$ and E as vectors at an instant, the first is transformed into the second by multiplying by Z exp. $(j\phi)$. Note that the instantaneous current $\dot{q}$ is not transformed into the instantaneous voltage e by multiplying by Z.

Probably because of the analogy of velocity to electrical current, the accepted definition of mechanical impedance is $Z = F/V$, where this is the ratio of force to velocity, both being complex quantities. More precisely, if the instantaneous

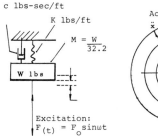

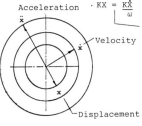

a. The system                    b. Motion Diagram

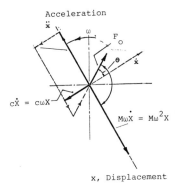

c. Force Phasors

Equation      $M\ddot{x} + c\dot{x} + Kx - F_0 \sin \omega t = 0$

Solution      $x = X \sin (\omega t - \phi)$         $\mathrm{Tan}\, \phi = c\omega/(K - M\omega^2)$

$$F_0/X = \sqrt{(K - M\omega^2)^2 + (c\omega)^2}$$

When $\mathrm{Tan}\, \phi$ is negative, force leads velocity.

In terms of velocity,      $M\omega\dot{X} + c\dot{X} + K\dot{X}/\omega = F_0$      (vectorially)

$$F_0/\dot{X} = \sqrt{(K/\omega - M\omega)^2 + c^2}$$         $\mathrm{Tan}\, \theta = (K/\omega - M\omega)/c$

Motion response is $X/(F_0/K) = \dfrac{1}{\sqrt{(1 - \omega^2/\omega_n^2)^2 + (2D\,\omega/\omega_n)^2}} = R_m$

$$D = c/(2\sqrt{KM}) = c/c_c$$

$$\mathrm{Tan}\, \phi = \dfrac{2D\,\omega/\omega_n}{1 - (\omega/\omega_n)^2}$$

Maximum motion response is $R_{m_{max}} = \dfrac{1}{2D\sqrt{1 - D^2}}$      (See Figs. 4.14 and 4.15)

Critical Frequency (maximum response)      $\omega_c = \omega_n \sqrt{1 - 2D^2}$      radians/sec

Damped Natural Frequency      $\omega_D = \omega_n \sqrt{1 - D^2}$      radians/sec

Undamped Natural Frequency      $\omega_n = \sqrt{\dfrac{K}{M}}$      radians/sec

Fig. 4.13   A forced mechanical system

force and velocity are envisaged as real sine functions $F_0 \sin \omega t$ and $V_0 \sin \omega t$, then in the accepted definition

$$Z = \frac{F_0 \, e^{j\omega t}}{V_0 \, e^{j(\omega t + \phi)}}$$

where $\omega$ is the radian frequency of vectors (phasors) $F_0$ and $V_0$. Thus, the same kind of consideration arises in the mechanical model as arises in the electrical model. Hence, V is the velocity at the point of application of the force F and Z is called the mechanical driving-point impedance.

Another impedance-type ratio is defined. $D = F/X$, the ratio of the complex peak force to complex peak displacement, is called the dynamic modulus, supposedly because it is the ratio of a stress-like quantity, namely a force excitation to a strain-like response. The topic of interest in all of the illustrations, from Fig. 4.1 through Fig. 4.22, is that of discovering for each system the ratios discussed above. Usually the required ratio of most interest is that of force to force or force to displacement or displacement to displacement. From these, the ratios of lesser interest, involving velocity or acceleration, are readily derived. The real and complex elements of the ratios are derived and expressed separately. The real part, of course, connotes the multiplier that transforms peak response into peak excitation as regards absolute value. The complex part connotes the multiplier that transforms the phase of the instantaneous response vector (phasor) into the phase of the instantaneous excitation vector.

The ratio sought in Fig. 4.13 is that of two displacements. The ratio in this instance is called motion response, and is given graphically in Fig. 4.14. Motion response is the real part of the ratio. Notice that the damped natural frequency is less than the undamped natural frequency and the critical is less than the damped natural frequency. Figure 4.15 gives the phase difference between excitation and response. Figures 4.14 and 4.15 taken together are, in fact, a graphical representation of a complex ratio analogous to $1/Z$.

Force transmissibility is considered in Fig. 4.16. Another ratio of interest, motion transmissibility, is considered in Fig. 4.17. The force and motion transmissibilities are given graphically in Fig. 4.18.

The motion response and force transmissibility diagrams for centrifugal excitation are given in Figs. 4.19 and 4.20. The exciting force is no longer constant but is proportional to the frequency of excitation. Just as before, the solutions are sought in terms of ratios such as $T_f$ and $R_m$, which are defined in the figures. The motion response ratio of Fig. 4.19 is the ratio that is traditionally used to describe the response of this system. Notice that $R_m$, as traditionally expressed, is no longer a ratio of response to excitation. Rather it is proportional to the absolute value of X, the displacement response.

Finally, Figs. 4.21 and 4.22 illustrate the response of a pallograph-like system. What is sought is the ratio of the displacement of the seismic part of the pallograph relative to the base, all divided by the absolute displacement of the base. This is of interest, for example, in determining the frequency response of a velocity-type vibration transducer. The complex relative displacement is denoted by $U - X$ which is a peak value. The real part of this, which is as before, the length of the instantaneous vector (phasor) which projects through the cosine of its angular position to give the instantaneous relative displacement, is denoted by $|U - X|$.

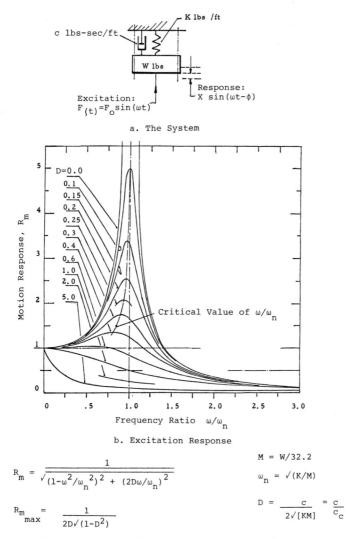

a. The System

b. Excitation Response

$$R_m = \frac{1}{\sqrt{(1-\omega^2/\omega_n^2)^2 + (2D\omega/\omega_n)^2}}$$

$$R_{m_{max}} = \frac{1}{2D\sqrt{(1-D^2)}}$$

$M = W/32.2$

$\omega_n = \sqrt{(K/M)}$

$D = \dfrac{c}{2\sqrt{[KM]}} = \dfrac{c}{c_c}$

(For Phasor Diagrams and Phase Angle Data, see Figs. 4.13 & 4.15)

Fig. 4.14   Response of a damped system to mass excitation where the damping force is c times the velocity

*Note:* The system is excited by a force $F_o \sin \omega t$ of radian frequency $\omega$ and of constant peak value $F_o$ lbs. The motion response $R_m$ is defined as $R_m = X/(F_o/K)$, which is the ratio of the peak displacement response to the static deflection that would be caused by $F_o$.

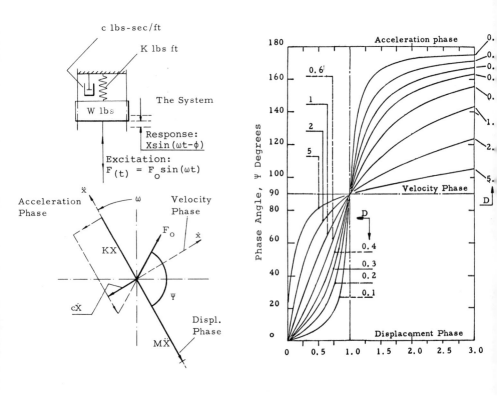

Force Diagram

Frequency Ratio     $\omega/\omega_n$

$\text{Tan } \phi = \dfrac{2D\omega/\omega_n}{1 - (\omega/\omega_n)^2}$

Critical Frequency
(maximum displacement)

$\omega_c = \omega_n \sqrt{1 - 2D^2}$     radians/sec

$M = W/32.2$

(For characteristic equations and motion response data
see Figs. 4.13 and 4.14.)

$\omega_n = \sqrt{\dfrac{K}{M}}$

$D = c/c_c$

$c_c = 2\sqrt{MK}$

Fig. 4.15   Phase relations as a function of damping and relative frequency

*Note:* The graph shows the phase relation of exciting force to mass displacement when the excitation frequency is varied in relation to $\omega_n$ for a
fixed value of $F_o$ and for various values of D, the damping ratio.

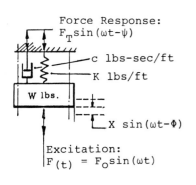

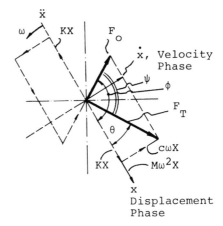

The System                              Phasor Diagram

$$F_T = X \sqrt{(c\omega)^2 + K^2}$$

$$F_0 = X \sqrt{(K - M\omega^2)^2 + (c\omega)^2}$$

Transmissibility $\quad T_f = \dfrac{F_T}{F_0} = \sqrt{\dfrac{K^2 + (c\omega)^2}{(K - M\omega^2)^2 + (c\omega)^2}}$

$$= \sqrt{\dfrac{1 + (2D\omega/\omega_n)^2}{(1 - \omega^2/\omega_n^2)^2 + (2D\omega/\omega_n)^2}} = T_m$$

$F_0$ Leads $F_T$ by angle $\psi$

$$\text{Tan } \phi = \dfrac{2D\omega/\omega_n}{1 - (\omega/\omega_n)^2} \; ; \text{Tan } \theta = \dfrac{c\omega}{K} \; ; \text{Tan } \psi = \dfrac{2D(\omega/\omega_n)^3}{1 - (\omega/\omega_n)^2 + 4D^2(\omega/\omega_n)^2}$$

$F_T$ is the force transmitted to the base

Graphs of $T_f$ and Tan $\psi$ are given in Fig. 4.18

$$\omega_n = \sqrt{K/M}$$

$$c = 2D \sqrt{KM}$$

$$K = c\omega_n/2D$$

Fig. 4.16   Forced vibration base response

*Note:* The numerical value for $F_T/F_0$ differs very little from that of $X/(F_0/K)$ (Fig. 4.14), except as D becomes large and when $\omega/\omega_n$ exceeds unity. When $\omega/\omega_n = \sqrt{2}$, $F/F_0$ is always unity for all values of D.

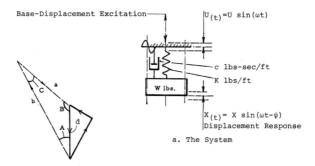

a. The System

b. Triangle Diagram

$\underline{U}$, $\underline{X}$, indicates Vector Quantities

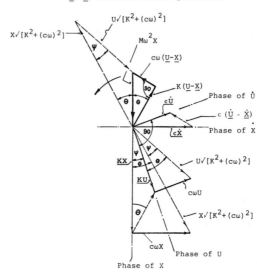

c. Phasor Diagram

Equation     $M\ddot{x} + c(\dot{u} - \dot{x}) + K(u - x) = 0$

From Fig. 4.17b     $d = rb; a^2 = b^2 + r^2b^2 - 2rb^2 \cos A$

Therefore     $r = \dfrac{M\omega^2}{\sqrt{K^2 + (c\omega)^2}}, \qquad \cos A = \dfrac{K}{\sqrt{K^2 + (c\omega)^2}}$

$\dfrac{b}{a} = \dfrac{X}{U} = \sqrt{\dfrac{1 + (2D\omega/\omega_n)^2}{(1 - \omega^2/\omega_n^2)^2 + (2D\omega/\omega_n)^2}} = T_m$     (same as $T_t$, Figs. 4.16 and 4.18)

$\operatorname{Tan} \psi = \dfrac{2D(\omega/\omega_n)^3}{1 - (\omega/\omega_n)^2 + 4D^2(\omega/\omega_n)^2}$     (same as $\operatorname{Tan} \psi$, Figs. 4.16 and 4.18)

Fig. 4.17   System for motion transmissibility

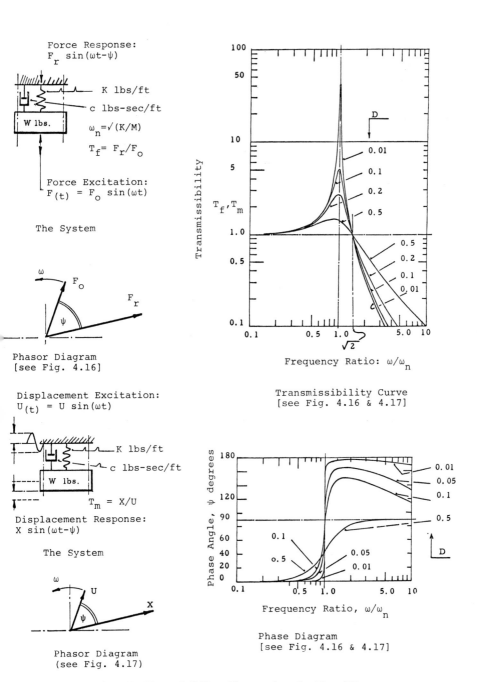

Force Response:
$F_r \sin(\omega t - \psi)$

K lbs/ft

c lbs-sec/ft

W lbs.

$\omega_n = \sqrt{(K/M)}$

$T_f = F_r/F_o$

Force Excitation:
$F_{(t)} = F_o \sin(\omega t)$

The System

Phasor Diagram
[see Fig. 4.16]

Transmissibility

$T_f, T_m$

Frequency Ratio: $\omega/\omega_n$

Transmissibility Curve
[see Fig. 4.16 & 4.17]

Displacement Excitation:
$U_{(t)} = U \sin(\omega t)$

K lbs/ft

c lbs-sec/ft

W lbs.

$T_m = X/U$

Displacement Response:
$X \sin(\omega t - \psi)$

The System

Phasor Diagram
(see Fig. 4.17)

Phase Angle, $\psi$ degrees

Frequency Ratio, $\omega/\omega_n$

Phase Diagram
[see Fig. 4.16 & 4.17]

Fig. 4.18   Transmissibility of force and motion $T_f$ and $T_m$

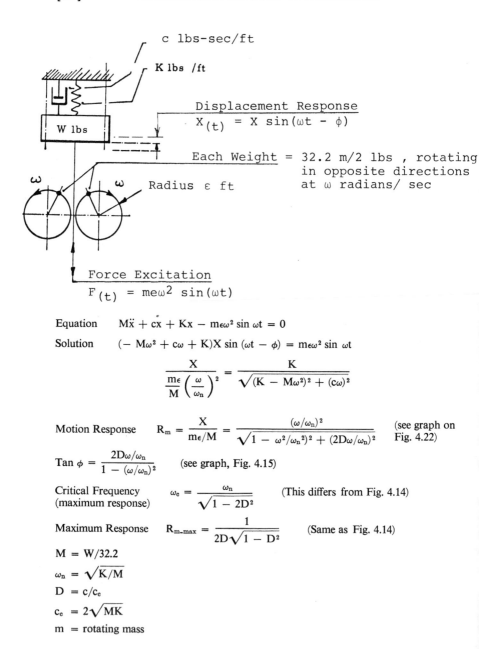

c lbs-sec/ft

K lbs /ft

Displacement Response

$X_{(t)} = X \sin(\omega t - \phi)$

Each Weight = 32.2 m/2 lbs , rotating in opposite directions at $\omega$ radians/ sec

Radius $\epsilon$ ft

W lbs

Force Excitation

$F_{(t)} = m e \omega^2 \sin(\omega t)$

Equation     $M\ddot{x} + c\dot{x} + Kx - m\epsilon\omega^2 \sin \omega t = 0$

Solution     $(- M\omega^2 + c\omega + K)X \sin (\omega t - \phi) = m\epsilon\omega^2 \sin \omega t$

$$\frac{X}{\dfrac{m\epsilon}{M}\left(\dfrac{\omega}{\omega_n}\right)^2} = \frac{K}{\sqrt{(K - M\omega^2)^2 + (c\omega)^2}}$$

Motion Response     $R_m = \dfrac{X}{m\epsilon/M} = \dfrac{(\omega/\omega_n)^2}{\sqrt{1 - \omega^2/\omega_n^2)^2 + (2D\omega/\omega_n)^2}}$     (see graph on Fig. 4.22)

$\text{Tan } \phi = \dfrac{2D\omega/\omega_n}{1 - (\omega/\omega_n)^2}$     (see graph, Fig. 4.15)

Critical Frequency
(maximum response)     $\omega_c = \dfrac{\omega_n}{\sqrt{1 - 2D^2}}$     (This differs from Fig. 4.14)

Maximum Response     $R_{m\text{-max}} = \dfrac{1}{2D\sqrt{1 - D^2}}$     (Same as Fig. 4.14)

$M = W/32.2$

$\omega_n = \sqrt{K/M}$

$D = c/c_c$

$c_c = 2\sqrt{MK}$

$m = $ rotating mass

Fig. 4.19   Centrifugal excitation—motion response

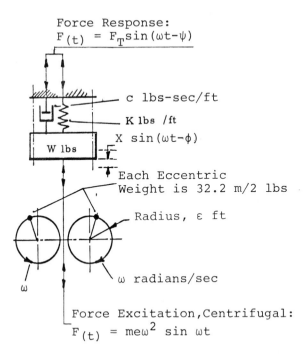

Force Response:

$$F_{(t)} = F_T \sin(\omega t - \psi)$$

c lbs-sec/ft

K lbs /ft

$X \sin(\omega t - \phi)$

W lbs

Each Eccentric Weight is 32.2 m/2 lbs

Radius, $\varepsilon$ ft

$\omega$ radians/sec

$\omega$

Force Excitation, Centrifugal:

$$F_{(t)} = m\epsilon\omega^2 \sin \omega t$$

$F_T = X\sqrt{(c\omega)^2 + K^2}$

$m\epsilon\omega^2 = X\sqrt{(K - M\omega^2)^2 + (c\omega)^2}$

$T_f = \dfrac{F_T}{m\epsilon\omega^2} = \sqrt{\dfrac{1 + (2D\omega/\omega_n)^2}{(1 - \omega^2/\omega_n^2)^2 + (2D\omega/\omega_n)^2}}$    (see Figs. 4.16 and 4.18)

$\text{Tan } \psi = \dfrac{2D(\omega/\omega_n)^3}{1 - (\omega/\omega_n)^2 + 4D^2(\omega/\omega_n)^2}$    (see Figs. 4.16 and 4.18)

Phase Angle    (see Fig. 4.16, replace $F_0$ with $m\epsilon\omega^2$)

$\omega_n = \sqrt{K/M}$

$c = 2D\sqrt{KM}$

$D = c/c_c$

$M = c/2D\omega_n$

$K = c\omega_n/2D$

Fig. 4.20   Force transmissibility $T_f$, centrifugal excitation

*Note:* For vector diagram, see Fig. 4.16 and replace $F_0$ $m\epsilon\omega^2$. For an exact solution of this system, M must be equated to $(W/32.2 + m)$. However, m is usually relatively small and is assumed here and in Fig. 4.19 to be negligible so that M is taken equal to W/32.2.

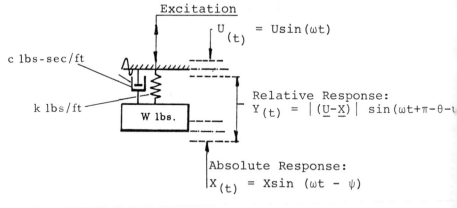

From the diagrams of Figs. 17bb & c:

$$\frac{d}{a} = \frac{|(U - X)|}{U} = \frac{d}{b} \cdot \frac{b}{a}$$

$$\frac{d}{b} = \frac{M\omega^2}{\sqrt{K^2 + (c\omega)^2}}; \text{ whence:} \qquad \text{Note: } \frac{b}{a} = T_m$$

$$\frac{d}{a} = \frac{|(U - X)|}{U} = \frac{T_m M\omega^2}{\sqrt{K^2 + (c\omega)^2}} \qquad M = c/2D\omega_n$$

$$\omega_n = \sqrt{K/M}$$

$$= \frac{(\omega/\omega_n)^2}{\sqrt{(\omega^2/\omega_n^2 - 1)^2 + (2D\omega/\omega_n)^2}}. \qquad \text{(see Graph, Fig. 4.22)}$$

Note also: $\dfrac{|(U-X)|}{U}$ is Relative Response Ratio. See Figure 4.17 for Absolute Response

Ratio $\dfrac{X}{U}$. The Three Relative Responses, given as follows, $\dfrac{|(U-X)|}{U}, \dfrac{|(\dot{U}-\dot{X})|}{\dot{U}}, \dfrac{|(\ddot{U}-\ddot{X})|}{\ddot{U}}$
are numerically equal for Sinusoidal Excitation.

Fig. 4.21    Response of pallographs and of transducers having a mechanical system as illustrated above

*Note:* The system as shown illustrates displacement relations. When
excitation is sinusoidal, the velocity, acceleration and displacement ratios
of response to displacement are all the same in absolute value. For
pallographs, velocity pick-ups and transducers that are truly seismic
such as in the illustrated system, it is desirable to compute the ratios:
$(U - X)/U$, $(\dot{U} - \dot{X})/\dot{U}$ and $(\ddot{U} - \ddot{X})/\ddot{U}$. All three are numerically
equal for sinusoidal excitation.

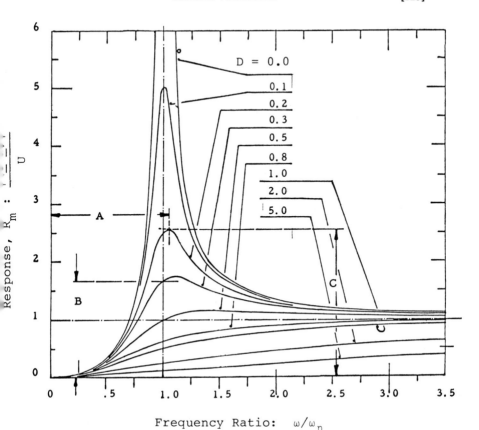

Fig. 4.22   Graphical response data for Figs. 4.19 and 4.21

$$R_m = \frac{X}{m\epsilon/M} = \frac{(\omega/\omega_n)^2}{\sqrt{(1 - \omega^2/\omega_n^2)^2 + (2D\omega/\omega_n)^2}}$$
(see Fig. 4.19
centrifugal excitation
and motion response)

$$\frac{|(U - X)|}{U} = \frac{(\omega/\omega_n)^2}{\sqrt{(1 - \omega^2/\omega_n^2)^2 + (2D\omega/\omega_n)^2}}$$
(see Fig. 4.21
Pallograph response)

Critical Frequency Ratio    $A = \dfrac{\omega_c}{\omega_n} = \dfrac{1}{\sqrt{1 - 2D^2}}$

Maximum Response            $C = \dfrac{1}{2D\sqrt{1 - D^2}}$

Response for $\omega/\omega_n = 1$    $B = \dfrac{1}{2D}$

## 7. ACTUAL EXAMPLES

Two simple examples are given to illustrate forced vibration and its control.
Diminishing the strength of the vibration source (for example, by mass balancing)

will often control it, as will altering the natural frequency, either by changing the mass or the stiffness (seldom by any attempt to alter the damping). Discrete dampers seldom appear in lineal systems whereas they very often appear in torsional systems.

For machines operating at speeds below about 60 cps, the natural frequency of response is usually higher than the exciting frequency. An example is a centrifuge with a horizontal shaft operating at 1500 rpm, where the frequency of the shaft and basket, as a cantilever, is say 4000 cpm. A system of this kind is sometimes called stiff. The effect on its response of altering stiffness or mass may be seen in Fig. 4.22. For rotating unbalance, for example, if stiffness is increased to increase $\omega_n$ so that the operating frequency is altered from 0.8 $\omega_n$ to 0.5 $\omega_n$, the response decreases from about 1.8 to 0.4. If mass balancing is carried out, then the response diminution will be directly proportional to the improvement in balance, other factors remaining the same. The effect of altering the operating speed may also be understood from Fig. 4.22.

For machines operating above about 60 cps, the natural frequency of response is often below the exciting frequency. Systems of this kind are sometimes called flexible. All rotors have some unbalance. As the speed of a flexible rotor is gradually increased, the unbalance force is more or less in phase with the radial displacement which it induces, as shown in Fig. 4.15. The large rotor mass with its small eccentricity exhibits a simple centrifugal tendency. When the speed is significantly above the critical speed, the dynamic behavior becomes quite different. Consider a disc so thin that the effect of centrifugal couple is negligible. The disturbing force leads the displacement more and more as the speed increases, until its phase is effectively 180 degrees ahead of the displacement, as in Fig. 4.15. In other words, the center of gravity exhibits a constant centripetal tendency to reach the axis of the bearings. It effectively achieves this so that the disc rotates around its center of gravity and the shaft whips at a radius about equal to the eccentricity of the center of gravity in relation to the geometrical (shaft) center of the disc. This may be partially understood from Fig. 4.22, wherein $R_m$ tends to unity as the exciting frequency increases without limit. The exciting mass is now M and since $XM/m\epsilon$ is unity, X is the same as $\epsilon$, the eccentricity of unbalance. In the treatment given in Fig. 4.19, it is assumed that m is small compared with M. This is not always so as, for example, in the case of the disc on the flexible shaft above, where M = m. For an exact treatment, the M of Fig. 4.19 defined as W/32.2 must be rewritten as M + m.

It is evident from the foregoing that, in order to control forced vibration, the nature of the machine system must be appreciated together with the effects of varying mass or stiffness. Whether the system be flexible or stiff, mass balancing, if properly done, always decreases vibration.

A fixed-speed 3600 rpm motor is illustrated in Fig. 4.23. Despite acceptable rotor balance, vibration caused harm to the coupling and bearings. A system such as this is usually arranged so that the natural frequency of the lateral vibration of the motor on its base is higher than the rotative frequency (forcing frequency). The usual rule, in this case to reduce vibration, is to increase stiffness so as to lower the ratio of forcing frequency to frequency of free response (see Fig. 4.14). If it seems desirable to determine the nature of the uncorrected

Fig. 4.23   Base plate for a 50 hp, 3600 rpm motor for a centrifugal pump,
before and after stiffening

motor a significant mass may be clamped to the motor. If this increases the
vibration it may be concluded that the uncorrected natural frequency is above
the forcing frequency. If it decreases the vibration, then it must be assumed that
the uncorrected natural frequency is below the forcing frequency. Similar considerations apply to alteration of stiffness. By using a large wooden or metal

Fig. 4.24   A pair of induced draft fans, turbine driven to a maximum speed of 900 rpm, each about 250 hp with an output of 60,000 cubic ft/min

lever to stiffen the system, if the vibration decreases, it may be assumed that the uncorrected natural frequency is above the forcing frequency. Thus, it is seen that there is no general rule as to what to do to diminish vibration. However, in general, all forced systems should be arranged to be definitely stiff or definitely flexible. Figures 4.14 and 4.22 make it clear that it is possible to operate at or near the critical frequency if there be sufficient damping. However, if any sudden change in mass balance then occurs, disaster may result.

A considerable stiffening of the motor base shown in Fig. 4.23 reduced the vibration to an acceptable level. But the case of variable-speed machines is more difficult as illustrated in Figs. 4.24 and 4.25. A pair of fans rest on a concrete floor about four levels above the boiler so that they are not within view of the operator as he alters the fan speed to suit his steam requirements. The wheels of these induced-draft fans have 70-in diameters by 18-in widths, of the squirrel-cage type. Vertical vibrations of the floor were usually about 0.020 in PP. The critical speed was found to be about 750 rpm for each, with a maximum operating speed of 900 rpm. This is an undesirable situation. Such fans preserve their balance well, despite considerable accretion of scale.

As illustrated in Fig. 4.25, careful balancing held the critical-speed vibration to less than one mil PP. But the accretion or loss of scale could lead to the disastrous response indicated by Fig. 4.25, curve A. It was therefore desirable to raise the natural frequency of the fan-on-floor to above about 1500 cpm. This was not done because of the large-scale structural alteration required. Arrangements were made to operate at 900 rpm or below 650 rpm.

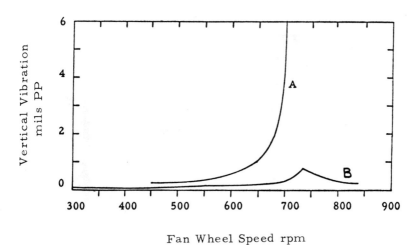

Fan Wheel Speed rpm

Fig. 4.25   The vertical vibration of the fan (Fig. 4.24) measured at various speeds, before and after balancing. Curve A is before balancing, and critical speed is about 730 rpm.

# REFERENCE

1. Jackson, C. Personal Communication, Monsanto Chemical Co., Inc., 1964, 1968.

# SECTION II

# ISOLATION AND DAMPING

# SECTION II CONTENTS

# SYMBOLS

| | | |
|---|---|---|
| IL | intensity level (re $10^{-12}$ watts/m²) | dB |
| W | acoustic power | watts |
| r | radius | feet |
| SPL | sound-pressure level (re 0.0002 $\mu$ Bar) | dB |
| PWL | sound-power level (re $10^{-12}$ watts) | dB |
| DI | directivity index | |
| A | propagation attenuation | dB |
| Hz | frequency | cps |

# 5

# SOUND AND NOISE IN AIR

*William S. Mitchell*

## 1. EQUIPMENT SOURCES

In recent years there has been a continuing effort to remove or reduce some of the more intense noise from our social environment. Such noise is harmful in that it may cause temporary and, in some cases, permanent hearing loss. Moreover, it is often considered a nuisance because of its subjective character of annoyance, its interference with communication and its effect on worker efficiency. The seriousness of the problem is reflected in existing noise regulations such as the Walsh-Healey Public Contracts Act (U.S. Government) and Standards for Noise Control, State of California. A graph of acceptable noise levels as defined by the latter is shown in Fig. 5.1.

The isolation and damping of sound and noise in air presents problems quite unlike those found in the treatment of mechanical vibration. In dealing with mechanical vibration, isolation of a system from its supporting structure is readily accomplished with compliant springs, rubber supports and air sacs. A vibrating piece of equipment can be isolated from its environment with specially constructed floating supports. All of the methods used are for one common purpose: to prevent the transmission of mechanical vibration from an operating device to its supporting structure or from the supporting structure to the sensitive device. What is considered here is how to isolate or damp mechanical vibrations after they are radiated into air as sound.

The source of any sound is either a mechanical vibration or an aerodynamic pressure fluctuation. The former is associated with mechanical surface motion, the latter with compressible gas flows of the type found in air jets and air-moving equipment. When a machine surface, or for that matter any surface, oscillates in some form of periodic or random motion, the motion generates alternating pressure waves that propagate from the moving surface at the velocity of sound. Those motions between 20 Hz and 20 kHz stimulate the hearing mechanism so that we are aware of their existence.

The mechanical motion can be the result of infinitely many effects. There is, for example, building resonance, improper lubrication or unbalance due to manufacturing errors. Consider the following few examples. Improper lubrica-

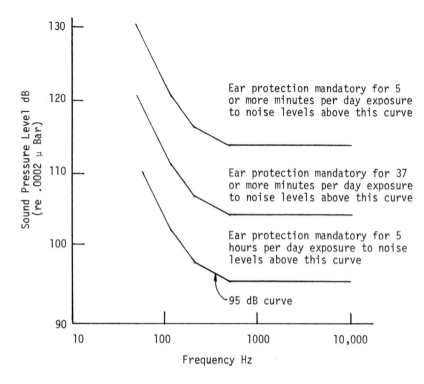

Fig. 5.1   Acceptable noise levels as recommended by the State of California, Division of Industrial Safety, Article 55, Standards for Noise Control

tion causes intermittent contact of dry bearing surfaces. The radiated sound has a high-frequency content and resembles a high-pitch screech. This is band-limited random noise. Air turbulence is most often produced in high-speed rotating machines such as fans, blowers and compressors. It is also produced by jet flows, nozzles and diffusers. Relative motion between the flowing air and the surrounding still air produces tiny cyclone-like swirls of air. These result in radiated sound. Relative motion between flowing air and compressor blades also causes vortex noise. As a vortex separates from a moving blade and is intercepted by a successive blade, aerodynamic noise is produced. Noise is also generated in piping systems by air turbulence caused by elbows and mechanical restrictions. With regard to electrical motors, the quietest motors are of the totally enclosed, nonventilated type. Open motors are noisier, but fan-cooled motors are the noisiest. Three-phase motors operate more quietly than single-phase machines because of the absence of switching mechanisms and lower magnetic densities.

Mechanical noise is caused by engines, bearings, transmission drives and power conversion equipment. A typical power engine will have aerodynamic noise generated by fan belts and fan blades in a frequency range of 500 Hz to 6 kHz. Engine air intakes are characterized by the hissing sound of air rushing through the inlet duct. There is also sound radiation from the duct due to internal mechanical valve noise with frequencies between 100 Hz and 3 kHz. At the exhaust outlet of an engine there is high-intensity noise whose frequency spectrum is in a range of 100 Hz to 400 Hz. Although this is relatively low-frequency sound and readily identifiable, it will increase in frequency with higher operational speeds of the engine. Sound is also radiated over a broad spectrum of frequencies from the transmission case and gear drives. Consideration must be given to all possible sources of sound when attempting to isolate or dampen sound and noise in air.

## 2. SOUND PROPAGATION

Sound propagates in air between the source and final receiver. Sometimes the path of radiated sound is direct; that is, it propagates along the line of sight. At other times it is reflected or diffracted by surfaces on adjacent equipment as well as walls, floors and ceilings. This results in complex sound paths which can proceed around corners or setup systems of standing waves.

The geometry of a noise source influences the manner in which sound is propagated through air. For example, a simple, small, low-frequency noise source creates a sound field which exhibits spherical spreading. Thus, at some distance, r, from the source of sound, the intensity level approximated by the inverse-square law is

$$IL = 10 \, Log_{10} \frac{W}{4\pi r^2 I_0} \, dB \qquad (re. \, 10^{-12} \, W/m^2)$$

where W is the acoustic power of the source in watts. This is also expressed as

$$IL = 10 \, Log_{10}W - 20 \, Log_{10}r + 99 \, dB$$

A line of these same sources (for instance a row of machines) perpendicular to the direction of r produces a higher sound intensity level at the same location r than that of a single machine. This type of response is found to obey more closely an inverse relation: IL is proportional to $1/r$, where a doubling of distance between source and receiver produces an intensity level change of $-3$ dB instead of a $-6$ dB change as predicted by the inverse-square law. Further, there is little directivity associated with low-frequency sounds which spread equally in all directions: forward, backward and around corners. Conversely, high-frequency sound is directional, often diffracted and absorbed in large part by air. It is also important to remember that airborne noise is not always generated from a primary source of vibration. The acoustic source may be nothing more than an efficient acoustic radiator excited into resonance by an internal machine vibration.

## 3. OUTDOOR SOUND PROPAGATION

Many times measurements must be performed out-of-doors. If there is any appreciable distance between the noise source and a measuring transducer, such as would occur when making environmental noise surveys around a plant site to satisfy local noise codes, then atmospheric conditions can influence greatly the accuracy of the measured data. For instance, sound waves are affected by density, temperature, wind and humidity. Thus, at the microphone the sound-pressure level will fluctuate with time, dependent on atmospheric conditions. For ideal outside conditions, that is, when there is no atmospheric attenuation of the radiated sound and spherical divergence takes place, the measured sound-pressure level above a hard-ground surface is

$$\text{SPL} = \text{PWL} + 10 \, \text{Log}_{10}\text{D}_\theta - 20 \, \text{Log}_{10}\text{r} + 2.5 \, \text{dB} \qquad (\text{re} \ .0002 \, \mu\text{B})$$

where the power level PWL is referred to (re $10^{-12}$ watts), $\text{D}_\theta$ is the directivity factor and r is the distance in feet from the source to a measurement transducer. In reality a lower sound-pressure level will be measured in actual environments because of sound absorption in the air, wind and temperature gradients, air turbulence, trees and shrubbery and the condition of the ground surface (grass is more absorptive than hard clay). Total attenuation due to the atmosphere can be calculated from two far-field measurements of sound-pressure level. For any two given distances $r_1$, $r_2$ from the source

$$\text{SPL}_1 - \text{SPL}_2 = 20 \, \text{Log}_{10}\frac{r_2}{r_1} + A \qquad (r_2 > r_1)$$

where A is the propagation attenuation in dB. Attenuation is dependent on frequency. Hence, it is necessary to average calculated values of A over the frequency range of interest. As sound propagates parallel to the ground, it passes through existing temperature gradients maintained by the temperature difference between the ground and air. But the velocity of sound is proportional to the square root of absolute temperature. So if warm air is sitting over cool ground (or water), the sound is diffracted downward. Conversely, when the ground is warm and the overlying air is cool, horizontal sounds are diffracted upward. This causes a shadow zone into which few sound waves travel. Wind gradients, which decrease in magnitude toward the ground, bend sound waves upward away from the ground, upwind of the sound source. However, downwind of the sound source, sound is diffracted towards the ground. Therefore, no shadow zone exists downwind of a sound source subjected to wind effects alone. A typical shadow zone is shown in Fig. 5.2. The combined effects of temperature and wind can cause a variety of shadow zone patterns which must be considered when making acoustical measurements in addition to considering sound source directivity. Sound-pressure level differences as great as 30 dB can exist between upwind and downwind measurements. Moreover, peak-to-peak SPL fluctuations at the receiver will be from 5 dB in a stable atmosphere (non-turbulent) to 20 dB in a turbulent one.

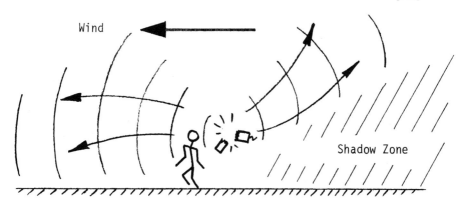

Fig. 5.2   Shadow zone, effect of wind on outdoor sound propagation

## 4. SOUND EFFECTS ON PERSONNEL

Noise is an industrial problem. When it is in excess of recommended limits, industrial noise will lead to temporary, partial or complete loss of hearing. In the latter case it can be accompanied by physical pain. Today's industrial workers are more aware than ever of this latent hazard and, consequently, are more inclined to file claims for compensation of hearing loss, especially in view of the favorable attitude of the nation's courts. Because of this potential loss, insurance companies are now devoting more of their energies and financial resources to identifying acoustically hazardous environments and notifying interested parties of their existence, thus making noise a significant economic problem. When noise levels are sufficient to mask audio communication, they become both an annoyance and a threat to human safety—the latter when it interferes with audible warning sounds. Intense background noise tends to isolate people from their environment in such a way that the noise of approaching vehicles and operating equipment is no longer discernible to their senses. Because noise results in a subjective response, some people respond differently to different types of noise. In general, high-frequency sounds are more annoying than low-frequency sounds, and prolonged high-level noise and repetitive, intermittent or modulated sounds are found to be more distracting. Noise creates a feeling of fatigue in those persons subjected to extended periods of high-intensity noise. This is precipitated by an increase in nervous activity brought on by increased muscular tension. Under these conditions, a worker becomes less efficient on his job and eventually careless in his duties. This inevitably manifests an economic loss.

## 5. EQUIPMENT ENCLOSURES

As noted previously, airborne sound is caused by either air turbulence or some form of mechanical motion. Since it spreads outward from the source, it is highly recommended that the source of the sound be determined first and every effort made to eliminate or reduce it at that location. However, since this is not always possible, the next approach is to isolate or dampen the sound waves as they propagate through the air. Noise can be isolated successfully by constructing either a partial or complete enclosure around the source of the acoustic disturbance. When the sources are air ducts, machines or other equipment not requiring continuous personnel attention, the problem is relatively simple. The source can be totally enclosed in a sound-isolating structure which is carefully sealed at all joints to prevent acoustical *leaks*. Sound will propagate through any opening in a structure no matter how small it may be. For example, some cinder block wall construction requires a covering coat of plaster or other surfacing material to obtain sufficient transmission loss through the wall since sound passes easily through the voids in the block material. Sound-attenuating enclosures are commercially available from firms such as Industrial Acoustics Co., New York and Eckel Industries, Inc., Massachusetts. Typical enclosures are capable of attenuating transmitted sound levels from 26 dB at low frequencies to 62 dB at high frequencies as shown in Fig. 5.3. Note that sound attenuation depends on frequency. Enclosures can be used to keep noise both in and out.

When an operating machine radiates intense noise into its surroundings and the noise is considered a sound hazard, a box-like structure can be constructed around the machine to contain the sound radiation. The construction of the enclosure can be quite simple or very complex, depending on the required amount of noise attenuation. If the required noise reduction is not too great, then an enclosure constructed of light-gauge metal with damping compound and sound-absorbent glass wool applied to the inside metal surfaces may prove to be adequate. For sources of higher acoustic power it may be necessary to isolate the

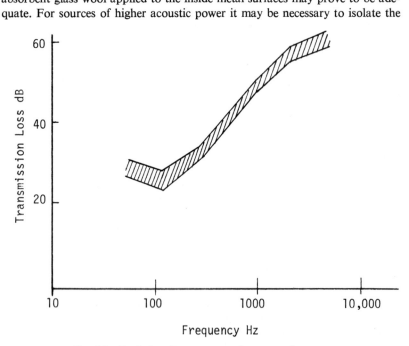

Fig. 5.3   Typical enclosure attenuation versus frequency

source from its supporting structure and provide an acoustically treated enclosure for the source. This isolated and enclosed system is subsequently enclosed in a second acoustically treated enclosure to effect a greatly increased sound transmission loss. Various treatments of the source are required to attenuate different parts of the frequency spectrum. For instance, vibration isolation of the source is most helpful in attenuating the low frequencies. Acoustically absorbent materials are most effective at the higher frequencies as are most enclosures without acoustical treatment. Thus, a combination of different treatments results in a partial attenuation of low-frequency sound and quite good attenuation of the highs. The resulting output becomes a moderate level low-frequency sound. Proper acoustical treatment can provide sound attenuation of 20 dB to 30 dB at frequencies below 100 Hz with 55 dB to 60 dB attenuation above 1 kHz.

## 6. WALL CONSTRUCTION

When consideration is being given to the construction of office areas or quiet rooms, an investigation should be made of the different types of wall constructions and their noise-controlling capabilities. Sound travels as readily through some building materials as it does through the air. For example, it has been previously noted that sound will pass through some forms of unsealed cement and cinder blocks. In addition, airborne sound impinging on a plasterboard-constructed wall is transmitted from one wallboard to the other by the coupling action of the wall studs. At the present time there are no required code standards for wall construction in the United States. (Such is not the case in Europe.) What is available in the United States is suggested construction methods for various type walls. Each type construction has an associated number such as STC-40 indicative of the ability of the wall to stop ordinary sound. An indication of the effectiveness of a given *sound transmission class* wall can be seen in the values for sound transmission through a wall given in Table 6. In each case the higher the number the more effective the sound barrier.

TABLE 6
SOUND TRANSMISSION OF STC WALLS

| STC NUMBER | SOUND TRANSMISSION |
|---|---|
| −25 | normal speech can be understood quite easily |
| −30 | loud speech can be understood quite easily |
| −35 | loud speech is audible but not understood |
| −42 | loud speech is audible as a murmur |
| −45 | loud speech is difficult to sense |
| −48 | some loud speech barely audible |
| −50 | loud speech not audible |

A fundamental rule for reducing the sound transmission of walls recommends using thick walls of dense material. But this is relatively expensive and in some cases creates structural problems. A more modern approach is to use sound-deadening insulation board in combination with standard gypsum or plasterboard. A typical installation is as follows. One-half-inch sound-deadening insulation board and 5/8-in thick gypsum wallboard are applied to both sides of 2-in x 4-in studs on a 2-in x 4-in wooden plate. The insulation board is nailed to the studs and the gypsum board is then laminated to the insulation board with an appropriate laminating compound. To be completely effective, the perimeter joints of the wall and all other wall joints are sealed to avoid acoustical flanking. This type of construction meets the requirements of STC-49 which is quite good when compared to the value of STC-38, associated with conventional 2-in x 4-in stud-framed walls.

Another construction rated as an STC-49 wall utilizes fiber glass instead of sound-deadening insulation board. First, 2-in x 4-in studs are staggered on a 2-in x 6-in plate. The spacing between the studs is filled with 1 1/2-in fiber glass building insulation to absorb sound energy in the wall and improve sound transmission loss. Then the gypsum wallboard is applied on each side of the wooden framing with the result that each side of the wall partition is supported by its own stud system. The resulting wall has an average STC-49 rating and is somewhat thicker than the previous construction. However, it uses standard construction materials and techniques with no appreciable increase in cost (see Fig. 5.4).

Good-quality construction is necessary to insure the effectiveness of a sound barrier partition. Walls must fit properly at the floor and ceiling. Further, solid core doors that fit closely in the door frames against tight-fitting perimeter seals

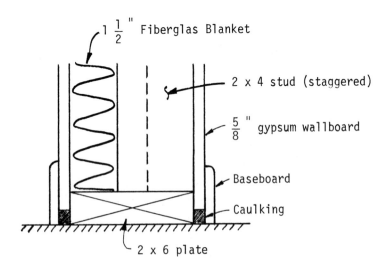

Fig. 5.4   Wall construction for sound isolation STC–49

and thresholds should be used to prevent bypassing of the sound-attenuating effect of the wall construction. This also means that construction should include caulking or sealing around all electrical, plumbing and mechanical penetrations into and from a partition.

## 7. FLOOR AND CEILING CONSTRUCTION

Floors require acoustical treatment for the same reason as the walls—to effect noise reduction. However, consideration must be given to the effect of impact noise. Two basic forms of construction are recommended in wood-framed structures. In the first a finished floor is floated on two layers of hardboard and one layer of Fiberglas Noise Stop Board. These are adhered in layers above the existing subfloor. Three-inch fibrous glass insulation is installed in the joist cavity to absorb sound. The Noise Stop Board must be caulked around the edges to prevent acoustical flanking. The finished floated floor, whether it be wood, vinyl, tile or sheet vinyl, is kept back away from the vertical walls to prevent impact noise from being transmitted into the walls and, consequently, to other areas of the structure.

The other type of construction utilizes a carpet and pad floor covering. Here, the carpet and pad are installed over the entire floor surface, composed of 1/4-in hardboard and 5/8-in plywood subfloor, to cushion the effect of impact. Fiber glass insulation is also installed in the joist cavity. This latter construction has the disadvantage of requiring considerable maintenance due to wear and clean-

TABLE 7
**RECOMMENDED SOUND TRANSMISSION CLASS AND IMPACT NOISE RATING VALUES FOR WALL AND FLOOR CONSTRUCTIONS**

| WALL SYSTEMS FOR LOW BACKGROUND NOISE | | |
|---|---|---|
| ADJOINING SPACE | BEDROOM | OTHER ROOMS |
| Living unit | STC-50 | STC-45 |
| Corridor | STC-45 | STC-40 |
| Public space (Avg. noise) | STC-55 | STC-50 |
| Public space (High noise) | STC-60 | STC-55 |
| Bedroom | STC-45 | N.A. |

| FLOOR SYSTEMS FOR LOW BACKGROUND NOISE | |
|---|---|
| Floors separating living units | INR  0 |
| Corridor floor above living unit | INR +5 |
| Living unit floor above public space | INR −5 |
| Public space above living unit | INR +5 |

ing of the carpet. Moreover, the impact performance depends on the quality of the carpet and pad which should be at least 65 oz/yd and 45 oz/yd, respectively. Further limitations of this type of construction are that holes cannot be randomly cut through the carpeted floor or lower ceiling. Neither can recessed light fixtures be used below; special installation of ceiling light fixtures is required.

The formerly mentioned floating-floor construction will have a sound-transmission value of STC-53. When the finished floor is a hard surface, the impact noise rating is six. But this is increased to thirty with the addition of a carpet and pad. The system with carpet and pad on the subfloor should have an STC value of fifty (depending on pad and carpet) and an impact noise rating of twenty-seven. Impact noise rating numbers are given in Table 7. Some typical constructions are shown in Fig. 5.5. This latter type of construction is used over concrete slabs.

The ceiling surface of an enclosure is also important in the process of noise reduction, not so much for transmission loss but for its ability to reduce environ-

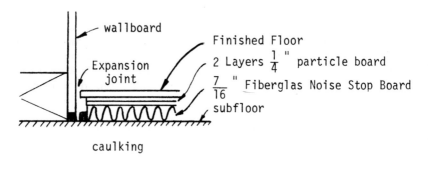

(a)

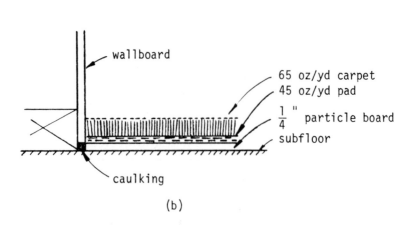

(b)

Fig. 5.5  Typical construction of impact noise rated floors

mental noise by absorption. Acoustical ceilings are applied either directly or suspended. Ceiling tile is available in various sizes and can be installed in one of many ways: for example, on furring strips or stapled directly to gypsum wallboard. It is also practical to cement the acoustical tiles in place with a special acoustic adhesive.

## 8. PERSONNEL ENCLOSURES

There exist times when it may be important to locate personnel in a hazardous area where the SPL exceeds 90 dB. In such cases as this, a sufficiently large enclosure or sound-isolation room with doors, windows and proper ventilation must be erected in the hazardous sound field. The enclosure provides necessary worker protection while allowing sufficient freedom to work. Sound-isolation rooms can be made with either single or multi-wall construction. Commercially available enclosures are made with prefabricated panels that have a solid sheet metal outside shell and an inner wall of solid or perforated metal. Acoustically absorbent material is inserted in the space between the walls. Doors, windows and structurally reinforced floors all form part of the enclosure structure. These latter items need to be mounted with proper acoustical seals and hardware; otherwise, acoustical flanking will result in an ineffective sound-isolation room. One manufacturer of sound enclosures furnishes a basic wall thickness of four inches; however, this can be increased for added noise reduction. In the case of cinder block or brick wall construction, mass would normally be added in the form of increased wall thickness to increase transmission loss. A common approach for obtaining maximum sound transmission loss through a wall is to use multi-wall construction. This amounts to constructing a room on isolators inside a room with an air space separating the inner and outer walls.

Sometimes machinery must be accessible to operators and service personnel. It may also be desired to acoustically isolate only a portion of a production system; therefore, it is not always possible to completely enclose a noise source. In these specific instances, partitions or panels can be placed in the vicinity of the source to alter and absorb some of the acoustic energy of the radiated sound. Panel construction can be relatively simple since it need only hold sound absorbing material in a given location. For example, Eckel Industries manufactures sound-attenuating panels in standard lengths of from 4 ft to 10 ft, by 30 in wide. These are constructed with a perforated metal facing behind which is a 2-in layer of fibrous glass. For adverse environmental conditions they are supplied encased in a thin plastic film. These units can be easily installed in existing structures. Typical sound-absorption values for the abovementioned panels are shown in Fig. 5.6. Panels such as these act as soft reflectors, but they do so far more effectively at high frequencies than they do in the lower range. This results in a substantial change in the directivity of the sound field. Because the sound waves from direct radiation impinge on absorbent surfaces, the pressure levels of reflected waves are attenuated far below those that are incident. This assists in lowering the level of the environmental sound field.

Consider a large factory space containing a variety of machines. If the interior walls and ceiling of the enclosure are acoustically treated, negligible benefit

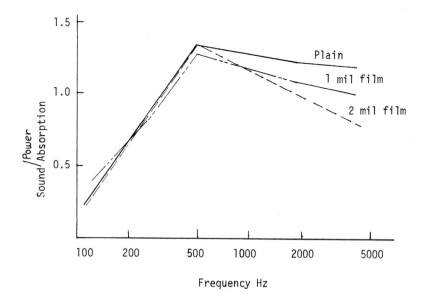

Fig. 5.6   Sound-power absorption of fibrous glass-packed, acoustically
absorbent panels (ASTM–C 423–66)

results for the worker who is not close to a wall but is in close proximity to a
noisy machine, since he is most affected by the direct, radiated sound. This direct
sound is far greater in level than sound in the reverberant field. As a further
example, consider a small room containing an intense sound source such as a
friction cutoff saw. If the room is acoustically untreated, the intensity level of
the room will increase to dangerous levels because of reverberation effects.
With adequate treatment, a lowered sound level can be maintained in the room
and the machine operator is subjected to line-of-sight noises only. However,
these may be as dangerous as the previously mentioned reverberant sound.
Now, if this same device is located in a large enclosure remote from any reflecting wall, then the effect of the enclosure on the machine and operator is negligible.
No acoustic treatment of such an enclosure will aid the hapless operator. This
situation can be improved with localized acoustical treatment and personal
safety devices, such as ear plugs and protective earmuffs. These examples should
dispel the belief that acoustical treatment of any environment automatically
results in noise reduction.

   Product noise is readily analyzed in a low-background-noise sound enclosure.
Sometimes, because of economic and manufacturing considerations, the enclosure must be constructed over a portion of a production line. In many cases
these are only partial enclosures since the products of manufacture need to pass
from one area of processing to the next without external interference. Such enclosures are used to analyze the acoustic radiation of a device, such as an electric motor, and compare its output with an electronically "good" device. The

result is automatic quality control. With respect to the nature of enclosures, then, the following observations should be noted. First, the effect of a high-level sound source on its local environment can be substantially reduced with proper enclosure treatment. Second, for vibration and acoustical testing in noisy areas, volumes can be effectively enclosed and maintained at a reduced sound level. Third, acoustically absorbent surfaces are ineffective in reducing noise levels in the vicinity of a source where direct radiation is the main concern. Sound enclosures and partitions afford a practical means of isolating and damping airborne noise in industrial environments when it is impractical to isolate or dampen the source.

Many acoustical emissions are the result of structural resonances in operating equipment excited by remote mechanical vibration. For example, gearbox and transmission drives will radiate sound due to the internal excitation of gears. The sound levels of these emissions are found to increase significantly with increasing gearbox power rating. In some instances high capacity gearboxes have been found to generate noise sound pressure levels in excess of 100 dB. Isolation of a noise source from its foundation results in a significant reduction of structure-borne noise accompanied by a decrease in acoustic radiation.

## 9. DAMPING AND DAMPING MATERIALS

One generally accepted method for reducing the amount of noise emitted by a vibrating surface is to reduce the amplitude of the vibratory motion through the application of a deadener or damping material. This is more effective when applied to thin, compliant panels than to rigid structures like gearbox housings. In the latter case sound attenuation is due mainly to sound-transmission-loss characteristics of the coating. Damping materials arrest surface motions by changing vibratory energy into heat. This heat is the result of particle motions in the viscous damping material.

Damping materials currently available cover a broad range of physical properties and characteristics. When selecting a damping material, consideration must be given to the manner by which it will be applied. In some cases it may be possible to coat a surface by brushing or spraying the material. In other cases it may have to be applied by troweling. This will depend on the characteristics of the vibrating surface and the weight and thickness of the material to be applied. With certain orientations, such as vertical surfaces or the undersides of panels, it often becomes necessary to load an open-mesh cloth or felt with the damping material and apply the composite to the vibrating surface. This is especially true with thick coatings which, without the holding effect of the cloth, tend to flow after application. Loading a material such as cloth (one or several layers) also results in ease of handling and application of the damping material. Selection of a deadener involves other considerations: flammability, aging properties, chemical resistance, water absorption, toxicity and cost. These are only a few of the many factors which determine whether or not a material is appropriate to a given situation. Many damping compounds are commercially available for sound deadening. They are applied to such devices as noisy machinery and the undersides of automobile bodies. However, an ordinary roofing compound sold by Sears, Roebuck and Co. is very effective in noncritical situations, and has the advantage of being much more economical. This latter material is a composition of asphalt and asbestos fibers. Damping compounds are generally applied directly to an acoustically noisy surface with a recommended cover of 40 percent of the surface area and with a coating thickness twice that of the thickness of the vibrating surface. An alternate application recommends a coating thickness of 20 percent by weight of the vibrating surface.

Recent developments in the application of damping coatings allow substantial reductions in radiated equipment noise. It has already been noted that coatings result in sound-transmission loss, but these coatings are usually quite thick in comparison to the thickness of the vibrating surface. This is necessary to satisfy the 20 percent by weight requirement. Efficient damping results when a thin, constraining layer of sheet metal or other suitable material is placed over the viscous damping coating. Constriction of the damping material results in increased viscous shearing action in the region between the metal surfaces during flexural vibration; thus, a thin coating can be applied and still achieve the same degree of attenuation.

Deadeners are also used in the construction of enclosure walls. Consider a high-intensity sound source located in a thin wall portable enclosure, such as shown in Fig. 5.7. A first approach to isolating or damping this source of sound and noise is to support the source on some type of resilient mounting and apply necessary localized acoustical treatment to the source. This prevents the transmission of structure-borne vibration to sympathetic members of the enclosure where it can be externally radiated as airborne sound. It also attenuates direct-sound radiation. However, it may not be economically feasible to decouple the source from its support or to decouple the support from the surrounding enclosure. Then damping material must be applied in some manner to the walls of the enclosure to reduce the effect of structure-borne and acoustically induced vibration.

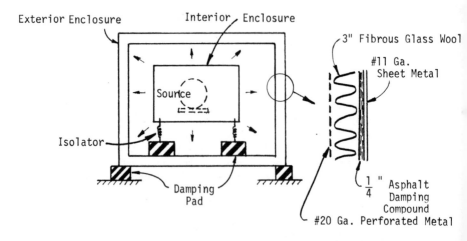

Fig. 5.7   Wall construction for sound isolating enclosure

Rubber is another type of useful damping material. One disadvantage, however, is that it cannot be applied easily to most nonplane surfaces, and this neces-
sitates that these latter be covered with one of the more recently available synthetic rubber compounds produced by such companies as General Electric
and Dow-Corning. This, however, is quite high in cost of application to broad
surface areas.

Rubber can be used in many forms. In sheet form it can be cut to shape and
cemented directly to a flat surface with a dependable brand of rubber-metal
cement. One common application is the insertion of rubber (neoprene) damping
pads beneath vibration isolation springs. An interesting application occurs when
machining an object such as a metal horn. If the horn is very large, vibration
induced by the cutting action of the tool is sufficient to cause the horn to resonate.
In a recent case involving the machining of a large, conical, aluminum horn,
resonance occurred during semi-finishing cuts. The resulting sound-pressure
level ten feet from the mouth of the horn near the axis was 110 dB, approximately,
with an accompanying degradation of the machined surface finish. This response
was satisfactorily eliminated by applying several wraps of 1/4-in diameter rubber
belting about the circumference of the horn.

Acoustical damping materials for architectural use are available in the form
of preformed tiles or plastic-like acoustic plasters. The latter, which can be applied by hand or machine methods, cure and become porous acoustical materials. Since the effectiveness of the acoustic plasters depends on final surface
porosity, great care must be exercised in finishing and applying decorative paints.
These must be nonbridging emulsions or casein paints preferably applied by
spraying. Oil base paints tend to seal the surface pores and this renders the
finished material nonacoustic. The absorptive power of the plaster coatings
increases with higher frequency, yet they are ineffective in the lower-frequency
range. In addition, they are fragile, easily damaged and quite troublesome to
refinish and repair.

Lead possesses excellent acoustical properties: it has very high density and a
comparatively low modulus of elasticity, two characteristics required for ideal
sound isolation of noise. Because in thick section it is too heavy for practical
application, it is widely used as thin sheeting and in combination with other
materials: in sheet form cemented to building surfaces; bonded to plastic foam
or plastic sheets; and bonded to vinyl films.

Consider now the damping which can be applied to the interior of an office-
type enclosure. Rugs, draperies and acoustical ceilings all contribute to the
total sound absorptive characteristic of the room. So do people and miscellaneous furnishings. By increasing the quantity of absorbent material in a room,
reverberation time can be lowered until the room is considered "dead." In this
condition each sound radiated in the room is absorbed rapidly by the interior
surfaces, and only direct-radiated sounds are of any consequence. This results
in a very low level of sound intensity in the room. When minimum amounts of
damping material are added to a volume, best results are obtained by placing
the material in the corners of the room and in the vicinity of the more intense
sound sources. Such positions correspond to regions of greatest acoustic activity.

## 10. NOISE IN AIR-MOVING SYSTEMS

Noise generated in air-moving systems is in large measure a function of the mechanical air mover or fan. This is most certainly true when the duct work of the system is short. On the other hand, when the distribution system is long, fan noise becomes masked by the sounds of air turbulence and radiations from the entrance grilles. Other sources of noise that concern structure-borne and airborne transmissions are wall and panel vibrations and the forces transmitted from the operating machine to the structure. Mention was made previously of methods for isolating and damping these various sources. Basic fan noise can be efficiently attenuated by enclosing the fan along with its auxiliary equipment, such as heating and cooling coils, filters and dehumidifying units, in a sound-isolating enclosure. This enclosure, or acoustic plenum as it is often called, prevents the radiation of airborne sound into the surroundings. If the fan base is isolated from the enclosure or structure with vibration supports and isolation pads, then structural transmissibility becomes nil. In addition, when the fan outlet is connected to the plenum structure or distribution duct work through a flexible coupling, that element of structure-borne noise is controlled. Consideration must also be given to the potential sound radiation of the plenum air inlet. This source is effectively suppressed by installing a duct sound attenuator in the inlet duct work. A typical attenuator is a section of duct work containing flow-forming baffling. Interior surfaces are made of wire mesh or perforated sheet behind which is installed acoustically absorbent material. These devices, which are referred to as packaged silencers, sound traps and duct silencers, are available from Koppers Co., Industrial Acoustics Co. and others. With a properly designed and constructed acoustic plenum, sound-attenuation levels of approximately 15 dB can be expected in the critical low-frequency range. Duct silencers are used effectively at or near outlets of the distribution duct work. These units attenuate noise generated in the airflow by the effects of turbulence as air passes through dampers, elbows or other internal restrictions. These silencers produce the added benefit of controlling cross-talk; that is, sound originating in one room propagating into an adjoining or nearby room through the air-distributing duct work. A general plan for silencing mechanical equipment is shown in Fig. 5.8.

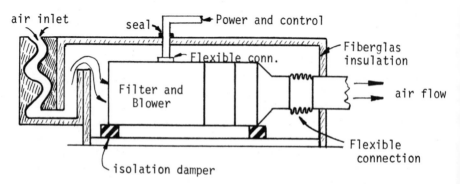

Fig. 5.8  Sound/vibration treatment for an air moving system

## 11. PERSONNEL PROTECTION

There are times when it becomes impractical or uneconomical to create a quiet environment by the methods noted above. For example, maintenance personnel servicing jet aircraft at an airport must have freedom of mobility; they cannot be restricted to the confines of a sound enclosure. Neither, with the present state of the art, is it practical to effect large noise reductions in the operating jet-aircraft engines. The same situation exists in manufacturing environments containing rolling, forging and riveting operations, or other machinery and processes that result in the generation of intense sound. In each of these cases, people are required to work in a hazardous environment, making it necessary to provide them with personal protection, rather than an altering of their surroundings.

The ear is sensitive to pressure pulsations of air. Such pulsations pass through the inlet canal of the ear and vibrate the membrane or eardrum at the end of the inlet canal. The mechanical vibration of the eardrum is amplified by the bone structure of the middle ear and this eventually stimulates the hearing mechanism in the inner ear. It follows that in order to protect the hearing mechanism, we could plug the inlet canal by some means and thus prevent the passage of air pulsations into the middle ear. Various materials have been used for this: candle wax, rolled cotton and even swimmers' ear plugs. However, these do not provide complete protection from hazardous noise. One type ear plug that has been recommended by people unfamiliar with the dangers of industrial noise is that type often used by hunters and gun enthusiasts. This latter-type plug contains a slit valve which supposedly remains open during periods of normal sound and closes quickly when subjected to a high impulsive pressure of the type originating with discharged firearms. However, if dust particles or other foreign matter become lodged on the surfaces of the slit, then closing of the valve becomes impossible and the ear is left completely unprotected. For the cost involved (less than one dollar) it is recommended that scientifically proven ear plugs, such as the "Ear Defender" which is available from the Mine Safety Appliances Co., be considered for more positive protection. Some sound fields have very high-intensity levels. When working in such sound fields, it is often found that even well-fitted ear plugs do not provide sufficient sound attenuation. Two factors contribute to this condition. First, the ear plug itself is capable of only a given level of noise reduction. Second, sound is transmitted by bone conduction. Thus, even with a well-plugged ear canal, sound impinging on a person's head in the area of the ear will find its way to the hearing mechanism although it will be substantially attenuated by the path. Acoustical earmuffs provide additional protection for the ear because of their greater sound attenuation and the larger head area covered. The choice of ear protection will depend on the individual as well as the environment. For instance, ear plugs are small, inexpensive and comfortable to wear in hot environments or for extended periods of time. However, they are a personal-type item and each user must be supplied with his own set. On the other hand, earmuffs are more expensive, provide more noise attenuation and are physically preferred by some to ear inserts. Since these latter protectors do not present hygienic problems, they can be distributed among different users and also used to provide protection for infrequent visitors.

The insertion of some materials into the ear canal can result in hearing problems. For example, low-level sounds will be sufficiently reduced by plugs made of candle wax so that it is not possible to hear ordinary conversation. With properly designed and fitted ear plugs only slight attenuation of conversation occurs. In these cases, work can be carried on without the distraction of a near-soundless environment. The same is true for earmuffs. In extreme cases it may be necessary for operating personnel to wear both earmuffs as well as ear plugs. In any case, the protection should reduce the perceived level of hazardous industrial noise without restricting the ability to converse in the environment.

# BIBLIOGRAPHY

Baade, P. K. "Sound Radiation of Air-Conditioning Equipment; Measurement in the Free Field Above a Reflecting Plane." *ASHRAE Trans.* 1882 (1964): 217-227.

Beranek, L. L. *Noise Reduction*, New York: McGraw-Hill Book Co., 1960.

Harris, C. M. and Crede, C. E. *Shock and Vibration Handbook*, 1st ed. New York: McGraw-Hill Book Co., 1961.

Kinsler, L. E. and Frey, A. R. *Fundamentals of Acoustics*, 2nd ed. New York: John Wiley & Sons, Inc., 1962.

Raybek, E. E. "Sound of AC Motors." *Design News* 18 (1963): 14-21.

Ruffini, A. J. "Bearing Noise, Part 1, Analysis of Rolling-Element Bearings." *Machine Design* 35 (1963): 232-255.

Sanders, G. J. "Noise Control in Air-Handling Systems." *Sound and Vibration*, February 1967, pp. 8-18.

Schrader, E. W. "The Sound of Ball Bearings." *Design News* 17 (1962): 12-15.

von Meier, A. "Application of Lead for Sound Insulating Partitions." *Sound and Vibration*, May 1967, pp. 14-19.

Zwikker, C. and Korsten, C. *Sound Absorbing Materials*, Amsterdam: Elsevier Publishing Co., 1949.

# SYMBOLS

| | | |
|---|---|---|
| c | damping coefficient | lbf·sec/ft |
| $c_c$ | critical damping | lbf·sec/ft |
| E(t) | forcing function | — |
| F(t) | forcing function | — |
| $F_T$ | transmitted force | lbf |
| $F_0$ | disturbing force | lbf |
| f, $f_n$ | disturbing frequency, natural frequency | $sec^{-1}$ |
| $\overline{G}(f)$, G(f) | transfer functions | — |
| H(t) | time-random function | — |
| k | spring rate | lbf/in |
| K(t) | time-random function | — |
| m | mass | lbm |
| n | integer | — |
| R(t) | response function | — |
| S(f) | spectral density | — |
| T | period | sec |
| $T_R$ | transmissibility | — |
| X | frequency dependent displacement | inch |
| $X_0$ | displacement amplitude | inch |
| $\overline{x^2}$ | mean-square displacement | inch |
| $x_s$ | static deflection | inch |
| x(t) | response function | — |
| x | displacement | inch |
| $\omega$ | angular frequency | radians/sec |
| $\omega_n$ | natural frequency (angular) | radians/sec |
| $\omega_d$ | damped natural frequency | radians/sec |

# 6

# RANDOM VIBRATION—ISOLATION AND DAMPING

*William S. Mitchell*

## 1. INTRODUCTION

The vibrational response of operating machines, vehicles and structures creates numerous problems in maintenance as well as environmental comfort. One is that sustained excessive vibration of rotating equipment will cause an eventual degradation of the mounting, mechanical connections and internal mechanisms; and this is accompanied by psychologically annoying sound. Such response should not be allowed to exist if only because of the economics associated with the problem: indications of failure of the supporting structure and mechanical connections necessitate considerable maintenance time. Aside from the inconvenience of reworking part of an operating system, losses can be relatively large when certain systems, such as those found in chemical plants and refineries, are inadvertently shut down due to a vibrational-related failure of a piece of critical equipment. Thus, some operations require duplication of pumping and process-equipment. To circumvent these high investment costs, preventative maintenance programs have been established in some industries. These programs allow operations personnel to monitor the condition of an operating machine and schedule maintenance accordingly. Another example in which vibrational response is influential in design is that of the automobile.

Consider the effect of having no isolation or damping in the finished product. Wheel and axle response would be transmitted directly to the vehicle body and this in turn would produce large accelerations of the passenger and driver seats. Consequently, the effect of crossing tar lines and roadway gravel would be sufficient at any speed to cause extreme discomfort. Current practice requires pneumatic tires to isolate and dampen the effects of the roadway surface. In addition, an isolation system of shock absorbers and springs is located between the tires and the vehicle body. Vibrations ultimately transmitted through the vehicle frame and body are attenuated by the construction of the passenger seats. Most seats are made of a combination of wire springs and thin sheet foam rubber, felt or other similar material. The finished automobile is comfortable for the passenger who experiences a minimum of vibration. An added benefit of the isolation and damping of the vibrational input from the tire-road interface is the resulting quietness inside the vehicle. Isolation and damping reduces the amount of energy available to excite acoustic response in various body panels and other structural components. The finished product is readily salable and highly desired by the general public.

Industrial machines are treated in a similar manner. All rotary equipment, such as air-moving blowers and pumps, vibrates because of hydro or aerodynamic effects as well as inherent unbalance. It is impossible to balance ordinary reciprocating machines. Therefore, when resulting vibratory forces are transferred to the structural support, the support responds to the excitation because of its own compliance and vibrates at the frequency of the exciting force. If structural resonance of the support is not far removed from the exciting frequency of the machine, floor displacements become large, possibly resulting in a serious condition. The structure may be locally overstressed, for example, or vibration may be sufficient to annoy inhabitants of the structure. Rotary unbalance and internal mechanisms generate the alternating forces that promote vibrational response. When this energy is transmitted to other parts of a structure that resonate at sympathetic frequencies, the surrounding environment is filled with noise. A first requirement then is to isolate a machine from its support. This includes the base and all mechanical connections. Only after this initial step should localized vibration and acoustical treatment be applied to the machine. A properly isolated machine will transmit negligible energy to a structure. Then it can be regarded as a source of direct sound only.

## 2. THEORETICAL CONSIDERATIONS—PERIODIC

The purpose of isolation and damping is to attenuate the transmitted forces of vibration so that the source is supported only statically by the structure. This decrease of force transmission is brought about by a proper selection of isolation components to make the source, in effect, a free body unrelated to the structure. Any isolation and damping should be done without interfering with the operation and maintenance of the machine.

Consider a typical simple vibrating system as shown in Fig. 6.1 (most systems can be approximated by a model of one degree of freedom having a resilient support with or without damping and subjected to a sinusoidal forcing function). The damped natural frequency of this type system is

$$\omega_d = \omega_n \left[ 1 - \left( \frac{c}{c_o} \right)^2 \right]^{1/2} \qquad \text{(radians per sec)}$$

where the undamped natural frequency, $\omega_n = 2\pi f_n$, is described by the system properties of mass and spring rate as

$$\omega_n = \left( \frac{k}{m} \right)^{1/2}$$

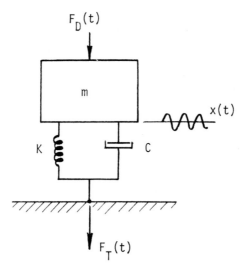

Fig. 6.1   Model—Damped, single-degree-of-freedom system with forced
excitation

For lightly damped systems the damping ratio of viscous damping to critical damping $c/c_c$ is much smaller than one. Thus, it becomes practical to approximate

$$\omega_d \doteq \omega_n = 2\pi f_n$$

By definition, $T_R$, the force transmissibility, is the ratio of transmitted force to applied force. For an input force of $F_0 \sin \omega t$ the transmitted force $F_T$ will be due to the sum of the forces induced by the spring and the damper. If the mass and spring rate of the system remain constant and only the forcing frequency is varied, then

$$T_R = \frac{F_T}{F} = \frac{\left[1 + \left(2\frac{c}{c_c}\frac{\omega}{\omega_n}\right)^2\right]^{1/2}}{\left\{\left[1 - \left(\frac{\omega}{\omega_n}\right)^2\right]^2 + \left(2\frac{c}{c_c}\frac{\omega}{\omega_n}\right)^2\right\}^{1/2}}$$

Transmissibility is plotted with respect to the ratio $f/f_n$ in Fig. 6.2, where f is the excitation frequency in cycles per second. From this figure the following can be noted. First, when $f/f_n$ is greater than 1.4, transmissibility is less than one. Thus, vibration isolation is possible only at those frequencies for which f is greater than 1.4 $f_n$. Further, when f is greater than 1.4 $f_n$, a system supported on spring isolators only is superior to one containing damping. This implies that in the higher frequencies damping is detrimental to a vibrating system. However,

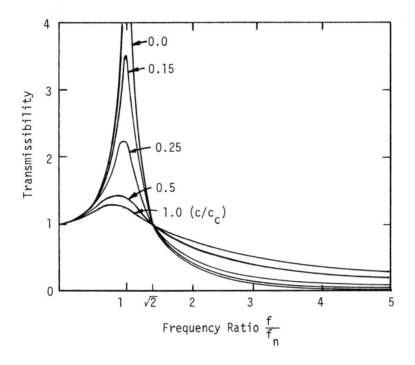

Fig. 6.2   Transmissibility curve of a simple, linear vibrating system with
various degrees of damping ratio (c/c₀)

if a system must pass through a resonant state at $f = f_n$ to attain its operating
frequency, then damping is very beneficial as it prevents excess displacement at
the resonant speed. The conclusion is that damping is required up to a value of
$f = 1.4\ f_n$, after which it adversely affects a system. Where desirable, displacement resonance can be limited by mechanical stops.

At low frequencies displacement of the mass in a lightly damped system is in
phase, approximately, with the excitation. Since phase is the implied angular
difference of time or position between forces and motions as noted in mathematical expressions or phasor diagrams, an "in phase" condition means zero
degrees. As the forcing frequency approaches $f_n$, the phase angle increases to
90 degrees, where the transmissibility with negligible damping becomes infinite.
At the higher frequencies, the phase angle may approach 180 degrees, depending
on the damping ratio.

If the value of the damping coefficient is small, transmissibility reduces to

$$T_R = \frac{1}{1 - \left(\dfrac{\omega}{\omega_n}\right)^2}$$

This expression is represented by the curve $c/c_c = 0$ in Fig. 6.2. It should be noted that it is a good approximation for many systems supported by springs only since in these cases the damping coefficient is approximately zero. When the system stiffness and damping coefficient remain constant, then the system is defined as linear.

For lightly damped linear systems there is a relationship between the natural frequency of the supported mass and $x_s$, the static deflection

$$f_n = \frac{3.14}{\sqrt{x_s}}$$

Figure 6.3 shows the effect of the forcing frequency on a system with respect to its static deflection. The broken line is the curve expressed by the above equation. Within the region of the graph below and to the left of the natural frequency line, magnification of system response occurs. Above and to the right of the broken line are those conditions for which vibration isolation is possible. For example, if a mass is supported on a very compliant spring system with a static deflection of 0.5 in, then at 33.5 cps 98 percent of the forces of vibra-

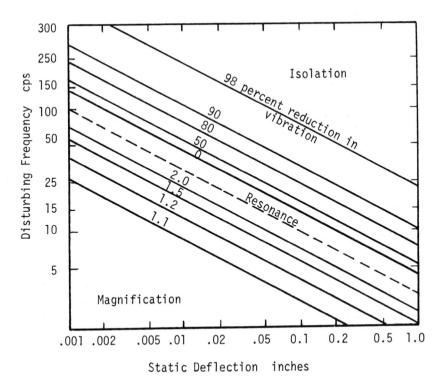

Fig. 6.3   Isolation efficiency curve for a simple, linear, vibrating system
with negligible damping

tion are prevented by the suspension from reaching the supporting structure. As frequency decreases, more of the disturbing force is transferred to the structure until at 4.5 cps the system is in a state of resonance. Two things can be implied from this. First, the graph is useful in selecting an isolator with a given static deflection when subjected to a specified mass load. Second, it shows that for a given forcing frequency, isolation of forces is possible if the suspension system is sufficiently compliant. If a system is operating in a frequency range above its resonant frequency, or the frequency at which it responds most strongly, then isolation requirements should be based on the lowest noted frequency of the system. This is particularly true when the disturbing frequency is the result of rotation. By isolating the lowest frequency, all higher frequencies are isolated at higher efficiencies. These concepts are generally true whether the excitation is sinusoidal, complex or random. When deflection rate is nonlinear, however, or damping is large or the random-frequency spectrum is broad, these approximations will not hold.

## 3. THEORY OF RANDOM VIBRATION

In the aforementioned theory it was assumed that impressed forces were periodic. Consider now the effect of a random signal as the input excitation. If the random excitation is narrowband, the vibrating system responds strongly when the band contains or is in the region of the natural frequency of the system. But if the limits of the narrowband input are far removed from the frequency of system resonance, the system reacts in a random way. Therefore, the above expressions are valid only qualitatively for describing lightly damped systems excited with narrowband random noise. A different situation exists with wideband random excitation. This type of input contains many frequencies. System response will depend on the power spectral density of the excitation and the transfer function of the system. For example, if the greater portion of the input energy is contained in a specific narrowband of the wideband random signal, that part of the system sensitive to this frequency band will respond strongly. One characteristic of wideband random excitation is that the response of the system is governed by interactions of its various modes. This response is not mathematically predictable. Because of filtering action by system components, particularly that of structural filtering, wideband random excitation is often altered to narrowband random response.

When a resilient system is excited by random forces, it will normally respond in random motion depending on the resonant frequency of the system and the power spectral density of the input. For an arbitrary forcing function $E(t)$, the response of a system will be $R(t)$; the ratio of output to input is called the transfer function

$$\overline{G}(f) = \frac{R(t)}{E(t)}$$

which is noted to be a function of frequency. If a generalized form of $\overline{G}(f)$ is found for a given physical system, then for any given input F(t) the response of the system x(t) can be calculated from the following.

$$x(t) = G(f)F(t)$$

However, this latter expression is very limited in practical use for several reasons. If excitation is wideband random, F(t) cannot be described beforehand. Also, G(f) is most conveniently determined only at a single location within a given system. The transfer function can be found experimentally by applying a constant amplitude, variable frequency, sinusoidal excitation to the base of the system and measuring the response at a given location. The transfer function is then calculated by dividing the measured response by the magnitude of the input excitation. Thus, the frequency dependent G(f) is obtained, but for one location only; and it has required considerable time and instrumentation, especially with the more complex systems.

The response of randomly vibrating systems is best described in terms of statistical values. For random vibration, the mean-square value of a response is by definition

$$\overline{x^2} = \lim_{t \to \infty} \frac{1}{T} \int_0^T x^2(t)dt$$

The time-varying response x(t) can be expressed in the form

$$x(t) = X \cdot H(t)$$

where X is the random-frequency amplitude of the response and H(t) is the time function of the motion. But the amplitude response X can be expressed in terms of a transfer function G(f) as a product $X_0 G(f)$. Hence, the mean-square response becomes

$$\overline{x^2} = \lim_{t \to \infty} \frac{1}{T} \int_0^T X_0^2 G^2(f)H^2(t)dt = \lim_{t \to \infty} \frac{1}{T} \sum_n X_n^2 G_n^2(f)H_n^2(t)\Delta t$$

where for small $\Delta t$

$$\overline{x_n^2} = X_n^2 G_n^2(f)H_n^2(t)$$

In the same manner, the excitation can be expressed as

$$F(t) = F_0 K(t)$$

where $F_0$ is the amplitude of the excitation and K(t) describes its time-random motion. The mean-square value is thus

$$\overline{F^2} = \lim_{t \to \infty} \frac{1}{T} \int_0^T F_0^2 K^2(t)dt = \lim_{t \to \infty} \frac{1}{T} \sum_n F_n^2 K_n^2(t)\Delta t$$

and for small $\Delta t$

$$\overline{F_n^2} = F_n^2 K_n^2(t)$$

Dividing the mean-square value of the excitation within a frequency interval $\Delta f_n$ by $\Delta f_n$, we obtain $S_n(f)_I$, the spectral density of the excitation

$$S_n(f)_I = \frac{\overline{F_n^2}}{\Delta f_n}$$

The same is true, for the response

$$S_n(f)_R = \frac{\overline{x_n^2}}{\Delta f_n}$$

If the number of frequency components is large, which is typical of random vibration, we note that spectral density curves will be continuous. The mean-square response is

$$\overline{x_n^2} = G_n^2(f)\overline{F_n^2} = G_n^2(f)S_n(f)_I\Delta f_n$$

But as previously noted $\overline{x_n^2}$ and $\Delta f_n$ are related by $S_n(f)_R$. Therefore

$$S_n(f)_R = G_n^2(f)S_n(f)_I$$

The transfer function for excitation and response also describes the transfer from input to output with respect to spectral density.

We note then the following. A random excitation possesses a definite spectral density which will result in response according to the transfer function of the system. For very simple systems it results in random response. Wideband random excitation is typically filtered by a structure to narrowband response. Here, the structural filtering is described by the transfer function. Narrowband excitation will cause a system to resonate when the band contains the resonant frequency of the system. Further, random input to a resilient system produces response composed of interacting effects between modes.

Isolation and damping of random forces and motions depends on a knowledge of the related power spectral density. For instance, if the random input to an isolator contains large amounts of energy in a given frequency band, $\Delta f_i$, then for satisfactory isolation the isolator must be most effective within that frequency range. It becomes obvious that if power spectral density shows large amounts of energy in two or more frequency bands, then one isolator may not be totally effective. Moreover, if an isolator is rated in terms of load versus static deflection, then it is quite possible that the most effective frequency region of the isolator will not coincide with the frequency region of maximum energy input. It is because of situations such as these that the application of isolators to some vibrating systems results in either no attenuation whatsoever or aggravation of the problem. In developing the state of the art for vibration isolators, it has been generally assumed that the excitation is sinusoidal or approximately so. In truth many vibrating systems do satisfy this latter condition of sinusoidal-like response. Therefore, it is reasonable to select isolators which perform well with sinusoidal excitation for application to a great many systems. But, when a system is known to have random response, application of these same devices without consideration of input spectral density becomes a *hit-and-miss* proposition.

## 4. SPECIFYING ISOLATOR SYSTEMS

The specification of proper isolators and damping materials is an important part of every installation of dynamic equipment. For example, it makes little sense to mount an expensive turbine or compressor on isolators where each one is made by a different manufacturer. This can result in the poor dynamic performance of the machine due to such effects as uneven settling, misalignment and uneven spring rates. Likewise, it is unsound practice to employ different brands of vibration control on individual units comprising an operating assembly. Manufacturers invariably rate isolation devices under differently controlled conditions. Hence, matching of isolators is best accomplished through the utilization of those obtained from the same source. This also avoids the problem of trying to define responsibility when an isolated system fails to meet isolation specifications, where assorted isolators have been used on a common support.

Various means have been proposed for specifying isolator performance. Theoretically, isolator efficiency is the more proper except that it requires a rigid base which is seldom realized; and because of floor deflection it is difficult to measure accurately after installation. Isolation efficiency is evaluated in the field by measuring the vibration at the base of a machine and comparing it with the vibration of the support. Another recommendation calls for the use of isolator deflection since it can be easily verified at the time of installation. Although it may be a convenient method for checking, it is meaningless when the excitation is random.

Helical-wound, metal spring isolators are very efficient in isolating high- and low-frequency forces. To increase the efficiency of isolation in terms of acoustical response it is necessary to install sound-isolation pads of rubber or other suitable material at the base of the isolator to isolate the high-frequency noise of the equipment which is transmitted to the structure through the coil of the metal spring. Where vibration isolation is not as critical as acoustical isolation, machines and structures can be set directly on isolation pads. Exposed isolation pads made of felt or compressed cork deteriorate with age and require periodic replacement. This is expensive because it requires that the system be disconnected from its environment and raised to replace the isolation pads. Isolation pads made from rubber or rubber-covered high density, precompressed glass fibers are more expensive than other bulk materials but their life expectancy and continued high performance often justifies the additional initial cost. In some long-term applications excellent results have been obtained by using composite pads of lead and asbestos fiber.

## 5. CONSTRUCTION PRACTICES

A common method of mounting equipment in critical locations requires the use of a concrete inertia base. Increasing use of this method is found today because of the improved techniques in building construction which produce buildings of lesser mass and more compliance and the current practice of installing larger, more powerful machines in equipment rooms located in the upper portions of a structure. These bases are used under individual pieces of equipment as well as for mounting an assemblage of operating machines. The latter case is exemplified by a floating floor.

Floating floors are constructed over partial or entire areas of a concrete structural slab. First, a large isolation panel is made from a sheet of high-quality exterior plywood. Load-bearing, fibrous glass isolation pads are properly distributed over its surface and bonded to it. The panel is positioned on the structural slab with the isolators bearing the weight of the panel. Then the perimeter of the floor is lined with a fiber glass isolation board, felt or cork to separate the floating floor from the walls; and the isolation panel and perimeter material are waterproofed with a plastic sheet or mastic compound in preparation for the pouring of concrete. Finally, steel reinforcement is added and the mass of concrete is poured. The finished floating floor is shown in Fig. 6.4. Because of the mass density of the concrete floor, it is an effective barrier to airborne sound which would normally propagate to the space below. In some cases fibrous glass wool is placed between the isolation pads for additional acoustic attenuation.

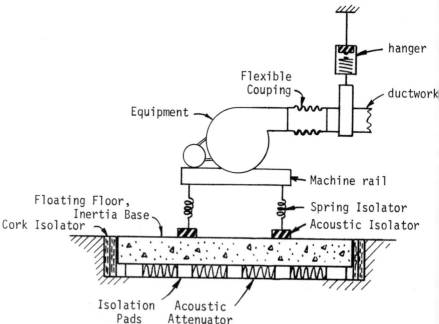

Fig. 6.4 Typical mounting construction elements for vibration and acoustic control

Inertia bases require steel reinforcement to provide sufficient stiffness for mounting and maintaining the alignment of operating machines. Minimum recommended concrete thickness is 8 percent of the longest base dimension which is increased for structural integrity and required weight. The base must also be large enough to support drive motors, piping elbows and other devices required for the successful operation of the major pieces of equipment. This construction will satisfy load requirements and serve as a motion-reducing mass as well as a discontinuity in the propagation path for noise and mechanical vibration.

Other methods are equally successful in constructing a floating floor. For instance, a construction recommended for acoustical control is to locate individual 2-in high neoprene isolators on the structural slab or have them pre-bonded to formwork panels. These panels can be of plywood, transite, hardboard, sheet metal or any suitable sheet material that is low in cost and will not deteriorate with age. This is important if all isolators are to be equally loaded over a long period of time. All seams and joints must be covered and secured in place to prevent the poured concrete from seeping through a joint and bridging the vibration-acoustical barrier. The upper surfaces of the forms are waterproofed with roofing felt and hot tar or polyethylene sheet. Finally, the perimeter is protected with 1-in thick, lightweight cork. This construction, having a 4-in-thick concrete slab, supposedly affords vibration isolation above 10 cps. Economical, high-frequency or more costly lower-frequency isolation can be obtained by using thinner or thicker pads, respectively. Properly loaded neoprene pads will deflect in an amount of 10 to 15 percent of overall thickness. Where vibration isolation is of major concern, metal springs are more effective then neoprene pads because of their larger deflections and negligible damping.

Spring-supported floors are formed in two ways. In the first, a plastic sheet is placed over the area on which the floating floor is to be located. Perimeter forms are positioned and reinforcing steel is set within the forms. Then, spring mountings in special canisters are located in and attached to the reinforcing steel. After concrete has been poured and cured, the springs are loaded by means of adjustment bolts, thus "floating" the floor. The other method requires a structural frame with isolator mounting brackets along its sides. Concrete is poured into the reinforced steel framework which is separated from the area by a plastic sheet. After the pour has hardened, spring isolators are installed at all the mounting brackets and the slab is raised into an operating position. These massive foundations and floating floors, when they are properly made, will isolate suitably most sinusoidal and random vibrations. An inertia mass provides a rigid base for those machines which will distort ordinary structural steel bases. In addition, they act as a mass barrier between the noise-generating equipment and the building structure. The weight of the base also resists reaction forces and torques produced by such equipment as fans, pumps and compressors. The additional mass of the concrete inertia base reduces the response motion of the equipment during transient disturbances as well as normal operation. Recommended weights of inertia bases are based on the magnitude of the forces to be resisted. But generally the minimum ratio of the base weight to equipment weight is approximately 1/1 for small fans, 1.5/1 for low-pressure pumps and fans (less than 5-in static pressure and 75 hp) and 2/1 for high-pressure fans and large motors in excess of 75 hp.

Single pieces of operating equipment are satisfactorily mounted on individual inertia bases. The inertia base is made of internally reinforced welded channel that is filled with a mass of concrete. Brackets are located along the sides of the frame for mounting the base on isolators. The equipment, including drive motor, is secured to the inertia mass. Any connection to this equipment, such as inlet or outlet duct work, piping or wire conduit, is made with a compliant joint. For example, a blower outlet can be connected to the distribution duct work by means of a flexible, neoprene connection. When the inertia base is finally supported on a system of high-deflection springs and damping pads, vibration disturbances are effectively isolated from the structure. It is good practice to require that the same manufacturer supply the welded base frame in addition to the isolators for each particular piece of equipment. This is to insure that the inertia mass is correctly sized for the installation. Moreover, it eliminates the expense of design and field fabrication. Many manufacturers of isolator equipment supply inertia frames of proper dimension that satisfy recommended weight ratios as well as minimizing space requirements. They also include mounting templates, stiffening members and mounting brackets for the mechanical isolators. In some installations, structural capabilities prohibit the use of large concrete inertia bases and floating floors. Then widely reinforced steel bases can be used, although they are not as effective as the more massive concrete bases.

Mechanical equipment with their associated drives should be mounted rigidly on a common base plate or, where the driver is integral with the equipment, on a set of rigid rails. These supports provide distinct advantages, the most important being alignment, which result in increased bearing and coupling life as well as reduced drive-belt wear. Installation time is reduced to a minimum since the equipment can be first accurately aligned on the base and then easily installed on the supporting isolators. In some cases isolators are supplied as an integral part of the base plate. The use of base plates results in an uneven load distribution. Consequently, they should be equipped with effective leveling arrangements and have lateral, adjustable snubbing. Snubbing is often necessary for those machines that must pass through resonance while accelerating to or decelerating from operating speed. Rail supports should be used when the drive is not an integral part of the machine or when the base is not sufficiently rigid for point support. They provide a low-cost means of support; in addition, they are effective in reducing cantilever effects resulting from the installed equipment. Special rails are available for suspending equipment from vertical walls. These work quite well in supporting air-conditioning and heating equipment. The built-in isolators and snubbers allow the higher wall surfaces to be used for mounting equipment while providing good isolation of vibratory forces.

The reason for applying mechanical isolators to a vibrating system is to attenuate those forces and motions of the excitation before they are transmitted to either the structure or to delicate equipment. In some cases the excitation is sinusoidal; in others it is random. Frequencies that are normally encountered range from 1 to 1500 cps. This range covers the response of most shipboard and aircraft motions as well as industrial equipment. In many cases, the response of an isolated system is too complex for accurate mathematical description. For example, there may be many degrees of freedom associated with a mechanical system and each degree of freedom will have its own natural frequency and modal response. Moreover, unsymmetrical mountings or skewed forces result in a coupling of the modes of vibration. A common method of analyzing resilient systems is to assume that the system has only one degree of freedom and that both excitation and response are simple harmonic. Then, the simplified system has but one associated natural frequency and displacement, velocity and acceleration can be conveniently approximated. However, most excitations are multi-frequency. When the excitation is random, physical testing of isolator systems, using limited simulated input spectra, provides the only accurate means of anticipating vibrational response. This can be an expensive and time-consuming process. Likewise, theoretical analysis requires preknowledge of input spectra and isolator characteristics, neither of which is often known.

## 6. SELECTION OF ISOLATORS

Isolator selection depends on the disturbance that requires isolation, whether it be from vibration or acoustic effects. In every case, an isolator must provide a resilient support, with or without damping, for the purpose of reducing the transmission of motion or exciting forces into or from a system. The natural frequency of the isolated system must not coincide with the frequency of the disturbance, or the system will respond in a state of resonance and either damage the individual isolators or the components of the isolated system. Selection must take into account life expectancy, environmental conditions, space limitations, weight, cost and dynamic response. Unfortunately, the latter is often neglected. Commercial literature rates individual isolators in terms of load-carrying capability and static deflection only. Thus, selection occurs without consideration of dynamic characteristics. Manufacturers supply a variety of components for vibration isolation and damping. These include pads of fibrous or cellular materials, solid elastomers, metal springs of varied design and rubber containers or bellows containing air. Each isolator responds to loading according to its deflection rate and damping properties.

Before selecting an isolator, it should be determined precisely what function it must perform; whether it must isolate or attenuate motion, force, shock or noise. It is necessary to note whether a product or instrument is to be protected from its supporting structures or the supporting structure is to be uncoupled from a mechanical disturbance. The weight of the system requiring support must be determined and consideration given to space limitation for the isolator. Additionally, the location of the center of gravity of the system should be noted with respect to the anticipated supports.

Coupling is a mechanically induced condition which results from the location of supporting elements. If isolators are positioned along coplanar lines passing through the center of gravity, then response modes are decoupled. It is essential that the frequency of the disturbing vibration be identified, and this is accomplished relatively easily when the excitation is very narrowband random, complex periodic or sinusoidal.

The fundamental frequency of a complex periodic response is usually the only frequency of interest since the harmonics are of higher frequencies and, generally, of decreasing amplitude. However, an energy density spectrum is required when the disturbance is random multi-frequency. Wideband random excitations produce system response in accordance with the energy distribution described by their input spectra. Hence, it is possible to install isolators in a randomly excited system only to have it respond in a state of resonance when the energy of the random motion is concentrated about the same frequency as the natural frequency of the isolated system. Conversely, if the energy is concentrated below the natural frequency of the system, magnification of the excitation occurs. When the dominant frequency of the excitation is known, consideration can be given to the transmissibility of the given system. It should be noted that decreasing the transmissibility implies larger, more expensive isolators.

Having defined frequency and transmissibility, it is then a matter of determining the required static deflection of an isolator and the value of its natural frequency. A "rule of thumb" suggests that the natural frequency of the supported system should be less than one-third of the lowest disturbing frequency. For example, if a disturbance is due to electromagnetic effects in a motor where the disturbing frequency is 60 cps, then the isolated motor should have a supported, undamped natural frequency of 20 cps or less which corresponds to a static deflection of 0.025 in and a reduction in vibration of approximately 87 percent.

Isolated systems manifest certain physical characteristics. First, isolation of motions and forces improves with a lowering of the natural frequency of the supported system. Second, as the natural frequency is reduced, isolators become larger and system deflection increases (the system support becomes very "soft"). Because of these inherent characteristics, isolator selection involves a compromise so that good isolation is obtained with reasonable deflection. After isolators are selected, friction or damping can be specified for those systems that must operate below or pass through a resonant frequency. This not only effectively reduces amplitude response at resonance, it also reduces isolation efficiency at higher frequencies. Where necessary, isolators with nonlinear deflection rates can be applied to limit amplitude response. Otherwise, mechanical stops or snubbers should be added to the system. When specifying helical springs it is important that they have a high aspect ratio (diameter/height) in order to insure horizontal stability. Tall, slender springs are inherently unstable in the lateral direction under load. Thus, they permit horizontal deflection and rocking motions. One manufacturer recommends a value of 0.8 with respect to the operating height of the spring (see Fig. 6.5).

If a system is subjected to shock or the supported weight can vary, then spring isolators should be specified with a minimum travel of 1.5 times their static deflection to prevent bottoming of the isolators. Typical minimum static deflections associated with various types of mechanical equipment are listed in Table 8.

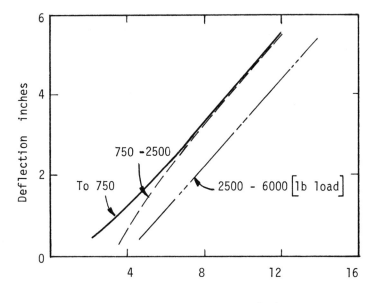

Fig. 6.5   Minimum recommended outside diameters for loaded helical
coil springs

## 7. COMMERCIAL ISOLATORS

Helical spring isolators are available in a wide assortment of ranges, styles and ratings. They can be obtained with internal or external adjustment bolts in addition to leveling screws. Externally adjustable isolators are used to support equipment when the isolator is accessible from above. Adjustable snubbers are recommended for machines such as punch presses and drop-hammer forges. Some isolators are provided with nonadjustable snubbers, and these are recommended where operating frequency remains constant; that is, under centrifugal fans and air-conditioning compressors and blowers. When acoustical isolation is necessary, damping material can be added to either the top or bottom of the isolator or both. The larger isolators utilize a system of two or more individual springs, thus providing for greater static loads with low deflection. Helical metal springs, when used without additional damping, have linear deflection rates. A linear deflection rate provides constant stiffness until the spring coils bottom. Although metal springs contain negligible damping, they are excellent transmitters of acoustical-frequency vibration. Hence, they are normally installed in series with damping pads. For special application, wire-wound forms can be produced to effect nonlinear spring rates. Response of various-type isolators is shown in Fig. 6.6. Spring isolators are supplied with appropriate housings for use in corrosive environments.

**TABLE 8**
**TYPICAL STATIC DEFLECTION IN INCHES FOR**
**RESILIENT, LOW-DAMPING ISOLATORS**

| EQUIPMENT | | LOCATION IN STRUCTURE (LEVEL) | | |
|---|---|---|---|---|
| | | BASEMENT | 20' | 30' |
| Refrigeration Machines | | 0.25 | 0.25 | 1.00 |
| Fans | rpm < 200 | 0.35 | 4.50 | 4.75 |
| | 200 – 300 | 0.35 | 3.50 | 3.75 |
| | 300 – 500 | 0.35 | 2.50 | 2.50 |
| | rpm > 500 | 0.35 | 1.00 | 1.75 |
| Cooling Blowers | rpm < 500 | 0.35 | 0.35 | 1.75 |
| | rpm > 500 | 0.35 | 0.35 | 1.00 |
| Pumps | | | | |
| Close–coupled | small | 0.35 | 0.35 | 1.00 |
| | large | 0.75 | 1.00 | 1.50 |
| Base Mounted | small | 0.35 | 0.35 | 1.50 |
| | large | 1.00 | 1.00 | 1.75 |
| Packaged Air-Conditioning Sets | | | | |
| Suspended | rpm < 500 | 1.25 | 1.25 | 1.25 |
| | rpm > 500 | 1.00 | 1.00 | 1.00 |
| Floor Mounted | rpm < 500 | 0.35 | 1.75 | 1.75 |
| | rpm > 500 | 0.35 | 1.00 | 1.00 |
| Internal Combustion Engines and Equipment | < 25 hp | 0.35 | 0.35 | 1.75 |
| | 25 – 100 hp | 0.35 | 1.75 | 2.50 |
| | > 100 hp | 0.35 | 2.50 | 3.50 |
| Reciprocating Compressors | rpm < 750 | 1.00 | 1.50 | 2.50 |
| | rpm > 750 | 1.00 | 1.00 | 1.50 |

Rubber, when properly used, provides satisfactory vibration control. It is available as a natural compound or as a synthetic substitute and can be molded in a variety of forms and compositions. It is readily bonded to metal; hence, it

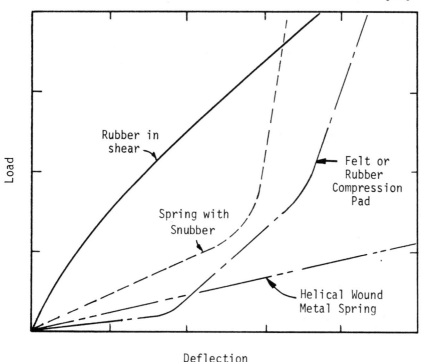

Fig. 6.6   Qualitative relationships of load versus deflection for various
type isolators

can be obtained with or without metal inserts. The design of a rubber isolator
can permit small or large static deflections. Moreover, rubber can be practically
loaded in compression or shear, although in shear it offers better isolation characteristics. When it is used in compression, it has less isolation efficiency, but it
does provide for overload protection and displacement limitation. Synthetic
neoprene rubber is used in applications requiring resistance to oil and other
hydrocarbon substances, whereas natural rubber is specified for best vibration
isolation properties. Rubber isolators are popular due to their low unit cost and
effectiveness over a wide range of frequencies. They are also easy to install.
Unlike metal coil springs, rubber isolators in shear have nonlinear response
characteristics; deflection rate is high with small and large deflections and low
over some range in between. A well-designed isolator that includes both shear
and compression loading can have linear response for limited values of deflection.
   Isolators were first used to support a mass from below. But as noise and vibration specifications became more stringent and floor space became more costly,
increasing numbers of systems were supported from above by specially constructed isolators called hangers. Typical examples of supported systems are
piping, air-conditioning and heating units and ceilings. Isolation hangers are
constructed with helical springs, rubber isolator elements or both. Typical con-

figurations are shown in Fig. 6.7. These units are inserted between the supporting structure and the system being isolated. They are used in conjunction with wire when supporting suspended ceilings or with steel rods when supporting piping. A major advantage of using overhead-supported equipment is that it makes available additional usable floor space.

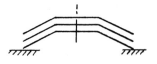

a. Belleville Springs

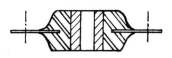

b. Bushing Isolator

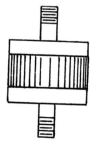

c. Stud Isolator

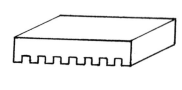

d. Compression Pad

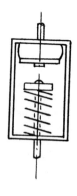

e. Hanger

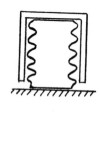

f. Air Spring

Fig. 6.7  Vibration and acoustic isolators

Vibration pads that are loaded in compression should be installed where minor noise and vibration must be reduced in noncritical situations and it is not necessary or desirable to bolt the system down. These pads seldom exceed 3/4 in in thickness except when stacked for increased deflection. Although the use of compression pads is generally a low-cost, convenient solution, they are not nearly as efficient as rubber in shear or metal coil springs. Isolation pads are manufactured from various materials, such as low- and high-density rubber foams and fibers that include wool and synthetic fibers, fibrous glass, hair and felt. These materials are generally processed into sheet material and molded parts. Many pad materials are preformed in some way and can be rapidly cut to size and installed at the supporting points of a machine. Some pads are molded of neoprene in ribbed and waffle patterns; these are supplied in hardnesses of from thirty to sixty durometer. Low durometer pads are preferred for reducing noise and vibration; the harder materials cost less, but they are also less efficient. Since neoprene rubber has a tendency to set, pad life is extended by using this material with light unit loads. Where supporting surfaces are small (which would result in high unit loads), metal or wooden plates can be bonded to a large compression pad before installing it under a machine, thus effecting low loading pressure. Pads can also be made of natural rubber when maximum isolation efficiency is required. In finished form elastomer sheets have a wide range of properties depending on the compounding and curing of the material. Some isolation pads are made by laminating a thick layer of granular composition cork between the inner and outer rubber surfaces. Such pads are recommended for acoustical rather than vibration attenuation. Another sandwich-type construction contains a high-density matrix of fibrous glass totally enclosed in a resilient neoprene jacket. The combination of trapped air and fibrous glass inside the jacket has properties which might be expected from a pneumatic volume containing high viscous damping.

In addition to being laminated with rubber, cork finds usefulness by itself, in the form of individual pads and as a continuous sheet for damping structural vibration. Typically, this material is used for the vertical support of vibration-isolated floating floors. Of the two types of cork mentioned for isolation and damping of vibration only one is totally satisfactory: pure natural cork. This is the actual bark of the cork oak tree. It is approximately 50 percent air by volume with the air trapped within the cellular structure of the bark, thus making it extremely resilient. Compressed cork and granular cork sheet are manufactured materials containing binders which deteriorate during prolonged vibration. Natural cork will prove completely satisfactory for isolating vibration when thicknesses in excess of two inches (three inches preferred) and surface loadings of ten to twenty pounds per square inch are used. Cork responds somewhat like an air-spring isolator in that moderate loading is required for optimum results. Operating within this range, large deflections occur with small changes in load. Aging of cork is evidenced by permanent set, that is, compression under load.

Felt pads are applied where isolation requirements are not too rigid and the isolator must be cemented in place. They are recommended for use when machinery movement must be closely controlled. Average surface loading is 50 psi. Felt exhibits high damping, thus it is effective in reducing the resonant response of a system. When vibration is not excessive, the compression pad need not be cemented to the machine. Felt is satisfactory for use above 40 cps in thicknesses of 1/2 to 1 in, where the softer systems (lower natural frequencies) require thicker pads.

Pads are also made by impregnating layer-on-layer of woven cotton duck with natural or synthetic rubber. This type pad is useful where alignment must be maintained under high loading conditions up to a maximum of 1500 psi. Typically, these are installed under drop-hammer anvil pads and printing press supports. Stainless-steel wire mesh is used for special applications requiring heat and corrosion resistance. This material is formed into knitted wire pads which have high internal friction. The characteristic property of all compression pads is that their stiffness increases with deflection.

Air mounts are a more recent form of vibration isolator which provide a high degree of isolation and stability at moderate cost. As mentioned above, the lower the natural frequency of the system the better the isolation. In this respect, air mounts are superior to other isolators since they require relatively small deflections to produce natural frequencies between 0.5 cps and 3 cps. This range of isolation is impractical for metal springs because of the required large static deflection. It should also be noted that these devices possess good isolation in the lateral as well as vertical direction. In some cases, air mounts are used as passive elements. Air is sealed in the elastomer bellows and internal pressure and deflection adjust themselves to the load impressed by the supported system. A more costly arrangement provides for introducing air into the bellows. This results in the supported system being elevated when necessary and internal air pressure can be varied to control the natural frequency of the system. The greatest advantage of this latter system is that during periods of nonuse the air can be released from the bellows, thereby relieving the stresses in the elastomer while the system is supported on mechanical stops. This produces longer life for the air mounts and results in savings in replacement and maintenance costs. When an air spring is properly loaded, large deflections will result from small changes in the load. This is characteristic of a soft suspension system.

There are other types of isolators commercially available, for example, metal Belleville springs. However, they can be considered as specialty items which do not fall into the more general category of isolators as described above. If consideration is given to the frequency spectrum of an excitation and isolators are selected according to isolation requirements, isolator type and characteristic response, then satisfactory results should be obtained for a long service life installation at reasonable cost.

# BIBLIOGRAPHY

Beranek, L. L. *Noise Reduction*. New York: McGraw-Hill Book Co., 1960.

Carson, R. W. "How to Select Vibration Isolators." *Product Engineering*, March 4, 1963, pp. 696-707.

Crede, C. E. *Vibration and Shock Isolation*. New York: John Wiley & Sons, Inc., 1958.

Eberhart, L. "The Specification of Products for Vibration Control." *Sound and Vibration*, May 1967, pp. 8-13.

George, D. "Noise and Vibration in Quality Control, Design, and Maintenance." *Sound and Vibration*, October 1967, pp. 21-26.

Harris, C. M. and Crede, C. E. *Shock and Vibration Handbook*. New York: McGraw-Hill Book Co., 1961.

Macduff, J. N. and Curreri, J. R. *Vibration Control*. New York: McGraw-Hill Book Co., 1958.

Rathbone, T. C. "How to Measure Product Vibration." *Product Engineering*, October 14, 1963, pp. 105-113.

Rieger, N. "Vibration in Geared Systems." *Machine Design*, September 16, 1965.

Wallerstein, L., Jr. "A New Approach to Vibration for Low Frequency Sources." Design Engineering Conf., Amer. Soc. of Mech. Engr., Paper No. 68-DE-S4, April 1968.

# SYMBOLS

| | | |
|---|---|---|
| A | area | inches-square |
| C | carbon, centigrade | — |
| cal. | calorie | — |
| D | deflection | inches |
| F | force, fahrenheit | pound |
| f | frequency | cps |
| g | gram | — |
| H | hydrogen | — |
| K | spring rate | pounds/inch |
| K | thermal conductivity | Btu/in-hr-°F |
| $l$ | linear deflection | inches |
| $\Delta l$ | unit thickness | inches |
| $M_{T,s}$ | modulus (tension, shear), elastic | pounds/inch-square |
| $\dot{Q}$ | heat flow rate | Btu/hr |
| S | stress | pounds/inch-square |
| t | thickness | inches |
| T | temperature | degrees |
| $\Delta T$ | unit temperature difference | degrees |
| $\epsilon_p$ | Poisson's ratio | — |
| $\epsilon$ | strain | — |

# 7

# PHYSICAL PROPERTIES OF RUBBER

*William S. Mitchell*

## 1. GENERAL PROPERTIES OF COMPOUNDS

The word elastomer is a modern-day generic term used to describe all rubber-like materials, including natural rubber, synthetic rubber and plastics. Many types of available rubber and plastic have similar characteristics. For instance, they soften when heated, have a large value of percent elongation and they are resilient. Only rubber has found wide application in the isolation and damping of vibrating systems. Because of the many variables involved in formulating and curing rubber compounds, the physical properties of a specific type of rubber may vary from one batch to another. In addition, when testing the properties of rubber, there are anomalies in individual tests. Thus, current literature contains what appear to be discrepancies in reported data. This applies to mechanical as well as to comparative properties. Values given in this chapter should be considered not as absolute values but as an average of current data.

Rubber is obtained in its natural state or synthesized by modern chemistry. The natural form is obtained principally from the rubber tree, *Hevea brasiliensis*, which is just one of many hundreds of tropical plants producing rubber-based gums. This tree is the major source of commercially processed natural rubber. It is found growing wild in Central America and is successfully cultivated on plantations in Southeast Asia and on the islands of the southwest Pacific. When the bark of the tree is cut, a milky emulsion of rubber globules and water containing one-third rubber seeps out. The liquid, which is commonly known as latex, is collected by plantation workers and later coagulated by the addition of a dilute acid. It is then rolled into thin sheets and cured over fire. The crude product is called smoked sheet rubber. In addition to being processed into this sheet form, a small percentage of the liquid latex is used in the manufacture of adhesives and commercial products such as rubber gloves.

Natural rubber is an unsaturated hydrocarbon ($C_5H_8$) composed of the polymer "isoprene." This molecule has the following structure.

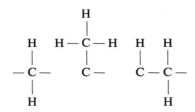

Unlike the metals that have rigid and well-ordered structures, an elastomer of natural or synthetic rubber can be described as a coherent, elastic solid which is a tangled mass of twisted and intertwined chain-like molecules. The material has a high molecular weight (68). At ordinary temperatures it is soft, gummy and sticky and at low temperatures it is hard and brittle. Unvulcanized natural rubber can be dissolved by organic solvents, such as gasoline, naphtha and turpentine, thus forming a sticky solution. It is very deformable, in some cases stretching to ten times its original length. Moreover, it will support a load of 11,000 psi in tension (based on actual cross-sectional area this is equivalent to 45,000 psi, approximately). Interestingly, the material does not obey Hooke's relation of

$$S = M\epsilon \quad \text{(stress = modulus} \times \text{strain)}$$

Rather, it exhibits a slight hysteresis. Original dimensions are not completely recovered when a deformed sample is subsequently unloaded. Sunlight, and to some degree air, is detrimental to natural rubber products as it causes them to become hard and brittle in time. These undesirable properties of rubber can be avoided with the addition of fillers, followed by vulcanization.

In 1840, Goodyear discovered that the properties of natural rubber are changed significantly by vulcanization. This is a heating of the rubber after it has been compounded with sulfur. The resulting product is superior to unvulcanized rubber in that it is elastic over a wide temperature range, possesses relatively high tensile strength and is resistant to the organic solvents. Whereas sulfur is required to alter the molecular structure of natural rubber, zinc oxide, carbon black and other fillers are added before vulcanization to enhance the physical properties of the isoprene-sulfur molecule. For example, carbon black increases the wear resistance of rubber. Other fillers add bulk, increase or decrease hardness and add strength and color to the finished product. They also increase toughness and improve aging resistance. Overall, vulcanized natural rubber is superior to all synthetic rubber, although for specific applications special synthetics have been developed that possess one or two superior properties. For example, rubber can be formulated to high tensile strength and elongation. The elastic property of natural rubber is better than that of any of the synthetics. Therefore, it is recommended for those applications requiring maximum resilience. Natural rubber has lower energy loss due to hysteresis than any of the synthetics. Consequently, it has a lower rate of heat buildup during rapid deformation.

One type of synthetic rubber which is not resistant to petroleum-based oils is currently being produced in large volume at low cost. This is a styrene-butadiene compound which is used for tires, hoses, tubes and molded products.

However, the nitrile-butadiene compounds are resistant to petroleum-based oils. A special polysulphide compound, having the trade name Thiokol®, has excellent resistance to such oils and solvents, but it also has low-temperature operating limitations and a noted lack of resilience. The chloroprene rubbers, such as neoprene, possess good resistance to organic oils. Additionally, they have good resilience and are resistant to the effects of sunlight as well as a wide range of chemical substances. Silicon-based rubber, which has moderate strength, has found wide application where extremes in temperature are encountered, as it can be formulated to retain its resilience in the temperature range of $-50°F$ to $520°F$. It finds limited general application, however, because of its high cost. These rubber compounds are available in liquid and paste form, some of which require two-part mixing. They are also available as RTV (room-temperature-vulcanizing) single-element compounds. The acrylic compounds have very poor low-temperature properties. However, they are oil resistant and do possess good high-temperature physical properties, thus making them an excellent choice where heat and oil resistance are required.

Rubber possesses great versatility because of its adhesion characteristics, moldability and inherent damping. It is highly resilient in many formulations and possesses excellent sound-deadening characteristics. All these factors make it readily acceptable for use as a sound and vibration isolator. Rubber has practically no elasticity by volume. For example, if a sample were confined in a rigid container or cylinder, it would not deflect noticeably when subjected to a load applied by a piston. If the sample were subjected to a load without peripheral restraint, the material would flow out or "bulge" from under the load. Rubber compounds are capable of significant energy storage, greater in proportion to size than any other material. The coefficient of friction between vulcanized rubber and steel is high, often exceeding a value of 1.0. Water acts as a lubricant for a rubber-metal interface, producing a coefficient of friction of approximately 0.02. Rubber is an excellent electrical insulator with a dielectric strength in the range of 100 to 500 volts per mil. This compares favorably with kraft paper (100 volts per mil) and glassine paper (300 volts per mil). Characteristics of various rubber compounds are shown in Tables 9, 10 and 11.

Synthetic rubber is produced mainly in four formulations: Neoprene, GR-S, Butyl and Acrylonitrile-Butadiene. Also, Polybutadiene is increasing in output because of the consumer's demand for a more wear-resistant tire. GR-S, which is also called BUNA-S, is a copolymerization of butadiene and styrene. The cheapest of all the synthetics, it constitutes approximately 80 percent of the synthetic rubber produced in the United States today, as its physical properties compare favorably to those of natural rubber, and its heat and abrasion resistance make it suitable for automobile tires. However, it has high hysteresis loss. Butyl, which has outstanding gas-retaining properties, has low resilience at room temperature while at 212°F it is comparable to gum natural rubber which has good resilience. Neoprene is an acetylene-chlorine polymer (chloroprene) which is far superior to natural rubber in its resistance to sunlight, ozone, oxygen, oil, grease and acids. Consequently, it is used for oil and gas service (hoses, tanks) and for containers in chemical service. Physical properties of typical synthetic rubber of various degrees of hardness are presented in Table 11.

## TABLE 9
## TYPES AND CHARACTERISTICS OF ELASTOMERS

| TYPE | CHARACTERISTICS |
| --- | --- |
| Natural Rubber (NR) | excellent mechanical performance, poor oil and chemical resistance, affected by heat, sunlight and oxygen |
| Synthetic Rubber (Isoprene) | excellent mechanical performance, poor oil and chemical resistance, affected by heat, sunlight and oxygen |
| Styrene-Butadiene (GR-S) | good physical properties when fiber reinforced, good wear resistance, affected by weather; poor oil and chemical resistance, poor electrical properties |
| Butyl | low gas permeability and damping properties; good chemical resistance; medium strength and poor oil resistance |
| Polybutadiene (BR) | better properties than Styrene-Butadiene, improved wear resistance, medium strength |
| Ethylene Propylene | weather resistant with good electrical properties; affected by solvents, low tear resistance |
| Chloroprene (Neoprene) | good mechanical properties, resistant to flame, weather and oils; affected by aromatic hydrocarbons |
| Acrylonitrile–Butadiene | excellent oil and chemical resistance, fair mechanical strength, poor low temperature properties |
| Polyurethane | exceptional abrasion and tear resistance, resistant to oxygen, sunlight and ozone; affected by aromatics and esters; good vibration damping and sound deadening |
| Silicon | wide useable temperature range; low strength, wear and oil resistance; good flex resistance |
| Polysulfide (Thiokol) | excellent resistance to oil, solvents, and water; poor physical properties, objectionable odor |
| Fluorocarbon (Viton) | excellent oil and chemical resistance; flame resistant at high temperature; poor low-temperature properties |
| Polyacrylate | excellent oxygen, ozone, heat, gasoline and oil resistance; poor physical strength at low temperature |

## 2. ADHESION

All of the common rubber compounds will adhere to ordinary cloth fabrics including cotton, rayon, wool and nylon if the fabric is first treated with a special liquid chemical "dip." The resulting bond is partly chemical, partly mechanical, and it will have excellent strength. Rubber can also be bonded to metal. For compounds having a hardness of fifty durometer and greater, adhesion values of 250 psi can be obtained, although 125 psi is more the average. Smooth metal surfaces give best adhesion results. Good design features large fillets at sections where the rubber material joins the metal surface. Care should be taken when applying bonded parts to specific installations involving wide

**TABLE 10**
**COMPARATIVE PROPERTIES OF RUBBER**

| PROPERTY | NATURAL RUBBER | NEOPRENE | GR-S | BUTYL | NITRITE | POLYSULFIDE | SILICON |
|---|---|---|---|---|---|---|---|
| Tensile | E | E | E | F | G | F | P |
| Elongation | E | E | G | E | G | F | P |
| Metal bonding | E | E | E | G | E | P | F |
| Adhesion to fabrics | E | E | G | G | G | F | G |
| Compression set | G | G | G | F | G | P | G |
| Resilience | E | G | G | P | G | G | E |
| Dielectric strength | G | F | G | G | F | F | E |
| Abrasion resistance | E | E | E | E | E | G | P |
| Tear resistance | E | G | F | E | G | P | P |
| Chemical resistance | F | G | F | E | F | G | G |
| Resistance to swelling | P | F | P | P | G | G | F |
| Resistance to aging | F | E | G | E | G | E | G |

## TABLE 11
### REPRESENTATIVE PHYSICAL PROPERTIES OF RUBBER AND OTHER MATERIALS

| MATERIAL | ULTIMATE STRENGTH PSI×10³ | ULTIMATE ELONGATION PERCENT | ELASTIC MODULUS PSI×10⁶ | SHEAR MODULUS PSI×10⁶ | POISSON'S RATIO |
|---|---|---|---|---|---|
| Steel, annealed | 56 | 30 | 29 | 11.5 | 0.30 |
| Aluminum, pure | 13 | 40 | 10.3 | 3.8 | 0.33 |
| Plastic, rigid | 10 | 20 | 0.4 | — | — |
| Rubber, hard 90H | 10 | 6 | 0.15 | 0.0625 | 0.20 |
| Rubber, stiff 60H | 3.5 | 450 | 0.005 | 0.00016 | 0.50 |
| Rubber, soft 40H | 2.5 | 800 | 0.00015 | 0.00005 | 0.50 |

temperature variations since the coefficient of expansion for rubber vulcanizates is generally higher than that for metal. Rubber can be adhered to stainless steel, carbon steel, aluminum, zinc, powdered metals, plastics and, in particular, brass. Usually adhesion surfaces are precoated with a special cement that is compatible with both the metal and the rubber compound. Adhesion to plated metals has been less than satisfactory except with platings of brass. Metal parts requiring finish plating or oil impregnation can be finish processed after rubber to metal pieces have been bonded.

## 3. SHAPE FACTOR

The shape rather than size of a rubber sample may have a pronounced effect on its stiffness when it is loaded in compression. This is in contrast to samples loaded in shear and tension which are not noticeably affected by shape. Samples with flat, parallel end surfaces under load and free perpendicular sides have a shape factor defined by the ratio of the area of one loaded surface to the total free surface area of the sample. By definition this ratio is limited to test samples having equal area-loading surfaces which include rectangular prisms, cubes and cylinders. Thin, flat samples have relatively large shape factor values; however, these values can be decreased by using samples that are either hollow or thick in section, or both. When rubber is so compressed that the loaded surfaces cannot move laterally, the effective stiffness of the sample will depend on the shape factor of the part. If the sample is thoroughly lubricated at the interface where the load is applied, then the compressive stress is independent of the shape factor. This phenomenon is shown in Fig. 7.1, and it holds true for those shapes mentioned above. It can be noted that lateral restriction of the loaded surface greatly stiffens a compressed rubber sample that has a high shape factor. With small values of shape factor, the effect of lateral restriction becomes negligible. Shape factors should be limited to a range of 0.25 to 1.0 for best results. Factors of less than 0.25 may result in column-type buckling. Unbonded samples with high shape factors are subject to large decreases in stiffness from accidental lubrication.

## 4. STRESS-STRAIN RELATION

A characteristic physical property of rubber is its low modulus of elasticity. At ordinary temperatures a low-carbon steel rod has a strain value of 0.2 percent at its elastic limit. A corresponding value for rubber is 300 percent. In addition, stress values for metals are of an order two greater than rubber. Hence, the resulting modulus of rubber as defined by the ratio of stress to strain is very low by comparison to steel. The result of this is that large deformations result from low levels of stress. With respect to a stress-strain relation, material modulus is defined from the shape of the linear portion of the stress-strain curve. However, no part of this curve is linear in the case of rubber; consequently, the modulus of rubber is a variable quantity. A qualitative stress-strain curve for a typical rubber compound is shown in Fig. 7.2. It can be seen that the stress-strain relation is approximately linear for low values of compressive strain and

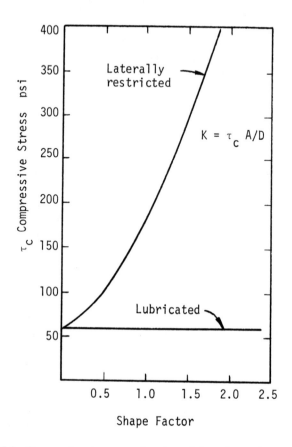

Fig. 7.1  Compressive stress as a function of shape factor for restricted and lubricated loaded surfaces on 40 durometer rubber at 20 percent deflection (K-spring rate, A-area, D-deflection)

high values of shear strain. These latter values correspond to relatively low levels of stress. This indicates the material has favorable compliance for application to the isolation of harmful vibration. Because rubber can be subjected to very large strains, the definition of modulus as applied to other materials will not suffice here. Hence, the modulus of rubber is computed from the slope of a chord line connecting the position of strain to the origin of the stress-strain diagram. Values for the different moduli are based on percent elongation and given in terms of an extension or compression modulus. The relation between the modulus in tension and the modulus in shear for rubber undergoing small strains is

$$M_T = 2 M_s(1 + \epsilon_p)$$

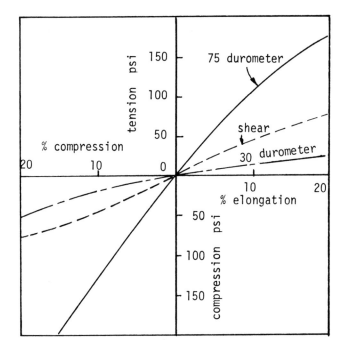

Fig. 7.2   Stress-strain relation for rubber in tension and compression
with a comparative curve for shear

where $M_T$ and $M_s$ denote the tensile and shear moduli, respectively, and $\epsilon_p$ is Poisson's ratio. With rubber subjected to a load, the original volume may increase or decrease depending on a specific compound. But assuming no change in volume, Poisson's ratio will approach a limit of 0.5, compared to a value of 0.3 for metals.

Whereas the elastic modulus implies a stiffness for a given material, bulk modulus is a measure of its resistance to volume compressibility. The bulk modulus of rubber is very high, ranging from 300,000 psi for gum natural rubber to 750,000 psi for the synthetics. By comparison, water has a value of approximately 300,000 psi. Thus, rubber is less compressible than water. An indication of the volume reduction of rubber under increasing stress levels is shown in Fig. 7.3.

## 5. DYNAMIC MODULUS

Properties of rubber are also expressed in terms of a dynamic modulus. This is necessitated by the fact that rubber stiffens when subjected to vibratory loads; thus, the effective modulus as measured from dynamic experimental data is termed the dynamic modulus. The ratio of dynamic modulus to static modulus

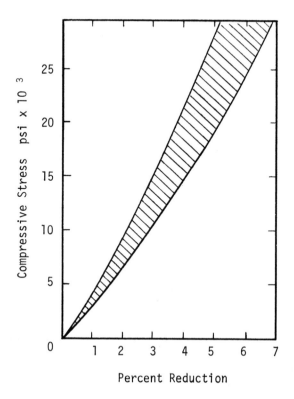

Fig. 7.3    Volume compressibility of rubber at ambient temperature

will range between 1 and 2, depending on the hysteresis property of the rubber. Since the effective modulus influences stiffness and the natural frequency of any elastic system equals the half-power of the ratio of stiffness to mass, resonant response of the system will be shifted toward a higher frequency. This implies that resulting isolation will be less than what is theoretically predicted. Consequently, softer than necessary suspension systems should be specified where isolation of dynamic systems is critical. If the elastomer is maintained at constant temperature over a moderate range of frequencies, then the dynamic modulus will be approximately independent of frequency. At the high frequencies, dynamic modulus increases. This undesirable increase is offset somewhat by the resulting increase of temperature within the rubber element which tends to reduce the dynamic modulus. High damping rubber, such as Thiokol RD or polyurethane, shows a rapid increase in the value of its dynamic modulus with increasing frequency while its damping factor varies relatively slowly. Displacement amplitude has no effect on the dynamic properties of gum natural rubbers. However, this is not the case with filled compounds; as displacement amplitude increases dynamic modulus is reduced, even if temperature of the elastomer is held constant. In some cases this effect is attributed, in part, to the temperature rise caused by increased viscous energy losses in the rubber. Thus, as temperature increases, resilience increases, but the dynamic modulus decreases. Moreover, this results when displacement amplitude increases.

## 6. HARDNESS

Rubber is classified both in terms of its formulation and its hardness. Hardness is defined as a relative resistance to surface indentation, implying a qualitative measure of the stiffness or rigidity of the material. A durometer scale of hardness based on ASTM specification D-676 is the most widely accepted. As examples of durometer hardness: a rubber band (35-40), tire tread (60-70) and bowling ball (95-99). Natural rubber can be processed to hardnesses of thirty to one hundred durometer; synthetic polymers are normally made in hardnesses in excess of forty-five.

## 7. RUBBER IN TENSION

Because of its low tear resistance, rubber is seldom used in tension for the isolation of vibratory forces. It does find use, however, in shock applications where extremely large deflections are required and for the isolation of ultra-low frequencies. In these instances rubber cord is used, consisting of many individual rubber strips encased in a braided covering. Suggested maximum values of working tensile stress are: 50 psi for durometers below fifty, 60 psi for durometer levels between fifty and sixty and 75 psi for all others. The modulus of elasticity with respect to elongation and hardness is shown in Fig. 7.4.

## 8. RUBBER IN COMPRESSION

Rubber is more widely applied in compression than it is in tension because of its high potential energy storage and effective stiffness. Moreover, it can support heavy loads without failure. These characteristics depend on the modulus or inherent stiffness of the rubber, the ability of its unloaded surfaces to expand (because of its low-volume compressibility) and the restraint offered by the loading surfaces to lateral motion of the material. In the latter case, bonding of the rubber to a metal support will insure maximum apparent stiffness. A roughened surface is also very effective, because any type of lubrication on a smooth metal plate in contact with rubber will result in a significant reduction of the effective stiffness. Rubber samples subjected to compressive loading obey the "similarity rule": similarly shaped samples undergo the same percentage of deflection when subjected to equal compressive loads. Therefore, the use of rubber models in experimental investigations can lead to useful and meaningful conclusions. The relationship between compressive stress and durometer hardness is shown in Fig. 7.5. Because of the inherent inaccuracies in measuring hardness, only a qualitative relationship exists; however, the trend is well defined.

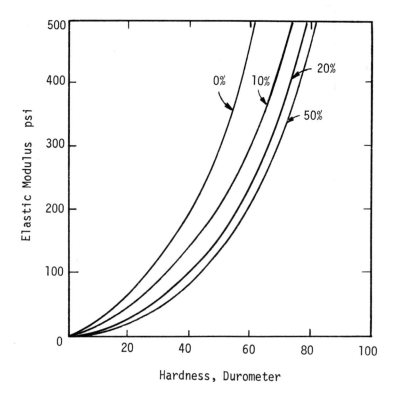

Fig. 7.4   Elastic modulus for rubber in tension as a function of hardness
and percent elongation

In proper application a rubber piece will react like a spring with a spring rate
K defined by the ratio of force to deflection or

$$F = K D$$

where F is the force applied and D is the deflection of the rubber in the direction
of the applied load. The spring rate can also be expressed in terms of the modulus
M as

$$K = \frac{MA}{D}$$

where A is the cross-sectional area under load. If several rubber pieces are
mounted in parallel, the equivalent spring rate is

$$K_E = K_1 + K_2 + \ldots K_n$$

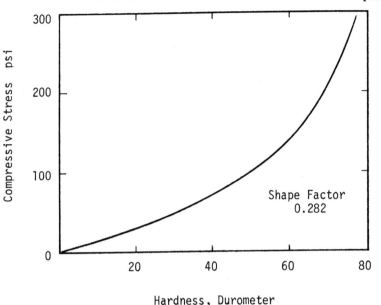

Fig. 7.5 Compressive stress of rubber samples with varying degrees of hardness at 20 percent deflection

But, if the pieces are arranged in series, then the equivalent spring rate becomes

$$K_E = \left[ \frac{1}{K_1} + \frac{1}{K_2} + \cdots \frac{1}{K_n} \right]^{-1}$$

## 9. RUBBER IN SHEAR

The majority of the rubber materials currently applied to the isolation of noise and vibration is subjected to loads which induce shear as the predominant stress. The reason for this is seen in Fig. 7.2. Rubber is more compliant when loaded in shear than it is when loaded comparatively in compression, although energy storage capacity is relatively smaller. Shear strain is defined as the ratio of the linear deformation, $l$, of a rubber sample to its thickness, t. This quantity is the tangent of the angular displacement of a sample in shear as shown in Fig. 7.6c. By comparison, shear strain for alloy steel is approximately 0.02, whereas for some rubber it is as high as 5. The early use of metal-backed rubber in shear, limited shear stresses to approximately 25 psi. Modern improvements in bonding techniques permit higher allowable stresses as indicated in Fig. 7.7.

The modulus of rubber in shear is dependent on shear strain as well as hardness. For example, the value of the shear modulus of a sample with a given hardness will pass through a minimum value as shear strain increases. Since spring rate is proportional to modulus, rubber in shear will exhibit a softening of its effective stiffness during periods of deformation. Variations of 30 percent are

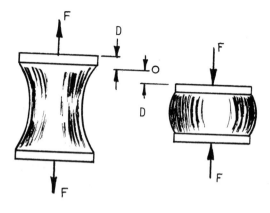

a. Tension        b. Compression

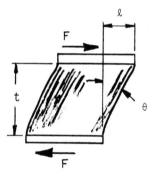

c. Shear

Fig. 7.6   Loading of rubber samples in tension, compression and shear

possible, depending on the rubber formulation and working hardness. Shear modulus as a function of hardness is shown in Fig. 7.8.

## 10. THERMAL PROPERTIES

Specially formulated rubber parts will function satisfactorily in a temperature range of −70°F to 520°F. However, increases in permissible working tempera-

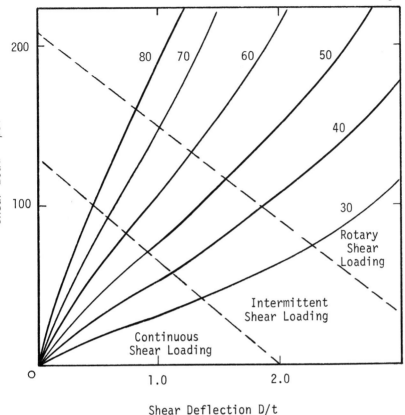

Fig. 7.7　Allowable shear loading for various shear deflections; natural
rubber-metal bond (with durometer hardness as a parameter)

ture are accompanied by reductions in other physical properties. At elevated
temperatures, the tensile strength of natural rubber is much greater than any
of the synthetics. Butyl rubber shows the least change as temperature increases;
neoprene shows the most. As material temperature increases, the elastic moduli
of all the previously mentioned rubber compounds, except neoprene, are found
to increase. In addition, Butyl rubber exhibits the greatest increase in resilience;
natural rubber and neoprene show practically none. Another result of increasing temperature is an increase in the hysteresis of natural rubber and GR-S.
Most other compounds show a decline. As temperature is lowered toward zero,
hysteresis as well as values of moduli increase; also, a crystallization of Butyl,
natural rubber and neoprenes occurs. At very low temperatures embrittlement
of the rubber's internal structure causes these materials to be brittle.

The specific heat of rubber is the amount of heat required to raise a unit mass
of this material by one degree of temperature. For natural rubber, the specific
heat varies with both temperature and elongation. Thermal conductivity K is
the rate at which heat Q is conducted through a material of area A and thickness

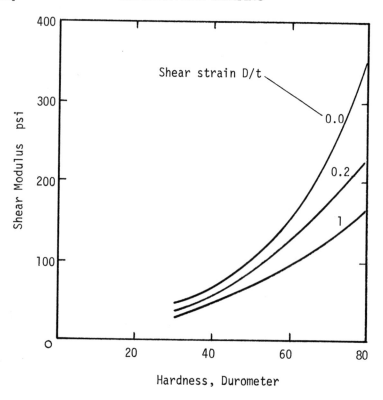

Fig. 7.8   Shear modulus as a function of hardness for various shear strains

$\Delta l$, exhibiting a temperature difference across the thickness of $\Delta T$. These quantities are related by the heat-conduction equation

$$\dot{Q} = KA \, \Delta T / \Delta l$$

The conductivity of rubber is relatively low. This is advantageous in that rubber can be used for heat insulation. But it is a great disadvantage in vibratory systems because heat cannot be dissipated rapidly and temperature rise can be considerable. Table 12 denotes typical values of thermal properties.

## 11. TIME-RELATED PROPERTIES

Creep occurs in rubber just as it does in metal: deformation of the material under load takes place over an extended period of time. Whereas creep in metal is associated with elevated temperatures, it is experienced in stressed rubber at

**TABLE 12**
**THERMAL PROPERTIES OF RUBBER**

| COMPOUND | SPECIFIC GRAVITY | THERMAL EXPANSION %VOLUME CHANGE/°F | SPECIFIC HEAT BTU/LB$_m$°F | CONDUCTIVITY BTU/INCH HR °F × 10$^{-2}$* |
|---|---|---|---|---|
| Natural Rubber | 0.91 | 0.037 | 0.50 | 0.69 |
| Neoprene | 1.23 | 0.04 | 0.52 | 0.93 |
| GR-S | 0.93 | —— | 0.44 | 1.26 |
| Butyl | 0.92 | —— | 0.47 | 0.44 |
| Nitrile | 0.99 | —— | 0.47 | 1.26 |

*For 0.69, read 0.0069, etc.

any temperature. Deformation due to creep is greatest during the initial period of loading. After an extended period of time, creep rate assumes a much lower value. Deformation continues until such time as the part fails by fracture or mass flow relieves the applied load and resulting stress. Creep due to tension is 30 percent greater than that for shear, while in compression creep is approximately 20 percent less. Soft natural rubber of about forty durometer hardness has a relatively low rate of creep. Compounds of sixty durometer hardness exhibit creep rates 30 percent above those for soft natural rubber. As the degree of hardness increases, creep rates are likewise increased. When a stressed part is unloaded, it tends to regain its original shape. That portion of a dimension not recovered is termed permanent set. The synthetics manifest more creep and permanent set than natural rubber; hence, they are less suited for long-term applications. Creep rate is reduced and possibly minimized by the proper application of rubber formulations, low working stresses and low working temperatures, usually below 150°F.

A changing of the physical properties of rubber due to the passage of time is defined as aging, and it occurs whether or not a piece of the material is in actual service. It can be minimized by storing the material in cold water; it is accelerated by air and sunlight. Proper compounding and curing of the natural and synthetic rubbers can result in rubber products showing good resistance to the effects of aging. In general, the synthetics tend to stiffen and become brittle. Natural rubber first hardens slightly and then becomes very soft. Neoprene and Butyl compounds exhibit the best resistance to aging, GR-S and natural rubber the least.

## 12. ENERGY LOSS

If a rubber sample is first loaded short of its breaking point and then unloaded, a resulting plot for its elongation versus load will resemble that shown in Fig. 7.9. This curve exhibits a hysteresis loop. The loop, which encloses a finite area, is a measure of the energy loss over a cycle of stretching of the rubber. Repeated stretching by a constant magnitude force results in a constant hysteresis area.

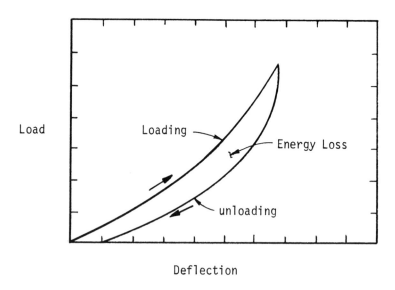

Fig. 7.9  Load-deflection diagram of a rubber sample in tension showing
hysteresis

Testing of rubber samples has shown that the energy loss per cycle increases with an increase in dynamic loading, while the ratio of energy loss to total energy input remains constant. Further, both the energy loss per cycle and the ratio of energy loss to total energy input increases with an increase of frequency. Energy loss per cycle increases with an increase in rubber volume while the ratio of energy loss to energy input decreases. The hysteresis area can be varied widely by altering the rubber compound. For instance, a highly pigmented rubber has high hysteresis loss, while gum natural rubber has almost none. The internal energy loss that takes place in the vibration of rubber is explained by the relaxation theory. Rubber is stretched without heat loss, thus causing a rise in temperature. The process continues until mechanical equilibrium is established. The resulting temperature rise is greater than the subsequent temperature drop during contraction. Since mechanical work is transformed into heat and radiated to the surroundings the relaxation process is irreversible (i.e. total input energy cannot be recovered).

## 13. VIBRATION-RELATED PROPERTIES

Rubber possesses physical characteristics which make it very suitable for anti-vibration mountings. For optimum vibration isolation, a rubber-like mount should be resilient, have a large damping factor for energy dissipation, and the dynamic modulus and damping factor should vary slowly with frequency. Since all of these requirements cannot be met with any one specific rubber compound, it has been suggested that two independent rubber pieces be applied in parallel so that each experiences the same dynamic strain. Since the damping factor of rubber increases with temperature, it follows that better damping results at higher temperature. Damping is also increased by the addition of a filler such as carbon black.

Transmissibility of an isolator can be defined as the ratio of output to input force or output to input displacement, and is an indication of the energy being conducted through the isolator. In the transmission of energy, consideration should also be given to wave effects within the rubber of an isolation mount. Predicted transmissibility will be in error when the physical dimensions of the elastomer coincide with half wavelengths of the elastic wave traveling in the material. This so-called "wave effect" occurs at many frequencies, the first of which is in the region of 200 cps for most commercial isolators. It will vary in frequency depending on dimensions of the mount and the hardness of the elastomer. Error also occurs when rubber material contains a substantial quantity of carbon black. Transmissibility for a simple mount is shown in Fig. 7.10.

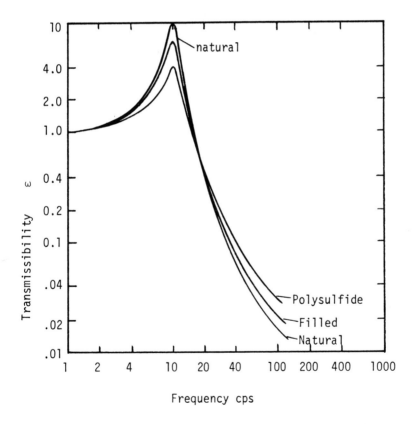

Fig. 7.10  Transmissibility curves for linearly acting isolators with various elastomer elements (10 cps system)

Here, the effect of substituting various elastomers of like durometer hardness into the isolator system can be seen. The less "natural" the material, the higher the degree of damping.

Many commercially available vibration isolators use rubber in compression, shear or both. When used in combination, loads producing small deflections are satisfactorily carried in shear. During large deflections produced by overload, the rubber becomes loaded in compression, thus acting as a snubber and compression spring to the motion. Typical isolator designs are shown in Fig. 7.11. These are indicative of only a few of the many types that are manufactured today.

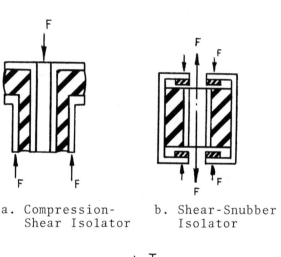

a.  Compression-        b.  Shear-Snubber
    Shear Isolator          Isolator

c.  Torsion
    Isolator

Fig. 7.11  Specialty vibration isolators

# BIBLIOGRAPHY

*Handbook of Molded and Extruded Rubber*, 2nd ed. Akron, Ohio: The Goodyear Tire and Rubber Co., 1959.

*Some Physical Properties of Rubber*. Akron: U.S. Rubber Co., 1941.

*Vanderbilt Rubber Handbook*. New York: R. T. Vanderbilt Co., 1958.

Beranek, L. L. *Noise Reduction*. New York: McGraw-Hill Book Co., 1960.

Burton, W. E. *Engineering With Rubber*. New York: McGraw-Hill Book Co., 1949.

Church, A. H. *Mechanical Vibrations*. New York: John Wiley & Sons, Inc., 1957.

Crede, C. E. and Harris C. O., *Shock and Vibration Handbook*. New York: McGraw-Hill Book Co., 1961.

Crede, C. *Shock and Vibration Concepts in Engineering Design*. Englewood Cliffs, N. J.: Prentice-Hall, Inc., 1945.

Fisher, H. L. *Rubber and Its Uses*. New York: Chemical Publishing Co., Inc., 1941.

Harris, C. O. "Some Dynamic Properties of Rubber." *J. Appl. Mech.* 9 (1942): 129, 135.

Mantell, C. L. *Engineering Materials Handbook*. New York: McGraw-Hill Book Co., 1958.

McPherson, A. T. and Klemin, A. *Engineering Uses of Rubber*. New York: Reinhold Publishing Corp., 1956.

Moeller, K. "Bearing Noise." *In Handbook of Noise Control*, edited by C. Harris, ch. 24. New York: McGraw-Hill Book Co., 1957.

Sack, H. S. and others. "Elastic Losses in Some High Polymers as a Function of Frequency and Temperature." *J. Appl. Phys.* 18 (1947): 450-456.

Snowdon, J. C. "Isolation From Vibration with a Mounting Utilizing Low and High Damping Rubberlike Materials." *J. Acous. Soc. Am.* 34 (1962): 54-61.

Snowdon, J. C. "Representation of the Mechanical Damping Possessed by Rubberlike Materials and Structures." *J. Acous. Soc. Am.* 35 (1963): 821-829.

Snowdon, J. C. "Rubberlike Materials, Their Internal Damping and Role in Vibration Isolation." *J. Sound Vibration* 2 (1965): 175-193.

Snowdon, J. C. "Occurrence of Wave Effects in Rubber Antivibration Mountings." *J. Acous. Soc. Am.* 37 (1965): 1027-1032.

Trelor, L. *The Physics of Rubber Elasticity*. Oxford: Clarendon Press, 1958.

Varga, O. *Stress-Strain Behavior of Elastic Materials*. New York: Interscience Publishers, 1966.

# SYMBOLS

| | | | |
|---|---|---|---|
| $\sigma$ | axial stress | R | dynamic resilience |
| $\sigma_0$ | stress amplitude under sinusoidal variation of stress | $\Delta$ | logarithmic decrement |
| $\tau$ | time | M | mass |
| $\omega$ | circular frequency | $K_1(\omega)$ | bulk storage modulus |
| i | $\sqrt{-1}$ | $K_2(\omega)$ | bulk loss modulus |
| $\epsilon$ | axial strain | $J(i\omega)$ | dynamic complex compliance (shear) |
| $\epsilon_0$ | strain amplitude under sinusoidal variation of strain | $G(i\omega)$ | dynamic complex modulus (shear) |
| $\delta$ | the phase angle between stress and strain | $K(i\omega)$ | dynamic complex modulus (volumetric deformation) |
| $E(i\omega)$ | complex dynamic modulus in tension | $B(i\omega)$ | dynamic complex compliance (volumetric deformation) |
| $E_1(\omega)$ | tensile storage modulus | $\bar{\lambda}$ | axial stretch ratio |
| $E_2(\omega)$ | tensile loss modulus | $\lambda_0$ | wavelength |
| $\dot{\epsilon}$ | strain rate | $T_g$ | glass transition temperature |
| $\eta$ | viscosity | $\omega_p$ | reduced frequency |
| $E(t)$ | relaxation modulus function | $\rho$ | density at temperature T |
| $D(t)$ | creep compliance function | $\rho_s$ | density at temperature $T_s$ |
| $T_s$ | reference temperature | $a_T$ | shift factor |
| $E(\tau)$ | relaxation spectrum | $Z_e(i\omega)$ | electrical impedance |
| $D(\lambda)$ | retardation spectrum | $Z(i\omega)$ | mechanical impedance |
| $\tau$ | relaxation time | $\psi$ | specific damping capacity |
| $\lambda$ | retardation time | $Q^{-1}$ | quality factor |
| $\bar{E}$ | voltage | C | wave velocity |
| I | current | $R_M$ | mechanical resistance |
| $\bar{R}$ | electrical resistance | $X_M$ | mechanical reactance |
| $\bar{X}$ | reactance | X | amplitude |
| $\omega_r$ | resonant frequency | $X_r$ | amplitude at resonance |
| $\bar{T}$ | transmissibility | $\doteq$ | approximately equal to |
| $G_1(\omega)$ | shear storage modulus | $>$ | is greater than |
| $G_2(\omega)$ | shear loss modulus | $<$ | is less than |
| $J_1(\omega)$ | storage compliance in shear | | |
| $J_2(\omega)$ | loss compliance in shear | | |

# 8

# DYNAMIC BEHAVIOR
# OF RUBBER

*M. G. Sharma*

## 1. INTRODUCTION

Rubber and rubber-like materials represent a class of materials that display highly elastic behavior. Under vibratory loading conditions, rubber-like materials are capable of absorbing energy which is partly stored (as in a spring) and partly dissipated (as in a dashpot) in internal friction like heat. Because of this character, rubber is mostly used as a cushioning and vibration isolating material in such applications as tires, belts and engine mounts. In these applications rubber may be subjected to rapidly repeated deformations which are small compared to ultimate breaking values. For the proper utilization of rubber and rubber-like materials in the above applications, it is very important to know their dynamic mechanical properties. Under small dynamic deformations, the mechanical properties of rubber and rubber-like materials are usually specified in terms of their response to sinusoidally varying forces. From a technological viewpoint, the consideration of behavior under sinusoidal loading conditions has limited scope. However, under linear response conditions, it is possible in most cases to relate the behavior under sinusoidal conditions to the behavior under nonsinusoidal conditions. But the same relationship does not exist when the response becomes nonlinear as in the case of rubber undergoing large deformations. There is no comprehensive theoretical treatment yet developed which can describe the dynamic mechanical behavior of rubber and rubber-like materials under nonlinear conditions. However, some attempts have been made in the light of experimental results to develop a theory to take into account the nonlinear effects.[1,2]

During the past two decades, the linear viscoelastic theory was considerably developed as a result of its bearing on the development of polymers.[3-7] As rubber and rubber-like materials are essentially viscoelastic in nature under dynamic loading situations, the application of linear viscoelastic theory to describe the dynamic mechanical behavior provides the background and understanding for interpreting dynamic experimental results. A literature search in the field of mechanical behavior of rubber indicates that considerable effort has been devoted to understanding the dynamic mechanical behavior.[8-12] In the following sections a brief review of the linear viscoelastic theory is given insofar as it applies to the general dynamic mechanical behavior of rubber. The dynamic mechanical properties of several rubber-like materials have been considered and their potentialities in vibration abatement and control are discussed. This is followed by the description of the experimental techniques for the evaluation of dynamic mechanical properties. Finally, some recent work on the non-linear dynamic mechanical behavior of rubber and rubber-like materials is discussed.

**Nature of Dynamic Mechanical Behavior in Rubber.** If a small sample of natural vulcanized rubber is pulled in tension slowly, it will show a response as indicated by the stress-strain curve in Fig. 8.1 plotted in terms of the nominal stress $\sigma$ (based on cross-sectional area of the undeformed specimen) and the stretch ratio $\lambda$ (ratio of extension at any stage of tensile loading to the original undeformed length). Figure 8.1 indicates that the natural rubber is essentially non-linear, displaying considerably large deformation. If the same test were conducted at several low, controlled strain rates, the stress-strain curves would coincide indicating that the behavior is essentially elastic in nature. However, if the strain rates are increased, the rubber behaves as a viscoelastic material indicating appreciable time-dependent behavior. These time effects may be displayed

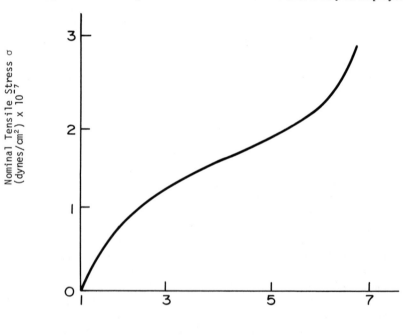

Fig. 8.1  Nominal tensile stress versus axial stretch ratio curve for natural rubber in simple tension (after Mooney [42])

at lower magnitude strain rates if the temperature is reduced. The viscoelastic behavior of rubber and rubber-like materials can better be studied if the magnitude of strains is restricted to anywhere between 1 to 10 percent. Under these circumstances rubber behaves linearly and the linear viscoelastic theory may be used to describe the dynamic mechanical behavior.

Suppose a specimen of a linear viscoelastic material is subjected to a periodic stress represented by

$$\sigma = \sigma_0 \, e^{i\omega t} \tag{1}$$

where $\sigma_0$ is stress amplitude, t is time, $\omega$ is circular frequency, and i is $\sqrt{-1}$. The strain is found to be also periodic but out of phase with stress as shown in Fig. 8.2a. It is given by

$$\epsilon = \epsilon_0 \, e^{i(\omega t - \delta)} \tag{2}$$

where $\epsilon_0$ is amplitude of strain and $\delta$ is phase angle between stress and strain. If the variation of stress and strain are both sinusoidal, a vector diagram involving these quantities can be drawn as shown in Fig. 8.2b. The response of the material can be specified by a complex modulus representing the ratio of stress to strain. This is given by

$$E(i\omega) = \frac{\sigma}{\epsilon} = \frac{\sigma_0}{\epsilon_0} e^{i\delta} = \frac{\sigma_0}{\epsilon_0} (\cos \delta + i \sin \delta) \tag{3}$$

where $E(i\omega)$ is the complex dynamic modulus corresponding to tensile deformation. The complex modulus, being a complex quantity, can be separated into a real part and an imaginary part as

$$E(i\omega) = E_1(\omega) + i \, E_2(\omega) = \left(\frac{\sigma_0}{\epsilon_0}\right) \cos \delta + i \left(\frac{\sigma_0}{\epsilon_0}\right) \sin \delta \tag{4}$$

Figure 8.2b and Eq. (4) indicate that the real part or the storage modulus represents the ratio of stress and strain in phase with the stress. Likewise, the imaginary part or the loss modulus represents the ratio of stress and the strain component, the latter being 90 degrees out of phase with the stress. The ratio of the loss modulus to the storage modulus is the tangent of the phase angle (usually called the loss tangent). It is evident, therefore, that two quantities are required to specify the dynamic behavior under sinusoidal stress, namely the storage modulus and the loss modulus.

The nature of dynamic mechanical behavior of linear viscoelastic materials in general, and rubber in particular, can be better understood by examining the variation of storage and loss modulus with frequency and temperature as shown in Fig. 8.3. As indicated by the figure, the storage and loss moduli are both small at low frequencies. The material is soft and rubbery in this region, displaying large deformation under loading. As the frequency is increased or the temperature is decreased the storage modulus increases rapidly until it reaches a high

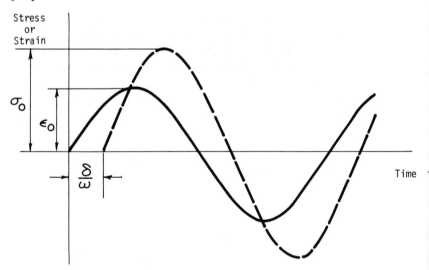

a. Sinusoidal Variation of Stress and Strain

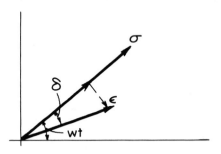

b. Vector Diagram for Sinusoidal Stress and Strain

Fig. 8.2   Sinusoidal stress and strain

constant value. The region where rapid increase in modulus takes place is usually termed as viscoelastic transition. For many materials this region extends over at least six to ten decades of the logarithmic frequency scale. During the transition the material displays considerable viscoelastic effect. As is seen from Fig. 8.3, the loss modulus assumes a small value at low frequencies and increases rapidly

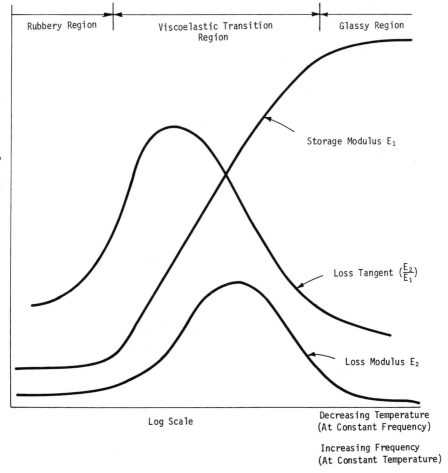

Fig. 8.3   Variation of storage modulus, loss modulus and loss tangent
with frequency and temperature

in the earlier part of the transition region until it reaches a maximum value.
During the later part of transition the loss modulus decreases rapidly and assumes again a small value at high frequencies. The frequency corresponding to
the maximum value of loss modulus corresponds to the maximum energy loss
in the material. Figure 8.3 also shows the variation of loss tangent with frequency
and temperature. Just as in the case of loss modulus, the loss tangent reaches a

maximum value in the transition region. However, the frequency at which this occurs is lower than the frequency at which the loss modulus becomes maximum. What has been said above about the variation of dynamic moduli and loss tangent with frequency at a specified temperature applies equally well to the variation of dynamic moduli with inverse of temperature at a specified frequency. This close relationship between temperature and frequency effects will be elaborated later in this article.

All viscoelastic materials within a linear range of response display the dynamic mechanical behavior with frequency as indicated in Fig. 8.3. The transition region for these materials is distinguished by a characteristic frequency or a characteristic temperature. The characteristic temperature, usually designated as the transition temperature, is a distinguishing property of elastomers and polymers. The harder (*glassy*) the polymer is under room-temperature conditions, the smaller will be its transition frequency or larger its transition temperature. Likewise, the softer (*rubbery*) the polymer or elastomer, the larger the transition frequency and the smaller will be its transition temperature. Rubber and rubber-like materials belong to the later category.

## 2. LINEAR VISCOELASTIC MATERIAL

For a perfect elastic material any imposed stress is proportional to the resulting strain (assumed small). If the imposed stress is removed, the strain is completely recovered. In addition, the work done during the application of stress is stored as strain energy which is completely recovered after the stress is removed. For a Newtonian fluid the imposed stress is proportional to the rate of strain (velocity). When the stress is removed the deformation is irrecoverable, and the work done by the stress during deformation is completely dissipated in the evolution of heat in the material. A linear viscoelastic material displays combined elastic and fluid material properties. As a result, if a uniaxial stress is suddenly imposed on a specimen of a linear viscoelastic material which is held for a certain duration of time and released thereon (Fig. 8.4a), the strain distribution for such a load history is found to be as shown in Fig. 8.4b. Figure 8.4b indicates that after the load is removed the strain does not recover immediately as in the case of perfect elastic material, but recovers partially after infinite time. The unrecovered portion of strain is the permanent deformation due to viscous flow. Rubber at room temperature does not display the behavior indicated above. It behaves as an elastic material, displaying large deformation and a nonlinear stress-strain relation (see Fig. 8.1). However, at low temperatures rubber does display creep and recovery. As shown in Fig. 8.4b, the total creep strain is made up of an instantaneous component, a retarded elastic component, which is recoverable after considerable time, and an irrecoverable viscous strain. The last component of strain has been found to be rather small for most rubber-like materials.

## 3. MODEL REPRESENTATION

As linear viscoelastic behavior can be considered to be a combined behavior of perfect elastic and ideal Newtonian fluid behavior, the response of linear visco-

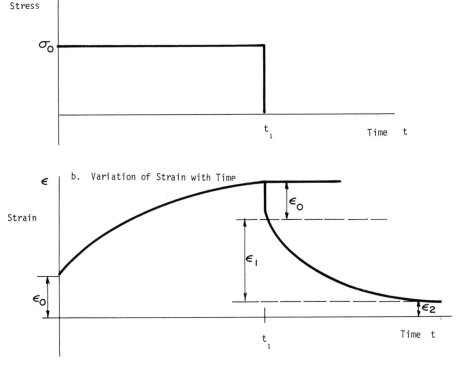

Fig. 8.4   Stress and strain time variation

elastic materials may be visualized from the response of mechanical models constructed from combinations of springs and dashpots. There are two basic types
of models which are customarily used for the representation of viscoelastic
behavior. These are the Voigt model and the Maxwell model (see Fig. 8.5).
Although these mechanical models do not represent the molecular processes
responsible for the observed viscoelastic behavior, they do give physical insight
into the phenomenological aspect of the process. For instance, to duplicate the
response of a linear viscoelastic material under creep and creep recovery, a
model constructed of three individual elements may be considered (Fig. 8.6).

In Fig. 8.6a the three-element model may be considered as made up of one
Maxwell element and one degenerate Maxwell element consisting of a spring
alone. Fig. 8.6b is an equivalent model made up of one Voigt element consisting
of a spring only. The response of the three-element models to various stress or

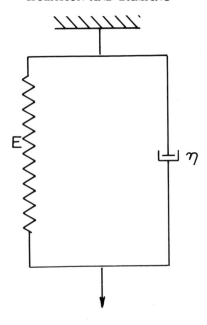

a.   Voigt Model Representation

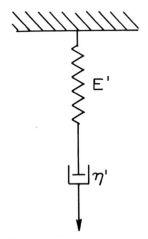

b.   Maxwell Model Representation

Fig. 8.5   Model representation

strain histories could be evaluated by formulating the differential equations relating the stress and the strain.

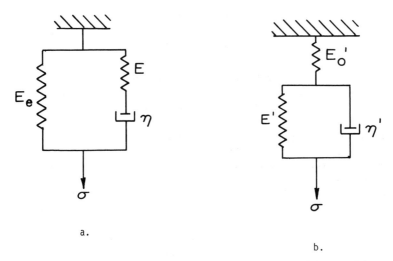

a.

b.

Fig. 8.6   Equivalent model representations for the standard linear solid

## 4. STRESS AND STRAIN RESPONSE OF THREE
## PARAMETER MAXWELL MODEL REPRESENTATION

Suppose a stress $\sigma$ is applied to the model in Fig. 8.6a, where the stress is divided into $\sigma_1$ and $\sigma_2$, respectively, between the spring and the Maxwell element. The total stress can be represented by

$$\sigma = (\sigma_1 + \sigma_2) \tag{5}$$

However, the strain $\epsilon$ resulting from the stresses $\sigma_1$ and $\sigma_2$ in the two components of the model must be the same. This leads to the construction of the following equations.

$$\epsilon = \frac{\sigma_1}{E_e}$$

$$\dot{\epsilon} = \frac{\dot{\sigma_2}}{E} + \frac{\sigma_2}{\eta} \tag{6}$$

where $\dot{\epsilon}$ = time rate of strain and $E_e$, $E$ and $\eta$ are constants in the model representation. Equations (5) and (6) can be reduced to one equation relating the total stress to total strain which one observes in a test involving the material. That can be shown to be[13]

$$\left(1 + \frac{E_e}{E}\right)\dot{\epsilon} + \frac{E_e}{\eta}\epsilon = \frac{\dot{\sigma}}{E} + \frac{\sigma}{\eta} \tag{7}$$

Calling $\tau = \eta/E$, and $\lambda = \eta[(1/E_e) + (1/E)]$, Eq. (7) becomes

$$\sigma + \tau\dot{\sigma} = E_e(\epsilon + \lambda\dot{\epsilon}) \tag{8}$$

where the dot on the symbols represents time rate of change, $\tau$ is relaxation time and $\lambda$ is retardation time. Equation (7) represents a differential equation relating stress and strain that describes the general behavior of a linear viscoelastic material for any kind of stress or strain history.

There are three types of stress or strain histories that are employed to study the viscoelastic behavior of rubber. The first corresponds to a creep experiment which involves the application of a stress of certain magnitude on a specimen. This stress is held constant during the experiment and the resulting strain is measured as a function of time. The ratio of strain to the constant value of stress yields the creep compliance function $D(t)$, that is a function of time alone, for a linear viscoelastic material.

In the second type of experiment, a strain of constant magnitude is applied suddenly to a specimen. Then stress necessary to keep the strain constant with time is measured. The ratio of stress to strain is evaluated. This ratio is usually called the relaxation modulus function $E(t)$.

In the third type of experiment, a sinusoidally varying stress or strain is applied and the resulting strain or stress is measured with time. For a linear viscoelastic material the resulting stress or strain due to the above histories is also found to be sinusoidal. The ratio of stress to strain obtained at any time is called the complex modulus $E(i\omega)$. Varying the frequency of sinusoidally varying strain, the complex modulus could be determined as a function of frequency.

Suppose an oscillating strain history as represented by

$$\epsilon = \epsilon_0\, e^{i\omega t} \tag{9}$$

is applied to a material obeying the stress-strain law in Eq. (7) (see Fig. 8.6a). Then the stress necessary to impose this strain history is found to be sinusoidal but leading the strain by a phase angle $\delta$. That is

$$\sigma = \sigma_0\, e^{i(\omega t + \delta)} \tag{10}$$

In Eqs. (9) and (10) the real and imaginary parts represent, each by themselves, oscillatory quantities with circular frequency $\omega$. Consideration of strain or stress in this form rather than as a purely sinusoidal variation introduces considerable mathematical simplicity.

Substituting Eqs. (9) and (10) into Eq. (8) and rearranging, the expression for complex modulus becomes

$$E(i\omega) = \frac{\sigma_0}{\epsilon_0}\, e^{i\delta} = \frac{(1 + i\omega\lambda)}{(1 + i\omega\tau)}\, E_e \tag{11}$$

where the ratio of amplitudes $\sigma_0/\epsilon_0$ represents the absolute value of the complex modulus $|E(i\omega)|$. The complex modulus $E(i\omega)$ can be separated into the following real and imaginary parts.

$$E_1(\omega) = \frac{(1 + \omega^2\tau\lambda)}{(1 + \omega^2\tau^2)} E_e \qquad (12)$$

$$E_2(\omega) = \frac{\omega(\lambda - \tau)}{(1 + \omega^2\tau^2)} E_e \qquad (13)$$

From Eqs. (12) and (13) the expression for loss tangent can be written as

$$\tan \delta = \frac{E_2(\omega)}{E_1(\omega)} = \frac{\omega(\lambda - \tau)}{(1 + \omega^2\tau\lambda)} \qquad (14)$$

To clearly represent the variation of the storage modulus and the loss modulus with angular frequency, the following geometric mean of the two characteristic times $\lambda$ and $\tau$ can be introduced.

$$\bar{\tau} = (\tau\lambda)^{1/2} \qquad (15)$$

From the three independent material constants $\lambda$, $\tau$ and $E_e$ in Eq. (8) a new constant can be derived which has considerable physical significance. Suppose within a short time, $\Delta t$, the stress is increased by a finite amount $\Delta\sigma$. Integrating both sides of Eq. (8) with respect to time over the interval $\Delta t$, the first term on both sides of the equation approaches zero as the interval $\Delta t$ is made smaller and smaller. As a result the following relation between increments of stress $\Delta\sigma$ and of strain $\Delta\epsilon$ may be obtained.

$$\tau\Delta\sigma = E_e\lambda\Delta\epsilon$$

or

$$\frac{\Delta\sigma}{\Delta\epsilon} = E_0 = \frac{E_e\lambda}{\tau} \qquad (16)$$

where $E_0$ is instantaneous modulus. Using the instantaneous modulus $E_0$ and the equilibrium modulus $E_e$, their geometric mean can be written as

$$\bar{E} = (E_0E_e)^{1/2} \qquad (17)$$

Using Eqs. (15) and (16), Eq. (14) can be rewritten as

$$\tan \delta = \frac{E_0 - E_e}{\bar{E}} \frac{\omega\bar{\tau}}{1 + (\omega\bar{\tau})^2} \qquad (18)$$

The first factor in Eq. (18) represents the relative difference between the instantaneous modulus and the equilibrium modulus. The second factor gives the frequency variation of the loss tangent. This factor has a maximum when the product $\omega\bar{\tau}$ is unity. If the factor $\omega\bar{\tau}/[1 + (\omega\bar{\tau})^2]$ is plotted against $\log(\omega\bar{\tau})$, it is seen to be a symmetrical function of $\omega\bar{\tau}$ with the general characteristic of an error curve. The maximum contribution of this factor can be seen to be half after substituting $\omega\bar{\tau} = 1$ in the expression [Eq. (18)]. Therefore, the maximum value of loss tangent becomes

$$(\tan \delta)_{max} = \frac{E_0 - E_e}{2\bar{E}} \tag{19}$$

The storage modulus expression [Eq. (12)] can be rewritten [using Eq. (16)] as

$$E_1(\omega) = E_0 - \frac{(E_0 - E_e)}{1 + (\omega\bar{\tau})^2} \tag{20}$$

In the limit of low and high frequencies, $E_1(\omega)$ reduces to the following.

$$E_1(\omega) = \begin{cases} E_e & \omega\bar{\tau} \ll 1 \\ E_0 & \omega\bar{\tau} \gg 1 \end{cases} \tag{21}$$

Figure 8.7 shows how $E_1(\omega)$ varies between the two limits represented by Eq. (21). It can be seen that $dE_1(\omega)/d(\omega\bar{\tau})$ reaches a maximum value when the

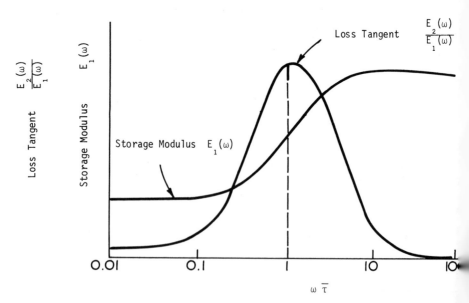

Fig. 8.7   Variation of storage modulus and loss tangent with frequency
for a standard linear solid

product $\omega\bar{\tau}$ reaches unity. This is just where tan $\delta$ is also maximum. Figure 8.7 also illustrates that $E_1(\omega)$ reaches its two extreme values [Eq. (21)] at frequencies where tan $\delta$ is still appreciable.

Rewriting the expression for the loss modulus, using Eq. (16)

$$E_2(\omega) = \frac{E_0 - E_e}{E_e} \cdot \frac{\omega\tau}{1 + \omega^2\tau^2} \tag{22}$$

Equation (22) indicates that $E_2(\omega)$ is maximum at $\omega\tau = 1$, and has a maximum value of $(E_0 - E_e)/2E_e$. This means that the peak occurs to the right of the loss tangent peak in Fig. 8.7. This is because $\omega\tau = 1$ corresponds to $\omega\tau = (E_0/E_e)^{1/2}$.

The model shown in Fig. 8.6a is equivalent to the model shown in Fig. 8.6b because the differential equation relating stress and strain is similar for both cases, as can be seen from the results that follow. The relationship between total stress and partial strain for the model of Fig. 8.6b can be given by

$$\sigma = E_0' \epsilon_1$$

and
$$\tag{23}$$

$$\sigma = E'\epsilon_2 + \eta'\dot{\epsilon}_2$$

where $\epsilon_1$, $\epsilon_2$ are the partial strains and $\sigma$ is total stress.

Rewriting Eq. (23) in terms of total stress and strain

$$E_0'E' \epsilon + E_0'\eta' \dot{\epsilon} = (E_0' + E')\sigma + \eta' \dot{\sigma} \tag{24}$$

Calling

$$\tau = \frac{\eta'}{E_0' + E'} , \; E_e = \frac{E_0'E'}{E_0' + E'} \text{ and } \lambda = \frac{\eta'}{E'} \tag{25}$$

Eq. (24) reduces to Eq. (8). Therefore, the two models (Figs. 8.6a and 8.6b) are equivalent. It is convenient to investigate the response of the model (Fig. 8.6b) to a sinusoidal variation of stress. That is

$$\sigma = \sigma_0 \, e^{i\omega t} \tag{26}$$

The strain response to the above stress will also be oscillatory but lags behind the stress by a phase angle $\delta$. It is given by

$$\epsilon = \epsilon_0 \, e^{i(\omega t - \delta)} \tag{27}$$

Substituting Eqs. (26) and (27) into Eq. (8) with the value of $\tau$, $\lambda$ and $E_e$ given by Eq. (25) and rearranging, the expression for complex dynamic compliance can be shown to be

$$D(i\omega) = \frac{\epsilon_0}{\sigma_0} e^{-i\delta} = \left(\frac{\epsilon_0}{\sigma_0}\right)(\cos\delta - i\sin\delta) = \frac{1 + i\omega\tau}{E_e(1 + i\omega\lambda)} \tag{28}$$

Calling $D(i\omega) = D_1(\omega)$, $- i\,D_2(\omega)$ Eq. (28) can be separated into real and imaginary parts as given by

$$D_1(\omega) = \frac{\epsilon_0}{\sigma_0} \cos \delta = \frac{1 + \omega^2\tau\lambda}{E_e(1 + \omega^2\lambda^2)}$$

$$D_2(\omega) = \frac{\epsilon_0}{\sigma_0} \sin \delta = \frac{\omega(\lambda - \tau)}{E_e(1 + \omega^2\lambda^2)} \tag{29}$$

and

$$\tan \delta = \frac{\omega(\lambda - \tau)}{1 + \omega^2\tau\lambda}$$

where $D_1(\omega)$ is storage compliance and $D_2(\omega)$ is loss compliance. By rewriting Eq. (29) in terms of geometric mean values of retardation time $\bar{\lambda} = (\tau\lambda)^{1/2}$ and the compliance function $\bar{D} = (D_0 D_e)^{1/2}$ where $D_0$, $D_e$ are the instantaneous and equilibrium compliances,[13] respectively, the variation of storage compliance and the loss tangent with $\omega\bar{\lambda}$ can be plotted just as in Fig. 8.7. It can be shown that the storage compliance decreases from a high value at low frequencies to a low value at high frequencies. The rate of decrease is high at the value of $\omega\bar{\lambda}$ equal to unity. This is just where the loss tangent is maximum.

Although the model shown in Fig. 8.6b is simple, in describing the viscoelastic behavior of rubber it represents the actual behavior only approximately. The reason for this is that the model gives rise to a transition that is sharp whereas in an actual case the transition is more gradual (extending over several decades of frequencies), indicating that several transitions characterized by differing relaxation times $\tau_j$ or retardation times $\lambda_j$ (where $j = 1, 2, 3, \ldots n$) occur simultaneously. This type of behavior can be described by a chain of Maxwell elements combined in series. This model representation is usually termed the generalized Maxwell model representation (see Fig. 8.8).

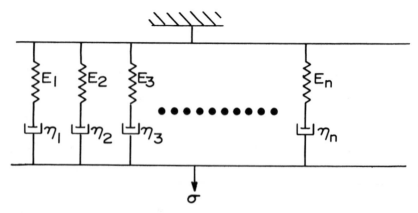

Fig. 8.8   Generalized Maxwell model representation

Depending upon the number of Maxwell Elements, there exists as many relaxation times. They are given by

$$\tau_1 = \frac{\eta_1}{E_1}, \qquad \tau_2 = \frac{\eta_2}{E_2}, \qquad \tau_3 = \frac{\eta_3}{E_3} \ldots \tau_n = \frac{\eta_n}{E_n} \tag{30}$$

The stress response of the generalized Maxwell model for an oscillatory strain history represented by $\epsilon = \epsilon_0 \, e^{i\omega t}$ leads to the following expressions for the complex modulus, the storage modulus and the loss modulus functions.

$$E(i\omega) = \sum_{j=1}^{n} \frac{E_j \, i\omega\tau_j}{1 + i\omega\tau_j}$$

$$E_1(\omega) = \sum_{j=1}^{n} \frac{E_j \, \omega^2\tau_j^2}{1 + \omega^2\tau_j^2} \tag{31}$$

$$E_2(\omega) = \sum_{j=1}^{n} \frac{E_j \, \omega\tau_j}{1 + \omega^2\tau_j^2}$$

Figure 8.9 shows the representation of the storage modulus distribution for polyisobutylene, a compound of high molecular weight, by means of a six-element (three Maxwell elements) generalized Maxwell model. The experimental data for the storage modulus is also shown in Fig. 8.9. It can be seen that the six-element model fits the experimental data for polyisobutylene better than the three-element model.

## 5. DISTRIBUTION FUNCTION OF RELAXATION TIMES

Suppose the number of elements in the generalized Maxwell model (Fig. 8.8) is increased without limit. There results an infinite number of units with spring stiffness E and dashpot viscosity $\eta$ varying continuously with retardation time $\tau$. The function $E(\tau)$, usually called the distribution function of relaxation times, specifies the viscoelastic response of relaxation times lying in the interval $(\tau + \delta\tau)$ and $\tau$. For the continuous Maxwell model representation, the complex dynamic modulus, storage modulus and the loss modulus expressions can be written down by changing the summations of the discrete number terms in Eq. (31) to integrals involving the continuous function $E(\tau)$. These expressions are[13]

$$E(i\omega) = \int_0^\infty \frac{E(\tau)i\omega\tau}{(1 + i\omega\tau)} \, d\tau$$

$$E_1(\omega) = \int_0^\infty \frac{E(\tau)\omega^2\tau^2}{1 + \omega^2\tau^2} \, d\tau \tag{32}$$

and

$$E_2(\omega) = \int_0^\infty \frac{E(\tau)\omega\tau}{1 + \omega^2\tau^2} \, d\tau$$

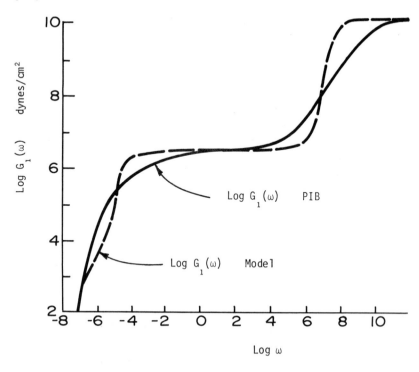

Fig. 8.9   Storage modulus of high molecular weight polyisobutylene
and an equivalent six-element model

The second and third integrals in Eq. (32) are functions of the relaxation spectrum. This indicates that $E_1(\omega)$ and $E_2(\omega)$ are not independent of each other.

## 6. GENERALIZED VOIGT MODEL REPRESENTATION

The model system discussed above is essentially of the Maxwell type. There is yet another type of model that is used to describe the viscoelastic behavior. This type of model is constructed from a series of Voigt (Kelvin) elements. The combination of three elements with a spring and dashpot in parallel (Voigt element) is shown in Fig. 8.6b. For this particular case the response is identical to the model shown in Fig. 8.6a. However, the model represents one transition. When there are a number of viscoelastic transitions occurring at various transition frequencies, the behavior can be described by a chain of Voigt elements as shown in Fig. 8.10.

The dynamic response of the model in Fig. 8.10 can be determined by subjecting it to an oscillating stress as represented by Eq. (26). From the strain response, which is also sinusoidal, it is possible to obtain the following expressions.

$$D(i\omega) = \frac{1}{E_0} + \sum_{j=1}^{n} \frac{1}{E_j(1 + i\omega\lambda_j)}$$

$$D_1(\omega) = \frac{1}{E_0} + \sum_{j=1}^{n} \frac{1}{E_j(1 + \omega^2\lambda_j^2)} \quad \text{(33)}$$

$$D_2(\omega) = \sum_{j=1}^{n} \frac{1}{E_j} \frac{\omega\lambda_j}{1 + \omega^2\lambda_j^2}$$

The generalized Voigt model represents a finite number of transitions as specified by various retardation times $\lambda_1, \lambda_2 \ldots \ldots \ldots \lambda_n$. As in the case of the generalized Maxwell model, the extension of the discrete Voigt model representation to a continuous one is immediate. The following expressions may be associated with the continuous Voigt (Kelvin) model representation.

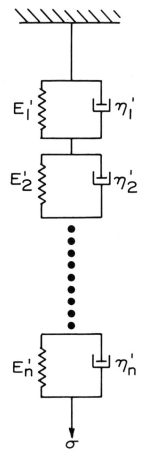

Fig. 8.10   Generalized Voigt model representation

$$D(i\omega) = \frac{1}{E_0} + \int_0^\infty \frac{D(\lambda)}{1 + i\omega\lambda} \, d\lambda$$

$$D_1(\omega) = \frac{1}{E_0} + \int_0^\infty \frac{D(\lambda)}{1 + \omega^2\lambda^2} \, d\lambda \tag{34}$$

$$D_2(\omega) = \int_0^\infty \frac{D(\lambda)\omega\lambda}{1 + \omega^2\lambda^2} \, d\lambda$$

where $D(\lambda)$ is the distribution function of retardation times or the retardation spectrum and $1/E_0$ is instantaneous compliance.

## 7. CORRELATION OF STATIC AND DYNAMIC PROPERTIES OF RUBBER-LIKE MATERIALS

The quasi-static properties of rubber-like materials are usually determined from creep and stress relaxation experiments. For a material with linear viscoelastic response, these experiments yield the creep compliance function $D(t)$ and the relaxation modulus function $E(t)$, both functions of time alone. However, the creep compliance function and the relaxation modulus function represent the behavior of rubber-like materials for specific stress or strain histories. In order to determine the response for a general stress or strain history it is necessary to consider an important property of the linear system, namely, the principle of superposition. The principle, usually termed the Boltzmann Superposition Principle, states that the effect of a sum of causes is equal to the sum of the effects of each of these causes.

## 8. BOLTZMANN SUPERPOSITION PRINCIPLE[14]

Suppose a stress history representing increments of stresses $\Delta\sigma_j$ of various times, $t_j$, where $j = 0, 1, 2, 3, \ldots$ is applied to a sample (see Fig. 8.11a). The strain at any time $t > t_j$ due to the previous application of increments of stress can be considered as the sum of the strains that would be observed at time, t, if each of the increments of stress had been applied independently. That is

$$\epsilon = D(t)\Delta\sigma_0 + D(t - t_1)\Delta\sigma_1 + D(t - t_2)\Delta\sigma_2 \ldots \tag{35}$$

Each term in the right-hand side of Eq. (35) represents the contribution to the strain at any time, t, from a particular stress increment.

Using the superposition principle, it is possible to determine the strain as a function of time when the stress history and the creep compliance function are known. Suppose the stress history as a function of time is given (see Fig. 8.11b), and the stress as a function of time is to be determined. This can be accomplished by determining the stress increments from the slope of the stress-time curve (Fig. 8.11b) and replacing the summation in Eq. (35) by an integration. Then, the strain as a function of time is determined by the integral

$$\epsilon(t) = \int_{-\infty}^{t} D(t - t') \frac{d\sigma}{dt'} \, dt' \tag{36}$$

where $t'$ is past time and $t$ is present time. Likewise, if the strain is given as a function of time and the stress necessary to produce the strain history is to be determined, this can be done by using the integral

$$\sigma(t) = \int_{-\infty}^{t} E(t - t') \frac{d\epsilon}{dt'} \, dt' \tag{37}$$

where $E(t)$ is the relaxation modulus function.

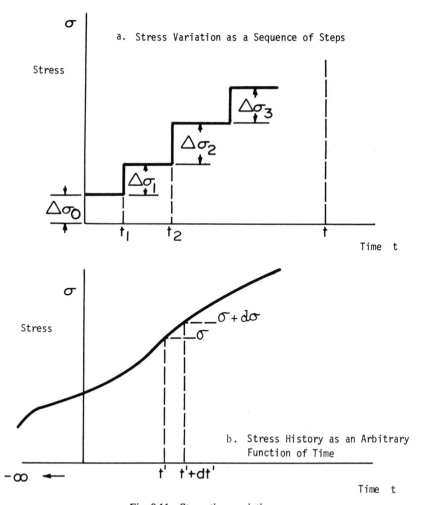

Fig. 8.11   Stress-time variation

## 9. RELATIONSHIP BETWEEN THE RELAXATION MODULUS AND THE COMPLEX DYNAMIC MODULUS

Suppose an oscillating strain history such as represented by Eq. (9) is imposed on a sample. The resulting stress as a function of time can be determined by using Eq. (37), representing the Boltzmann Superposition Principle. Substituting Eq. (9) for $\epsilon$ in Eq. (37), the expression for strain at any time, t, becomes

$$\sigma(t) = \int_{-\infty}^{t} E(t - t') \, \epsilon_0 i\omega e^{i\omega t'} dt' \qquad (38)$$

Calling $(t - t') = \theta$, Eq. (38) can be reduced to

$$\sigma(t) = \epsilon(t) \int_0^{\infty} E(\theta) e^{-i\omega\theta} d\theta$$

or

$$\frac{\sigma(t)}{\epsilon(t)} = E(i\omega) = i\omega \int_0^{\infty} E(\theta) e^{-i\omega\theta} d\theta \qquad (39)$$

Noting $E(i\omega) = E_1(\omega) + iE_2(\omega)$, Eq. (39) becomes

$$E_1(\omega) = \omega \int_0^{\infty} E(\theta) \sin \omega\theta d\theta$$

and

$$E_2(\omega) = \omega \int_0^{\infty} E(\theta) \cos \omega\theta d\theta \qquad (40)$$

Equation (40) indicates that the storage and the loss moduli are the one-sided Fourier sine and cosine transforms, respectively, of the relaxation modulus function E(t).

Similarly, the following relationship between the dynamic compliance functions and the creep compliance function can be established. They are as follows.

$$D(i\omega) = \int_0^{\infty} \frac{dD(\theta)}{d\theta} e^{-i\omega\theta} d\theta$$

$$D_1(\omega) = \int_0^{\infty} \frac{dD(\theta)}{d\theta} \cos \omega\theta d\theta \qquad (41)$$

$$D_2(\omega) = \int_0^{\infty} \frac{dD(\theta)}{d\theta} \sin \omega\theta d\theta$$

## 10. DETERMINATION OF THE RELAXATION SPECTRUM

In the foregoing sections two response functions, the storage and loss moduli (or compliance), were introduced for the description of dynamic mechanical behavior of linear viscoelastic materials. The relationship between these functions and the response functions obtained from creep and relaxation experiments was established. In addition, two new functions, namely the distribution function of relaxation times (relaxation spectrum) $E(\tau)$ and the distribution function of retardation times (retardation spectrum) $D(\lambda)$ were introduced. It must be quite clear that these functions exist as a result of consideration of mechanical models for the description of viscoelastic behavior. Evaluation of distribution functions or spectra for a particular elastomer is known to provide insight into molecular processes responsible for the observed viscoelastic response. As a result the knowledge of spectra makes it possible for chemists to improve or change the chemical condition of a material to provide a particular viscoelastic response needed. Mathematically speaking, the relaxation spectrum or the retardation spectrum can be obtained, in principle, by the inversion of the integrals in Eqs. (32) and (34). Such inversion equations for the relaxation and retardation spectra have been given by Gross.[15] These equations can be used provided analytical expressions for dynamic modulus or compliance functions in terms of frequency can be evaluated from the experimental data. As the experimental data, generally, can be obtained only for a finite range of frequencies and the inversion equations call for analytical expressions for dynamic modulus (or compliance functions) applicable for an infinite frequency range, it is not practicable to use the inversion equations. Therefore, several approximation methods have been developed for the determination of spectra. These approximation methods require, more or less, the knowledge of derivatives of dynamic response functions plotted in terms of log $\omega$. It has been shown[16] that the first-order approximation for the relaxation spectrum is given by

$$[E(\log \tau)] \simeq -\left.\frac{dE_1(\omega)}{d \log (\tau)}\right]_{\tau = 1/w} \tag{42}$$

Equation (42) indicates that the value of the distribution function for $\tau = 1/\omega$ is approximately equal to the negative slope of the storage modulus versus log $(1/\omega)$ curve. Similarly, it has been shown that the zero order approximation for the relaxation spectrum obtained from the loss modulus data is given by

$$[E(\log \tau)] \simeq \frac{4.606}{\pi} E_2(\omega) \tag{43}$$

Comparing Eqs. (42) and (43), it is evident that the loss modulus at any frequency is related to the slope of $E_1(\omega)$ versus log $1/\omega$ curve at that frequency.

It can be shown that the relaxation spectrum can also be obtained to a first approximation from the slope of the stress relaxation modulus versus log t curve. That is

$$[E(\log \tau')] \simeq -\frac{dE(t)}{d \log t} \tag{44}$$

where in this case $\tau'$ is time t. Equations (42) and (44) indicate that for a linear viscoelastic material there exists a simple relationship between dynamic and quasi-static properties.

## 11. TEMPERATURE DEPENDENCE OF DYNAMIC MECHANICAL BEHAVIOR

The examination of dynamic mechanical data for most rubber-like materials and several polymers obtained at various temperatures indicates that the behavior under elevated temperatures corresponds to the dynamic behavior at lower frequencies under room conditions. Likewise, the dynamic mechanical response at low temperatures corresponds to the behavior at higher frequencies observed at room temperature conditions. This observation has led to the so-called time- (or frequency) temperature equivalence principle. By this principle the storage modulus or the loss modulus curves obtained for a certain frequency range and at several temperatures when plotted in terms of log $\omega$ can be displaced by a proper amount to obtain a composite curve for storage modulus and loss modulus referred to a standard temperature for an extended frequency range. The amount by which each curve is to be shifted depends on the temperature difference $(T - T_s)$, where $T_s$ is the standard temperature and T the temperature at which a particular dynamic response curve (storage or loss modulus) is obtained. The time-temperature equivalence principle holds good also in the case of creep and stress relaxation behavior of various elastomers and polymers.[17]

The time-temperature equivalence principle has been found to be very useful for extending either the frequency or the time range of variation of modulus and compliance functions of various rubber-like materials. Although there exist many types of experimental setups for studying the dynamic properties of elastomers and polymers, the frequency range covered by these at a given temperature is limited. This limitation can be overcome by obtaining the data for a limited frequency range at various constant temperatures and reducing the data to a reference temperature. Several reduction schemes for constructing composite curves have been proposed.[18-20] These depend on the manner in which the shifting of curves is performed and the way the expression for the shift factor (representing the exact amount the curves need to be displaced) is arrived at. However, the reduction scheme can be expressed in general terms, by the following equations for the storage and loss moduli.[18]

$$\frac{T_s \rho_s}{T \rho} E_1(\omega, T) = E_1(\omega_p, T_s)$$

$$\frac{T_s \rho_s}{T \rho} E_2(\omega, T) = E_2(\omega_p, T_s) \tag{45}$$

and

$$\omega_p = \omega a_T$$

where T is the temperature at which the $E_1(\omega)$ and $E_2(\omega)$ curves are known;

$T_S$ is the standard temperature to which the data for temperature T is to be reduced; $\omega_p$ is reduced frequency, $a_T$ is the shift factor, a function of temperature alone and $\rho_S$, $\rho$ are the densities at temperatures $T_S$ and T, respectively. Equation (45) indicates that to reduce the data at temperature T to correspond to the data at $T_S$, the modulus values at T are multiplied by a factor $(T_S\rho_S/T\rho)$. This multiplication factor represents a correction to take into account the effect of temperature on modulus values that follows from the kinetic theory of rubber elasticity. The reduced modulus curves are then shifted along the log $\omega$ scale to give composite curves for the storage and loss moduli covering many decades of frequency. This is affected by shifting all curves, for storage modulus for instance, one at a time with respect to the reference curve at temperature $T_S$ until portions of curves superimpose to give a composite curve. The same procedure is followed for the composite curve for the loss modulus. The shift factor $a_T$, representing the amount each modulus curve is shifted along the low $\omega$ axis, is a monotonically decreasing function of temperature as shown in Fig. 8.12. Williams, Landel and Ferry[6] have found that the temperature dependence of the viscoelastic behavior of most elastomers in a range of temperature $T_g$ to $T_g + 100°C$, where $T_g$ = glass transaction temperature, can be represented by the following expression for the shift function.[6]

$$\log a_T = -\frac{C_1^g(T - T_g)}{C_2^g + (T - T_g)} \tag{46}$$

where $C_1^g$, $C_2^g$ are universal constants with values 17.44 and 51.6, respectively.
The time-temperature equivalence principle has been used to extend the range

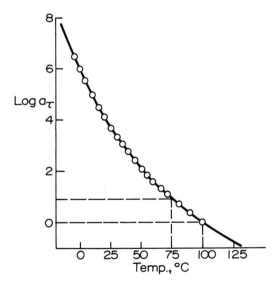

Fig. 8.12   Temperature dependence of the shift factor $a_T$ (after Ferry[6])

of variation of various response functions to twenty decades of frequency or time. The principle has been applied to most rubber-like materials, although there are exceptions.

## 12. ELECTRO-MECHANICAL ANALOGY AND MECHANICAL IMPEDANCE

It is well known that there exists a close mathematical similarity between the mechanical response of a viscoelastic material and the electrical response of a two-terminal network. The correspondence between various electrical and mechanical quantities is shown in Table 13. Table 13 indicates that the series coupling in a mechanical model must be equivalent to parallel coupling in an electrical network to achieve mathematical identity. Similarly, parallel coupling in the mechanical model must correspond to series coupling in the electrical network.[21] Using the electro-mechanical analogy, it is possible to develop the concept of impedance which has been used in the analysis of electrical circuits to describe the dynamic mechanical behavior of viscoelastic materials.

TABLE 13
CORRESPONDENCE BETWEEN ELECTRICAL AND
MECHANICAL QUANTITIES

| | |
|---|---|
| Quantity of Electricity (Q) | Strain $\epsilon$ |
| Electro-Motive Force (E) | Stress $\sigma$ |
| Current (I) | Strain Rate ($\dot{\epsilon}$) |
| Resistance (R) | Viscosity ($\eta$) |
| Capacitance (C) | Compliance (D) |
| Parallel Coupling | Series Coupling |
| Series Coupling | Parallel Coupling |

Suppose a sinusoidal voltage is applied to an electric circuit consisting of resistances and capacitances. The current generated will also be sinusoidal but will lead the voltage by a phase angle $\phi$, as indicated by the following equations.

$$\bar{E} = R_e[\bar{E}_0 e^{i\omega t}] \tag{47}$$

and

$$I = R_e[I_0 e^{i(\omega t + \phi)}] \tag{48}$$

where $R_e$ represents the real part of the complex function. The electrical impedance is defined as the ratio of voltage to current. That is given by using Eqs. (47) and (48).

$$Z_e(i\omega) = \frac{\bar{E}_0}{\bar{I}_0} \frac{e^{i\omega t}}{e^{i(\omega t + \phi)}} \tag{49}$$

where $Z_e(i\omega)$ is the electrical impedance, a complex quantity.

The mechanical impedance, which is analogous to the electrical impedance, is also a function of frequency. Using Eq. (49) and the correspondence between electrical and mechanical quantities (Table 13) it is possible to write the following expression for the mechanical impedance.

$$Z(i\omega) = \frac{\sigma_0\, e^{i\omega t}}{\epsilon_0\, e^{i(\omega t + \phi)}} = Z_1(\omega) - i\, Z_2(\omega) \tag{50}$$

where $Z(i\omega)$ is the mechanical impedance, $Z_1(\omega)$ the real part of the complex impedance and $Z_2(\omega)$ the imaginary part of the complex impedance. The real part of the complex impedance represents the ratio of stress in phase with the rate of strain to the rate of strain itself. It can be shown to be equal to $E_2(\omega)/\omega$. Likewise, the imaginary part of the complex impedance is equal to the ratio of the stress component 90 degrees out of phase with the rate of strain to the rate of strain. The imaginary part of the complex impedance can be shown to be equal to $E_1(\omega)/\omega$.

The electro-mechanical analogy described earlier can be used to analyze the steady-state response of various mechanical models, consisting of springs and dashpots, to sinusoidal loading. Assume the dynamic mechanical behavior of a rubber-like material is adequately represented by the three-element model representation (Fig. 8.6a). By the electro-mechanical analogy, the equivalent electric circuit can be constructed (see Fig. 8.13).

By using the elementary rules for alternating current networks, the impedance of the circuit in Fig. 8.12 can be shown to be

$$Z_e(i\omega) = \left(\frac{\bar{E}}{\bar{I}}\right) = \frac{1}{iC_1\omega} + \frac{1}{\dfrac{1}{r} + iC_2\omega} \tag{51}$$

Noting $Z_e(i\omega) = \bar{R} - i\bar{X}$, Eq. (51) can be separated into real and imaginary parts as given by

$$\bar{R} = \frac{r}{(1 + r^2 C_2{}^2 \omega^2)}$$

and $\tag{52}$

$$\bar{X} = \frac{1 + r^2 \omega^2 C_2 (C_1 + C_2)}{c\,\omega (1 + r^2 C_2{}^2 \omega^2)}$$

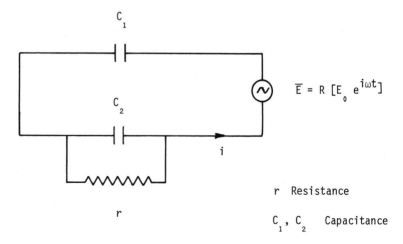

$$\overline{E} = R\,[E_0\,e^{i\omega t}]$$

r   Resistance

$C_1$, $C_2$   Capacitance

Fig. 8.13   Equivalent electric circuit for a three-element mechanical model

where $\overline{R}$ is the total resistance in the network and $\overline{X}$ is the total reactance in the network.

Making use of the correspondence between the electrical constants in the network and the mechanical constants in the model (see Table 13) the following relations hold.

$$C_1 \longrightarrow \frac{1}{E_e} \qquad\qquad \overline{R} \longrightarrow \frac{E_2(\omega)}{\omega}$$

$$C_2 \longrightarrow \frac{1}{E} \qquad\qquad \overline{X} \longrightarrow \frac{E_1(\omega)}{\omega}$$

$$r \longrightarrow \eta$$

Substituting the above into the correspondence Eq. (52), the expressions for the storage and the loss modulus are

$$\frac{E_1(\omega)}{\omega} = \frac{\left[1 + \dfrac{\eta^2\omega^2}{E}\left(\dfrac{1}{E_e} + \dfrac{1}{E}\right)\right]E_e}{\omega\left(1 + \dfrac{\eta^2\omega^2}{E^2}\right)} \tag{53}$$

and

$$\frac{E_2(\omega)}{\omega} = \frac{\eta}{\left(1 + \dfrac{\eta^2\omega^2}{E^2}\right)} \tag{54}$$

Equations (53) and (54) can be shown to be equivalent to Eqs. 12 and 13 after substituting $\tau = \eta/E$ and $\lambda = \eta(1/E_e + 1/E)$.

Previously, the mechanical impedance was defined in terms of the steady-state response of the material to sinusoidal loading [Eq. (50)]. In the case of a general dynamic stress or strain history the mechanical impedance function can be expressed in terms of a Fourier transform as

$$Z(i\omega) = \frac{F[\sigma(t)]}{F[\epsilon(t)]}$$ (55)

where $F[\sigma]$ represents the Fourier transform

$$F[\sigma(t)] = \int_{-\infty}^{\infty} \sigma(t)e^{-i\omega t}dt$$ (56)

## 13. BASIC TYPES OF DEFORMATION AND DYNAMIC MECHANICAL PROPERTIES

The dynamic behavior of viscoelastic materials depends upon the type of deformation imposed on them. There are two basic types of deformation that may be considered in studying the dynamic mechanical properties of rubber. These are the shear deformation and the volumetric (dilatational) deformation. The shear deformation involves distortion (change in shape) without any change in volume; the volumetric deformation involves the change in volume without any distortion. Any type of deformation, including the tensile or compressive deformations, may be derived from a combination of shear and volumetric deformations. In the foregoing sections the dynamic behavior of rubber-like materials was discussed mainly in terms of tensile modulus or compliance functions. The same discussions apply equally well in the cases of shear and volumetric deformations. However, it is necessary to emphasize that viscoelastic behavior is more pronounced in shear deformation than in the case of tensile or volumetric deformation. This is particularly important in the case of rubber as it is used primarily for vibration isolation and control.

As in the case of tensile deformation, the dynamic modulus or compliance functions in shear and volumetric deformation can be specified.[13] They are

Shear deformation—

$$G(i\omega) = G_1(\omega) + i\, G_2(\omega)$$

$$J(i\omega) = J_1(\omega) - i\, J_2(\omega)$$

$$G(i\omega) = \frac{1}{J(i\omega)}$$ (57)

$$|G(i\omega)| = \sqrt{[G_1(\omega)]^2 + [G_2(\omega)]^2}$$

$$|J(i\omega)| = \sqrt{[J_1(\omega)]^2 + [J_2(\omega)]^2}$$

$$\tan \delta_8 = \frac{G_2(\omega)}{G_1(\omega)} \text{ or } \frac{J_2(\omega)}{J_1(\omega)}$$

where $G_1(\omega)$, $G_2(\omega)$ are the storage and loss moduli in shear, $J_1(\omega)$, $J_2(\omega)$ the storage and loss compliances in shear, $G(i\omega)$ the complex dynamic modulus in shear, $J(i\omega)$ the complex dynamic compliance in shear; $\delta_s$ the phase angle between shear stress and shear strain and $|G(i\omega)|$ represents the absolute value.

Volumetric deformation—

$$K(i\omega) = K(i\omega) + iK_2(\omega)$$

$$B(i\omega) = B_1(\omega) - iB_2(\omega)$$

$$K(i\omega) = \frac{1}{B(i\omega)}$$

$$|K(i\omega)| = \sqrt{[K_1(\omega)]^2 + [K_2(\omega)]^2}$$  (58)

$$|B(i\omega)| = \sqrt{[B_1(\omega)]^2 + [B_2(\omega)]^2}$$

$$\tan \delta_V = \frac{K_2(\omega)}{K_1(\omega)} \text{ or } \frac{B_2(\omega)}{B_1(\omega)}$$

where $K_1(\omega)$, $K_2(\omega)$ are the storage and loss moduli in volumetric deformation, $B_1(\omega)$, $B_2(\omega)$ the storage and loss compliances in volumetric deformation, $K(i\omega)$ the bulk complex dynamic modulus, $B(i\omega)$ the bulk complex dynamic compliance and $\delta_V$ the phase angle between volumetric strain and hydrostatic stress. The dynamic compliance functions for tensile deformation can be expressed in terms of the dynamic compliance functions for shear and volumetric deformations, as follows.

$$D(i\omega) = \frac{1}{3} J(i\omega) + \frac{1}{9} B(i\omega)$$

$$D_1(\omega) = \frac{1}{3} J_1(\omega) + \frac{1}{9} B_1(\omega)$$  (59)

$$D_2(\omega) = \frac{1}{3} J_2(\omega) + \frac{1}{9} B_2(\omega)$$

Rubber and most rubber-like materials are incompressible under room temperature conditions. As a result $B_1(\omega)$ and $B_2(\omega)$ become zero and dynamic compliances in tension and shear are simply related as

$$D(i\omega) = \frac{1}{3} J(i\omega)$$

$$D_1(\omega) = \frac{1}{3} J_1(\omega)$$  (60)

$$D_2(\omega) = \frac{1}{3} J_2(\omega)$$

## 14. RELATIONS BETWEEN THE DYNAMIC MODULUS AND MEASURED QUANTITIES

There are several experimental techniques to measure the dynamical properties of rubber-like materials. Each experimental setup yields different measured quantities. The experimental data obtained by different methods needs to be reduced to either dynamic modulus or compliance functions. Various experimental techniques used to measure the dynamic mechanical properties of rubber-like materials will be covered in a later section. However, the interrelationship between the dynamic modulus or compliance functions and the experimentally measured quantities is established below.

**Logarithmic Decrement.** The logarithmic decrement is defined as

$$\Delta = \log_e \frac{X_1}{X_2} \tag{61}$$

where $X_1$ and $X_2$ are the successive amplitudes in a free vibration experiment (see Fig. 8.14). In a free vibration experiment the logarithmic decrement and the frequency of oscillation $\omega_f$ are measured. These experimental quantities can be related to the dynamic properties of the material by means of the following equations.

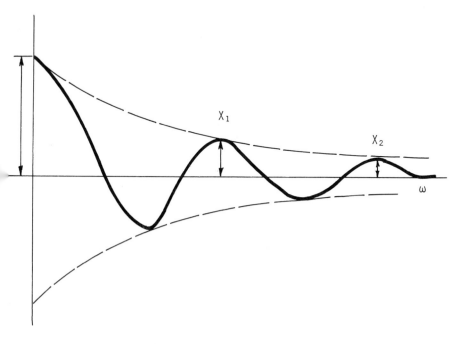

Fig. 8.14  Free-damped vibration

$$E_1(\omega_f) = M\omega_f{}^2\left(1 + \frac{\Delta^2}{4\pi^2}\right) \tag{62}$$

$$E_2(\omega_f) = M\omega_f{}^2\frac{\Delta}{\pi} \tag{63}$$

$$\tan\delta = \frac{E_2(\omega_f)}{E_1(\omega_f)} = \frac{\Delta/\pi}{1 + \Delta^2/4\pi^2} \tag{64}$$

where $\omega_f$ is circular frequency of oscillation in a free vibration experiment and M is mass of the vibrating system. When the damping (internal friction) in the material is small, $\Delta$ is small compared to 1. Equation (64) may then be simplified to

$$\frac{E_2}{E_1} = \frac{\Delta}{\pi} \tag{65}$$

**Bandwidth.** Suppose the resonance curve (see Fig. 8.15) representing the amplitude ratio plotted against frequency is obtained from a forced vibration experiment. From the frequency corresponding to a particular resonance and the half-width of the half-power width, it is possible to obtain the expressions for the storage and loss modulus. They can be shown to be[22]

$$E_1(\omega) = M\omega_r{}^2$$

$$\frac{E_2(\omega)}{E_1(\omega)} \simeq \frac{1}{\sqrt{3}}\frac{\Delta\omega}{\omega r} \tag{66}$$

or

$$\frac{E_2(\omega)}{E_1(\omega)} \simeq \frac{\Delta\omega'}{\omega_r}$$

where $\omega_r$ is the resonant frequency. Further

$$\frac{\Delta\omega}{\omega_r} = \frac{\omega_2 - \omega_1}{\omega_r}$$

where $\omega_1$ and $\omega_2$ are the frequencies at which the amplitude is half the maximum amplitude at resonance. And

$$\frac{\Delta\omega'}{\omega_r} = \frac{\omega_2' - \omega_1'}{\omega_r}$$

where $\omega_1'$ and $\omega_2'$ are the frequencies at which the amplitude is $1/\sqrt{2}$ or 0.707 of the maximum amplitude.

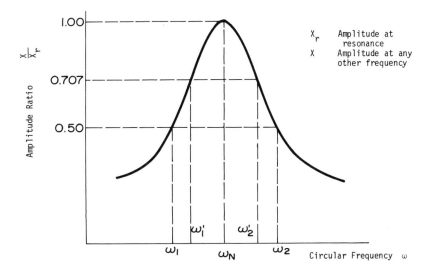

Fig. 8.15 Vibration-amplitude ratio versus circular frequency curve

**Quality Factor $Q^{-1}$.** The quality factor represents a parameter analogous to that used in an electric circuit and indicates the sharpness of a resonance peak and amplification produced by the resonance.[23] The quality factor is equal to the loss tangent, that is

$$\frac{1}{Q} = \frac{E_2(\omega)}{E_1(\omega)} = \tan \delta \qquad (67)$$

**Specific Damping Capacity.** The specific damping capacity is defined as the ratio of energy dissipated per cycle to the maximum strain energy stored per cycle expressed as

$$\psi = \frac{\Delta W}{W} \qquad (68)$$

where $\Delta W$ is the energy dissipated per cycle in a free or forced vibration experiment and $W$ is the maximum strain energy stored during the cycle. Consider a specimen of rubber-like material subjected to a uniaxial oscillating stress represented by

$$\sigma = \sigma_0 \sin \omega t \qquad (69)$$

The resulting oscillating is given by

$$\epsilon = \epsilon_0 \sin(\omega t - \delta) \qquad (70)$$

The energy loss in the material during a complete cycle can be evaluated from the following integral that includes Eqs. (69) and (70).

$$\Delta W = \int_0^{2\pi/\omega} \sigma \frac{d\epsilon}{dt} dt = \int_0^{2\pi} \sigma_0 \epsilon_0 \sin \omega t \cos(\omega t - \delta) \, d(\omega t) \tag{71}$$

Solving the above integral, the expression for energy loss becomes

$$\Delta W = \pi \, \sigma_0 \epsilon_0 \sin \delta = \pi \, E_2(\omega) \epsilon_0{}^2 = \pi D_2(\omega) \sigma_0{}^2 \tag{72}$$

where $\sigma_0 \sin \delta = \epsilon_0 \, E_1(\omega)$ and $\epsilon_0 \sin \delta = \sigma_0 \, D_1(\omega)$. The maximum strain energy stored in the specimen during a complete cycle can be shown to be

$$W = \frac{1}{2} E_1(\omega) \epsilon_0{}^2 \tag{73}$$

From Eqs. (72) and (73) the specific damping capacity is

$$\psi = \frac{\Delta W}{W} = 2\pi \cdot \frac{E_2(\omega)}{E_1(\omega)} = 2\pi \tan \delta \tag{74}$$

**Dynamic Resilience.** The dynamic resilience R is usually defined as the ratio of the energies stored in two cycles of a free damped vibration experiment. Since the energy stored is proportional to the square of amplitude, the expression for the dynamic resilience can be given as

$$R = \frac{X_1{}^2}{X_0{}^2} \tag{75}$$

where $X_1$, $X_0$ are successive amplitudes (see Fig. 8.14). Noting $\Delta = \log_e X_1/X_0 = [E_2(\omega)/E_1(\omega)]$, the dynamic resilience, the expression for the dynamic resilience finally becomes

$$R = e^{-2\Delta} = e^{-[2\pi E_2(\omega)/E_1(\omega)]} \tag{76}$$

If the energy loss is small for a material the dynamic resilience expression simplifies to

$$R = (1 - 2\Delta) \tag{77}$$

## 15. DISSIPATION OF ENERGY AND HEAT GENERATION IN RAPID OSCILLATING DEFORMATION

Rubber-like materials dissipate energy when subjected to rapid cyclic deformation, causing evolution of heat. In the case of small sinusoidal deformation, the rate at which the energy is being dissipated can be shown to be

$$U = \omega \, E_2(\omega) \, \frac{\epsilon_0{}^2}{2} = \omega \, D_2(\omega) \, \frac{\sigma_0{}^2}{2} \qquad (78)$$

where U is energy dissipated per unit volume per sec in sinusoidal uniaxial deformation and $\epsilon_0$, $\sigma_0$ are the amplitudes of strain and stress, respectively.

The temperature rise caused by energy dissipation may reach a steady value for continuous sinusoidal deformation, depending on the heat transfer to the surroundings. Suppose that natural rubber with $J_2(\omega) = 1.0 \times 10^{-8}$ cm$^2$/dyne was subjected to a cyclic deformation at 10 cycles/sec with an amplitude of shear stress of $10^6$ dynes/cm$^2$. The rate of energy dissipation for this material can be calculated by Eq. (78) to be 0.0038 calorie/cc./sec. Noting the heat capacity of the material is about 0.5 calorie/deg/gm, the temperature rise is about 0.008 deg/sec if no heat loss takes place due to conduction.

## 16. RELATIONSHIP OF DYNAMIC MECHANICAL PROPERTIES TO TRANSMISSIBILITY

Rubber and rubber-like materials have been extensively used to reduce and control vibration in machine components under periodic or transient loading. It is a common practice to use anti-vibration mountings to reduce the transmission of energy from a vibrating source to the system to be isolated. The quantity that determines the effectiveness of vibration isolation is usually called the transmissibility T, and it is defined as the ratio of the amplitude at the output of the isolator to the imposed amplitude at the input. The transmissibility is also defined as the ratio of transmitted force to the externally imposed force. Considering a one-dimensional vibrating system (see Fig. 8.16) in which a mass M is subjected to sinusoidal oscillations of frequency $\omega$ and isolated by a rubber isolator R from the massive foundation F, the following expression for the transmissibility can be obtained.[24,25]

$$T = \frac{F_2}{F_1} = \left[ \frac{1 + (\tan \delta)^2}{\left(1 - \frac{\omega^2 M}{K E_1(\omega)}\right)^2 + (\tan \delta)^2} \right]^{1/2} \qquad (79)$$

where K is a constant depending on the geometry of the mounting and $F_1$ and $F_2$ are the amplitudes of the imposed and the transmitted forces, respectively. Resonance will occur in the vibrating system when $\omega^2 M / K E_1(\omega) = 1$, that is, when T reaches a maximum value. In Fig. 8.17 the variation of T with log $\omega$ is plotted. Calling the resonant frequency $\omega_r$, the ratio M/K can be expressed in terms of the storage modulus and resonant frequency as follows.

$$\frac{M}{K} = \frac{E_1(\omega_r)}{\omega_r{}^2} \qquad (80)$$

where $E_1(\omega_r)$ represents the value of the storage modulus at a resonant frequency $\omega_r$.

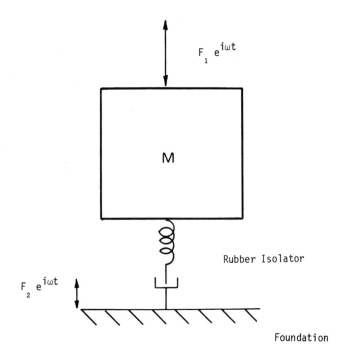

Fig. 8.16    Forced vibration of a mass with rubber isolator

Using Eq. (80), the expression for the transmissibility can be rewritten as

$$\bar{T} = \frac{1 + (\tan \delta)^2}{\{1 - r_f^2[E_1(\omega_r)/E_1(\omega)]\}^2 + (\tan \delta)^2} \qquad (81)$$

where $r_f$ is the frequency ratio $(\omega/\omega_r)$.

In order to illustrate the relative importance of the loss tangent (representing damping) and the storage modulus $E_1(\omega)$ (representing stiffness) or rubber-like materials in transmissibility, four hypothetical cases of a variation of tan $\delta$ and $E_1(\omega)$ are considered and the transmissibility variation with the frequency ratio is shown plotted in Fig. 8.18. Figure 8.18 indicates that only vibration beyond the resonant frequency $\omega_r$ of the mounting system can be isolated. In order to make the resonant frequency $\omega_r$ smaller, a rubber-like material with a low value of $E_1(\omega)$ may be selected for a given geometry of mounting. However, considerations such as space and load bearing capacity set a lower limit to $E_1(\omega)$. An examination of transmissibility curves in Fig. 8.18 shows that rubber-like materials with minimum rate of change of the storage modulus with frequency provide a maximum amount of vibration isolation beyond resonance. Although rubber-like materials with high damping reduce amplification at resonance to a great extent, they are bad vibration isolators at higher frequencies.

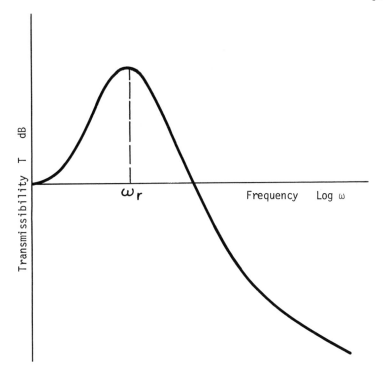

Fig. 8.17 Variation of transmissibility with frequency

It is clear from the previous discussion that the performance of rubber-like materials in vibration isolation and control applications depends on their dynamic mechanical properties. As the dynamic mechanical properties of a particular material are generally affected by temperature in addition to the frequency range, the operating temperature of a vibration isolator is of paramount importance in the selection of particular material for vibration control applications. To discuss the relative importance of various elastomers in vibration isolation applications, the transmissibility curves for Neoprene W, Nordel, Hevea, polyisobutylene and SBR are shown in Fig. 8.19. These curves were originally computed by Yin and Pariser[26] at 25°C (77°F) from reduced dynamical data using Eq. (81). While arriving at these curves, the resonant frequencies were chosen as 200 cycles/sec for all systems. Figure 8.19 indicates that, except for polyisobutylene, all the materials have approximately the same isolation characteristics. However, if the temperature is changed to −18°C (0°F), the performance of various materials differs appreciably, as is shown in Fig. 8.20. This spread in the transmissibility curves may be attributed to different temperature effects on the storage modulus of materials. It is interesting to note from Fig. 8.20 that Hevea rubber and Nordel have better isolation characteristics at this temperature than the other materials considered.

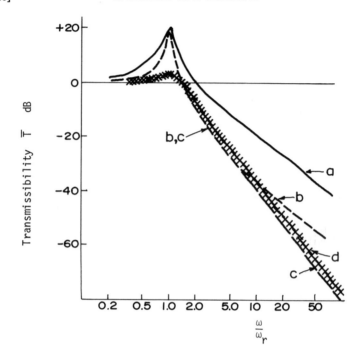

Fig. 8.18   Four illustrative Transmissibility (T) curves versus normalized frequency $(\omega/\omega_r)$ where T (db) $= 20 \log$ (fraction of force transmitted): (a) Tan $\delta = 0.1$, E′ $\doteq \omega$; (b) E′ constant, tan $\delta \doteq \omega$ or 0.1 at resonance frequency $\omega_r$; (c) E′ constant, Tan $\delta = 0.1$; (d) E′ constant, Tan $\delta = 1.0$ (after Yin and Pariser [26])

As opposed to vibration isolation, a rubber-like material may be chosen for damping vibration. The criterion for selection of an adequate material for the purpose is that the material must have a high loss tangent in the temperature and the frequency region of interest. For the effective selection of rubber-like materials as vibration dampers in a dynamic system operating in a narrow frequency range and high temperature range, it is very essential to know how the storage modulus and the loss tangent for various materials vary with temperature. Figures 8.21 and 8.22 show such curves plotted at an arbitrarily chosen frequency.

## 17. METHODS OF MEASURING DYNAMIC MECHANICAL PROPERTIES

To describe the dynamic mechanical behavior of rubber for an extended frequency scale and temperature, several widely different types of measurements are necessary. Although most of the experimental techniques customarily employed involve sinusoidal excitation of the material, at times nonsinusoidal periodic changes are imposed on the material to obtain the dynamic mechanical

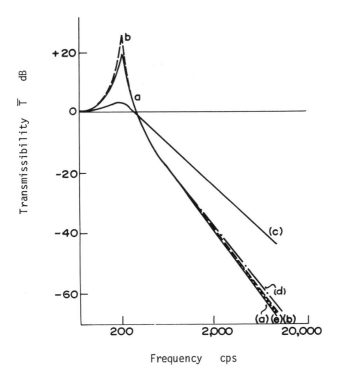

Fig. 8.19  Computed T curves for (a) Nordel, (b) Hevea rubber, (c) PIB, (d) SBR, (e) Neoprene W at 25° C (77° F) with resonance frequencies arbitrarily chosen as 200 cps (after Yin and Pariser [26])

properties. Whatever technique is used, the purpose of each test is to obtain sufficient information that will lead to the evaluation of the storage modulus (or compliance) and the loss modulus (or compliance) for an extended frequency range. In order to be able to obtain the two response functions, all dynamic experimental methods must lead to the measurement of at least two quantities.

Various dynamic experimental techniques are classified according to the relationship that exists between the geometric dimensions of the sample and the wavelength of the stress waves propagated at the frequency of measurement. If the critical dimension of the specimen $l$ (for tensile deformation the length in the direction of stress, and for shear deformation the dimension perpendicular to the plane of shear) is small compared to the stress wavelength

$$\lambda = \frac{2\pi}{\omega} \left( \frac{G_1}{\rho} \right)^{1/2} \qquad \text{in shear}$$

or

$$\lambda = \frac{2\pi}{\omega} \left( \frac{E_1}{\rho} \right)^{1/2} \qquad \text{in tension}$$

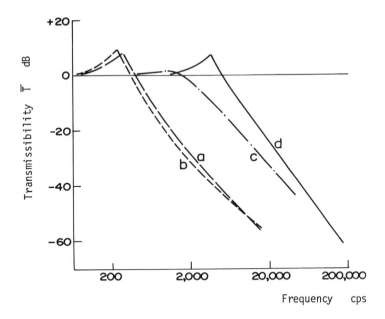

Fig. 8.20   Performance at—18° C (0° F) predicted for the same systems as
in Fig. 8.19 (after Yin and Pariser [26])

where $G_1$ and $E_1$ are the shear storage and tensile storage moduli, respectively;
the inertia of the sample can be disregarded. If the critical length of the specimen is comparable to the stress wavelength, a standing wave phenomenon is
observed and the specimen vibrates with various modes, depending on the fre-

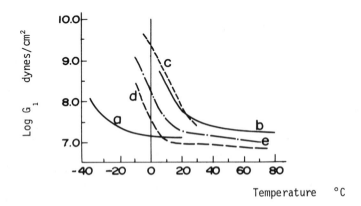

Fig. 8.21   Elastic shear modulus ($G_1$) at 100 cps versus temperature for:
(a) Nordel, (b) Viton A, (c) Neoprene ILA, (d) Neoprene W, (e) Hypalon 20
(after Yin and Pariser [26])

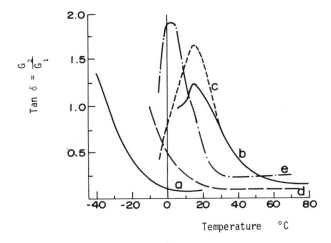

Fig. 8.22   Mechanical loss tangent (tan δ) at 100 cps versus temperature
for the same systems as in Fig. 8.21 (after Yin and Pariser [26])

quency of excitation. However, when the critical length of the specimen is large
compared to the stress wavelength, viscoelastic waves are generated in the medium and the inertia of the medium influences the dynamic mechanical behavior
appreciably.

Based on the above criterion, the testing techniques for studying the dynamic
mechanical properties of rubber-like materials may be broadly classified as
follows.

1. The direct measurement of stress and strain method.

2. A method involving the free and forced vibration of a dynamic system
with added inertia.

3. Free and forced vibration of specimens without any added inertia.

4. Wave propagation method—in this category there are two kinds, namely,
the continuous wave method and the pulse propagation method.

In the direct method of measurement, the specimen size is small compared to
the wavelength of the stress waves generated. Therefore, the inertia of the sample
can be neglected. The specimen is subjected to an oscillating (sinusoidal) stress
and the resulting oscillating strain which lags behind the stress is measured.
From the ratio of amplitudes of strain to stress and the measured phase angle,
the storage compliance and the loss compliance corresponding to the chosen
frequency of the oscillating stress are evaluated. By conducting the same test
at various frequencies of operating stress, it is possible to evaluate the storage
and loss compliance functions of the test material. Instead of a measurement of
phase angle and the ratio of amplitudes of strain and stress, sometimes a hysteresis loop is traced. From the area of the hysteresis loop and its orientation to

the stress and strain axes the dynamic properties may be evaluated. The frequency range covered by the above experimental technique is small.

In the free vibration of a dynamic system a specimen, whose dimensions are small compared to the stress waves that may be generated at the selected frequency of the vibrating system, is chosen and is connected in series with a mass M. The specimen acts as a spring and dashpot of a one-degree-of-freedom vibrating system with the mass providing the inertia. A slight disturbance of an otherwise equilibrium system will lead to the system executing free vibration. By measuring the frequency at which the oscillations occur and the ratio of successive amplitudes of oscillations the dynamic properties of the material under investigation can be obtained. By varying either the dimensions of the specimen or changing the mass M it is possible to allow the system to execute free vibrations at various frequencies. The frequency range covered in these experiments is small as in the previous method, but the range of frequencies is higher than that achieved by the previous method.

The forced vibration method uses essentially the same experimental procedure as the free vibration method, but with certain differences. The dynamic system, consisting of the specimen in series with a mass M, is subjected to a periodic stress or strain at various selected frequencies. The response of the dynamic system is observed under these conditions. The ratio of the stress amplitude to the strain amplitude and the phase angle are measured, which leads to the determination of the dynamic properties corresponding to a selected frequency of fluctuation. A variation of this method involves the continual variation of frequency until the system experiences resonance. From the amplitude ratio curve at resonance and the frequency associated with resonance, storage and loss moduli values corresponding to the resonant frequencies can be evaluated. Varying the size of the sample and the mass allows the variation in resonant frequencies, leading to the determination of dynamic properties at various discrete values of frequencies. Limitations in the sizes of mass and sample do not allow the determination of dynamic properties for an extended frequency range. However, this method allows for the measurement of dynamic properties for a larger frequency range than those described earlier.

Suppose the size of the sample is comparable to the generated stress wavelength, and the sample is subjected to either an impulse or a steady fluctuating stress. The specimen would execute various vibration modes, depending on the frequency at which the oscillating system vibrates under free vibration conditions or the frequency selected under steady fluctuating stress. Either way, the inertia of the sample needs to be considered, By this method it is possible to evaluate the dynamic properties of the test material for an extended frequency scale. The measurement of dynamic properties by this technique closely follows that of the free and forced vibration method considered earlier.

If the sample size chosen is large compared to the stress wavelength propagated, viscoelastic waves are generated in the medium. These waves would be attenuated completely if the length of the specimen were large. By measuring the velocity and attenuation of the wave, the dynamic properties of any material can be determined.

In any experimental investigation for the measurement of dynamic properties, the following three steps need to be considered.

1. Selection of a proper experimental technique and the dimensions of the sample to determine the dynamic mechanical properties for a certain range of frequencies.

2. Mathematical formulation of the dynamic system consisting of either the specimen and added inertia or the specimen alone.

3. Evaluation of dynamic mechanical properties from the observed data.

Steps 2 and 3 will be considered subsequently, followed by a brief description of various experimental techniques.

## 18. MATHEMATICAL FORMULATION OF DYNAMIC SYSTEMS

We present first the formulation for cases where the characteristic length of a specimen is much smaller than the stress wavelength.

**Case** (i). In this case the inertia of the sample is neglected. The specimen is subjected to a sinusoidal stress or strain. The ratio of amplitude of stress to strain and the phase angle between stress and strain at a certain frequency lead to the determination of dynamic properties. This method is suitable for dynamic property measurements at low frequencies up to 100 cps.

Suppose a sinusoidal fluctuating stress is induced in the specimen shown in Fig. 8.23a. Then the differential equation for the dynamic system is given by

$$\left(\frac{E_2}{\omega}\right) \frac{d\epsilon}{dt} + E_1\epsilon = \sigma_0 \cos \omega t \tag{82}$$

where $E_1$, $E_2$ are the storage and loss moduli in tension and $\epsilon$ is uniaxial strain at any time t. Hence

$$\epsilon = \frac{\sigma_0 \cos(\omega t - \delta)}{(E_1{}^2 + E_2{}^2)^{1/2}} \tag{83}$$

where $\delta = \tan^{-1}(E_2/E_1)$, the loss angle. Equation (83) and the fluctuating stress distribution defines the hysteresis loop shown in Fig. 8.23b.

From Eq. (83) $\epsilon = 0$, where

$$(\omega t - \delta) = (2n + 1) \frac{\pi}{2} \tag{84}$$

or

$$\omega t = (2n + 1) \frac{\pi}{2} + \delta \tag{85}$$

and where n = 1, 2, 3 . . . . Stress corresponding to Eq. (85) is

$$\sigma_1 = \pm\sigma_0 \sin \delta \tag{86}$$

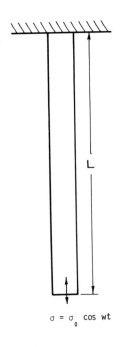

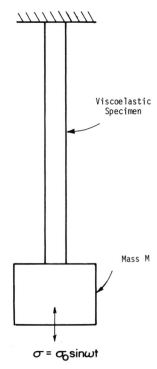

a.  A Specimen of a Viscoelastic Material
    Subjected to Oscillatory Stress

b.  Longitudinal Vibration of
    Viscoelastic Material

Fig. 8.23   Oscillitory deformation of a viscoelastic material

A maximum strain of

$$\pm\epsilon_2 = \frac{\sigma_0}{(E_1{}^2 + E_2{}^2)^{1/2}} \qquad (87)$$

occurs at $(\omega t - \delta) = n\pi$, or

$$\delta = \omega t + n\pi \qquad (88)$$

The stress corresponding to Eq. (87) is

$$\sigma_2 = \sigma_0 \cos(\omega t + n\pi) = \pm\sigma_0 \cos \delta \qquad (89)$$

where the loss angle

$$\delta = \tan^{-1}\left(\frac{\sigma_1}{\sigma_2}\right) = \tan^{-1}\frac{E_2}{E_1} \qquad (90)$$

Equation (90) shows that the ratio (OA/EC) in Fig. 8.23b gives the loss angle. Using Eqs. (87) and (90) it is possible to calculate $E_2$ and $E_1$ for the material at any selected frequency.

**Case (ii).** In this case also the inertia of the sample is neglected, but an external mass M is provided to account for the inertia of the vibrating system. The system is given a disturbance by a sudden application and removal of the load in an axial direction (see Fig. 8.24), and the resulting motion of the system is observed.

The differential equation of motion for the system can be given as

$$M \frac{d^2\epsilon}{dt^2} + \frac{E_2}{\omega_f} \frac{d\epsilon}{dt} + E_1\epsilon = 0 \tag{91}$$

The solution of Eq. (91) is

$$\epsilon = \epsilon_0 e^{-[E_2 t / 2\omega_f M]} \cos(\omega_f t + \beta) \tag{92}$$

where

$$\omega_f^2 = \frac{E_1}{M} - \left(\frac{E_2}{2\omega_f M}\right)^2 \tag{93}$$

is the frequency of free vibration and $\epsilon_0$ and $\beta$ are initial boundary conditions. Equation (92) represents damped oscillation of the system as shown in Fig. 8.14.

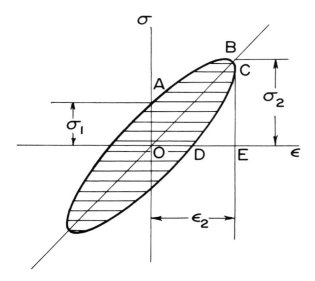

Fig. 8.24 A hysteresis loop in a sinusoidal experiment

The logarithmic decrement can be determined from successive amplitudes as follows.

$$\Delta = \ln\left(\frac{\epsilon_1}{\epsilon_2}\right) = \frac{\pi E_2}{M\omega^2} = \pi\left(\frac{E_2}{E_1}\right) \tag{94}$$

Using Eqs. (93) and (94) it can be shown that

$$\frac{E_1}{M} = \omega_f^2\left(1 + \frac{\Delta^2}{4\pi^2}\right) \tag{95}$$

When the damping is small, $\Delta^2/4\pi^2$, being much smaller than 1, Eq. (95) then leads to

$$\frac{E_1}{M} = \omega_f^2$$

**Case** (iii). The vibrating system is essentially the same as Case (ii), except that a sinusoidal stress is induced and the steady-state vibration of the system is studied.

The governing equation for the system subjected to a steady sinusoidal stress $\sigma = \sigma_0 \sin \omega t$ (see Fig. 8.24) is

$$M\frac{d^2\epsilon}{dt^2} + \frac{E_2}{\omega}\frac{d\epsilon}{dt} + E_1\epsilon = \sigma_0 \sin \omega t \tag{96}$$

The steady-state solution for Eq. (96) can be shown to be

$$\epsilon = \frac{\sigma_0 \sin(\omega t - \phi)}{\left[M^2\left(\frac{E_1}{M} - \omega^2\right)^2 + E_2^2\right]^{1/2}} \tag{97}$$

where $\tan \phi = E_2/(E_1 - M\omega^2)$ or

$$\phi = \tan^{-1}\frac{E_2/E_1}{1 - \dfrac{M\omega^2}{E_1}}$$

Calling $\omega_r^2 = E_1/M$ the natural frequency of oscillation, Eq. (97) becomes

$$\epsilon = X \sin(\omega t - \phi) \tag{98}$$

where

$$X = \frac{\sigma_0}{\sqrt{M^2(\omega_r^2 - \omega^2) + E_2^2}} \tag{99}$$

At resonance $\omega = \omega_r$; Eq. (99) then reduces to

$$X_r = \frac{\sigma_0}{E_2} \tag{100}$$

From Eqs. (99) and (100)

$$\frac{X}{X_r} = \frac{E_2}{[M^2(\omega_r^2 - \omega^2) + E_2^2]^{1/2}} \tag{101}$$

Plotting Eq. (101) (see Fig. 8.15) the dynamic properties can be obtained from the shape of the resonance curve (amplitude ratio plotted against the frequency of fluctuation). For a linear viscoelastic material, the frequency corresponding to the maximum value in the amplitude ratio can be used to calculate the storage modulus at that frequency and half-width [or the power width see Eq. (66) of the resonance curve], that is, the difference in the values of $\omega$ at which the magnitude of $X/X_r$ is lower by one-half can be related to the loss modulus $E_2$. Assuming $\omega_1$ and $\omega_2$ are the values of frequency at which $X/X_r$ is reduced by one-half, it can be shown[27] that

$$\omega_1^2 - \omega_2^2 = 2\sqrt{3}\,\frac{E_2}{M} \tag{102}$$

When the viscous losses in the material are small

$$\frac{\Delta\omega}{\omega_r} \approx \frac{\omega_1 - \omega_2}{\omega_r} \approx \frac{\omega_1^2\,\omega_2^2}{2\omega_r^2} \tag{103}$$

Substituting Eq. (31) in Eq. (32)

$$\frac{\Delta\omega}{\omega_r} \doteq \sqrt{3}\,\frac{E_2}{E_1} = \sqrt{3}\,\tan\delta \tag{104}$$

*When the characteristic length of a test specimen is comparable to stress wavelength,* corresponding to a chosen frequency, standing waves are generated in the test sample and the inertia of the specimen cannot be neglected.

Consider a vibrating reed clamped at one end and performing vibration of constant amplitude and frequency in the direction of thickness of the reed (see Fig. 8.25). The vibrating end of the reed is free. The dynamic mechanical behavior of the test material can be determined by studying the motion of the reed. The equation of motion of the vertical displacements of the viscoelastic reed, as shown in Fig. 8.25, is given by[28]

$$I\,E(i\omega)\,\frac{\partial^4 y}{\partial \chi^4} + m\,\frac{\partial^2 y}{\partial t^2} = 0 \tag{105}$$

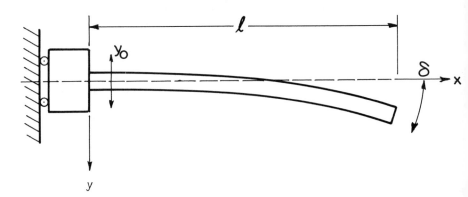

Fig. 8.25   A vibrating reed

where I is moment of inertia of the cross-sectional area of reed, $m$ is mass per unit length, y is vertical displacement and $\chi$ is horizontal coordinate. The above differential equation is based upon the usual beam bending theory which assumes lateral dimensions are small compared with its length.

In a vibrating reed test a steady forced oscillation at a selected frequency is considered. Suppose the forced oscillation is assumed to be of the following form.

$$y = y \, e^{i\omega t} \tag{106}$$

Substituting Eq. (106) into Eq. (105) and rewriting

$$\frac{d^4\bar{y}}{d\chi^4} - \left[\frac{M^2}{I} E(i\omega)\right] \bar{y} = 0 \tag{107}$$

Equation (107) is similar to the equation for forced oscillation of an elastic beam with the difference that the amplitude y is complex in this case and depends on the complex modulus $E(i\omega)$.

Equation (107) can be solved to satisfy the boundary conditions

$$\chi = 0, \qquad y = y_0, \qquad \frac{d\bar{y}}{d\chi} = 0$$

$$\chi = l, \qquad \frac{d^2\bar{y}}{d\chi^2} = \frac{d^2\bar{y}}{d\chi^3} = 0 \tag{108}$$

Solving Eq. (107) for the boundary conditions represented by Eq. (108) and putting $\chi = l$ in the resulting solution, the motion of the free end of the reed can be shown to be

$$y(l) = \bar{y}(l)\, e^{i\omega t} = y_0 \left( \frac{\cosh\theta + \cos\theta}{1 + \cosh\theta \cos\theta} \right) e^{i\omega t} \tag{109}$$

where

$$\theta = \left[ \frac{M\omega^2}{I}\, E(i\omega) \right]^{1/4} l \tag{110}$$

From Eq. (109) the amplitude ratio between the free end and the clamped end of the reed $(y/y_0)$ is a complex quantity indicating that the motions of the two ends of the reed are, in general, out of phase. For a particular length of the reed, amplitude ratio versus frequency curves are obtained which give a series of maxima at several resonant frequencies of the system. From the distribution of amplitude ratios around a particular maximum, the dynamic modulus values can be evaluated as described earlier.

*When the characteristic length of a test specimen is much larger than stress wavelength*, the waves are propagated in the specimen. Considering only one-dimensional wave propagation in the medium, the governing equation for the system is given by

$$E_1 \frac{\partial^2 U}{\partial \chi^2} + \frac{E_2}{\omega} \frac{\partial^3 U}{\partial t \partial \chi^2} = \rho \frac{\partial^2 U}{\partial t^2} \tag{111}$$

where $E_1$, $E_2$ are the storage and loss tensile moduli, respectively, U the longitudinal displacement in the medium and $\rho$ the material density. Suppose a sinusoidal displacement of $U = U_0 \cos \omega t$ is applied at one end $(\chi = 0)$ of the specimen which is in the form of a thin, long strip. The solution of Eq. (111) for this boundary condition can be shown to be

$$U(\chi, t) = U_0\, e^{-[2\pi r_0/\lambda_0]\chi} \cos\left( t - \frac{\chi}{C} \right) \tag{112}$$

where $\omega$ is angular frequency of the propagated wave, C is velocity of wave propagation, $\lambda_0 = 2\pi C/\omega$, the wavelength and $2\pi r_0$ is attenuation per wavelength. The attenuation per wavelength can be determined by the ratio of two consecutive amplitudes at any instant. Consider the motion of the wave at time $t = 0$. The displacement U as a function of $\chi$ can be determined from Eq. (112) as

$$U(\chi, 0) = U_0\, e^{-[2\pi r_0 \chi/\lambda_0]} \cos \frac{\omega \chi}{C}$$

or

$$U(\chi, 0) = U_0\, e^{-[2\pi r_0 \chi/\lambda_0]} \cos \frac{2\pi \chi}{\lambda_0} \tag{113}$$

The variation of displacement with the coordinate $\chi$ is shown in Fig. 8.26.

From Eq. (113) attenuation per wavelength can be shown to be

$$2\pi r_0 = ln \frac{U_0}{U_1} = ln \frac{U_1}{U_2} \tag{114}$$

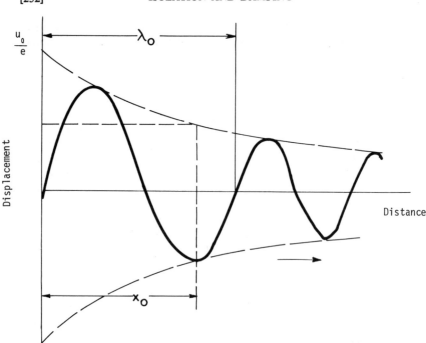

Fig. 8.26  Attenuation of a propagated wave

Substituting Eq. (112) for Eq. (111) gives

$$E_1(\omega) = \rho C^2 \frac{1 - r_0^2}{[1 + r_0^2]^2}$$

$$E_2(\omega) = 2\rho C^2 \frac{r_0}{[1 + r_0^2]^2} \qquad \qquad \textbf{(115)}$$

and

$$\tan \delta = \frac{E_2(\omega)}{E_1(\omega)} = \frac{2r_0}{(1 - r_0^2)}$$

## 19. DYNAMIC EXPERIMENTAL TECHNIQUES

A survey of literature[28] indicates that a wide variety of dynamic test techniques have been used. In Table 14 is a comparative summary to show their respective limitations and advantages. The experimental techniques classified in Table 14 will be considered briefly.

**Direct Measurement of Stress and Strain.** When the size of the specimen and the frequency are such that inertia can be neglected, the simple way to determine the dynamic properties is to measure both stress and strain as functions of time in sinusoidal deformations. These measurements may be made by straining the specimen under investigation with a spring of some kind. If the spring is stiffer than the specimen, a prescribed sinusoidally varying strain applied to the pair will be practically equal to the strain of the specimen. The spring deflections are therefore proportional to stress. Various devices for recording the sinusoidal variation of stress and strain have been used. [29-32] An optical device for recording the two sinusoidal graphs of stress and strain is due to Leathersich. [29] In an apparatus developed by Phillipoff, [32] the two displacements proportional to stress and strain are recorded by differential transformers. The amplified outputs of the differential transformers are applied to the horizontal and vertical plates of a cathode ray oscilloscope, and an ellipse is traced. The two components of the complex modulus are evaluated from the ellipse geometry.

**TABLE 14**
**COMPARISON OF DYNAMIC TEST TECHNIQUES**
**(AFTER FERRY[28])**

| CLASS | $\dfrac{l}{\lambda_0}$ | $\dfrac{E_2}{E_1}$ or $\dfrac{G_1}{G_2}$ | FREQUENCY RANGE (CPS) |
|---|---|---|---|
| Direct Measurement of Stress and Strain | small | not limiting | continuous $10^{-6}$ to $10^2$ |
| Resonance Devices with Added Inertia | small | small | discrete $10^{-2}$ to $10^5$ |
| Transducer Measurements of Stress and Strain Ratios | small | not limiting | continuous 10 to $10^4$ |
| Wave Propagation | large | small | continuous $10^2$ to $4 \times 10^4$ |
| Characteristic Impedance Measurements | large | not limiting but usually large | discrete $10^3$ to $10^7$ |
| Resonance Vibrations | order of unity | small | discrete 1 to $10^4$ |

**Resonance Devices with Added Inertia.** When a fluctuating force not only deforms a specimen of material under investigation but also oscillates a mass M, whose inertia is large compared with that of the specimen, the dynamic mechanical properties may be evaluated by exciting mechanical resonance. Though this technique involves less complicated apparatus, any single test provides only single values for both the storage and the loss moduli. Measurement is made only at the resonant frequency, and any alteration in resonant frequency is possible by either readjusting the added inertia or changing the sample geometry only.

Assuming the oscillating motion is linear, the driving force may be represented as a complex quantity with the real part in phase with the rate of displacement and the imaginary part 90 degrees out of phase. Accordingly, the mechanical impedance is defined as follows.

$$Z = \frac{f}{v} = R_M + i\,X_M = \frac{aE_2}{\omega} + i\left(\omega M - \frac{S_M}{\omega} - \frac{aE_1}{\omega}\right) \qquad (116)$$

where $R_M$ is mechanical resistance, $X_M$ is mechanical reactance, M is moving mass, a is constant depending on sample geometry and $S_M$ is elastance associated with the apparatus exclusive of the specimen. When the peak constant force f is attained, either M or $\omega$ is adjusted until maximum velocity v or amplitude X is observed. The expression of real and imaginary parts of the modulus can be given as

$$E_1 = \left(\omega^2 \frac{M}{a} - \frac{S_M}{a}\right) \quad \text{and} \quad E_2 = \frac{|f|}{aX} \qquad (117)$$

Another method of obtaining $E_2$ is to measure the amplitude X at a series of frequencies near resonance and construct a response curve from which the half-width $\Delta\omega$ is evaluated [see Fig. 8.15 and Eq. (66)].

Instead of the resonance method described above, a free vibration method can be used. The moving system is displaced from its equilibrium and then allowed to oscillate freely. From the trace of the exponentially damped sinusoidal function, the logarithmic decrement $\Delta$ is evaluated. The knowledge of the logarithmic decrement makes it possible to determine either the modulus or compliance functions [Eqs. (62) and (63)].

**Transducer Measurement of Stress and Strain Ratios.**[33] An electromagnetic transducer represents a device for converting electrical energy to mechanical energy or vice versa. Utilizing this principle, quantities such as stress or strain rate (velocity), which are necessary to describe a mechanical system, can be measured very accurately. The transducer method has the advantage over the other methods in that neither stress nor strain need be measured directly and that the frequency range can be continuously varied. This makes it possible to determine the variation of complex modulus over a large frequency range. The experimental arrangement for the electromagnetic transducer is described elsewhere. However, the expressions for mechanical resistance and mechanical reactance, expressed in terms of the measured electrical quantities, can be shown to be as follows.

$$R_M = \frac{(\beta l)^2\,(\overline{R} - \overline{R}_0)}{[(\overline{R} - \overline{R}_0)^2 + (\overline{\chi} - \overline{\chi}_0)^2]}$$

$$\qquad (118)$$

$$X_M = \frac{(\beta l)^2\,(\overline{\chi} - \overline{\chi}_0)}{[(\overline{R} - \overline{R}_0)^2 + (\overline{\chi} - \overline{\chi}_0)^2]}$$

where $\overline{R}$ and $\overline{R}_0$ are the electrical resistances of the coil in motion and at rest, respectively, and $\bar{\chi}$ and $\bar{\chi}_0$ are the corresponding reactances. From $R_M$ and $X_M$ the moduli $E_1$ and $E_2$ can be calculated by Eq. (117).

**Wave Propagation Method.** When the critical dimensions of a specimen under investigation are large compared to the wavelength of stress waves which are propagated in the medium at the frequency of measurement, the inertia of the sample needs to be considered. In addition, because of the wave effect, a simple relation between the force and the resulting displacement cannot be found. By observing the velocity and attenuation of these viscoelastic waves, the dynamic properties of the material can be ascertained. Two kinds of stress waves may be propagated in rubber-like materials. These are the waves of distortion and the waves of dilation. The velocity of waves of distortion depends on the shear storage modulus $G_1(\omega)$ alone, but the velocity of waves of dilation depends on the shear storage modulus, the bulk storage modulus $K_1(\omega)$ and also on the geometry of the specimen.[34]

Numerous experimental techniques[35-37] have been developed to study the dynamic mechanical properties of rubber and rubber-like materials by wave propagation methods. Most of these techniques deal with one-dimensional wave propagation. This can be achieved by using a long thin strip of the material under investigation as a specimen. The lateral dimensions (thickness, etc.) must be very small compared to the stress wavelength. Under these conditions the generated waves approximate to a one-dimensional disturbance, and a complete attenuation prevents any reflection from the far side of the sample opposite to the source of excitation. The exponentially damped wave is characterized by its wavelength $\lambda_0$ and the critical damping distance $\chi_0$ (see Fig. 8.26). The dimensionless ratio $\lambda_0/\chi_0$ is a convenient index of severity of damping.

**Characteristic Impedance Measurement.** This is an alternate method for the determination of the complex modulus in the case of wave propagation in a viscoelastic medium, which completely decays before reaching the opposite boundary. By this method the complex modulus is evaluated by measuring the complex ratio of force to velocity at the surface from which the wave is propagated. For detailed description of this method and measuring techniques, the reader is referred to Mason.[38]

**Resonance Vibrations.** When the critical dimension of a specimen under investigation is of the same order of magnitude as the stress wavelength at the frequency of measurement, resonance vibrations can be executed in the specimen. In this case the inertia required for resonance is provided by the material itself instead of an added mass M. The calculations for the dynamic properties resemble those discussed in the case of resonance devices with added inertia in that the real part of the complex modulus is determined by the resonant frequency of a particular vibration mode. The imaginary part of the complex modulus is obtained by one of the following observations.

1. Measurement of amplification at resonance
2. Half-width of the response curve against frequency
3. Logarithmic decrement in free vibration

A commonly used experimental arrangement for this purpose is a vibrating reed, which is strained either in tension or in bending. Vibration is usually excited electromagnetically. The displacement can be measured by visual observations with a microscope. This method has been described in great detail by Nolle. [39]

## 20. NONLINEAR DYNAMIC MECHANICAL BEHAVIOR

The foregoing treatment of dynamic mechanical behavior of rubber-like materials applies to the linear range representing about a 10 percent strain. The linear response makes it possible to apply the linear theory of viscoelasticity to describe the dynamic mechanical behavior of rubber-like materials. It is well known that linear response is displayed by rubber-like materials at either high frequencies or low temperatures. However, the low-frequency or the high-temperature dynamic response of rubber-like materials has been found to be nonlinear. This nonlinearity in the dynamic response of rubber-like materials is generally due to strain (deformation) becoming large. As a result a sinusoidal stress on the material does not lead to a sinusoidal response as in the case of linear behavior. This introduces difficulties in the specification of dynamic modulus and compliance functions. Although a considerable amount of work has been reported on small-strain behavior, the dynamic mechanical properties of rubber in the nonlinear region of large strains has received scant attention. Much of the existing literature on nonlinear dynamic behavior of rubber-like materials is concerned with the stress-strain behavior observed for a range of low-strain rates. In some cases the energy loss has been evaluated from hysteresis loops. It was Smith[40] who conducted experiments on polyisobutylene up to about 100 percent of extension and on Styrene-Butadiene rubber (SBR) up to about a 500-percent strain at various strain rates and temperatures. Similar work has been reported by Landel and Stedry.[41] The uniaxial tension results of these authors indicate that the stress can be expressed as the product of two functions, one of which depends on the strain and the other on the time taken to reach the strain, and that the time-temperature superposition principle can be used to reduce the data to a standard temperature. Schallamach, Sellen and Greensmith[2] have studied the dynamic mechanical behavior of two different vulcanized elastomers, natural rubber (NR) and Acrylonitrile-Butadiene rubber (ABR) by subjecting these materials to strain cycling at various strain rates.

The energy losses incurred during strain cycling, which were calculated from the areas of hysteresis loop, are shown plotted in Fig. 8.27 on a logarithmic scale as a function of the reduced rate. The loss ratios representing the ratio between energy dissipated and the maximum energy input are shown in Fig. 8.28 for ABR and NR. Harwood and Schallamach[11] have investigated the dynamic mechanical behavior of natural rubber by strain cycling up to a maximum elongation of 530 percent for various temperatures and strain rates. Their results are shown in the form of three-dimensional plots for two maximum extensions, $\epsilon = 2.8$ and $\epsilon = 5.3$ in Figs. 8.29 and 8.30. Further, the effect of ampli-

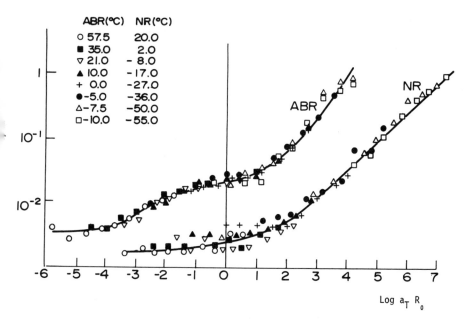

Fig. 8.27 Energy losses per unit strain (after Schallamach, Sellen and Greensmith [2])

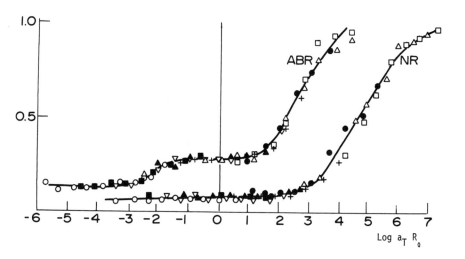

Fig. 8.28 Loss ratios during extension cycles as a function of the rate $a_T R_0$ (after Schallamach, Sellen and Greensmith [2])

tude, temperature and frequency on the dynamic mechanical behavior of rubber under compressive loading has been studied by McCallion and Davies.[1]

The foregoing discussion indicates that a general nonlinear viscoelastic theory for describing the nonlinear dynamic mechanical behavior of rubber-like materials needs to be developed. In the absence of such a theory the material scientist or engineer must rely solely on experimental evidence.

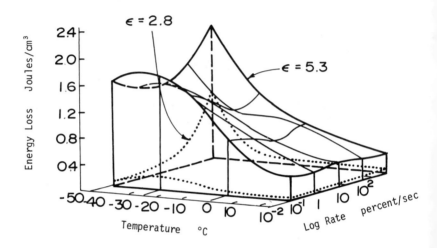

Fig. 8.29  Energy loss per cycle versus temperature and strain rate; two maximum extensions, $\epsilon = 2.8$ and 5.3 (after Harwood and Schallamach[11])

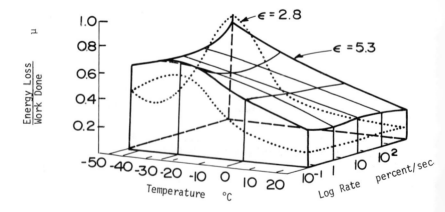

Fig. 8.30  Energy loss ratio versus temperature and strain rate; two maximum extensions, $\epsilon = 2.8$ and 5.3 (after Harwood and Schallamach[11])

# REFERENCES

1. McCallion, H. and Davies, D. M. "Behavior of Rubber in Compression Under Dynamic Conditions." *Proceedings of Institute of Mechanical Engineers* (London) 169 (1955): 1125.
2. Schallamach, A.; Sellen, D. B.; and Greensmith, H. W. "Dynamic Behavior of Rubber During Moderate Extensions." *British Journal of Applied Physics* 16 (1965): 241.
3. Gross, B. *Mathematical Structure of the Theories of Viscoelasticity.* Paris: Herman and Co., 1953.
4. Tobolsky, A. V. *Properties and Structure of Polymers.* New York: John Wiley & Sons, Inc. 1960.
5. Bland, D. R. *The Theory of Linear Viscoelasticity.* New York: Pergamon Press, 1960.
6. Ferry, J. D. *Viscoelastic Properties of Polymers.* New York: John Wiley & Sons, Inc. 1961.
7. Staverman, A. J. and Schwarzl, F. "Linear Deformation Behavior of High Polymers." In, *Die Physik der Hochpolymeren*, vol. IV, edited by H. A. Stuart. New York: Springer-Verlag, 1956.
8. Gehman, S. D. "Dynamic Properties of Elastomers." *Rubber Chemistry and Technology* 30 (1957): 1202.
9. Marvin, R. S. "Measurement of Dynamic Properties of Rubber." *Industrial and Engineering Chemistry* 44 (1952): 696.
10. Gehman, S. D.; Woodford, D. E.; and Stambaugh, R. B. "Dynamic Properties of Rubber." *Industrial and Engineering Chemistry* 33 (1941): 1032.
11. Harwood, J. A. C. and Schallamach, A. "Dynamic Behavior of Natural Rubber During Large Extensions." *Journal of Applied Polymer Science* 2 (1967): 1835.
12. Sharma, M. G. and St. Lawrence, W. F. "Dynamic Mechanical Behavior of a Composite Viscoelastic Material." *Fiber Science and Technology* 1 (1969): 171.
13. Sharma, M. G. *Mechanics of Polymers.* Proceedings of a short course. University Park: The Pennsylvania State University, June 1967.
14. Boltzmann, L. *Pogg. Ann. Physick* 7 (1876): 624.
15. Gross, B. "On Creep and Relaxation." *Journal of Applied Physics* 18 (1947): 212.
16. Andrews, R. D. "Correlation of Dynamic and Static Measurements on Rubberlike Materials." *Industrial and Engineering Chemistry* 44 (1952): 707.
17. Leaderman, H. *Elastic and Creep Properties of Filamentous Materials.* Washington, D.C.: Textile Foundation, 1943, p. 175.
18. Ferry, J. D.; Fitzgerald, E. R.; Grandine, L. D., Jr.; and Williams, M. L. "Temperature Dependence of Dynamic Properties of Elastomers; Re-

laxation Distributions." *Industrial and Engineering Chemistry* 44 (1952): 703.

19. Ferry, J. D. "Mechanical Properties of Substances of High Molecular Weight. V1 Dispersion in Concentrated Polymer Solutions and its Dependence on Temperature and Concentration." *Journal of American Chemical Society* 72 (1950): 3746.

20. Hopkins, I. L. "The Ferry Reduction and the Activation Energy for Viscous Flow." *Journal of Applied Physics* 24 (1953): 1300.

21. Alfrey, T., Jr. *Mechanical Behavior of High Polymers*. New York: Interscience Publishers, 1948.

22. Nielsen, L. E. *Mechanical Properties of Polymers*. New York: Reinhold Publishing Corp., 1962.

23. Ruzicka, J. E., ed. *Structural Damping*. Colloquium on Structural Damping, ASME, 1959.

24. Myklestead, N. O. *Fundamentals of Vibration Analysis*. New York: McGraw-Hill Book Co., 1956.

25. Snowdon, J. C. "Choice of Resilient Materials for Anti-Vibration Mountings." *Rubber Chemistry and Technology* 32 (1959): 1209.

26. Yin, J. P. and Pariser, R. "Dynamic Mechanical Properties of Several Elastomers and Their Potentialities in Vibration Control Applications." *Journal of Applied Polymer Science* 8 (1964): 2427.

27. Hillier, K. W. "The Measurements of Dynamic Elastic Properties." In *Progress in Solid Mechanics*, vol. II, edited by I. N. Snedden and R. Hill, Chap. V. New York: Interscience Publishers, 1961.

28. Ferry, J. D. "Experimental Techniques for Rheological Measurements on Viscoelastic Bodies." In *Rheology Theory and Applications*, edited by F. R. Eirich, Chap. II. New York: Academic Press, 1958.

29. Leathersich, W. "Apparatus for the Study of Creep of Dielectric Polymers and Their Dynamic Rheological Properties." *Journal of Scientific Instrumentation* 27 (1950): 103.

30. Davies, D. M. "The Effect of Frequency on the Behavior of Rubber Under Cyclic Deformation." *British Journal of Applied Physics* 3 (1952): 285.

31. Roelig, H. *Proceedings of Rubber Technical Conference*, London, 1938, p. 821.

32. Phillipoff, W. J. "Mechanical Investigation of Elastomers in a Wide Range of Frequencies and Further Investigations on Polymers." *Journal of Applied Physics* 25 (1954): 1102.

33. Smith, T. L.; Ferry, J. D.; and Shremp, F. W. "Measurements of Mechanical Properties of Polymer Solutions by Electro-Magnetic Transducer." *Journal of Applied Physics* 20 (1969): 144.

34. Kolsky, H. *Stress Waves in Solids*. New York: Oxford University Press, 1953.

35. Nolle, A. W. and Sieck, P. W. "Longitudinal and Transverse Ultrasonic Waves in Synthetic Rubber." *Journal of Applied Physics* 23 (1952): 888.

36. Ivey, D. G.; Mrowca, B. A.; and Gerth, E. "Propagation of Ultrasonic Bulk Waves in High Polymers." *Journal of Applied Physics* 20 (1949): 486.

37. McSkimin, H. J. "A Method for Determining the Propagation Constants of Plastics at Ultrasonic Frequency." *Journal of Acoustical Society of America* 23 (1951): 429.

38. Mason, W. P. "Measurement of Viscosity and Shear Elasticity of Liquids by Means of a Torsionally Vibrating Crystal." *Transactions of the American Society of Mechanical Engineers* 69 (1947): 359.

39. Nolle, A. W. "Methods of Measuring Dynamic Mechanical Properties of Rubberlike Materials." *Journal of Applied Physics* 19 (1948): 753.

40. Smith, T. L. "Nonlinear Viscoelastic Response of Amorphous Elastomers to Constant Strain Rates." *Transactions of Society of Rheology* 6 (1962): 61.

41. Landel, R. F. and Stedry, P. J. "Stress as a Reduced Variable: Stress Relaxation of SBR Rubber at Large Strains." *Journal of Applied Physics* 31 (1960): 1885.

42. Treloar, L. R. G. *The Physics of Rubber Elasticity*. Oxford: Clarendon Press, 1949.

# SECTION III

# ANALYTICAL METHODS

# SECTION III CONTENTS

# 9

## VELOCITY CRITERIA FOR MACHINE VIBRATION*

*Steve Maten*

Measuring and reporting equipment vibration severity is reduced to a simple skill at Shawinigan Chemicals Plant in Varennes, Québec, Canada now that velocity standards (Table 15) have been developed.

**TABLE 15**
**VELOCITY STANDARDS**

| DIRECTLY MEASURED MAX. VELOCITY IN/SEC PEAK | CLASSIFICATION | SEVERITY RATING |
|---|---|---|
| Above 0.5 | AA | *Extremely Rough*, dangerous, shut down machinery |
| From 0.3 to 0.5 | A | *Very Rough*, correct within a few weeks since major damage may occur |
| From 0.2 to 0.3 | B | *Rough*, correct to save wear |
| From 0.1 to 0.2 | C | *Fair*, minor fault, correction uneconomical |
| Up to 0.1 | D | *Smooth*, well-balanced and well-aligned equipment |

(Values are for bolted-down and steady-rotating equipment.)

The velocity measurements are taken in three directions (vertical, horizontal and axial) on the bearing housings of rotating machinery and recorded as a vertical bar on the severity report (Fig. 9.1). Depending on the reported vibra-

---

*Reprinted by permission from Hydrocarbon Processing Magazine, January 1967.

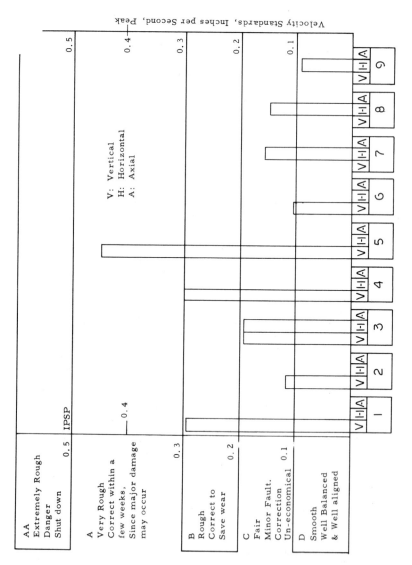

Fig. 9.1   Vibration severity report

tion severity, a follow-up by thorough vibration analysis is made where faults are indicated.

**Meter.** A simple portable, battery-operated vibration meter is used for the measurement of velocity, acceleration and displacement amplitude, having a pickup with flat response from 2 to 2000 cps and a resonant frequency of 3200 cps. The read-out is directly in inches per second.

**Warning.** For those who use vibration meters or analyzers measuring displacement amplitude and/or velocity at shaft speed only, Table 15 velocity standards should *not* be used. It is strongly recommended that instead they use the haystack chart by Michael P. Blake (see Fig. 9.2).

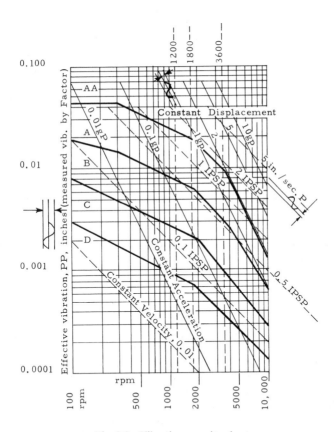

Fig. 9.2   Vibration severity chart

**Service Factors of Fig. 9.2**

Single-stage centrifugal pump, electric motor, fan............................. 1
Typical chemical processing equipment, noncritical.......................... 1

Turbine, turbo-generator, centrifugal compressor............................ 1.6
Centrifuge stiff-shaft; * multi-stage cent. pump............................... 2
Miscellaneous equipment, characteristics unknown............................ 2
Centrifuge, shaft-suspended, on shaft near basket............................ 0.5
Centrifuge, link-suspended, slung.......................................... 0.3

* Horizontal displacement on basket housing.

Effective vibration = measured peak to peak vibration, inches, multiplied by the service factor.

Machine tools are excluded. Values are for bolted-down equipment: when not bolted, multiply the service factor by 0.4 and use the product as a service factor. Caution: Vibration is measured on the bearing housing, except as stated.

It is important to note that the new velocity standards allow, in most instances, higher measured velocities than those obtained at shaft speed, particularly for low-speed equipment. This is quite acceptable, however, since most direct velocity readings are based on harmonics or bearing rumbles resulting in the higher readings. A typical case is described in Example 9. This compressor measured an amplitude of 1 mil at 327 rpm, corresponding with a velocity of 0.017-in/sec. The measured velocity is actually 0.08-in/sec or about five times as high and at approximately the tenth harmonic of the shaft speed.

In summary, the following vibration readings are important.

**Velocity.** When one is concerned with wear and internal machine failure, velocity measurements are important.

**Displacement Amplitude.** When one is concerned with studying unbalance and deflections, measure amplitude.

**Acceleration.** When one is concerned with designing for adequate strength, acceleration is important.

**Frequency.** When one is concerned with analyzing the source of vibrations and the study of environmental problems, measure the vibration frequency.

**Velocity Readings.** The energy generated by vibration is dissipated inside machinery as heat and impact between components causing internal wear and failure. The energy is proportional to the square of the mean velocity with which some of these components are chattering back and forth. Therefore, measuring velocities on the bearings of operating machinery will give us a measure of how much wear we may expect, regardless of what component it is that causes the measured velocities. For maintenance purposes, therefore, this appears to be a most valuable type of reading to obtain.

Anti-friction, journal and thrust bearings, thrust collars, shaft sleeves, mechanical seals, gear teeth, couplings, etc. along with their lubrication oil films are the built-in shock absorbers that dampen the machine's vibrations at the source. Failure of these parts keeps maintenance and operating costs high. In normal operating machinery, these failures usually occur here before any structural or environmental failures.

It is these items, therefore, with which the velocity measuring method is concerned.

The severity classifications can be interpreted approximately as follows.

**Class AA.** The oil film will be destroyed; therefore, metal-to-metal contact and bearing seizure or breakage of teeth can occur at any moment.

**Class A.** The oil film will break if oil viscosity or temperature are not controlled. Rapid wear is expected.

**Class B.** Gradual wear over a period of time is expected.

**Class C.** Little or no appreciable wear is expected—minor fault in installation.

**Class D.** Normal trouble-free installation. Components should last for several years.

**Displacement Amplitude Readings.** An unbalance in rotating machinery is one of the most common sources of vibration. Since the amplitude of vibration is proportional to the amount of unbalance, measuring of this quantity will aid chiefly in balancing when its timing is checked simultaneously with a stroboscope. A mean velocity reading is of no use here. These velocity readings are frequently not proportional to the unbalance anyway and they can be influenced by other components in the machine. Velocity can also be affected by the harmonic response of the materials of construction. For balancing, therefore, amplitude readings are essential.

Note: Since most amplitude readings obtained are at shaft speed (usually the source of the largest displacement), the higher harmonic velocities are being ignored when using amplitude as a measure of the machine's operating roughness. Therefore, these conditions must be compensated for in a haystack-shaped chart of allowable amplitudes and corrected with service factors for different types of machinery when using such readings.

**Acceleration Readings.** The dynamic force produced by machinery vibration is proportional to the mean acceleration of the vibration. This force can be very high if not controlled and can cause support or structural failures and other major catastrophes. Acceleration measurement is essential to equipment designers and for those specialists who troubleshoot environmental problems in the field to solve design errors or omissions. Acceleration readings can vary greatly with the environmental conditions, and the highest values measured are usually associated with the higher-frequency harmonics. As a measure of expected wear, the velocity can be calculated from acceleration readings assuming that the corresponding frequency is known.

**Frequency Readings.** For complete vibration analysis, the sources of various vibrations in rotating equipment components must be known. These sources are associated with shaft speeds and other component speeds. Therefore, frequencies must be measured. It can happen that different machines pulsate with one another and produce new so-called *beat*-vibrations. Support structures may oscillate at their natural frequencies sometimes caused by the primary exciting forces and sometimes by the beat-vibrations. Torsional vibrations or oil whirl may occur inside machines, etc. Therefore, frequency readings are most essential to the analyst. The most critical frequency analyzers are capable of measuring acceleration at frequencies from 2 to 20,000 cps. These analyzers provide sufficient measurements to cover analysis of some of the most complex environmental problems associated with mechanical vibrations. Piezoelectric pickups (accelerometers) are used with resonant frequencies in excess of 35,000 cps. For the simple analysis associated with machinery wear only, a pickup response between 2 and 2000 cps is generally quite adequate, assuming that the original installation did not have an environmental problem.

Standards—Are They Valid? The writer is greatly indebted to Mr. M. P. Blake's article, "New Vibration Standards for Maintenance."[1] On many occasions the included vibration standards were used here at Shawinigan's plant, and the new velocity standards have been extensively field-tested against the haystack chart.

For the sake of uniformity, the use of similar severity rating nomenclature is used in this paper.

In order to substantiate the usefulness of the velocity standards, the following examples have been taken at random from vibration records on various machines at all speed levels and recorded over the past three years.

The examples are not intended as complete case histories of vibration analyses. The information is limited to depicting a reading in one direction only and comparisons are made with the haystack chart for the purpose of evaluating the standards.

In some cases the comparison is not realistic, since the scope of Mr. Blake's article "excludes the problems of high-frequency vibration, and noise, stemming from rolling bearings, and miscellaneous high-frequency vibrations and noise, such as hydraulic and pneumatic" and these *are* included in the velocity method. Also in some cases extrapolation of the haystack chart was assumed when frequencies were beyond 10,000 cpm.

**Example 1.**
*Machine:* First casing of a large motor-driven triplex centrifugal compressor train.
*Readings:* (By analyzer)

|                     Primary                      |                     Secondary                      |
| ------------------------------------------------ | -------------------------------------------------- |
| rpm        = 10,800                              | Freq. cpm  = 5400                                   |
| Calc. ampl. = 0.065 mils                         | Calc. ampl. = 0.51 mils                            |
| Calc. vel.  = 0.036 in/sec                       | Calc. vel.  = 0.144 in/sec                         |
| Meas. acc.  = 0.114-g.                           | Meas. acc.  = 0.22-g.                              |

*Readings:* (By vibration meter)

Meas. vel.   = 0.29 in/sec
Meas. acc.   = 0.67-g.
Calc. ampl.  = 0.65 mils
Calc. cpm    = 8400

*Comments:* A misalignment associated with oil whirl caused by looseness in the journal bearings was indicated. The bearing ran at a high temperature (about 170 to 175°F, normal about 140 to 150°F).
*Check:* Velocity standard (meas. vel. 0.29 in/sec). Classification B (to be corrected).
*Haystack chart (1.6 service factor applied):* 0.065 mils at 10,800 rpm, Class D (no fault); 0.51 mils at 5,400 cpm, Class B (to be corrected); 0.65 mils at 8400 cpm, Class A (correct immediately).
*Note:* This compressor ran for seven months under the above conditions. Routine inspection revealed some fretting wear on the teeth of the oil-flooded gear-flex drive coupling. The coupling was replaced. Also the journal bearing babbits were damaged and replaced. A small alignment correction was made and the compressor bearing now runs with fewer vibrations, as can be seen in Example 2.

**Example 2.**
*Machine:* Same as per Example 1.
*Readings:* (By vibration meter)

Meas. vel.   = 0.12 in/sec
Meas. acc.   = 0.26-g.
Calc. ampl.  = 0.29 mils
Calc. cpm    = 7800

*Comments:* Some misalignment and oil whirl is still indicated. The bearing temperature, however, dropped to the normal of about 150°F.
*Check:* Velocity standard (meas. vel. 0.12 in/sec). Classification C (minor fault).
*Haystack chart (1.6 service factor applied):* 0.29 mils at 7800 cpm, Class B (to be corrected).
*Note:* The less serious rating of the velocity standard is acceptable because of satisfactory experience with other similar machines over the past three years.

**Example 3.**
*Machine:* 900 hp steam turbine.
*Readings:* (By vibration analyzer)

rpm          = 5400
Meas. acc.   = 0.285-g.
Calc. ampl.  = 0.7 mils
Calc. vel.   = 0.195 in/sec

*Comments:* The readings indicate an unbalanced or misaligned shaft. Pipe strain was actually distorting this turbine.

*Check:* Velocity standard (calc. vel. 0.195 in/sec). Classification C (minor fault).

*Haystack chart (1.6 service factor applied):* 0.7 mils at 5400 rpm, Class B (to be corrected).

*Note:* This turbine operated for more than a year before correction was made to reduce vibration. The vibration had not affected the turbine bearings in that period, but wear of the gear-flex coupling occurred to the extent that several teeth broke off. The coupling was replaced after a year and a half of service. Wear of this coupling was considerably accelerated by the loss of its lubricant caused by turbine shaft heat. A heavier grease is now used in the coupling. The "minor fault" classification of the velocity standards is a borderline case, since an unforeseen breakdown of the coupling could have occurred.

## Example 4.

*Machine:* 40 hp horizontal overhung centrifugal pump.
*Readings:* (By vibration meter)

| | | | |
|---|---|---|---|
| rpm | = 3600 | Meas. vel. | = 0.3 in/sec |
| Meas. ampl. | = 1.6 mils | Meas. acc. | = 0.3-g. |
| Calc. vel. | = 0.3 in/sec | Calc. ampl. | = 1.6 mils |
| Calc. acc. | = 0.3-g. | Calc. cpm | = 3600 |

*Comments:* The readings indicate an unbalanced or misaligned shaft but a smooth ball bearing.

*Check:* Velocity standard (meas. vel. 0.3 in/sec). Classification B (to be corrected).

*Haystack chart:* 1.6 mils at 3600 rpm, Class B (to be corrected).

*Note:* Alignment of this pump has repeatedly been difficult and the abovementioned vibrations have periodically resulted in bearing failures.

## Example 5.

*Machine:* 1250 hp synchronous motor with inboard fan bell housing.
*Readings:* (By vibration meter)

| | | | |
|---|---|---|---|
| rpm | = 1800 | Meas. vel. | = 0.45 in/sec |
| Meas. ampl. | = 4.5 mils | Meas. acc. | = 0.31-g. |
| Calc. vel. | = 0.42 in/sec | Calc. ampl. | = 3.5 mils |
| Calc. acc. | = 0.22-g. | Calc. cpm | = 2520 |

*Comments:* The motor required dynamic rebalancing after a previous major repair. Because of appreciable flexibility in the vertical direction of the end bell and looseness in the roller bearing, an additional bounce of the housing occurred every time the rotor dropped back. This resulted in a higher measured acceleration and an increase in calculated frequency.

*Check:* Velocity standard (meas. vel. 0.45 in/sec). Classification A (correct immediately).

*Haystack chart:* 4.5 mils at 1800 rpm, Class B (to be corrected).

*Note:* The more severe rating of the velocity standards is advocated, since it was only possible to keep this motor running for a few days after the bearing housing was provided with external cooling and cooler air was circulated through the motor. Without these precautions bearing seizure would have occurred at any moment, since the temperature tended to rise very rapidly.

The following example confirms the present quietness of this motor after rebalancing.

**Example 6.**

*Machine:* Same as Example 5.

*Readings:* (By vibration meter)

| | | | |
|---|---|---|---|
| rpm | = 1800 | Meas. vel. | 0.055 in/sec |
| Meas. ampl. | = 0.25 mils | Meas. acc. | = 0.22-g. |
| Calc. vel. | = 0.023 in/sec | Calc. ampl. | = 0.07 mils |
| Calc. acc. | = 0.011-g. | Calc. cpm | = 14,400 |

*Comment:* It is interesting to note that the measured velocity is more than double the amount at shaft speed and the acceleration is twenty times that at shaft speed.

*Check:* Velocity standard (meas. vel. 0.055 in/sec). Classification D (no fault).

*Haystack chart:* 0.25 mils at 1800 rpm, Class D (no fault); 0.07 mils at 14,400 cpm, Class D (no fault).

**Example 7.**

*Machine:* Small steam turbine.

*Readings:* (By vibration meter)

| | | | |
|---|---|---|---|
| rpm | = 1800 | Meas. vel. | = 0.16 in/sec |
| Meas. ampl. | = 1.6 mils | Meas. acc. | = 0.285-g. |
| Calc. vel. | = 0.15 in/sec | Calc. ampl. | = 0.47 mils |
| Calc. acc. | = 0.075-g. | Calc. cpm | = 6600 |

*Comments:* Slight looseness in the ball bearing and the 1.6 mils unbalance or misalignment vibration causes rattling of the bearing at higher frequency.

*Check:* Velocity standard (meas. vel. 0.16 in/sec). Classification C (minor fault).

*Haystack chart:* 1.6 mils at 1800 rpm, Class C (minor fault); 0.47 mils at 6600 cpm, Class C (minor fault).

**Example 8.**

*Machine:* Small steam turbine.

*Readings:* (By vibration meter)

| rpm | = 1200 | Meas. vel. | = 0.15 in/sec |
| Meas. ampl. | = 0.3 mils | Meas. acc. | = 0.57-g. |
| Calc. vel. | = 0.016 in/sec | Calc. ampl. | = 0.3 mils |
| Calc. acc. | = 0.0052-g. | Calc. cpm | = 13,800 |

*Comments:* These readings indicate a very smooth shaft but a somewhat faulty bearing.

*Check:* Velocity standard (meas. vel. 0.15 in/sec). Classification C (minor fault).

*Haystack chart:* 0.3 mils at 1200 rpm, Class D (no fault); 0.3 mils at 13,800 cpm, Class B (to be corrected).

*Note:* The lighter components of a faulty bearing are not as destructive as the heavier weights of rotors. The "minor fault" classification is therefore appropriate in this case.

**Example 9.**

*Machine:* 2000 hp–4-stage dynamically opposed reciprocating compressor.

*Readings:* (By vibration meter)

| rpm | = 327 | Meas. vel. | = 0.08 in/sec |
| Meas. ampl. | = 1 mil | Meas. acc. | = 0.0675-g. |
| Calc. vel. | = 0.017 in/sec | Calc. ampl. | = 0.5 mil |
| Calc. acc. | = 0.0016-g. | Calc. cpm. | = 3120 |

*Comments:* The 0.5 mil vibration at 3120 cpm is a harmonic shock-wave component resulting from the reciprocating action of the machine and is associated with slight unbalance and bearing clearances.

*Check:* Velocity standard (meas. vel. 0.08 in/sec). Classification D (no fault).

*Haystack chart:* 1 mil at 327 rpm, Class D (no fault); 0.5 mil at 3120 cpm, Class C (minor fault).

*Note:* Harmonic shock waves are less damped by bearing oil films than the vibrations at shaft speed. Therefore, checking the 0.5 mil harmonic vibration against the haystack chart is not a fair appraisal and the "minor fault" classification can be ignored.

**Equations.**

Energy $= \frac{1}{2} mv^2$

Frequency (cps) $= \dfrac{v}{2\pi A}$ , where A and v are peak values

Amplitude (PP) $= \dfrac{a}{2\pi^2 f^2}$ , where a is peak, and f is cps

Force $= ma$

Dynamic relationship of vibrating system $= ma + cv + kA = 0$.
*In which:*

| | |
|---|---|
| m  = mass | A  = amplitude |
| a  = acceleration | c  = damping factor |
| f  = frequency | k  = stiffness or spring factor |
| v  = velocity | $\pi$  = 3.14 |

*Note:* With two known values, either the displacement amplitude, velocity, acceleration or frequency can easily be computed with a vibration slide rule designed for this purpose.

# REFERENCE

1. Blake, M. P. "New Vibration Standards for Maintenance." *Hydrocarbon Processing and Petroleum Refiner* 43 (January 1964): 111–114.

# SYMBOLS

| $T_p$ | impulse duration | seconds |
|---|---|---|
| T | natural period of a system | second |
| $\omega$ | natural radian frequency of a system | radians per second |
| $f_n$ | natural frequency of a system | cycles per second |
| X(t) or x | instantaneous displacement | inches |
| F(t) | instantaneous force | lbs |
| X | maximum value of X(t) | —— |
| $X_i$ | maximum value of initial displacement, achieved at a given instant or during the preceding time | —— |
| $X_r$ | maximum value of residual displacement, achieved at a given instant or during the succeeding time | —— |
| W | weight | lbs |
| K | stiffness | lbs/inch |

# 10

## SHOCK AND IMPACT
## IN MASS EXCITATION

### *Michael P. Blake*

### 1. SCOPE

The transient and steady-state behavior of a mass and spring system, undamped and with one degree of freedom, are discussed in this chapter. Given the system of Fig. 10.1, the task is to find its response to a variety of force excitations applied to the mass. The case wherein the excitation is applied to the base is considered in Chapter 11. The phase plane method is introduced and applied to finding the response to a rectangular impulse and an impulse of generalized form. Next, the effect of varying the force, F, while the duration of the impulse remains the same is considered. Finally, attention is directed to shock spectra, as they are usually called. The task here is to determine the effect upon the system response of varying the impulse duration. In discussing shock spectra, four cases are considered.

An impact is a transient transmission of mechanical energy to or from a system, the duration being notable for its brevity in relation to other transient transmissions in the same system.

### 2. INTRODUCTION

Shock and impact are practically the same thing. There is a USASI standard

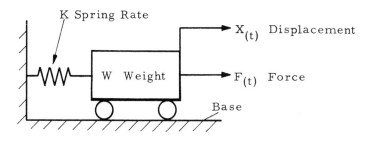

Fig. 10.1 The system

definition, where the concept of impact is defined as pertaining to the more rapid transients which are impactive without saying how rapid they must be. The term impulse connotes an impactive event, a force-time event, of any duration measured in terms of the time integral F dt. Pulse connotes a series of impulse-like events, such as a short series of sinusoidal oscillations or a short-time-duration group of impulsive forces.

Figure 10.2a shows a square impulse of unlimited duration. Here it is natural to consider the entire event as an impulse rather than as an impact, where the instantaneous generation of the force F is the impact proper. Likewise, in Fig. 10.2b the impulse may be considered as being flanked by two instantaneous impacts. A generalized impulse is shown in Fig. 10.2c. If the time, $t_1 - t_0$, is relatively long, the impulse has a small impactive content; whereas if the time is short the impulse has a large impactive content. Consider the analogous case of lifting and carrying a bucket without impact, and then dropping the bucket into a well. The impulse in Fig. 10.2c has superimposed on it five rectangular impulses,

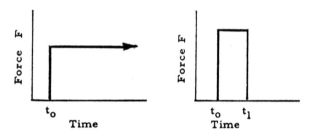

a. Square-Wave Impulse     b. A Time Limited
   of Unlimited Duration       Square-Wave
                               Impulse

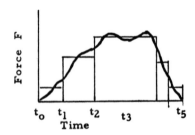

c. Generalized Impulse

Fig. 10.2   Various impulse waveforms

such as might be expected to induce the same response as does the generalized impulse. If the number of rectangles is sufficiently great, their effect is identical with that of the generalized impulse. In practice, by dividing the impulse into about five or more rectangles, very little response error is introduced regarding identical response.

Figure 10.3a shows a rectangular impulse of duration $T_p$ and the system response $X(t)$. Here $T_p$ is relatively long. The initial impact generates a transient response followed by a steady state, again followed by a transient and then a steady state. Figure 10.2b shows an impulse of duration far less than the natural period. The case (10.2b) is the more typical of practical examples.

Although the discussion stops short at the single-degree-of-freedom-undamped system, the given methods, with some modifications, are applicable to damped or bilinear systems as shown in Figs. 10.4a and 10.4b. In some instances they are applicable to nonlinear systems. A clear grasp of the system that is discussed affords a foundation for understanding the majority of practical cases that arise. The phase plane method has been chosen as a method of analysis because it appears the simplest to understand and the method that is likely to be most self-corrective regarding error.

## 3. BASIS OF THE PHASE PLANE METHOD

For convenience the equations of motion are given below, although it is possible to use the method without understanding the equations. The system is shown in Fig. 10.5. The task is to discover the system response to a rectangular impulse. Knowing this response, the response to other impulses may be derived by fitting a set of rectangles to other impulses; then the effect of varying the impulse duration becomes clear.

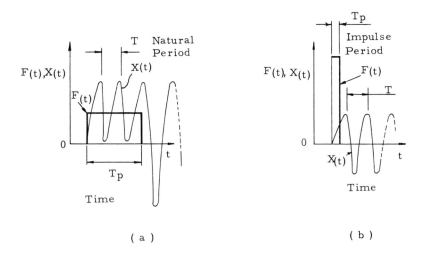

( a )          ( b )

Fig. 10.3   Response of an impact-excited spring-mass system

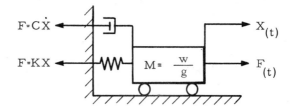

a. Damped Spring-Mass
   System

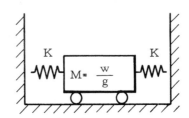

b. Bi-Linear Spring-Mass
   System

Fig. 10.4   Spring-mass systems

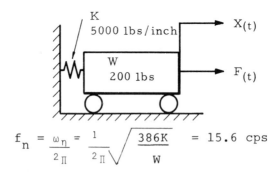

$$f_n = \frac{\omega_n}{2\pi} = \frac{1}{2\pi}\sqrt{\frac{386K}{W}} = 15.6 \text{ cps}$$

$$T = \frac{1}{f_n} = 0.064 \text{ sec.}$$

Fig. 10.5   System for discussion of the phase-plane method

Consider the behavior of the system during the interval when it is subjected to a constant amplitude force, F. The equation of motion is as follows.

$$X(t) = A \sin(\omega t) + B \cos(\omega t) + F/k$$

Writing x for $X(t)$ as a convenience, the above may be rewritten in either of the two following forms.

$$x = C \cos(\omega t - \alpha) + F/K$$

$$x = C \sin(\omega t + \beta) + F/K$$

where $C = (A^2 + B^2)^{1/2}$, $\alpha = \tan^{-1}(A/B)$, and $\beta = \tan^{-1}(B/A)$ as shown in Fig. 10.6.

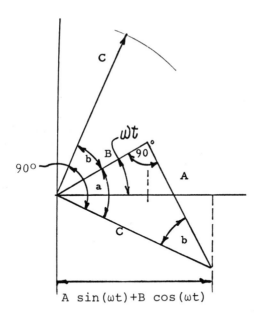

Fig. 10.6   Phasor diagram for the system shown in Fig. 10.5

Using the first form, $x = C \cos(\omega t - \alpha) + F/K$, and differentiating with respect to time we have

$$\frac{\dot{x}}{\omega} = -C \sin(\omega t - \alpha)$$

When these latter two expressions are squared and added, we get

$$\left(x - \frac{F}{K}\right)^2 + \left(\frac{\dot{x}}{\omega}\right)^2 = C^2 = A^2 + B^2$$

The above equation may be called the characteristic equation of the phase plane method. It rests on the assumption that $F/K$ is constant. The equation is that of a circle, with center at $F/K$ and radius $(A^2 + B^2)^{1/2}$. Next, we evaluate A and B. For B, let time, t, be zero in the equation for $X(t)$. The sine then becomes zero, and the cosine unity, giving:

$$B = X_0 - F/K$$

where $X_0$ is the value of x at time zero. And likewise, differentiating $X(t)$ to obtain velocity and letting time equal zero, we have

$$\frac{\dot{x}}{\omega} = A \cos(\omega t) - B \sin(\omega t)$$

so that

$$A = \frac{\dot{X}_0}{\omega}$$

where $\dot{X}_0$ is the velocity at time zero. For any time interval i, wherein the value of F is constant, the system behaves harmonically. Furthermore, the preceding analysis gives the following information, for the interval duration.

1. Velocity and displacement are the sine and cosine projections of a vector C, rotating at radian frequency $\omega$.

2. The center of rotation is at $F/K$.

3. The angle swept by C, during the $i^{th}$ interval is the interval duration multiplied by the degrees per second, which is

$$\text{Angle}_i = (t_i - t_{i-1})\left(360 \frac{\omega}{2\pi}\right)$$

**Response of a System to a Given Rectangular Pulse.** The natural frequency of the given system in Fig. 10.7 is 15.6 cps. If the force F were induced by mass, say $W_1/32.2$, then the natural frequency would be

$$f_{n1} = \frac{1}{2}\pi\sqrt{\frac{5000 \times 386}{200 + W_1}}$$

In many instances the mass of the impacting component is small as compared with that of the machine or foundation and so does not change the natural frequency.

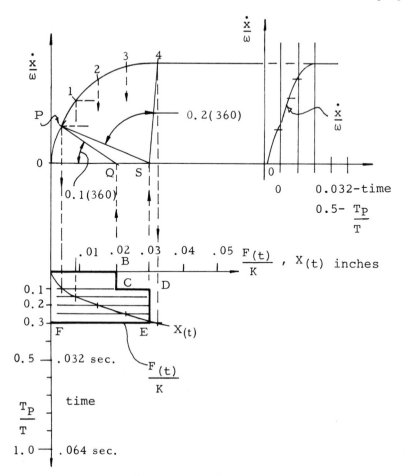

Fig. 10.7   Use of the phase-plane method to determine the displacement and velocity response of the given system to a rectangular impulse ABCDEF. The pulse amplitude is drawn in units of F/K. Point C is projected to the upper diagram to give center Q and an arc is drawn with radius QO. The angle of this arc is 0.1 (360) degrees. Point P is projected horizontally and vertically to give points on the velocity response curve and the displacement response curve. Point D is projected to upper diagram to give center S of arc radius SP and arc angle 0.2 (360) degrees. The method shown permits the determination of system response to a stepped rectangular pulse of any proportions, with minimum computation and small chance of error.

The elements of phase plane graphics are shown in Fig. 10.7. Included are two consecutive impulses of time durations $T/10$ and $T/5$, with $F/K$ equal to 0.02 and 0.03 in, respectively. It is convenient to draw the impulse in terms of $F/K$ instead of in terms of F, since this allows the impulse to be plotted on the scale of x and because the required centers for the construction are derived from the

F/K values. Likewise, it is most convenient to plot duration in terms of $T_p/T$ which is the relation of pulse duration to Natural Period. When this is done the relative pulse duration for any rectangle, multiplied by 360, gives the angle swept by vector C during the duration of the rectangle.

The two impulses shown in Fig. 10.7 are 100 lbs for 6.4 milliseconds and 150 lbs for 12.8 milliseconds. These are drawn as 0.02 by 0.1 and 0.03 by 0.2. In the upper part of the figure, the vertical axis is graduated in units of velocity divided by radian frequency. The horizontal axis is in units of $T_p/T$. Assuming that x and ẋ are both zero at time zero, an arc is drawn at center 0.02 and radius 0.02, with an angle of 0.1 times 360. If ẋ and x have values other than zero, these other values would first be plotted as a point in the upper figure. The correct radius would be the distance from the center to the point. Again, the angle would be 36 degrees. The vertical projection of the radius QP during the arc interval OP gives the displacement response, and the horizontal projection gives the velocity response.

For the second part of the impulse, F(t)/K equal to 0.03 is projected vertically to give a new center and a new arc is now drawn starting at P. It is valid to start from P because the new radius must be $(A^2 + B^2)^{1/2}$. This is so because, in fact, the new radius is one side of a right-angle triangle, wherein the other sides are $\dot{X}_0/\omega$ and $(F/K - X_0)$, and $\dot{X}_0$ and $X_0$ are, respectively, the initial velocity and initial displacements for the new impulse. For this second impulse, the arc is 72 degrees. Point 4 is projected vertically downward to give the displacement at the end of the second impulse. The displacement during a rectangular impulse (and the velocity) may be plotted as accurately as we please by dividing the arc into several divisions, such as 1, 2, 3, 4, and by dividing the vertical time interval into the same number of divisions to obtain the desired number of projected points.

**Response of a System to a Generalized Impulse.** A generalized impulse F(t)/K is shown in Fig. 10.8. The task is to find the system displacement response. The velocity response is readily obtained, as already noted, by horizontal projection from the upper diagram. In order to make the figure clear, only three rectangles are used to form an equivalent of the generalized impulse. The generalized impulse is made amenable to the phase plane method by replacing it with a series of rectangular impulses.

An arc of 36 degrees is first drawn where PQ is intersected at center 0.01, and then arcs of 108 and 72 degrees at centers 0.02 and minus 0.01. The center of the next arc now shifts to F/K equal to zero, which is the point 0. The transient state has ended and the system now executes free vibration. The time at the end of the third arc is 0.038 sec. The time duration of the arc that has a center at 0, until point P is reached, is about 121 degrees or ⅓ full cycle or 64/3 milliseconds, which is say 21 milliseconds, giving the time at P as 59 milliseconds. To plot the displacement, point P is projected vertically to a value of $T_p/T$ of 59/64. From here on the free displacement is plotted as a cosine curve with peak amplitude OP.

Because F/K is the static displacement for force F and because X(t) is the

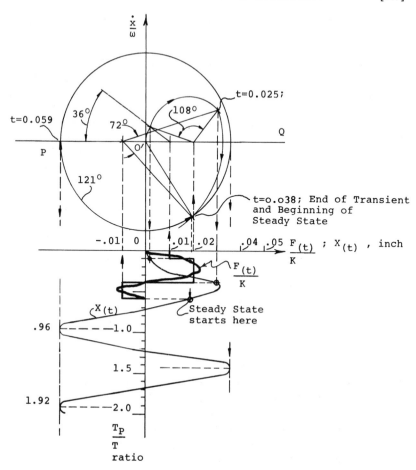

Fig. 10.8   Determination of displacement response X(t) when a generalized impulse $F_{(t)}$ is applied to the system. The given impulse $F_{(t)}$ is replaced by three consecutive rectangular impulses which have respective durations of 0.1, 0.3 and 0.2 expressed in dimensionless units $T_p/T$ and which have amplitudes of .01, .02 and −.01, respectively, expressed in units of F/K. The angular displacements of the system during the three impulses are, respectively, 36, 108 and 72 degrees, amounting to 0.6 T or 38 milliseconds. The angular displacement during an interval is given by the ratio of the interval to T, all multiplied by 360.

dynamic displacement, the graphic treatment readily illustrates the magnification that results from dynamic application of the force.

**How Response Is Affected by Variation of Force F, Other Things Being Equal.**
To determine the effect of varying F, suppose that the force F for a rectangular impulse is magnified by a factor m. Then F/K is also magnified and the radius of the phase plane arc is likewise so magnified. It follows that displacement and

velocity response, in general, are magnified by a factor m when the force is magnified by a factor m for the special case of the pulse duration being fixed.

**How Response Is Affected by the Duration of Impulse.** It is usual to consider response as being composed of a transient part and a later steady part. The distinction is not always as clearly made as might at first be supposed. The impulse of Fig. 10.2b engenders a cosine response. In Fig. 10.8 there are three such responses, all consecutive, which together make up the transient response. After that, at time 38 milliseconds, the steady state commences and endures until altered by another impulse. Here is a clear distinction. However, in Fig. 10.2a the duration of the impulse is not limited and the response is a steady cosine function from the very beginning, passing through as many cyclical vibrations as we direct. It seems proper, however, to describe this response as transient, which forces the definition that the steady vibration is that which occurs after the cessation of the pulse.

The transient part of the response is conveniently called initial, and the steady part is here conveniently called residual. During both the initial and residual intervals of time, there occur maximum values of displacement which we denote by $X_i$ and $X_r$, respectively. Similar remarks apply to velocity. When it is desired to control the response of a system, this may be done by altering the duration of the impulse or by altering the natural frequency of the system. For this reason it is of interest to determine $X_i$ and $X_r$ in terms of $T_p/T$, as $T_p$ is varied and T is held constant. A plot of $X_i$ and $X_r$ as a function of $T_p/T$ is known as a shock spectrum. This term seems misleading, and seems to have arisen from the fact that the plot is sometimes made as a function of frequency, or a frequency relation, as is done in plotting the response of a system to a forcing vibration. However, in this chapter the ratio $T_p/T$ shall be used as the independent variable against which the dependent $X_i$ and $X_r$ are plotted.

Shock spectra are derived below for three types of impulse.

1. Rectangular impulse, constant force F
2. Sinusoidal impulse (half-sine), and sawtooth impulse
3. Rectangular impulse having constant impulse value, such that F is inversely as $T_p$

**Rectangular Pulse, Constant Force F.** The derivation of the shock spectrum is shown in Fig. 10.9. The values of $T_p/T$ range from zero to one and maximum initial and residual displacements are plotted for $T_p/T$ = 0.1, 0.2, 0.3, etc. When $T_p/T$ exceeds 0.5, then $X_i$ is always twice F/K. Otherwise, $X_i$ is equal to $R(1 - \cos \alpha)$, where R is F/K and $\alpha$ is 360 $T_p/T$ degrees. For values of $T_p/T$ less than 0.5, $X_r$ exceeds $X_i$. It is evident from Fig. 10.9 that $X_r$ is equal to the square root of the sum of the squares of $X_i$ and R sin $\alpha$. Thus, for R = F/K

$$X_i = R(1 - \cos \alpha) \text{ when } T_p/T \text{ is less than } 0.5$$
$$X_i = 2R \text{ when } T_p/T \text{ equals or exceeds } 0.5$$
$$X_r = \sqrt{[R(1 - \cos \alpha)]^2 + [R \sin \alpha]^2} = R\sqrt{2(1 - \cos \alpha)}$$

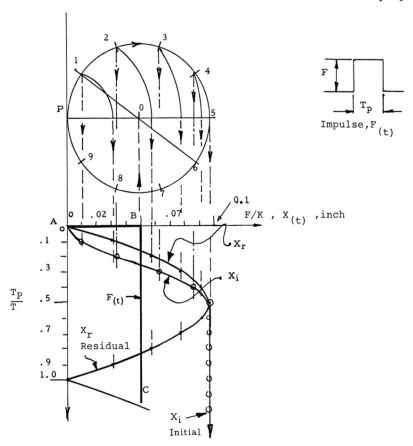

Fig. 10.9   Shock spectra for a rectangular impulse. The spectra give the maximum initial and maximum residual displacements induced by a rectangular impulse of relative duration $T_p/T$. The impulse ABC of amplitude $F/K$ = 0.05 is applied to the system for various values of $T_p/T$ up to one. The spectra are not response curves, although the two do coincide in part. For example, for a relative duration 1, the spectrum gives a value for $X_i$ (the maximum initial displacement) of 0.1; whereas at the instant corresponding to $T_p/T$ = 1, the actual value of displacement is zero. At the instant of relative duration 0.2, the residual displacement is about 0.033; but the spectrum value is about 0.058 at a later instant. To plot the spectra for any relative duration, say $T_p/T$ = 0.2, the plot of $X_i$ is found by projecting down from a point 2(2/10 way around the upper circle) to the line $T_p/T$ = 0.2. To plot $X_r$, draw an arc with center P and radius P2 and project down to $T_p/T$ = 0.2.

It is seen that $X_r$ may be diminished toward zero as $T_p/T$ approaches n, where n is zero or any integer. However, $X_i$ may be diminished only by making $T_p/T$ less than 0.5. It will be evident that in practical cases this may be achieved by increasing T or decreasing $T_p$ or both.

Instead of plotting X, the ratio $X/(F/K)$ may be plotted. The resulting dimensionless number is seen to vary from 0 to 2.

The two concepts, response curve and shock spectrum, are to be carefully distinguished. The first is a plot of actual displacement on the time ratio base $T_p/T$. The second is a plot for any given value of $T_p/T$ of the maximum values of $X_i$ and $X_r$ in the time interval from zero to, and including, $T_p/T$ and all the periods thereafter. This is, in fact, a dimensionless interval, although the term "time interval" has been used for purposes of explanation.

**Sine and Sawtooth Impulses.** The shock spectra (spectra of maximum response) for the half-sine and sawtooth impulses are given in Fig. 10.10 without derivation. The term "maximax" is sometimes used to denote the curve that indicates the greater of $X_i$ and $X_r$ for all values of $T_p/T$. For example, the maximax for a rectangular pulse is given by $X_r$ for values of $T_p/T$ up to 0.5 and is given by $X_i = 2F/K$ thereafter.

**Rectangular Impulse of Constant Value and Variable Duration Where F/K Is Inversely Proportional to $T_p$.** Referring to Fig. 10.11, an impulse of fixed value $(F/K)(T_p/T) = 0.01$ units is applied to the system. These units are convenient for drawing diagrams. However, impulse is usually measured in terms of the integral F dt. And this integral for the case of a rectangular pulse is simply $FT_p$, which in this example is $0.01KT$ or $0.01(5000)(0.064) = 3.2$ lbs-sec units.

So that the impulse area may be constant, $F/K$ diminishes as $T_p$ increases, and the following corresponding values are noticed in Fig. 10.11.

| $T_p/T$ | $F/K$ | $(T_p/T)(F/K)$ |
|---|---|---|
| .05 | .2 | .01 |
| .1 | .1 | .01 |
| .2 | .05 | .01 |
| .5 | .02 | .01 |
| 1.0 | .01 | .01 |

Using the phase plane method, the radius in each case is $F/K$ which is now variable and always equal to $0.01T/T_p = 0.00064/T_p$. The arc angle in each case is $360\,T_p/T$ degrees, or, say, $5600\,T_p$ degrees.

It is evident from the Fig. 10.11 that $X_i$ and $X_r$ shall have the same equations as given above in the section on Rectangular Pulse, Constant Force F of this chapter. But now it is to be borne in mind that R is variable. From the above:

$$R\alpha = \frac{0.00064}{T_p}(5600\,T_p) = 3.6$$

or

$$R = \frac{3.6}{\alpha}$$

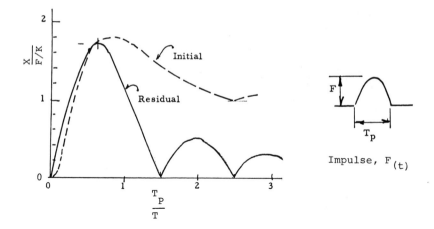

a. Half-Sine Impulse,
   Response Spectra

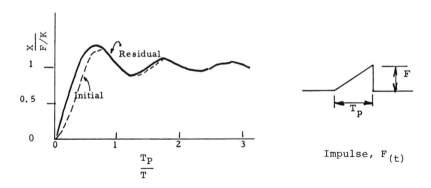

b. Saw-Tooth Impulse,
   Response Spectra

Fig. 10.10  Displacement response spectra

where R is the radius and $\alpha$ the angle. From the section noted above

$$X_i = R(1 - \cos \alpha) = \frac{3.6(1 - \cos \alpha)}{\alpha}$$

where $\alpha$ is less than 180 degrees, and

$$X_i = \frac{7.2}{\alpha}$$

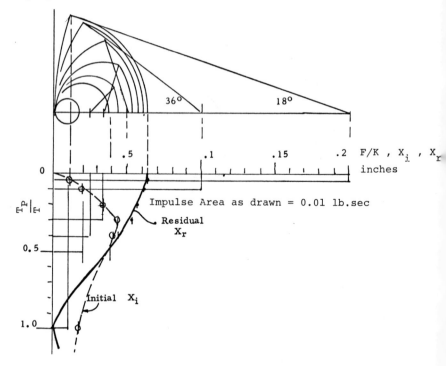

Fig. 10.11  Shock spectra for a rectangular impulse of variable duration, the impulse having a constant value so that its amplitude is inversely as its duration (Displacement spectra $X_i$ and $X_r$ are drawn.)

when $\alpha$ exceeds 180 degrees. Likewise

$$X_r = \frac{3.6\sqrt{2(1 - \cos \alpha)}}{\alpha}$$

It is seen that $X_i$ reaches a maximum before $T_p/T$ reaches the value 0.5 and then declines steadily, tending to zero as $T_p/T$ and $\alpha$ increase without limit. $X_r$ behaves differently and declines to zero for $\alpha = 360$ degrees. Thereafter, it increases again, only to reach zero once more each time $T_p/T$ passes through the values 2, 3, 4 and so on, the peak value of each successive undulation being lower than that of the preceding one.

## 4. CONCLUSIONS

The phase plane method is seen to be capable of determining the response of the simple system to various types of impact. From the phase plane diagram of the time response of the system, either displacement or velocity are readily derived, as are the shock spectra.

The particular utility of the shock spectra lies in the fact that they give the effects of varying the ratio of pulse duration to system natural frequency which is $T_p/T$.

Referring to the rectangular pulse, it is seen in Fig. 10.9 that the residual vibration may be eliminated if $T_p/T$ is zero or any integer. However, the zero value is not practicable. The initial (transient) vibration may be decreased only by arranging that $T_p/T$ is less than 0.5. When this is done, both the initial and residual response are decreased (amplitude response) as $T_p/T$ is decreased toward zero. In general, because impulses in machinery tend to be fixed in form and magnitude, the task of decreasing the response amounts to the task of increasing the natural period of the machine on its supports. The machine here corresponds to the system that has been analyzed in this article.

Using Fig. 10.10 it can be seen that the remarks of the preceding paragraph apply, with some modification, to both the half-sine impulse and the sawtooth. It is of interest to note that the greatest possible displacement amplitude for the rectangular impulse is twice the static displacement which is $2F/K$, whereas for the half-sine it is about 1.8 and for the sawtooth it is less than 1.5.

Referring to the Section about a rectangular impulse of constant value and variable duration, this example throws some light on the response due to impulses that diminish in amplitude as their duration increases. This author has not seen this example elsewhere. We may imagine a falling weight as yielding a constant impulse irrespective of its impulse duration, so long as the height of fall is constant and so long as the deformation at the point of contact is elastic. It is seen that the spectra in this example are quite different from those relating to impulses of constant amplitude in that it is necessary to increase, rather than diminish, the ratio $T_p/T$ in order to diminish the response. In order to effect a significant diminution in displacement response, it is necessary to increase the ratio $T_p/T$ considerably above about 0.25.

# SYMBOLS

| $T_p$ | impulse duration | seconds |
|---|---|---|
| T | natural period of a system | seconds |
| $\omega_n$ | natural radian frequency of a system | radians per second |
| $f_n$ | natural frequency of a system | cycles per second |
| X(t) or x | instantaneous displacement | inches |
| X | maximum value of X(t) | —— |
| $X_i$ | maximum value of initial displacement achieved at a given instant or during the preceding time. When describing the basis of the phase plane method, expecially in Fig. 11.5, the subscript i refers to the $i^{th}$ period; and it is used there in a special sense that is evident in the figure. | —— |
| $X_.$ | maximum value of residual displacement achieved in all of the residual period | —— |
| $x_i$ | instantaneous value of x during the initial period | —— |
| $x_r$ | instantaneous value of x during the residual period | —— |
| U(t) or u | instantaneous value of displacement excitation | —— |
| $\alpha$ | $(T_p/T)360$ | degrees |

# 11

## SHOCK AND IMPACT
## IN BASE EXCITATION

*Michael P. Blake*

### 1. SCOPE

The transient and steady behavior of an undamped mass and spring system with one degree of freedom is considered in this chapter. Given the system of Fig. 11.1, the task is to find its response to a variety of displacement excitations applied through the base. The phase plane method is described and applied to finding the response and shock spectra for a rectangular impulse and to finding the response to a generalized impulse. Displacement response is given chief consideration, while it is made clear that velocity response is readily determined using methods analogous to those used for displacement. The response and

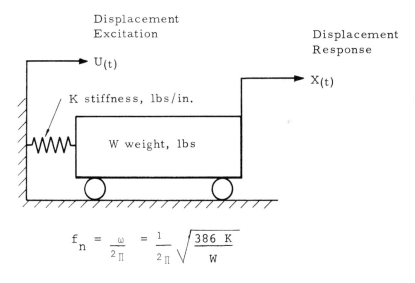

Displacement
Excitation

$U_{(t)}$

K stiffness, lbs/in.

W weight, lbs

Displacement
Response

$X_{(t)}$

$$f_n = \frac{\omega}{2\pi} = \frac{1}{2\pi}\sqrt{\frac{386\ K}{W}}$$

Fig. 11.1  The system

spectra for impulses, applied to the mass, rather than to the base, are discussed in Chapter 10.

Shock and impact are analogous and, in practice, equivalent terms defined as being a transient transmission of mechanical energy to or from a system, the duration of the transmission being notable for its brevity in relation to other transient transmissions in the same system.

## 2. INTRODUCTION

This chapter deals with just a few of the endless analytical considerations that relate to shock in trucks, automobiles, trains and the like. When rail cars collide or when a truck with a stiff wheel-suspension passes over a hump, they are excited through the medium of a spring. On the other hand, the machine mass in a drop hammer is excited through the mass. It is of interest here to determine, for example, the effect of the duration of an impulse or step due to the road surface upon the displacement induced in an automobile body. Although this chapter is confined to a single spring with undamped mass, nevertheless the discussion throws a useful light on the behavior of actual vehicles, which are usually comprised of several springs and damping elements.

It is instinctively known that a soft (long natural period) automotive suspension limits the body vibration. Likewise, it is known that if an automobile is driven over a long hump or, in an extreme case, over a hill, the body movement will follow the contour, irrespective of how soft the suspension may be. The body response has a very long period so that it is not physiologically troublesome, particularly because automobile speeds are low. Corresponding accelerations relating to low-frequency vibrations are low. On the other hand, aircraft may sustain a significant acceleration for long periods so that, in practice, it is not possible to isolate either passenger or freight from the full value of the overall acceleration.

The effect on a vehicle of a road surface producing regular wave-like disturbances is readily determined using methods applicable to forced vibration. These methods, however, are not applicable to shock analysis, wherein a typical task is to find the response when a truck wheel encounters a sudden step, followed by a smooth surface of·variable length.

In practice, vehicle suspensions are made about as soft as fashion or engineering considerations dictate, usually with some loss in overall stability while at the same time roads are constructed so that the durations of bumps at typical vehicle speeds are small or extremely long, such as for a hill, but not intermediate in duration.

Shock absorbers, so called, are dampers. They do not, in the ordinary sense of the words, absorb shock. In fact, they may increase the vertical velocity of the vehicle body during an ascent or descent of the wheels. Shock absorbers do, however, tend to limit the overshoot displacement of the body when the wheels perform transient ascents or descents and they do increase the overall stability. Their greatest effect is to quickly eliminate the steady-state vibration at suspension frequency which would endure for a long time without shock absorbers. This frequency is usually below about 10 cps and is experimentally known to be most physiologically disturbing. If freight, rather than the person, is the matter of interest, the shock absorber saves the freight from undue durations of vibration and, to an extent, relieves it of the wear and tear that may be connected with self-damping. In relation to some kinds of freight, it is conceivable that shock absorbers may harm rather than improve the ride.

The shock spectra are a most enlightening key to a host of problems connected with shock, just as the response curves for forced vibrations are a key to a host of problems connected with steady vibration.

In the following discussion, the same system, namely 200 lbs of weight and a stiffness of 5000 lbs/in, is used throughout. The rectangular impulse forms the basis of the analytical approach, not because it typifies what is found in nature, but rather because it permits the use of the phase plane method and because any kind of impulse may be replaced by a series of rectangles. This leads to a very small error. The reader can assure himself that this is so by comparing a phase plane computation for an impulse such as the half-sine with an exact mathematical result, as given in vibration texts.

In general, the forms of impulse measured in practice do not correspond to classical forms such as the sine or sawtooth; and they may not therefore be amenable to mathematical analysis, as regards response, except expression in the form of an equation, is possible. By replacing the measured impulse with a series of rectangles, it is immediately amenable to response determination, using the phase plane method.

## 3. A COMPARISON OF BASE EXCITATION AND MASS EXCITATION

Referring to Fig. 11.2, both kinds of excitation lead to similar results. A rectangular impulse of displacement amplitude U applied to the base leads to a vibration of semi-amplitude U and frequency $1/T$ so long as the pulse duration $T_p$ exceeds $T/2$. Likewise, a rectangular impulse of force F applied to the mass leads to a vibration of semi-amplitude $F/K$, with frequency $1/T$. Later, the application of the phase plane method shall make it clear that the system behavior is quite similar for base excitation and mass excitation.

**The Basis of the Phase Plane Method.** The method rests on the fact that the system of Fig. 11.1 behaves harmonically when left alone or free. Two classes of free vibration arise here: that in which the base is in its normal position, and that in which the base is in some other fixed position, due to the intervention of a rectangular impulse. The first class is called a residual or steady vibration. The second is called an initial or transient vibration. Both vibrations have the same radian frequency, although the initial class may not be allowed enough time to execute more than a fraction of a cycle.

Thus, referring to Fig. 11.3, the transient response to a step-like excitation must simply comprise a series of fragments of cosine curves, where the fragments differ in phase and curve amplitude but not in curve frequency. Fig. 11.3a shows three, among the unlimited number of free vibrations. Figure 11.3b shows a step-like excitation. The first step might induce, for example, a cosine response OA, the second CD and the third EF. These are all assembled in

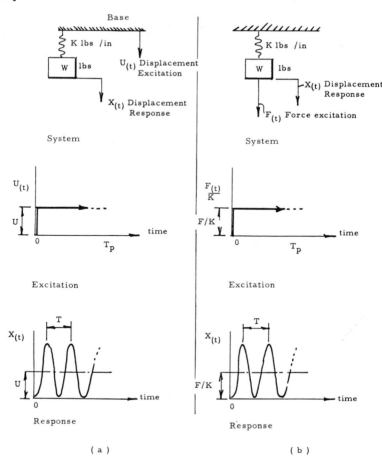

Fig. 11.2 Comparative effects of a rectangular impulse of duration $T_p$ greater than T, when applied to the simple system either as a base displacement or a mass impulse

Fig. 11.3c to illustrate the composition of a possible response. These considerations make it possible to use the phase plane method with some understanding and without recourse to mathematical equations. The equations are given below, together with a derivation of the phase plane method.

We consider the $i^{th}$ rectangle of a series of rectangular impulses 1, 2, 3, 4, . . . , as shown in Fig. 11.4. The equation of motion is

$$M \ddot{x} = K(U_i - x)$$

where x is the same as X(t) and $\ddot{x}$ is the acceleration. The solution of the equation of motion is

$$x - U_i = A_i \cos \omega(t - t_{i-1}) + B_i \sin \omega(t - t_{i-1})$$

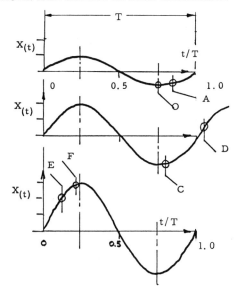

a. Some Free Response Curves

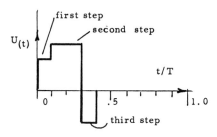

b. Rectangular Displacement
Excitation of Base

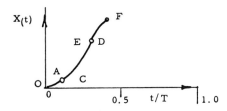

c. Transient Response, Displacement

Fig. 11.3   How the transient response for a rectangular excitation comprises successive fragments of cosine curves. During any step, the base is fixed so that the system executes some part of a (free residual) cosine curve. For the three steps of 3b, the corresponding cosine responses are, for example, OA, CD and EF. The parts OA, CD and EF are parts of three cosine curves, all having the same natural period, T and all differing in phase and amplitude.

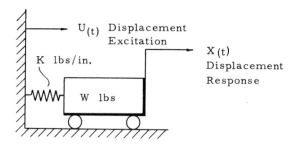

The System

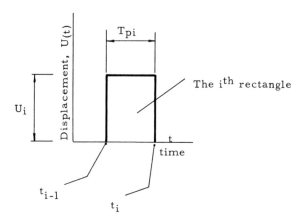

The Excitation

Fig. 11.4   Rectangular impulse excitation

Differentiating the solution we get the velocity $\dot{x}$, where

$$\frac{\dot{x}}{\omega} = -A_i \sin \omega(t - t_{i-1}) + B_i \cos \omega(t \quad t_{i-1})$$

Squaring and adding the two equations and remembering that $\sin^2 + \cos^2 = 1$, we have

$$\left(\frac{\dot{x}}{\omega}\right)^2 + (x - U_i)^2 = A_i^2 + B_i^2$$

which is the equation of a circle of radius $R = (A_i{}^2 + B_i{}^2)^{1/2}$. When $t = t_{i-1}$ and $x = x_{i-1}$, then $A_i = x_{i-1} - U_i$; and further $B_i = \dot{x}_{i-1}/\omega$, where $\omega = (K\ 368/W)^{1/2}$. From the term $(x - U_i)^2$ we note that the center of the circle is at $U_i$.

Figure 11.5 is a graphic presentation of the above results and considerations.

**Shock Spectra and Response, Rectangular Impulse.** Although a truly rectangular impulse cannot occur in practice since it connotes an interchange of energy taking place in a period of zero duration, nevertheless the rectangular pulse is perhaps the most tractable as regards analysis. For this reason it is profitable to understand the effects of the rectangular impulse. Familiarity with these effects enables us to analyze a variety of practical impulses by substituting a family of rectangular impulses to take the place of the real impulse and be equivalent to it, before analysis proceeds. The response of the system to the rectangular impulse is shown in Fig. 11.6.

In this figure, the subscript i refers to initial or transient conditions, and r refers to residual or steady conditions. The i has now no relation to the i that was previously used in connection with the $i^{th}$ interval. X and U denote maximum values of these functions, whereas x and u denote instantaneous (undefined) values. For example, $X_i$ is the greatest value of $x_i$ during the initial period, and $X_r$ is the greatest value of $x_r$ during the residual period. For the initial period

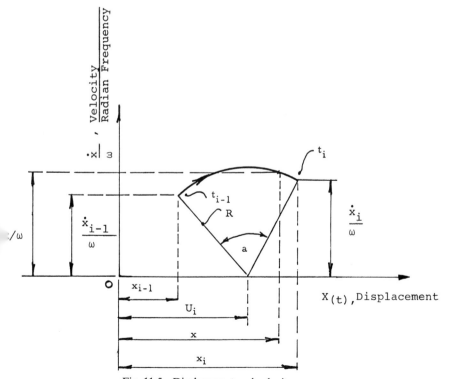

Fig. 11.5   Displacement and velocity response

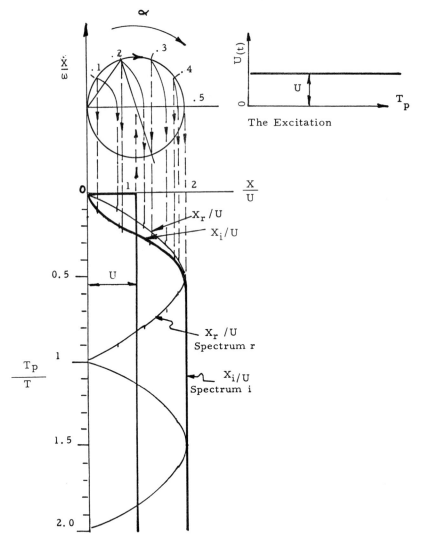

Fig. 11.6   Shock spectra for mass displacement, due to base displacement
by a rectangular impulse excitation

$x_i$ and $X_i$ are the same for values of $\alpha$ through 180 degrees; thus, the response and maximum response, or response and spectrum, are all the same thing. When $\alpha$ exceeds 180 degrees, $X_i$ is continually 2u, whereas $x_i$ equals $u(1 - \cos \alpha)$, for $0 < \alpha < 180$. Both the initial response and initial spectrum are shown in the figure, whereas only the spectrum is shown for the residual period.

**Response for a Generalized Impulse.** A generalized pulse $U(t)$ is shown in Fig. 11.7. A series of seven rectangular steps are substituted so as to be about equiva-

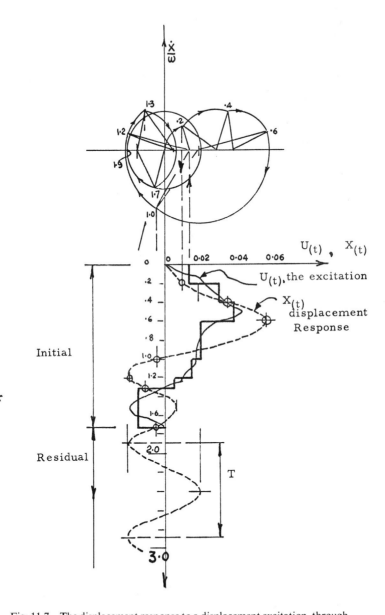

Fig. 11.7  The displacement response to a displacement excitation, through the base, using the phase-plane method. The rectangular steps are substituted to replace the generalized excitation $U_{(t)}$. For each rectangular step an arc is drawn with center at step height and angle equal to step duration $(360/T)°$.

lent to the pulse U(t), and the phase plane method is used to find the displacement response X(t) induced by the displacement pulse U(t).

## 4. SOME CONCLUSIONS

It is seen that the considerations and results relating to a displacement excitation applied through the base are largely similar to those relating to a force excitation applied through the mass, as discussed in Chapter 10. Again, the phase plane method is seen to be most useful, informative and not readily conducive to gross error.

The response spectra for the rectangular impulse gives an insight into the probable behavior of vehicle suspensions when the vehicle encounters a bump. Response spectra for sine and other impulse forms are available in the literature of shock analysis. The treatment given in this article is, of course, applicable to problems not connected with transit, such as the problem of isolating a machine from an impulse delivered through its base, which base might be, in fact, an elastic girder or spring mounting.

Although the given treatment is severely confined to a very simple system, it can afford sufficient insight into a variety of problems to enable a desirable correction to be made.

# BIBLIOGRAPHY

Harris, C. M. and Crede, C. E. *Shock and Vibration Handbook*, 1st ed. New York: McGraw-Hill Book Co., 1961.

Jacobsen, L. S. and Ayre, R. S. *Engineering Vibrations*, New York: McGraw-Hill Book Co., 1958.

Bruel & Kjaer Technical Review No. 3, 1966.

# 12

# DISCOVERY AND
# IDENTIFICATION OF FAULTS

*Michael P. Blake*

## 1. SCOPE

This chapter discusses current practice in discovering faults through the reading of common vibration survey type meters. It is pointed out that the typical velocity meter is adequate for simple vibrations over the frequency range from about 30 to 1000 cps. For complex vibrations, a true *peak* meter is recommended for discovery and classification in the given frequency range. Outside this range the wave analyzer is recommended for simple and complex vibration classification and also for fault identification. The response of typical meters to complex vibration is discussed together with the topic of classifying complex vibrations. Several conclusions and recommendations are included, along with relevant examples.

## 2. ELEMENTS OF DISCOVERY

Noise and vibration have always been the signals that reveal the presence of a mechanical fault. Thus, discovery has been a personal art. But with the existence of vibration meters it becomes possible to make discovery quite scientific. The vibration of a machine is measured and then compared with standards which, in turn, enable the observer to judge whether or not a fault is present, and if so, to classify it as insignificant or severe. Typical oscillographs of machine vibrations are illustrated in Figs. 12.1 and 12.2.

A fundamental sine wave of 30 cps is shown in Fig. 12.1a. A fifth harmonic wave, having an amplitude of 50 percent of that of the fundamental is shown added to it in Fig. 12.1b. When the amplitude of the harmonic is increased to 100 percent, we have the result shown in Fig. 12.1c. In Fig. 12.2, tenth harmonics with 50 percent and 100 percent fundamental amplitudes are added to their fundamental vibration waveform. When a vibration meter is subsequently used to measure these as machine vibrations, it yields a single reading, such as 0.9-in/sec peak. Referring now to Fig. 12.3, which is a typical vibration standard,

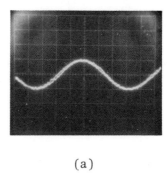

(a)

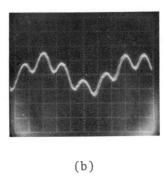

(b)

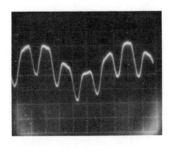

(c)

Fig. 12.1   A fundamental sine-wave signal and fifth harmonic effect

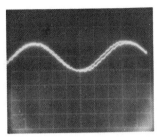

a. 30 cps. Fundamental

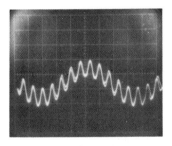

b. With 50% of Tenth Harmonic

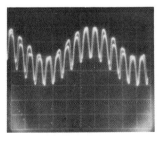

c. With 100% of Tenth Harmonic

Fig. 12.2   A fundamental sine-wave signal and tenth harmonic effect

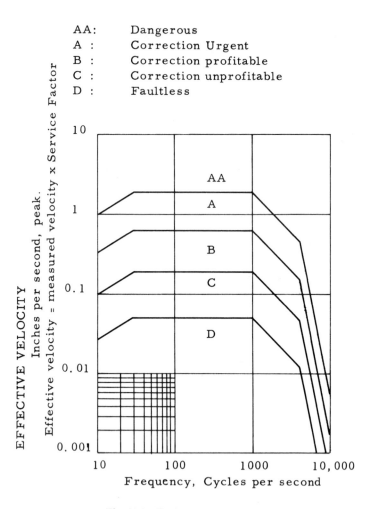

AA: Dangerous
A : Correction Urgent
B : Correction profitable
C : Correction unprofitable
D : Faultless

Fig. 12.3  Typical velocity standard

it is noticed that 0.9-in/sec peak is classified as being almost dangerous. Thus, the machine exhibits a fault that requires immediate attention.

The elements of fault discovery are, therefore, relatively simple and can be given as three steps:

1. A vibration measurement;

2. The measured vibration is multiplied by a service factor that takes account of the criticality of the machine. This yields what is called effective vibration;

3. The effective vibration is then classified from a standard, such as that shown in Fig. 12.3.

## 3. THEORY OF DISCOVERY

Through the use of a standard such as that of Fig. 12.3, discovery and classification arise at the same time. An existing fault is discovered, and it is automatically classified as lying in the region AA, A, B and so on. Simple as this procedure may appear, it required years of field experience—from the time of the early workers such as T. Rathbone to the present—to develop the procedure and foster its acceptance. Many difficulties still await resolution but before concentrating on these, we consider some further aspects of elementary classification.

It is noticed that Fig. 12.3 is arranged in terms of velocity as a criterion. There are two reasons for the growing use of velocity instead of displacement as the unit of measurement for purposes of classification. First, experience has shown that vibrations of various frequencies may be taken as having the same severity if they exhibit the same velocity. That is to say, if the frequency is doubled the fault is no worse provided the amplitude of the vibration is halved. Second, concerning the matter of severity judgments, national and international bodies, such as ANSI and ISO (American National Standards Institute, formerly United States of America Standards Institution and the International Organization for Standardization) are developing standards that are written in terms of velocity.

When velocity is used as a unit of measurement, it is not necessary to be concerned about frequency; whereas if displacement or acceleration is used, it is not possible to classify the severity of the vibration without knowing the response frequency. This will become clear by reference to Fig. 33.1 of this book. If the frequency exceeds the limits of about 30 to 1000 cps, then it must be taken into account. Some authorities prefer to contend that velocity is proportional to severity only in the range to about 600 cps, but the writer has based the standards of this handbook on a limit of 1000 cps, which coincides with proposed national and international standards.

When the dominant frequency of the measured vibration exceeds the given (flat) range, then it is not possible to use the noted standards without reference to the response frequency. In order to determine the dominant frequency, a wave analyzer or oscilloscope may be used, or the vibration may be measured both in terms of acceleration and velocity on an appropriate meter in order to derive frequency from the following relation. Velocity (in/sec peak) = 0.5 times displacement (inches peak to peak) multiplied by radian frequency.

$$V_{IPSP} = \frac{1}{2} \omega X_{PP}$$

Knowing the frequency of a simple vibration, such as shown in Figs. 12.1 or 12.2, the equivalent velocity is entered in Fig. 12.3 at the appropriate frequency in order to classify the severity.

## 4. CLASSIFICATION OF SIMPLE VIBRATIONS

Examples of simple vibration are shown in Figs. 12.1a and 12.2a. Such vibration is best measured in terms of velocity, and it is of no consequence whether measurements are expressed in terms of P, PP, average, or rms. It is usual, however, to measure velocity in terms of peak, although rms values have been predominant in discussions which might eventually lead to international standards. Some meters measure peak and some rms and so on. Some measure *true* rms and some read out an rms value from a meter that has an average response by using a factor to correlate average and rms values.

The classification of simple vibrations poses little difficulty once an understanding has been gained of the related units and procedures. However, when complex vibrations, such as illustrated in Figs. 12.1b, 12.1c, 12.2b and 12.2c, are considered the state of the art affords little help, and the vibration engineer must direct his attention to matters that are not only not at all well settled, but are such that by their nature it is unlikely that easy answers to related problems will ever appear. On the other hand, semi-arbitrary standard procedures are likely to emerge as national standards after discussion and research progresses to a new technical level.

## 5. MEASUREMENT OF COMPLEX VIBRATION

The simple vibration is such that a meter reading of its motion is proportional to its amplitude, irrespective of whether the meter reads average or peak or other. When the amplitude doubles, the reading doubles. For the complex vibration matters are not so simple. First, well-informed procedures are required for measurement. Second, classification does not result automatically when the measurement has been achieved. We first dismiss the relatively easy matter of measurement and then try to grasp the nettle of classification. Although measurement and classification are separated here for convenience of understanding, they are in fact inextricably related, just as are classification and identification.

In order to exemplify the response of various measuring instruments to complex vibration signals, two sets of signals have been generated and recorded on tape for the purposes of this chapter. They are illustrated in Figs. 12.1 and 12.2. The various signals were connected to measuring instruments, while an oscilloscope was used as a monitoring device. The arrangement is shown in Figure 12.4.

Because the relative amplitudes of Figs. 12.1a, 12.1b and 12.1c are 100, 150 and 200, respectively, it is desirable that a measuring instrument give readings that reflect these different values. The relative amplitudes of Figs. 12.2a, 12.2b and 12.2c are the same, i.e. 100, 150 and 200, except that the amplitude increase in this case is achieved with a tenth harmonic instead of a fifth harmonic. Table 16 illustrates the fact that a wave analyzer does reflect the increasing amplitude of the signal in all instances. The important conclusion here is that the oscilloscope and the wave analyzer faithfully reflect the change of the peak amplitude of the complex signal.

Coming now to simpler measuring instruments, it is instructive to consider how they respond to amplitude variations. Table 17 gives the response for three different typical meters, namely a meter capable of average and peak indication, a wave analyzer used as an all-pass meter and a simple vibration velocity meter. While the signal-peak amplitude ranges from 100 to 200, the average-response meter almost ignores the change, increasing 24 percent with a fifth harmonic increase and only 18 percent with tenth harmonics. The velocity meter shows a 33 percent increase in reading, the wave analyzer yields 47 percent and 45 percent and the peak meter shows 76 percent and 69 percent, respectively. All of the meters failed to reflect what was happening. (The rms meter may be set aside for the moment as it is not expected to follow peak variations.) The average meter may be rejected for this measurement because all six signals have the same average rectified value, so that a perfect average-response meter and a perfectly generated complex wave might together be expected to read 100 for all six signals.

## TABLE 16
### OSCILLOSCOPE AND WAVE ANALYSIS READ-OUT FOR SIGNALS OF FIGS. 12.1 AND 12.2

| SIGNAL | | | READ-OUT | | | |
|---|---|---|---|---|---|---|
| | | | WAVE ANALYZER, GENERAL RADIO 1564-A, ONE-TENTH OCTAVE ANALYSIS, QUASI RMS | | | |
| | OSCILLOSCOPE | | 30 CPS | | 150 CPS | |
| FIGURE | mVPP | RATIO | mV | RATIO | mV | RATIO |
| 1 a | 400 | 100 | 215 | 100 | ——— | ——— |
| 1 b | 600 | 150 | 215 | 100 | 110 | 51 |
| 1 c | 800 | 200 | 220 | 102 | 220 | 102 |
| | | | 30 CPS | | 300 CPS | |
| 2 a | 400 | 100 | 215 | 100 | ——— | ——— |
| 2 b | 600 | 150 | 215 | 100 | 115 | 53 |
| 2 c | 800 | 200 | 220 | 102 | 220 | 102 |

The above table illustrates the read-out of an oscilloscope and a wave analyzer of the signals that are shown in Figs. 12.1 and 12.2. The arrangement of the instrumentation is shown in Fig. 12.4. The important conclusion is that the wave analysis faithfully reflects the components of the complex waves of Figs. 12.1 and 12.2. On the other hand, an all-pass reading of the analyzer does not reflect the extent of the PP increase that occurs when harmonics are added to the fundamental signal.

Consider, however, the response of the two peak instruments. One ranges from 100 to 176 (or 169) and the other ranges from 100 to 133 when both should range from 100 to 200. No doubt this is because the meters are arranged to indicate a peak value that is derived from some kind of average or rms detection. Both peak meters would have ranged from 100 to 200 if the signal had been a simple one (one frequency) and had ranged from 100 to 200. The fact remains, however, that these meters do not reflect the true change in the peak amplitude of the complex signal.

Oscilloscope,                    Measuring Instrument,
used as a Monitor              Meter Type

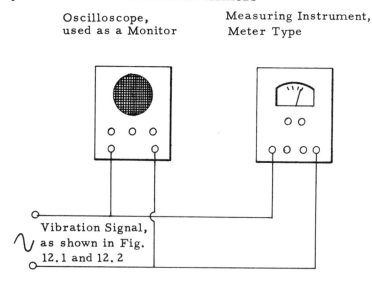

Fig. 12.4    Measurement system in which an oscilloscope is used to monitor
a meter read-out

## 6. CLASSIFICATION OF COMPLEX VIBRATION

It has been noted that some meters do not respond to a complex signal in the
true way that an oscilloscope responds, and that a wave analyzer analysis can be
almost as accurate and revealing as that derived from an oscillograph. Moreover,
sound-level meters do not have a response like that of an oscilloscope, but are
considered, nevertheless, to be very useful. In order to determine, therefore, what
kinds of meters or instruments are adequate and, indeed, what kind of detection
is desirable, it is first necessary to answer one vital question: "Is the complex
wave, with its peak amplitude of 200 twice as severe as the simple wave with its
peak amplitude of 100; or is the complex wave in fact any more severe than the
simple wave?" In other words, what is the severity criterion for complex signals?
We can answer this question in part. Fig. 12.5 illustrates the sort of amplitude-
frequency spectrum that derives from Fig. 12.1c with its equal components
at 30 and 150. The present state of the art accepts the fact that both components
in Fig. 12.5, the fundamental or the harmonic, are equally severe. In addition,
if either component is doubled in amplitude, the severity is doubled. Without
going into a detailed discussion it seems almost unavoidable to conclude that
the severity of a signal is arithmetically proportional to its peak displacement
amplitude.

## 7. METERS FOR COMPLEX VIBRATION

Taking the conclusion given above that the severity of a signal is directly proportional to its peak displacement amplitude, there is no difficulty as regards measur-

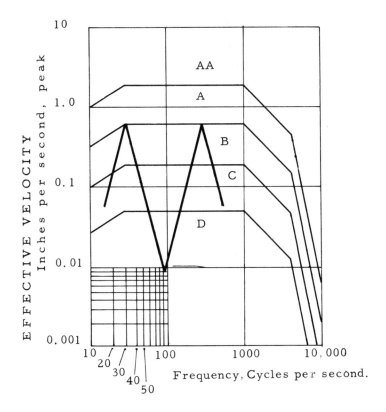

Fig. 12.5   Amplitude-frequency spectrum corresponding with the signal
of Fig. 12.1c

ing and classification, except when some frequency component falls outside the
frequency range of 30 to 1000 cps.

**The frequency band 30 to 1000 cps.** The oscilloscope indicates true peak. At the
same time it indicates if the frequency band is exceeded so that it is totally satisfactory, except for the limitation that it is not always easy to determine the frequency components. The wave analyzer shows the true peak by simply adding
the components that are within the stated 30 to 1000 cps band. An averaging
meter does not reflect the true peak value. The rms meter makes a better attempt
but the peak meter gives the best result. The most desirable meter for classification is, therefore, the peak meter. Next is the rms meter, while the averaging meter
is misleading and unavailing.

It can be deduced from these considerations that a graph that determines a
particular severity, such as any of the graphs in Fig. 12.3, is not a limiting envelope for amplitude-frequency spectra. On the contrary, it is the criterion graph

## TABLE 17

### THE RESPONSE OF FOUR DIFFERENT READ-OUTS TO THE SIGNAL WAVEFORMS ILLUSTRATED IN FIG. 12.1 AND 12.2.

| SIGNAL | | | READ-OUT | | | | | | | |
| --- | --- | --- | --- | --- | --- | --- | --- | --- | --- | --- |
| | OSCILLOSCOPE | | VIBRATION METER GENERAL RADIO 1553-A | | | | WAVE ANALYZER GENERAL RADIO, 1564-A, QUASI RMS DETECTION | | VIBRATION METER IRD., VIBRATION SELECTOR, #320 | |
| FIGURE | mVPP | RATIO | IPS² AVE. | RATIO | IPS² PEAK | RATIO | mV, ALL-PASS | RATIO | IPSP | RATIO |
| 1 a | 400 | 100 | 710 | 100 | 1250 | 100 | 220 | 100 | 0.165 | 100 |
| 1 b | 600 | 150 | 750 | 105 | 1750 | 140 | 250 | 113 | 0.180 | 109 |
| 1 c | 800 | 200 | 880 | 124 | 2200 | 176 | 325 | 147 | 0.220 | 133 |
| 2 a | 400 | 100 | 720 | 100 | 1300 | 100 | 220 | 100 | 0.165 | 100 |
| 2 b | 600 | 150 | 750 | 104 | 1800 | 138 | 250 | 113 | 0.180 | 109 |
| 2 c | 800 | 200 | 850 | 118 | 2200 | 169 | 320 | 145 | 0.220 | 133 |

for the arithmetic sum of individual components of a vibration signal. This is most important and is illustrated by the fact that whereas both of the spectral peaks in Fig. 12.5 are of the same severity, the spectrum as illustrated is nevertheless twice as severe as that wherein only one of the components is present. Thus, when a spectrum begins to swell, so to speak, as a machine deteriorates, matters become worse even if the swelling does not take the spectrum above the graph line that corresponds to the original component of greatest amplitude.

**The frequency band below 30 cps and above 1000 cps.** The frequency components as determined by a wave analyzer are added in any convenient way in order to determine the total peak value of the signal. The oscilloscope is valuable in that it gives the phase relations of components and a quick impression of the peak value. However, the components that lie outside the band of flatness (30 to 1000 cps) must be weighted before adding them to those that lie inside in order to arrive at a final computed peak value (see Fig. 12.3). On the whole, the wave analyzer is the most useful analytical instrument.

## 8. STANDARDS

For classification, the standard that has been quoted is that of Fig. 12.3, which is a typical standard adapted mainly to simple signals. Several difficulties have been noted in the matter of classifying complex signals. Many of the difficulties that exist could be avoided to an extent if a spectral-envelope standard were available. A spectral envelope is simply a curve drawn in the same units as those of Fig. 12.3 and such that a machine that is measured may be considered as faultless, rough, etc. Classification such as faultless, rough and so on would depend on whether the measured amplitude-frequency spectrum falls below a given standard spectral envelope which is based on experience or on a maker's recommendations. This involves more difficulties than may appear at first sight. Nevertheless, such envelopes can be useful in many applications. But other difficulties could arise from such a standard, as the reader may imagine, so that a standard of the type shown in Fig. 12.3 is generally the most useful. Figure 12.6 gives an impression of how the spectrum of an electric motor vibration may vary in practice. This should be interpreted with caution since it involves the vibrations of rolling bearings which are somewhat random and, therefore, beyond the scope of this discussion.

## 9. IDENTIFICATION

Identification is usually a simple procedure. It follows naturally after the fault has been discovered and classified. Again, here the wave analyzer is of great help. The vibration meter of the usual, simple kind is of little avail. Frequency is the key, sometimes helped by phase measurement when it is possible. Frequencies corresponding to rpm usually connote simple unbalance; frequencies corresponding with gear-tooth or other computed frequencies invariably identify the fault as a gear fault, a blade fault or some other internal fault. Various faults, identified in terms of frequency, are given in Table 18.

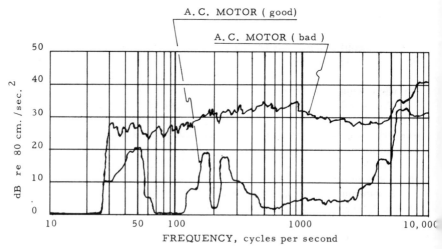

Fig. 12.6   Vibration spectra for an electric motor

**TABLE 18**
**FAULT IDENTIFICATION**

| FREQUENCY MULTIPLE (RELATIVE TO ROTATIVE OR MACHINE FREQUENCY) | PROBABLE IDENTIFICATION |
|---|---|
| 1 | Unbalance |
| 1 (sometimes 2) | Bent shaft |
| 1 | Cyclic fault in sleeve bearings |
| 2 | Misalignment |
| 2 | Looseness |
| 1 to 5 (belt rpm × 2) | Belt faults |
| 10 to 100 | Gear teeth |
| 1 | A single gear fault |
| 20 to 100 | Rolling bearings (frequency unsteady) |
| Less than 1 | Oil whip |
| Beat frequency | Two sources of vibration |

After Reliance Electric Co., *Instruction Manual*, TM–132–63, June 1967.

## 10. CONCLUSIONS

### Periodic Simple Vibration

1. Any vibration meter whether measuring in peak, peak-to-peak, rms or average is suitable for classification provided that the frequency is within

the range of 30 to 1000 cps (and this range includes the majority of machine vibrations).

2. For identification at any frequency, or for classification at frequencies outside of this range of 30 to 1000 cps, a wave analyzer is required.

### Periodic Complex Vibration

1. For classification in the range of 30 to 1000 cps the true peak meter is recommended, since it gives the greatest indication of increasing amplitude. The true rms meter is second best. The meter having an average response is not recommended.

2. For the most accurate classification in the range of 30 to 1000 cps, the wave analyzer is recommended, possibly supported by the oscilloscope.

3. For identification at any frequency, or for classification in the range outside of 30 to 1000 cps, a wave analyzer must be used.

### Other Considerations

1. The true peak meter is recommended for survey work, i.e. for the work of discovery and sometimes of classification; the wave analyzer is recommended for either discovery, classification or identification, since it is the more accurate and penetrating measurement instrument.

2. For discovery, classification and identification, and mainly for the first two because of the foregoing discussion, a peak or peak-to-peak mode of measurement and detection is recommended with second preference being given to rms. No recommendation is given at all to average detection. The rms meter is not recommended if it is, in fact, a meter having average detection and merely expresses its read-out in terms of an ostensible rms.

3. It is to be borne in mind that every signal, every machine condition, is a random signal in the sense that irrespective of a long history of periodicity or steadiness there is no guarantee whatever that it shall not become otherwise with little or no notice. The same remark applies to the problem of stating what is the next term in the series $1 + 2 + 4 + 8 + \ldots$. In fact, we have no way of knowing the next term, other than by mathematical stipulation. In the same vein, the unheralded 'blip' or other change in a vibration signal may be a significant warning or a significant event. We may think of the seismic history of a territory that has had an earthquake. A slow-response-long-term rms history virtually ignores the quake. A peak history does not. Likewise, an rms response virtually ignores steady harmonics except they exceed say 30% of fundamental amplitude. The point that is here therefore advocated is that if we may only have one detection mode, the peak is more likely to guarantee acceptable monitoring than is the rms. But the rms is a valuable additional information item and capability.

# BIBLIOGRAPHY

*Instruction Manual*, TM–1326–3. Cleveland: Reliance Electric Co., June 1967.

Blake, Michael P. "Vibration Standards for Maintenance." *Hydrocarbon Processing and Petroleum Refiner* 43 (January 1964): 111–114.

*Technology Interchange*. Chicago: Lovejoy, Inc., 1969.

*Instructions and Applications Manual*, Accelerometers 4308/09/10/11. Cleveland: Bruel and Kjaer, 1960.

# 13

## PHASE MEASURING
## FOR VIBRATION ANALYSIS

### Lionel Lortie

### 1. INTRODUCTION

Phase measuring, which is one of the more recent techniques in practical vibration analysis, is also one of the most powerful and most useful techniques. New applications and possibilities are indicated in almost each new vibrations problem that is studied. In brief it is recommended that phase measuring be used for analysis in all new problems that arise, no matter how simple they may seem at first and no matter how evident the solution may appear to be. Where the solution is evident, phase measuring is likely to afford new and unsuspected insight. It encourages the observer to think in terms of phase as well as in the more usual idiom of frequency. Besides this, phase measuring is a necessary key for solving several of the more elusive situations, such as those that arise in the vibration of structures and problems wherein it is desirable to identify a troublesome source among a number of possible sources, when none of the offending machines may be shut down for purposes of trial and error.

It is then recommended without reservation that the standard field equipment of a vibrations analyst contain a phase measuring system which, in general, will be the primary measuring instrumentation for the first look at a problem. If phase measuring is not included in the first look, it is probable that half of the problem will remain undisclosed. The problem may later reveal itself by accident, where it has not been discovered through analysis. Much measurement time and labor is likely to be saved if phase measurement information is available from the outset. The usual instrumentation comprises two identical measurement channels, including pickup, amplifier and band-pass filter, each leading to a dual trace oscilloscope.

Phase measuring is used in practical applications to discover a source of vibration and to disclose the mode of transmission; to disclose whether a component is behaving in a resonant way, in a stiff way or in a flexible way; to analyze the behavior of elements of various impedances that together make up a path of transmission; and to discover the points of translation of vibration in a structure wherein vibration occurs, and by this means to locate points of maximum stress.

Phase measuring nomenclature is not yet standardized; so when nonstandard terminology is used every effort shall be made to clarify what is connoted. In this chapter we aim at describing the measurement approach and hope to make clear the great possibilities of phase measuring by giving some elementary examples that illustrate principles and techniques. The word translation as noted above connotes a change in direction or a change of mode. For example, the vertical excitation of a pylon or flagstaff induces a vertical response. However, in suitable circumstances, the vertical mode of response gives way to a lateral mode toward the upper end of the pole. A point of maximum stress occurs at or near the point of translation. If this stress is measured by means of strain gages, the pylon or flagstaff structure will be considered safe if the measured stress is within allowable limits.

The more complex problems that arise in analyzing buildings and structures are beyond the scope of this chapter. However, the article points the way to a most valuable technique that is often overlooked. It is hoped that once the reader grasps the basic principles he can extend his own techniques to cover problems that arise from time to time.

## 2. INSTRUMENTATION

Although six, twelve or even more channels, each with its own pickup and filter, are occasionally used, the basis of all phase measuring is the two-channel system. The more complicated multiple-channel systems are merely extensions of this. They are used more for purposes of convenience and dramatization or illustration than for basic survey and discovery.

For rotating machinery, a stroboscopic index on the shaft may be used as a basis of reference. As the pickup is moved from here to there the index moves, revealing phase shifts. This is mentioned because stroboscopic equipment is often available for maintenance purposes. The most versatile instrumentation and that which is recommended is illustrated in Fig. 13.1. The pairs of transducers, amplifiers and filters are identical. If they are not, they must be at least near identical in phase shift effects that may be induced by variations of signal frequency or signal amplitude. Using this instrumentation, it is usual to fix one transducer as a phase reference. It may be fixed at the suspected source of vibration, at the point of vibration reception or at an intermediate point, depending on the nature of the problem.

When, for example, a simple sinusoidal vibration occurs at a bearing it is completely described if its amplitude and frequency are determined, supposing that its direction is also known. But if this vibration changes into a rocking mode or some other mode as it moves through the structure, then measurements around the structure will not reveal what is happening except when phase is measured as well as amplitude. Without phase measuring the mechanism of transmission is not appreciated. Thus, it becomes difficult or impossible to appreciate the vibration situation and, consequently, difficult or impossible to offer solutions.

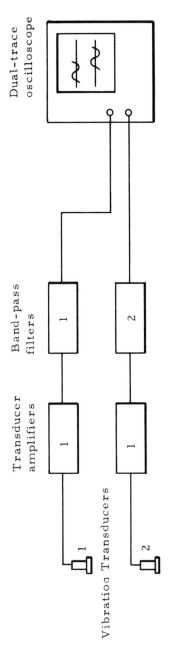

Fig. 13.1   Basic phase-measuring instrumentation

Returning to Fig. 13.1, any two-channel oscilloscope may be used providing it offers two vertical inputs and the time base is common to both traces. Also, it must be sufficiently accurate to reveal, for example, a phase shift of five degrees at a frequency of 1000 cps. Most of the usual problems lie in the frequency range of zero to 1000 cps so that even the typical dual trace that is obtained by alternating or chopping methods at a frequency of say 400 kcps is adequate for phase measuring work.

The band-pass filters are used, as may be supposed, for isolating the frequency of interest. For example, it might be that of a rotating machine, or that of an oscillating flagstaff or other structure. As noted before, the phase distortion (lead or lag) of each filter must be the same. In many applications it is of no consequence if this distortion varies with frequency, so long as the variation is the same for both filters. As may be imagined, applications sometimes arise in which it may be necessary to measure the relative phase of two signals of different frequency. In such an application, the phase distortion must be independent of frequency. For the same reasons, the above remarks apply to the amplifiers. Furthermore, if a signal waveform of variable frequency is displayed on an oscilloscope, the relative phases of the signal in the oscilloscope display shall only be accurate provided that the phase distortion of the oscilloscope amplifier or amplifiers is independent of frequency.

## 3. EXAMPLES

**Example 1: The Source of a Vibration and Mechanism of Transmission.** This example brings to light the principle that when vibration is transmitted via an acoustic coupling, the phase of the disturbed system is the same as that of the source of excitation. The term acoustic coupling is used here to include cases where there is not a simple transmission through a continuous medium but rather transmission through a first structure, then through air, water or some other medium of vastly different density or characteristic when compared with the characteristic of the source.

In this example (see Fig. 13.2), several compressors could have been the source of the offending vibration. Without phase measuring, the actual source could not have been located except by a laborious process of trial and error. The problem concerns that of a paper sheet which exhibits a thickness variation that relates to a frequency of 6 cps, taking into account the paper speed and the wavelength of the variations. A reference pickup located on the baffle of the head box revealed a vibration frequency of 6 cps. A pressure pickup in the paper emulsion revealed a pulsation frequency of 6 cps, in phase. A pressure pickup in the air space of the basement revealed an air pulsation at 6 cps. A point of interest is that the phase of the air vibration does not change as the transducer is moved from place to place, although the amplitude does change. Also, the air pulsation is in phase with the source vibration. A vibration signal from the base of one compressor in the basement exhibited a frequency of 6 cps. The fact that this compressor was in phase with the disturbed system revealed it as the offender. This was verified by turning it off. Notice that the problem appears solved without shutting down any of the machinery during measurement. Later on, a closer

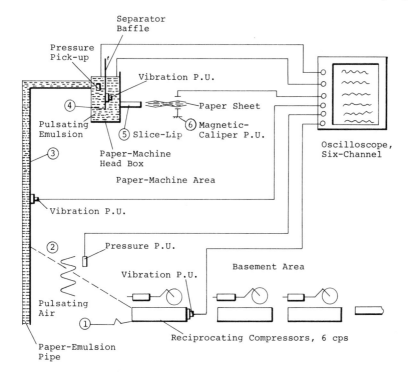

Fig. 13.2  Schematic arrangement of vibration transmission through a mechanical system. 1. Foundation of reciprocating compressor oscillating at 6 cps. 2. Fundamental frequency of room at 6 cps favoured coupling. 3. Emulsion carrying pipe which tuned to organ effect. 4. Pulsating emulsion entrained baffle in its first mode at 6 cps. 5. As baffle moved away from slice lip, emulsion thinned out and got denser as baffle moved in. 6. The magnetic caliper gauge displayed this end result as well as all the previous functions.

look at the offending compressor revealed one loose hold-down bolt. Tightening of this one bolt brought the problem to an end.

Another point of interest about acoustic coupling in a reverberant space is that the vibration of the receiving system depends instant by instant on the source vibration, so that the disturbance vanishes instantaneously if the source is brought to rest at any chosen part of its cycle. This may be demonstrated by exciting a rubber diaphragm via acoustic coupling with a loudspeaker source in a small room. The loudspeaker can be energized by means of a signal generator and amplifier. The oscilloscope is connected to a pickup on the speaker and one on the receiving diaphragm. When the loudspeaker is switched off, the signal trace becomes a straight line. At the same instant the signal trace from the diaphragm will also become a straight line. There is no period of delay.

Two useful principles have been described. In acoustic coupling the source and receiver vibrate in phase. When the source is brought to rest instantaneously, the receiver is brought to rest without delay. In this example, one operating compressor among several was discovered to be the source by determining which compressor vibrated in phase, in particular at its foundation, with the pressure in the emulsion headbox.

**Example 2: Mode of Vibration Transmission.** When a component such as a beam is excited, it is often of interest to know whether the transmission is *solid*, *resonant* or *elastic*. These terms, which are in practical use, are further described in Table 19.

TABLE 19
TRANSMISSION OF VIBRATION

| MODE OF TRANSMISSION (SEE FIG. 13.3) | PHASE DIFFERENCE, r-e | AMPLITUDE RATIO: $X_r/X_e$ |
|---|---|---|
| Solid S | Less than 20° | Little more than one |
| Resonant R | About 90° | Much more than one, as much as five or ten, depending on the amount of damping. |
| Elastic E | Approaching 180° | Less than one |

When a lightly damped system is excited at increasing frequencies, the phase relation (excitation phase minus response phase) gradually increases from zero to 90 degrees near the critical speed and to 180 degrees when the excitation frequency is far greater than the natural frequency. When damping is negligible or zero, the phase relation tends to be zero. It changes from zero to 180 degrees in one jump near the critical speed, as the frequency ratio $f_e/f_n$ increases.

Referring to Fig. 13.3, let it be supposed that $X_r$ is a troublesome vibration arising out of vibration $X_e$. It is desired to determine immediately and easily whether the mode of transmission is solid, resonant or elastic. The determination is readily achieved, as explained in Table 19 and Fig. 13.3, once the phase relation r-e is known.

Let it be supposed that the transmission is "solid." Softening of the beam or component stiffness only increases response, and increased stiffening can only give a negligible diminution. If the transmission is termed "elastic," stiffening of the beam makes matters worse ($X_r/X_e$ increases) and softening of the beam stiffness decreases the vibration response, but only by a negligible amount. Thus, when the transmission is solid or elastic, the nuisance response must usually be decreased and the problem solved through treatment of the disturbing excitation, $X_e$. It may be diminished or isolated. On the other hand, if the transmission is resonant, that is, if the phase relation is in the vicinity of 90 degrees, diminution of the response may be achieved by a modification of the beam configuration. By making it more flexible or by adding to its mass, a con-

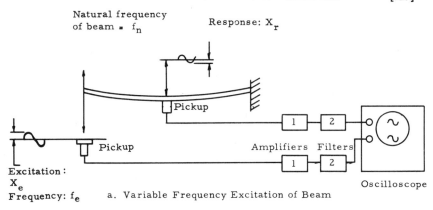

Natural frequency
of beam = $f_n$

Response: $X_r$

a. Variable Frequency Excitation of Beam

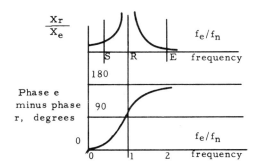

b. Phase Measurement as a Key to Mode
of Transmission, Whether "Solid, Resonant or Elastic".

Fig. 13.3   Mode of transmission of an excited beam

siderable diminution in the ratio of $X_r/X_e$ is possible. By making it stiffer (or lighter for the same stiffness), a significant diminution is possible, but not as great as that achieved by providing greater flexibility (decreasing stiffness), as is made clear in Fig. 13.3b. Briefly, a single-phase measurement makes it clear whether to devote attention to the source or to the responding system.

**Example 3: Systems with Multiple Components.** Figure 13.4 illustrates three beams excited in common by a displacement amplitude $X_e$. The response of A is solid, whereas B is resonant and C is elastic. Amplitude A is a little greater than $X_e$ and the phase lag A is near zero. Amplitude B is far greater than $X_e$ and the phase lag is about 90 degrees. Amplitude C is less than $X_e$ and the phase lag is about 180 degrees. This is not a practical example of the usual kind, but it does lead to an example of the composite cantilever and the practical cases of vibration transmission through series elements of varying impedance and varying natural frequencies. The three-component cantilever of Fig. 13.4

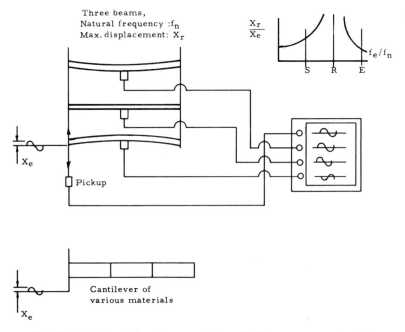

Fig. 13.4   Mode of vibration transmission, "solid, resonant or elastic," in systems having several components

is of constant cross section. In this connection it is worth remarking that for constant geometry and different materials, natural frequency often changes but little, because the relation of density to Young's modulus is often about the same as we go from one material to another. A notable exception is the metal lead. The illustrated cantilever may be almost quiescent at one location, violently vibrating at another and vibrating significantly at still another. Simple phase measurements will usually reveal what is occurring.

**Example 4: Multiple Impedances.** The vibrating flagstaff illustrated in Fig. 13.5 is indicative of a not uncommon situation wherein a nuisance vibration exists at a particular place in an office or apartment building. In the same building there may be two or as many as ten or more sources, all having the same frequency as the measured frequency of the nuisance vibration. The problem is to identify the source machine without shutting down the others, or better still without any shutdown. For example, in summer the stoppage of one fan may lead to more trouble than is caused by the nuisance vibration. Since the usual frequency method of identification fails, phase measuring is then the key.

As before, the flagstaff or the building in which the nuisance vibration exists is used as a reference for phase measuring. With another pickup various machines or possible sources are surveyed until one is found that is in phase with the reference phase; and that will be the source of the nuisance. As in the case of

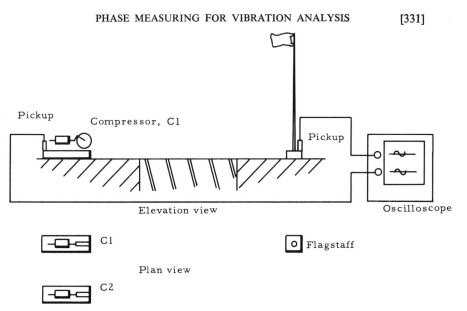

Fig. 13.5   The vibrating flagstaff illustrates identification of source of
a nuisance vibration

acoustic coupling, the same sort of principle arises here: when a vibration is
transmitted through a variety of media, the source that causes the nuisance
vibration will be in phase with that vibration. Also, the phase of the nuisance
vibration and of all the intervening media is the same as that of the source. In
this case it is measured at the base of the flagstaff. When the phase difference
between the flagstaff base and a compressor base is 90 degrees, that particular
compressor can induce no response in the flagstaff. Response occurs only when
the phase difference is near zero.

In summary when the phase difference is nearly zero, the source and response
are correlated. On the other hand, when the phase difference is nearly 90 degrees,
the source and the response are not correlated. So long as the flagstaff base
and a compressor base vibrate in phase, that particular compressor is the
offender, irrespective of how many impedances intervene between the two bases.
The impedances may consist of wood, soil, sand, concrete, etc. Further, if an
in-phase relation exists and a new impedance, such as a structural member or
earth trench, is introduced into the transmission path, then the phase relation is
likely to change and the flagstaff become quiescent. It is more difficult to approach quiescence by altering the soil path than by altering a structural member.
A structural member is sometimes, but not always, the total path, whereas the
soil at any given place is only part of the transmission path. The surface wave in
the soil may not be the causative vibration so that the use of trenches or holes or
bodies of water may have little or no effect.

**Example 5: Translation of a Vibration.** Consider that the flagstaff base vibrates
vertically (see Fig. 13.6). If vertical and horizontal vibration of the flagstaff is

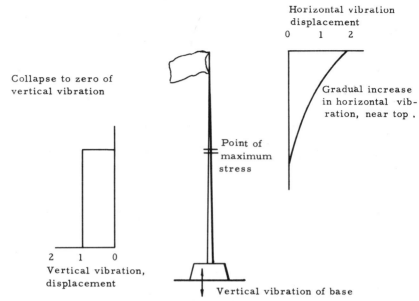

Fig. 13.6   A flagstaff used as an illustration of the translation of direction of vibration, such as occurs in building structures, leading to locations of maximum structural stress

measured along its length, the vertical component shows an almost total collapse at some point. Above this point, horizontal vibration shows a gradual increase. The transition point is a point of maximum stress, and strain gauges are best used to determine if the stress is acceptable. Between the base and the point at which vertical vibration amplitude almost vanishes, the horizontal amplitude is negligible. After translation it increases constantly or increases and then decreases, depending on which mode of vibration is taken up by the flagstaff. The mode depends on the relation of the disturbing frequency to the natural frequencies of the staff.

Now consider a building structure having a central elevator shaft and a surrounding independent structure tied to the center region by beams. Suppose local rail traffic causes vertical vibration of the structure. Vibration amplitude and phase, both horizontal and vertical, are measured on the structure and elevator shaft at various levels. In the lower areas, there is no phase difference between the central shaft and the surrounding structure. But higher up, one or the other is likely to lag. The lagging member is that which is receiving vibration while the leading member is that which is transmitting it. Such knowledge is necessary in order to formulate proposals for structural alteration when such is indicated.

At the point in the structure where translation of vibration occurs, strain-gauge measurements should be made to determine if the stress is acceptable. Sometimes it will be found that translation leads to a torsional mode in building structures as well as to a horizontal mode.

# MAINTENANCE MEASUREMENT EXAMPLES

# SECTION IV CONTENTS

# 14

# BALANCING OF
# FANS AND BLOWERS

*D. G. Stadelbauer*

Balancing of fans and blowers presents specific problems to the balancing machine because of horsepower requirement, windage fluctuation, tooling, wheel deflection, etc. This chapter will define these problems and suggest ways and means to overcome them.

## 1. BALANCING MACHINE DRIVE

When a fan or blower wheel is run in the open, that is, outside of its normal housing, a considerable amount of air turbulence results, which in turn causes all kinds of disturbances. Furthermore, the horsepower requirement generally increases proportionately to the third or fourth power of rotational speed. Since horsepower costs money, and since the drive mechanism in a given size balancing machine is limited in the horsepower it can transmit, various ways have been found to reduce the power requirement.

A low balancing speed helps. The simplest solution and the one most commonly used is to reduce the balancing speed to a minimum. A difficulty arises, however, because with decreasing speed the balancing machine's sensitivity is also decreased, thereby adversely affecting minimum achievable residual unbalance. The question then is, can a given machine indicate a small enough residual unbalance to meet the required balancing tolerance? If no information about the tolerance for a specific wheel is available, a recommended value can be found in ISO and USASI documents on balancing criteria. In the rotor category for fans, a balance quality grade should be sought which allows approximately the following residual unbalance in each of two correction planes.

$$U = \frac{20W}{N}$$

where U is residual unbalance in oz-in, W is weight of the fan in lbs and N is operating speed in rpm.

The problem of decreased sensitivity at low balancing speeds can quite often be overcome by equipping the machine with two speeds. The lower speed is used for heavy wheels on which large tolerances are acceptable. The higher speed, where the machine has greater sensitivity, is used for light wheels which require close tolerances. Since the horsepower requirement of a fan decreases proportionately to the third or fourth power of rotational speed, there is quite an advantage gained by reducing the balancing speed.

If the machine is equipped with a gear transmission, or if the lower speed is obtained by putting a larger drive pulley on the balancing mandrel, the torque available at the lower balancing speed is actually increased since the transmitted horsepower remains approximately constant. Eventually, of course, there will be a limit to the amount of torque that the machine can put into the fan, usually because of the size of the transmission, the universal joint driver or the belting.

When balancing armatures, paper rolls or sheaves, maximum drive power is used only during the short acceleration period. Once the work piece has come up to balancing speed, the torque requirement to keep it going drops considerably. Thus, the motor gets a chance to cool off. Not so on fans. Windage drag continues to demand high torque from the drive until the unbalance reading has been taken and the drive is shut off again. Therefore, the drive on a machine which is to balance fans should be sized so that its rated horsepower will accommodate the largest fan on a continuous duty basis.

Most of the time an operator is not aware of how much horsepower a fan really requires at a given balancing speed. He may eventually notice that something is wrong when the thermal overload on the drive motor kicks out, or some other component in the drive mechanism overheats or the unbalance readings seem to be incorrect.

Whereas the first two examples furnish a rather unmistakable warning, the matter of incorrect readings is more subtle and may not be as readily identified. Incorrect readings due to slight speed changes usually occur when the electronic instrumentation utilizes a fixed speed filter which requires a specific balancing speed for proper operation. When this speed is not attained, the amount of unbalance reading is generally smaller than it should be, and the angle reading may be ten or more degrees off.

Then, after an unbalance correction is applied in exact accordance with the machine's indication, a subsequent check run will show that another weight—although smaller—is required approximately 90 degrees away from the previous correction. The reason for this will be shown later. At this point suffice it to say that the apparent shift in readings may occur several times in succession until the desired tolerance is finally reached. Overall balancing time is thus considerably increased. Machines using instrumentation that is unaffected by speed variations avoid this problem.

**Balancing at Operating Speeds.** It is often asserted that fans or blowers should be balanced at their operating speed. Variations in blade size or position supposedly cause different quantities of air to be drawn into the wheel at different speeds, thereby causing first-order vibrations that could be corrected by balance weights. Experience does not bear this out, except in some isolated cases. These include certain propeller-type fans such as kitchen exhaust fans or window ventilating fans which are often of such lightweight design that the individual blades deflect at operating speed and produce an unbalance. It will then be necessary to balance them at their operating speed.

To rotate medium-size or large fans at their operating speed in a balancing machine would make for a very expensive installation, not only because of the tremendous amount of horsepower that would be required to overcome the air drag, but also for reasons of safety of the operating personnel. Blowers for blast furnaces or fans to ventilate tunnels may require anywhere from 300 to 10,000 hp at their operating speed so that the cost for the balancing machine drive would become prohibitive.

**Choosing the Direction of Rotation.** Certain fan wheels use more horsepower rotating forward, others rotating backward. If the balancing machine drive cannot be reversed, then perhaps the wheel can be reversed on the balancing mandrel. If the wheel still uses more horsepower than the machine can deliver, or if the windage causes erratic unbalance indication, a cover or shroud must be used.

## 2. SHROUDS

Centrifugal fans with cylindrical outside diameters can be covered very simply by wrapping a strip of heavy paper around the blower and fastening it with tape or cord. However, such a makeshift arrangement is usually too inconvenient for production balancing and, in any case, difficult or downright impossible on fans with tapered outside diameters or with propeller-type blades.

If a large number of the same type of fans are to be balanced, it may be more practical to make a clamshell shroud. The lower half rests on legs which generally are fastened to the balancing machine bed. The upper half is attached to the

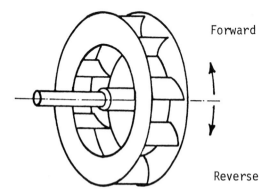

Fig. 14.1 Shrouded wheel. When shrouded, wheels may be rotated in either forward or reverse direction. When not shrouded or when a disc is used to block its axial intake opening, wheels should be run in reverse direction to reduce windage drag.

lower one by hinges at the rear so that the top part can be opened easily for loading, correcting and unloading the wheel.

Another way to reduce air drag is to mount a hub with a large disc on the balancing mandrel. The axial intake opening of the fan is then pushed up against the disc and the fan rotated in the direction in which it would normally suck in air through the axial opening.

## 3. FLEXIBLE SQUIRREL CAGE BLOWERS

On light squirrel cage blower wheels initial unbalance sometimes distorts the wheel. Indicated unbalance, therefore, is larger than the correction weight actually

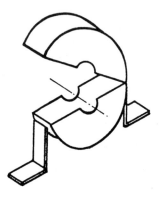

Fig. 14.2   Clamshell shroud. A clamshell shroud with hinged cover serves to reduce windage and hazard. One shroud may accommodate several wheel sizes. Air rotates with the wheel inside a sheet metal or plexiglas enclosure.

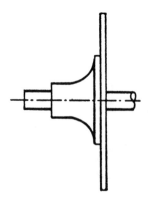

Fig. 14.3   Balancing mandrel with hub and disc. Disc covers air intake of centrifugal fans, reducing horsepower requirement and windage fluctuations.

necessary. In such cases it is advisable to stiffen the wheel by diagonal struts or reinforcement of the end rings. At the same time this will cut down handling damage, because if a wheel is bent after balancing it will be out of balance again.

## 4. MANDRELS AND ADAPTERS

Mandrels should be straight, balanced and, for production work, hardened and ground. When resting in the balancing machine supports, the area where the fan wheel will be mounted should have as little runout as possible. Multiply one-half of the total indicated runout (in inches) by the weight of the fan (in ounces) to determine the unbalance introduced by eccentricity of the mandrel. This tooling error ideally should be less than 10 percent of the balancing tolerance, but in no case should it exceed it.

For example, assume that the mounting surface of a mandrel has 0.001 in TIR and that it is to be used for balancing a wheel weighing 100 lbs (1600 oz). The mandrel eccentricity will displace the wheel by $\frac{1}{2}$ of 0.001 in, e.g. 0.0005 in from the mandrel's rotational axis, inducing a static unbalance of 0.0005 in $\times$ 1600 oz $= 0.8$ oz-in. This is acceptable if the balancing tolerance for the wheel is 8 oz-in. The operator simply has to balance the wheel to 7.2 oz-in so that, even when the angular position of the residual unbalance coincides with the tooling error, the total of the two is still within the allowable tolerance of 8 oz-in.

However, if a balancing tolerance of 1.6 oz-in is prescribed for the same wheel, the tooling error absorbs 50 percent of the tolerance. The operator is now forced to balance every wheel down to 0.8 oz-in, so that the aggregate of residual unbalance and tooling error will never exceed the balancing tolerance of 1.6 oz-in. Thus the operator's job has been made unnecessarily difficult.

Should a tolerance of less than 0.8 oz-in be prescribed for the 100 lb wheel, the mandrel with 0.001 in TIR is no longer usable. No matter how carefully the operator applies correction weights to the fan to obtain a zero indication on his balancing machine instrumentation, the wheel will be out of balance again when he mounts it on the presumably true running final shaft.

Similar difficulties arise if, although the wheel is balanced on a perfectly true mandrel, it is then mounted on a shaft that runs out. An obvious way to avoid all these problems is to do away with the mandrel altogether and run the wheel on its own shaft in the balancing machine. However, quite often this is not possible as, for instance, in mass production, on replacement fans, for reasons of shaft design or balancing machine limitations.

Another way to overcome mandrel problems is to balance the fan after assembly with the help of field balancing equipment. This, however, takes considerable time and special skill, neither of which may be in ample supply. In most cases, then, it is best to have true-running mandrels and final shafts. Sometimes fan wheels are provided with rather small bores. In such cases the mandrel should be kept as short as possible so that it does not deform during the balancing operation.

**Mounting the Wheel.** Common practice is to hold a wheel on its mandrel by tightening a setscrew in the wheel hub. A small flat area should be filed into the mandrel where the setscrew will contact it. Otherwise, ridges caused by the setscrew on the mandrel surface make removal of the wheel difficult. In fact, burrs in the mandrel's locating surface may damage the bore in the wheel hub.

For hubs with setscrew and keyway, a somewhat longer flat area should be filed on the mandrel. A key of half thickness but full length can then be inserted into the hub and pressed against the mandrel's flat area by means of the set-screw.

For production balancing a mandrel with keyway and *full* key is preferred. Again, the wheel's setscrew should be tightened on the key. Before using such a mandrel it should be balanced with a *half* key (see also, Balancing the mandrel and adapter, p. 344).

**Mandrel Fits.** Do not use size-on-size or interference fits, since these may distort a thin mandrel and make it difficult to pull the wheel off later. A snug push fit between mandrel and wheel usually gives the best results.

A loose fit, on the other hand, results in a displacement of the wheel when the setscrew is tightened. If the wheel is balanced in this offset position, it will be out of balance again when removed from the mandrel.

In other words, a wheel is balanced about the rotational axis as determined by the mandrel running in the balancing machine. This rotational axis is a straight line through the mandrel's journal centers. If the mounting surface diameter of the mandrel and that of the final shaft are not identical, or if either surface has a runout, the rotational axis of the wheel will shift when it is moved from mandrel to shaft. Any such shift of the rotational axis causes considerable unbalance, as the previous example in this section for runout and the example below for a loose fit show.

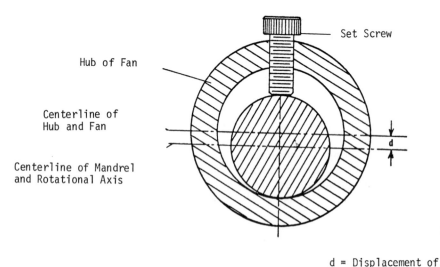

Set Screw

Hub of Fan

Centerline of
Hub and Fan

Centerline of Mandrel
and Rotational Axis

d

d = Displacement of
Hub and Fan

Fig. 14.4   Hub location on a shaft

**Example.** d = 0.002-in on a fan weighing 200 lbs (3200 oz). Unbalance caused by loose mandrel fit = 3200 oz × 0.002 in = 6.4 oz-in.

The same kind of error results if the mandrel fit is good but the fit between wheel hub and final shaft has too much clearance.

If a wheel has two setscrews offset by 90 degrees, both of them should be tightened in a given sequence on the mandrel. When the wheel is later mounted on its shaft, the identical sequence should be used so that the wheel is displaced approximately the same way. This procedure reduces any unbalance that may be brought about by a slight amount of clearance in mandrel and shaft fits.

One should also make certain that the shaft or, as the case may be, the armature to which the fan is to be assembled has previously been balanced, and that such balance was obtained with a key of full length but of half the normal key's thickness. This is current American practice but does not necessarily hold true for equipment coming from European countries. There, mandrels, shafts, armatures, etc. are generally balanced with a *full* key (see also footnote, p. 344).

**Type of Drive.** The most frequently used method to transmit torque to a fan wheel during the balancing operation is the universal joint driver, as shown below.

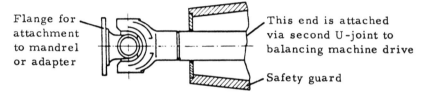

Flange for attachment to mandrel or adapter

This end is attached via second U-joint to balancing machine drive

Safety guard

Fig. 14.5   Universal joint driver

On machines equipped only with belt drives it often becomes necessary to provide a pulley on the mandrel over which the belt can drive. If the belt is run directly over the mandrel the belt may not be able to transmit the necessary torque. It continues to slip on the shaft, resulting in excessive belt wear and possibly fluctuating or incorrect unbalance readings. The end band drive is not recommended for fans that produce an appreciable amount of air drag, since this drag puts too much pressure on the band, causing fluctuations in the unbalance indication.

**Attaching the Drive Adapter.** To reduce the number of required adapters and to facilitate changeover from one size of wheel to another all mandrels within a certain range of sizes should be turned down at one end to a standard diameter. Two keyways, 180 degrees opposite each other, should be machined into the same end. * Thus, a single-drive adapter, hardened and ground, with mating bore

---

* Note: Although a single key may be sufficient to transmit drive torque, the second key, nevertheless, is recommended because it makes the balancing of mandrel and adapter much easier.

and permanently mounted keys, can be used to provide the connection between balancing machine drive and many different mandrels. The adapter remains attached to the machine's U-joint flange so that no bolt connection needs to be made when loading and unloading a mandrel with fan.

Looseness in the fit between adapter and mandrel causes unbalance, since not only the adapter but also half of the U-joint drive will be displaced. A sliding fit without noticeable play is recommended.

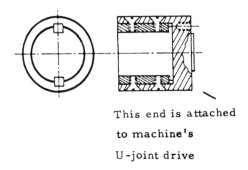

This end is attached

to machine's

U-joint drive

Fig. 14.6   Standardized drive adapter for balancing mandrels

The adapter shown above has no provision for preventing the mandrel from slipping out during balancing. No such provision is necessary if the fan, when rotated, creates its own thrust toward the drive.

Machines equipped with end band drive generally have built-in thrust by means of slightly canted crowned rollers. This thrust or feed toward the machine's headstock is sufficient to hold the mandrel in the adapter even if the fan generates no thrust of its own. However, on all other machines an end stop should be used to prevent the mandrel from sliding out of the adapter. Even on machines with a built-in thrust arrangement an end stop is required if the fan creates a negative thrust. A setscrew in the adapter will also give some protection, but an end stop is more positive and therefore preferred.

**Balancing the Mandrel and Adapter.** If the mandrel has a key for driving the fan, the key should be temporarily replaced by a half key.* Mandrel and adapter together are then dynamically balanced by adding balancing clay at each end of the mandrel. Next, the mandrel is pulled out of the adapter, rotated 180 degrees and reinserted. One half of the unbalance that is now indicated must be corrected in the adapter, the other half in the mandrel. At this time a permanent weight correction can be made by drilling or grinding. Then the mandrel is turned once more by 180 degrees in relation to the adapter, and again half of the indicated unbalance corrected in the adapter, the other half in the mandrel. Repeat this

---

* Note: When the mandrel substitutes for a final shaft, armature, etc., balanced with a full key, the mandrel should likewise be balanced with a full key.

procedure until the reversal error is considerably smaller than the balance tolerance allowed on the fan for which mandrel and adapter are to be used.

**Special Adapter for Vertical Machines.** For high-production balancing of small
blower wheels, machines with a vertical spindle are generally employed since
they somewhat facilitate the loading and unloading of wheels. Overall balancing
time may be far less than a minute, so every second counts. To eliminate the time
used up by the usual tightening (and subsequent loosening) of a wheel's setscrew,
an adapter as shown in Fig. 14.7 may be used. The central mandrel locates the
wheel, while the surrounding collar does the driving. The wheel is simply set on
the mandrel in such a way that the protruding part of the setscrew fits into a
slot in the adapter's collar. Opposite the slot a spring-loaded ball has been incorporated into the collar, pressing against the outside of the wheel hub. Thus,
any play that may exist between mandrel and hub is located on the setscrew
side, where it would normally be had the setscrew been tightened. Mandrel as
well as final shaft may both be slightly under size for easy loading without
detrimental effect on the balance condition of the wheel.

**Running Large Wheels in Their Own Pillow Block Bearings.** By using the final
shaft the need for a separate balancing arbor is eliminated. Since many types of
fans run in pillow block bearings, it may be desirable to use these same bearings
during the balancing process, thereby circumventing possible balance errors from
bearing eccentricities. One prerequisite for this approach is that the balancing

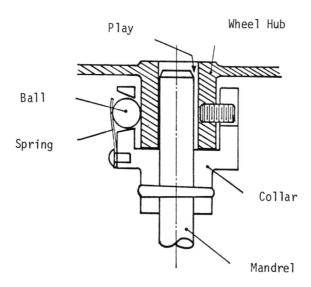

Fig. 14.7  Special adapter for high-production balancing of small fan
wheels on vertical balancing machine

machine supports be equipped with suitable carriages on which pillow blocks can be mounted.

In case of pillow blocks with spherical bearings, the required support carriages may be of comparatively simple design. For nonspherical pillow block bearings, however, the balancing machine carriages or supports must provide the possibility for spherical adjustment, or the balancing machine cannot indicate correctly. Such carriages or supports are somewhat more complicated and expensive.

## 5. METHODS OF CORRECTING UNBALANCE

Correction methods depend largely on the type and size of wheel that is to be balanced.

**Small and Medium-Size Squirrel Cage Wheels.** Most convenient is the use of spring-loaded clips which are forced over the inside edges of the blades. Such clips are commercially available in different sizes and weights, contoured to fit a certain blade curvature and equipped with a small, sharp spur to prevent falling off.

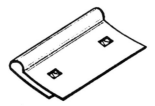

Fig. 14.8    Weight clip for squirrel cage blower wheel

In production balancing it works out well if the smallest clip is equivalent to the balancing tolerance and larger clips are available weighing exactly two, five and ten times as much. The unbalance indication can then be calibrated directly in tolerance units so that the operator does not have to work with a weighing scale. Correction with clips can be carried out in a matter of seconds.

If no clips are available, correction can be accomplished by flowing solder on the concave side of the galvanized blades. On large, ungalvanized wheels correction is usually made by bolting, and on even larger wheels by welding weights to the end rings.

When painting such wheels after balancing, with blades in a horizontal position, care should be taken that an excess amount of paint does not collect in the troughs formed by the bottom blades. An accumulation of paint could throw the wheel out of balance again.

If the wheel is small enough to be corrected with clips or bolt-on weights, painting should be done *before* balancing.

**Centrifugal Fans.** These are generally corrected by riveting, welding or bolting on weights. On very large wheels correction weights may weigh several pounds and are individually cut and weighed. They are generally tack welded first to make sure they are in the right position. A certain amount of underweight should be allowed to provide for the final welding seam.

A note of caution about welding in a balancing machine: Always attach the ground wire directly to the fan, never to the balancing machine. Otherwise, the current will go through the support rollers and bearings and ruin them very quickly. Use an asbestos cloth to shield, at least, the machine's roller bearings and bed-ways and the mandrel's running surfaces from welding spatter.

Fans made of thin gauge steel distort from welding heat. Bolting or riveting of weights is then recommended. Weights can be precut and stored in tray compartments marked with their weight. Bolts of different lengths and washers may be used to make up the difference. Care should be taken that bolts and nuts are well secured with lock washers and by peening or staking so that they will not come loose in operation.

Thin gauge wheels are sometimes distorted by large initial unbalance causing similar difficulties as those described in Section 3 of this chapter for flexible squirrel cage wheels without cross bracing. To keep initial unbalance to a minimum, care should be taken during the manufacturing process to build a wheel with as little runout as possible. Welding seams should be evenly distributed and of equal length and thickness.

If the initial unbalance is so large that it exceeds the measuring capability of the balancing machine or even causes the wheel to lift off the support rollers, a preliminary static correction should be made. Either distribute the static correction evenly between the two outside faces of the wheel or place the major correction in that side which shows the largest runout. This procedure will hold the induced couple unbalance to a minimum.

A small, spring-loaded weighing device (fish scale) may serve to ascertain the approximate value for the preliminary correction. With the wheel in the balancing machine but the drive disconnected, let the heavy spot come to rest at the bottom and mark its location. Then rotate the wheel by 90 degrees and hook the spring scale under the blade that is closest to the heavy spot. Hold the scale up and let it balance out the downward pull of the unbalance. The value for the correction weight can now be read off the scale.

If the correction weight is to be added at a smaller radius than the one at which the scale was hooked to the wheel, multiply the scale reading by the radius at which it was attached ($r_1$ in Fig. 14.9) and divide the result by the radius at which the weight is to be added ($r_2$).

Another type of temporary correction weight is shown below. It consists of a small C-clamp with setscrew and is frequently used for trial-and-error or field balancing.

**Miniature Blower Wheels.** Such wheels are frequently used for cooling electronic devices and often need be checked only for runout. By closely controlling the manufacturing process, the need for unbalance correction is eliminated. Wheels are graded for different applications into two or three unbalance categories.

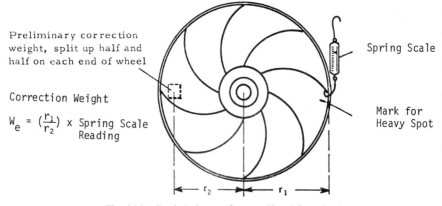

Preliminary correction weight, split up half and half on each end of wheel

Correction Weight

$$W_e = \left(\frac{r_1}{r_2}\right) \times \text{Spring Scale Reading}$$

Spring Scale

Mark for Heavy Spot

Fig. 14.9   Static balance of a centrifugal fan wheel

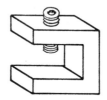

Fig. 14.10   Temporary clamp-on weight for blower wheels

**Small Cast Blower Wheels.** This type of wheel is sometimes drilled or ground for correction on single-plane, vertical balancing machines. For large production, mass centering is used. This is a process whereby the wheel's bore is finish-machined on the balancing machine so that the bore center coincides with the wheel's principal axis of inertia. No further unbalance correction is necessary.

**Single Resultant or Component Correction.** In describing the various methods of correcting unbalance it has been assumed so far that the correction weight would always be applied exactly 180 degrees opposite the heavy spot. However, in practical application it will be found that quite often a wheel ends up with weights at more than one angular position per plane before the desired balancing toler-ance is reached. Such scattering of weights comes about because, due to a wrong balancing machine indication, operator error or fan design, the initial correction weight was not placed at precisely the right angular position.

   A comparatively small angle error—even though the correct-size weight is used—will cause a surprisingly large residual unbalance approximately 90 degrees removed. Figure 14.11 shows how a 10-degree inaccuracy in applying a correction weight will result in a residual unbalance of 17.4 percent at an angular position 85 degrees removed from the initial unbalance.

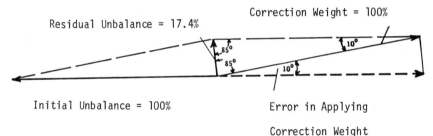

Fig. 14.11   Vector diagram for unbalance

In Table 20 are listed the residuals, expressed in percentages, of initial unbalance which result from applying unbalance correction of proper amount but at various incorrect angular positions.

**T A B L E   2 0**
**RESIDUAL UNBALANCE ERROR**

| DEGREE ERROR | PERCENT RESIDUAL UNBALANCE | DEGREE ERROR | PERCENT RESIDUAL UNBALANCE |
|:---:|:---:|:---:|:---:|
| 1 | 1.7 | 6 | 10.5 |
| 2 | 3.5 | 8 | 14.0 |
| 3 | 5.2 | 10 | 17.4 |
| 4 | 7.0 | 12 | 21.0 |
| 5 | 8.7 | 15 | 26.0 |

If the residual unbalance exceeds the balancing tolerance, a second correction weight must be applied. Since application of the second weight will most probably be affected by the same error as the first weight, it will leave a residual unbalance which then must be corrected by a third weight. Successive applications continue until the specified balancing tolerance is reached—a rather costly process.

Errors in machine indication can be corrected by repair or proper adjustment, provided the indicating system incorporates the feature of *plane separation*. This is the ability of a machine to indicate unbalance in two transverse planes at right angles to the rotational axis and preselected anywhere in the work piece where unbalance correction is convenient.

An incorrect operating procedure can, of course, also be corrected, but what is to be done when a weight is required between two blades on a five-bladed paddle wheel? The only solution is to split up the weight between the two adjacent blades.

Depending on the angles between machine indication and neighboring blades, one blade will need a larger weight than the other. A skilled man will eventually develop some feel for the splitting of weights, or he can consult a diagram or table. However, the most convenient and precise way to overcome the problem is to have a balancing machine that will read out unbalance directly in components. One such component meter is shown below (Fig. 14.12). It has interchangeable overlays which can be furnished for fans with any number of blades.

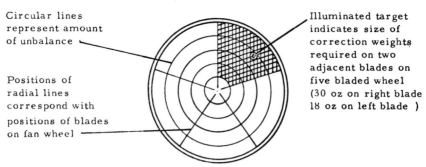

Circular lines represent amount of unbalance

Positions of radial lines correspond with positions of blades on fan wheel

Illuminated target indicates size of correction weights required on two adjacent blades on five bladed wheel (30 oz on right blade 18 oz on left blade )

Fig. 14.12    Balance test indication

# 15

## ROTOR BALANCING
## WITHOUT PHASE MEASUREMENT*

*Michael P. Blake*

Rotors may be balanced using only a dial indicator or simple vibration meter. The method is applicable to in-place or machine balancing. Conventionally, this is done by phase measuring using a stroboscope, a wattmeter circuit or perhaps an oscilloscope or even optical methods. The purpose of phase measuring is to find the location of the unbalance. However, it is often desirable to be able to balance without phase measuring, by using a dial indicator only. Mechanics often balance rotors in this way; but in this writer's experience their methods are usually hit and miss, resting on no theoretical base. Here is a method which appears to be generally unknown, although it can hardly be novel. It is satisfactory in practice and is very simple. Four total trial runs are made and the vibration is measured with a dial indicator or simple vibration meter. A simple graphical procedure gives the correct phase and the proper weight for the final correction.

## 1. EQUIPMENT REQUIRED

A common dial indicator and a few trial weights are all that is required for most problems. A typical trial weight is shown in Fig. 15.1, useful for rotors of about fifty pounds weight and upwards. Dial indicators will read up to about 10 mils TIR at 2000 rpm, and 5 at 3600 so that they are suitable for a wide range of speed and vibration. Any other form of vibration meter is also acceptable. Sometimes, a vibration meter is mandatory. For example, when measuring the vibration of a cooling tower in order to balance its fan, the whole tower moves so that there is nothing to which to attach the dial indicator. The indicator is what is often called a *driven* pickup and must be abandoned in favor of a meter with a *seismic* pickup. At speeds below about 900 rpm, most vibration meters exhibit large errors, say 50 percent at 300 rpm. This is of no consequence to this balancing method as long as the error is consistent.

---

*Reprinted with permission from *Hydrocarbon Processing*, March 1967.

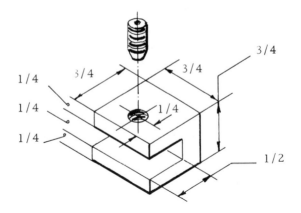

Fig. 15.1   A typical balancing weight. Usually about 12 are on hand.
Use a 1/4–28 hollow point set screw.

## 2. PROCEDURE

Four runs are made. First, the rotor is run without a trial weight and the vibration is measured horizontally at a bearing. Then three more runs are made, all at the same original speed. For each run of the three, a trial weight (always the same and always at the same radius) is placed successively at three different places around the rotor. The 120-degree positions are most convenient, but any position may be used. The three corresponding vibrations are measured. The data for the four runs might appear as those shown in Table 21.

### TABLE 21
### TYPICAL ROTOR BALANCING DATA

| RUN | AMOUNT OF TRIAL WEIGHT (OUNCES) | POSITION OF TRIAL WEIGHT ON ROTOR | REVOLU-TIONS PER MINUTE | VIBRATION MILS PP | VIBRATION SYMBOL |
|-----|------|------|------|------|------|
| 1 |  | —— | 300 | 44 | E |
| 2 | 10 | 12 o'clock | 300 | 50 | M1 |
| 3 | 10 | 4 o'clock | 300 | 42 | M2 |
| 4 | 10 | 8 o'clock | 300 | 96 | M3 |
| 5 | 8.8 | 2 o'clock | 300 | 15 | E1 |

Now draw a circle of radius E, to scale (see Fig. 15.2). Locate points 1, 2, 3 on the circle at the same angular positions as were used on the rotor. Using center 1 and radius M1, draw an arc to scale and likewise for center 2 and M2 and center 3 and M3. The arcs will all intersect at the same point or nearly so. Take P as being the average point of intersection, and join OP. The points 12, 4 and 8 o'clock on the rotor are indicated by 1, 2, 3, in Fig. 15.2. The proper position for the final correction weight is parallel to T, that is, at $\theta$ degrees clockwise from 1 on the rotor. The amount of the final correction weight is E/T times the amount of the trial weight used. In this example, it is 44/50 times 10 oz or 8.8 oz.

If the rotor is not now considered to be smooth enough while running, make the 8.8-oz correction permanent and start four more runs as if approaching a new problem. Usually a correction of about 50 percent or more is obtained for each four runs. This is about as much as is achieved per correction when phase measuring equipment is used.

Persons not familiar with balancing procedures are advised to work with the greatest forethought and safety. It is advisable to proceed with care, using very small weights, making sure that they are well fastened and taking care to stay out of the line of flight. There is no useful rule for the proper amount of a trial weight. That shown in Fig. 15.1 weighs about 2 oz, and is adequate for rotors from about 50 to 1000 lb or more. Rotors weighing say one pound usually rotate fast so that about a ⅛-oz weight is suitable. From one pound to 50 lb, about ⅛ oz multiplied by the square root of the rotor weight in pounds is usually safe. Very seldom does a trial weight exceed four ounces.

## 3. THEORY

Fig. 15.3 is a diagram of vibration amplitude. It is assumed that amplitude is proportional to the amount of unbalance. Suppose a rotor has an amount of

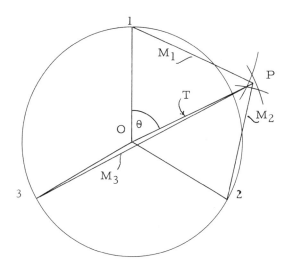

Fig. 15.2   Draw a circle and locate trial weight positions (120° apart).

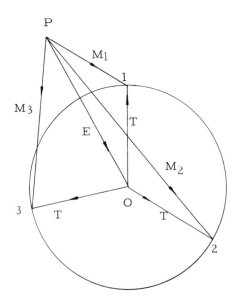

Fig. 15.3   A diagram of vibration amplitude

unbalance, E, and that a trial weight is placed in three angular positions at the same radius, successively 1, 2 and 3, and that its vector contribution to vibration is T. The measurable vibrations are E, M1, M2, M3, namely the original unbalance of the rotor and then the unbalance (vibration) with a trial weight successively at phases 1, 2 and 3. T cannot be measured directly so that Fig. 15.3 cannot be directly constructed from actual measurements.

Fig. 15.4 may be constructed from actual measurements. It contains three triangles, the same found in Fig. 15.3. Imagine the triangles of Fig. 15.3 rotated until the three vectors, marked T, all coincide. Fig. 15.5 is more convenient than Fig. 15.4 because it is more meaningful. It is derived from Fig. 15.4 by mirror reversal around the line 1-0. Now the numbers 1, 2, 3 appear in the same successive phase sequence as in Fig. 15.3, which represents the actual phase sequence of the trial weights. It is evident then that if in Fig. 15.5 a circle is drawn with the radius proportional to E, and that if points 1, 2, 3 are set out in the same phase relation as on the rotor, and that if arcs proportional to M1, M2 and M3, respectively, are drawn from these points, the arcs will meet at the same point or thereabouts. Then 0-C1 will represent, to scale, the vector contribution of the trial weight, and the correct phase for the final corrective weight shall be that of 0-C1.

**Advantages.** A great advantage is freedom from relying on phase measuring instrumentation. This is an asset if one cannot afford or cannot justify the instrumentation or if one is working in a remote location, far from the usual plant facil-

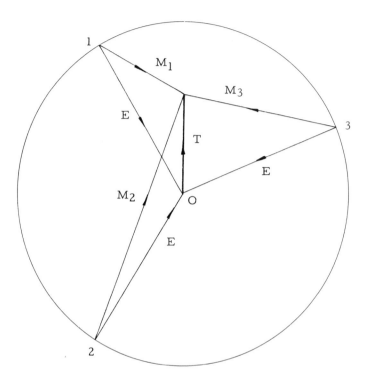

Fig. 15.4   Diagram contains same three triangles as Fig. 15.3

ities. The method is simple and is readily mastered by anyone who is interested. It often works with greater speed and certainty than the phase measuring method in difficult-to-balance machines or in the case of machines rotating below about 400 rpm. In bright sunlight where strobe signals are difficult to see, the method is superior. It is hardly overrating the four-run method to say it is a valuable enough technique to justify a brief consideration by anyone connected with mechanical maintenance or balancing procedures.

**Scope.** The four-run method appears to be applicable to any balancing problem that is usually done by the phase measuring method. The limitation is that it is generally a slow method rather than that it will not give satisfactory results. It is not always slower, and in many practical cases it gives surer and quicker results than the phase measuring method. But it does require four runs. When the phase measuring method is used over and over again on the same machine, the phase of unbalance in relation to the pickup phase is already known. There appears to be no reason to suppose that the four-run method cannot achieve the same precision of balance as is achieved by the phase measuring method.

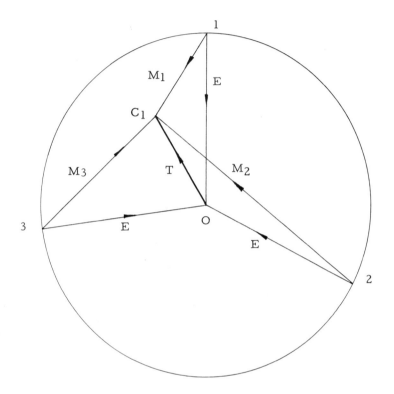

Fig. 15.5   Circle is drawn with radius proportional to E

Neither method can extend the precision beyond the level of the residual harmonics or external disturbances, and both can go just that far.

This writer has used the method in the balancing machine, to verify it; for boiler fans, for cooling tower fans and, even to balance automobile wheels by jacking up one rear wheel at a time. It should be remarked that a jacked-up automobile, with the engine running, is potentially dangerous.

For large cooling tower fans this writer prefers the four-run method. It seems to require less mental consideration, and it seems to proceed with more certainty than the phase measuring method. Perhaps this preference is based on conjecture rather than on fact. It is a fact, however, that the four-run method does work well, and it has been used to balance one fan at a time in a five-fan tower, with all fans running. This author has never tried this with phase measuring methods.

**Case Histories.** Two typical case histories are given and illustrated by Fig. 15.2 and Figs. 15.6–15.11. These are helpful in answering questions that would arise in the reader's mind regarding details of procedure and the practical utility of the method. By way of explanation, the "mil" is one-thousandth of an inch, and PP means peak-to-peak, which is the usual way of measuring vibration and is analogous to TIR for a dial indicator.

**Case History 1.** At 300 rpm the vibration of a six-blade cooling tower fan was measured horizontally and found to be 44 mils PP (0.044-in peak-to-peak). Ten-ounce weights were then attached successively at a 6.5-ft radius on blades 1, 2 and 3 (see Fig. 15.6). The corresponding vibration was 50, 42 and 95 mils (always measured at the same place and in the same direction). To find the required correction draw a circle, Fig. 15.2, with a radius of 0.044 in to scale, and mark off points 1, 2 and 3 in the same angular positions as they were on the fan (they are usually symmetrical but need not be). Using 1 as a center draw an arc of radius 0.050 in to scale. Using 2 as a center, draw an arc with a radius of 0.042 in to scale. And with 3 as a center, draw an arc with a radius of 0.095 in. The arcs should all intersect at a common point, or nearly so. Select a point P (see Fig. 15.2) which is an average location of the intersection. Join points O and P and label this line T. The angular location of the proper correction weight is at T, and its proper value is E/T (44/50) times 10 oz or 8.8 oz. If the final correction weight is attached to a radius of say one foot then the proper weight is 6.5/1 times 8.8 oz or 57 oz. If the weights may be attached only at blade positions, draw RC (Fig. 15.7) parallel to T of Fig. 15.2. This is the proper direction of the required correction. Then draw A parallel to blade A and 2 parallel to blade 2. If RC represents the required single-correction weight to scale, then A is the weight to be placed on blade A and 2 the weight to be placed on blade 2, all to scale.

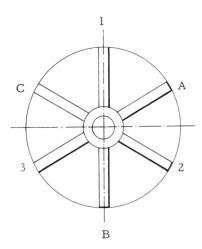

Fig. 15.6   Weights were attached to blades 1, 2 and 3 of this 6-bladed cooling tower fan.

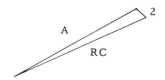

Fig. 15.7   Draw RC parallel to T (Fig. 15.2) if weights can only be attached to blades.

Fig. 15.8   Balancing machine for two rotors

### TABLE 22
### TRIAL RUNS FOR CASE HISTORY 1

| WEIGHT/POSITION | VECTOR | INDICATOR TIR |
|---|---|---|
| Fan as tested | E | 44 mils |
| 10 oz at 1 | M1 | 50 mils |
| 10 oz at 2 | M2 | 42 mils |
| 10 oz at 3 | M3 | 96 mils |
| 8.8 oz at T | E2 | 15 mils* |

* Percent correction achieved in one series of runs: $(44-15)/44 = 66$ percent

**Case History 2.** A small balancing machine, shown in Fig. 15.8 was used to balance two rotors. Each rotor weighed about ¾ lb. The vibration was measured at A and B with a dial indicator (see Fig. 15.9).

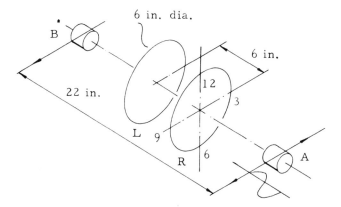

Fig. 15.9   Vibration measured at A and B

The method used was to make trial runs as tabulated in Table 23. The trial weights were placed at the 12, 4 and 8 o'clock positions on the rotor. A circle was drawn with radius E and points marked off at 12, 4 and 8 o'clock. Then, arcs were drawn with radii M1, M2 and M3 from centers 12, 4 and 8. A line was drawn from the intersection of the arcs and the center of the circle and labeled T (see Fig. 15.10). Line T is the correct phase position for final weight correction. The value of the final correction is E/T times the trial weight amount (see Fig. 15.11).

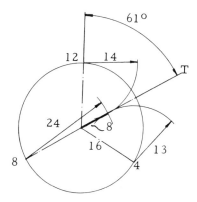

Fig. 15.10   Line T is the correct phase position for final weight correction

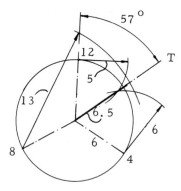

Fig. 15.11   Final correction is E/T times trial weight

TABLE 23
TRIAL RUNS FOR CASE HISTORY 2

| RUN | INDICATOR POSITION | INDICATOR READS: MILS TIR | REVOLUTIONS PER MINUTE | TRIAL WEIGHT, GRAMS | PHASE OF TRIAL WEIGHT | PLANE OF TRIAL WEIGHT | CORRECTION |
|---|---|---|---|---|---|---|---|
| 1 | A | 16 E | 950 | — | — | — | |
| 2 | A | 14 M1 | 950 | 4 | 12.00 | R | |
| 3 | A | 13 M2 | 1000 | 4 | 4.00 | R | |
| 4 | A | 24 M3 | 920 | 4 | 8.00 | R | |
| 5 | A | 3 E1 | 950 | 8 | 2.00 | R | $\dfrac{16\text{-}3}{16}$ : 82% |
| 6 | B | 6 E | 900 | — | — | — | |
| 7 | B | 5 M1 | 920 | 2 | 12.00 | L | |
| 8 | B | 6 M2 | 900 | 2 | 4.00 | L | |
| 9 | B | 13 M3 | 900 | 2 | 8.00 | L | |
| 10 | B | 2 E1 | * | 1.85 | 1.54 hr-min | L | $\dfrac{6\text{-}2}{6}$ : 66% |

* Not recorded, nominally 900 rpm

# 16

# ROTOR BALANCING
# WITH PHASE MEASUREMENT*

*Michael P. Blake*

About half of the vibration problems occurring in rotating equipment may be corrected by balancing. And most balancing can be done in-place, with the exception of small electric motors where the rotors are inaccessible for correction. Since the rotors are corrected in their usual environment, the result is often more satisfactory than that achieved by shop balancing and the labor and downtime saving is usually considerable.

The quickest method of balancing in-place is the phase measuring method. The instruments used are a stroboscope and a vibration analyzer, costing from about $1500 to $10,000. This cost is usually saved in about six months. Although the method is easily understood, proficiency comes only with practice because balancing in-place is no more than an art, receiving but little support from theory. For this reason several recommended procedures are given which are helpful in avoiding pitfalls, and the discussion of dynamic principles is limited to that which is known to be useful to the nonspecialist.

## 1. EQUIPMENT

A typical vibration analyzer contains a velocity pickup which fires a strobe lamp at some fixed point in its excursion back and forth. The pickup also serves to measure the vibration by generating a voltage proportional to velocity amplitude. This signal is amplified and usually passes through a wave analyzer or *filter* system, the purpose of which is the rejection of all frequencies except that of rotation so that balancing may proceed in the presence of other vibrations as if they were not there. In field practice most filters are of doubtful use because of their phase distortions, and because they are generally incapable of doing what is wanted of them (for example: separating frequencies that are close to each other). However, the filter is of considerable use in discovering the composition of the

---

*Reprinted with permission from *Hydrocarbon Processing*, August 1967.

vibration waveform, which is done by noting how much vibration amplitude is measured at various frequencies as the filter is run through its tuning range. The simplest equipment is recommended for field use. The more expensive instruments that may even read phase and amplitude on the same dial are more suitable for shop work.

The outstanding advantage of the strobe system is that it requires no connection to the machine other than the pickup, whereas the wattmeter system or oscilloscope or optical system requires some kind of generator connected to the rotating shaft. Often such a connection is physically impossible, and is usually inconvenient even if a photoelectric connection is used. Thus the wattmeter, perhaps the best phase measuring system, is usually confined to the shop where the generator connection is not a problem.

Besides a vibration analyzer, all that is required are a few dozen balancing weights ranging from about $\frac{1}{2}$ oz to perhaps 3 oz. An oscilloscope, when available, is most helpful in discovering faults such as loose-bearing pedestals or other troubles that may hinder the balancing effort and which must be corrected before balancing begins. For slow-speed rotors, say below 500 rpm, the oscilloscope gives no useful message to the eye. A DC paper strip recorder is used in this application.

## 2. METHOD

The method advocated here is that of taking the results of two trial runs and solving a simple vector construction problem. There are several more complicated methods but these are generally unsuited to the needs of in-place balancing.

**Single-Plane Correction.** For rotors that are corrected in one plane only (kinetically balanced), two runs and one correction should suffice; but it does not and about eight to ten total runs are normally required. However, a given correction is computed from only two runs at a time. For rotors that are corrected in two planes (dynamically balanced), perhaps fifteen or more runs are required. Happily, these rotors are not so common as might be suspected.

It appears that relatively wide wheels such as shown in Fig. 16.1 seldom seem to require anything more than kinetic balancing in-place, if originally dynamically balanced by the manufacturer.

**Two-Plane Correction.** Two-plane balancing proceeds in the same way as single-plane: the worst plane is corrected using measurements on the adjacent bearing; then the other plane is corrected using measurements on its adjacent bearing. The procedure is repeated, if necessary. Reference to a case history will make the basic correction procedure clear.

## 3. CASE HISTORY

In Fig. 16.1, an induced-draft fan is drawn approximately to scale. The pickup is attached rigidly, in the horizontal radial direction, to the bearing of interest. Two pickups and a switch are a welcome convenience. Occasionally machinery

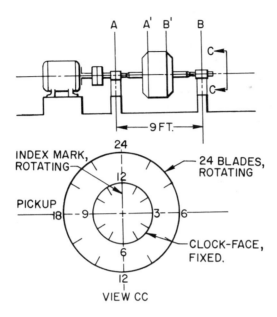

Fig. 16.1   Induced-draft fan described in case history

is situated on a floor or support that deflects considerably so that the vertical vibration is greater than the horizontal. In such cases, the pickup is attached vertically.

Figure 16.2 shows an induced-draft boiler fan being readied for balancing. After the first trial run, a weight is placed on both shrouds at blade No. 24 (see Table 24). This is a mere guess because there is no way of knowing where to place it, except the machine has been balanced before with the same instrumentation. If so, the phase relation of light spot (opposite to heavy spot) to pickup position is known and does not usually change too much. But even this cannot be relied upon. The phase relation is used as follows: in order to save a run or perhaps several runs, suppose that it is known that the light spot is three hours from the pickup, in the direction of rotation, when the strobe flashes. A trial run is made and the rotor is brought to rest. It is rotated until the strobe index is brought to the same phase position it was in when the strobe flashed. Three hours are counted off in the direction of rotation from the pickup position and a trial weight is placed there. If the placing is correct, either the index position during the next run will not change, and the amplitude will change; or the index will change phase by 180 degrees with or without an amplitude change. Except when the trial weights are too light to have any effect, a nonchanging index phase or a change of 180 degrees is the surest sign of progress. Until the characteristic reac-

Fig. 16.2 A view of an induced-draft boiler fan, showing clockface drawn on bearing housing. An index mark is drawn on the end of the shaft, opposite the 12 o'clock position on the fan wheel. A pick-up, strobe lamp, vibration analyzer and oscilloscope, are also seen.

tions of the rotor become apparent, it is desirable to proceed with unremitting caution using a mere fraction of the amount of weight that seems appropriate.

After the second trial run all of the weight is moved over to shroud B' because bearing B appears to react more than A, having increased in amplitude from 17 to 23 while A increased from 17 to 18 (see Table 24). Since vibration has worsened, it is also indicative of the weights being 180 degrees from their proper trial position. Hence, the weight is added at blade 12. A vector is now drawn to scale (Fig. 16.3) 17 mils at 7:30; and one of 13 is drawn, not at 8:30 but rather at 6:30, which is the mirror reflection position of 8:30 in relation to 7:30. This is because the index always moves contrary to the movement of the *light* spot. Sometimes a rotating clock face with a fixed index is preferred to a rotating index and fixed face. If a rotating face can be arranged on a coupling or shaft end, the apparent movement of the index in relation to the face is in the same direction as the movement of the light spot. In such an instance, the vector 13 would be drawn at 8:30.

Referring to Fig. 16.3, the rotor exhibited 17 mils and then 13. The vector that has the effect of transforming 17 at 7:30 into 13 at 6:30 is labeled TW and is about 9.5 units in length. On the same basis of argument, the vector that would have the effect of transforming 17 at 7:30 into zero would be 17 units long at phase 1:30. The trial weight would accomplish this if increased in the ratio 17/9.5 and moved counterclockwise through the angle S, which is about 45 degrees or 1.5 hours or three blades. This is done, and run No. 4 proceeds using 8 oz, which is the nearest available weight to 17/9.5 times 4 oz. It is futile to seek weights more

## TABLE 24
### CASE HISTORY DATA

| RUN NO. 1175 rpm | TRIAL WEIGHT AMOUNT AND LOCATION | | | | VIBRATION AMOUNT AT BEARINGS AND PHASE ANGLE | | | |
|---|---|---|---|---|---|---|---|---|
| | A′ | ANGLE BLADE A′ | B′ | ANGLE BLADE B′ | VIB. MILS PP A | PHASE CLOCK A | VIB. MILS PP B | PHASE CLOCK B |
| 1 | —— | | —— | | 17 | 7.30 | 17 | 7.30 |
| 2 | 2 oz. | 24 | 2 | 24 | 18 | 7.30 | 23 | 7.30 |
| 3 | —— | | 4 | 12 | 13 | 8.30 | 13 | 8.30 |

Now increase trial weight in the ratio 17/9.5; and, because angle S is about 45°, move trial weight counterclockwise 3 blades or 1.5 hours.

| 4 | —— | | 8 | 9 | 5 | 9.00 | 5 | 9.00 |
|---|---|---|---|---|---|---|---|---|

Now increase trial weight in the ratio 17/14, and because angle S is about 16 degrees, move trial weight counterclockwise by one blade.

| 5 | —— | | 11 | 8 | 2.5 | 6.00 | 1.5 | 5.00 |
|---|---|---|---|---|---|---|---|---|

Now weld the 11-oz weight at B′, 8th blade, and start a new problem.

| 6 | —— | | —— | | 5 | 3.00 | 7 | 3.00 |
|---|---|---|---|---|---|---|---|---|

Note: From runs 1 and 5, if blade 8 is proper correction position for index at 7.30 and pickup at 9, then this position is 2.5 hours clockwise from pickup, when strobe flashes; so now place 4 oz at blade 17.

| 7 | —— | | 4 | 17 | 1.2 | 12.30 | 3 | 12.30 |
|---|---|---|---|---|---|---|---|---|

Now use same trial weight and rotate through angle S, 30° clockwise.

| 6 | —— | | 4 | 19 | 0.4 | 12.00 | 0.7 | 7.00 |
|---|---|---|---|---|---|---|---|---|

Now weld on and make final check.

accurate than those available unless the latter are in error by more than about 50 percent at this stage. Had 17/9.5 times 4 oz been used, it *should have eliminated vector 17* so that the use of 8 oz, which is a trifle greater, should lead to overcorrection and a vibration of perhaps one mil at phase 1:30. But it is seen

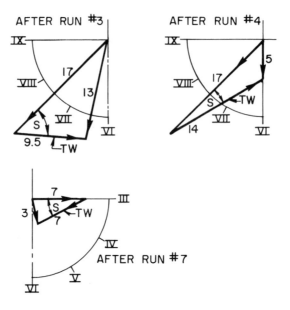

Fig. 16.3   Graphic approach to solution of case history problem

that no such thing occurs, which bears out the point that precise computation is a waste of time. Over and over this remark will be found applicable both in the case history and in general practice.

The vibration is now 5 mils and a second vector diagram is drawn (Fig. 16.3). The angle S is about 15 degrees or one blade. The trial weight is increased to 11 oz and moved to blade 8. It is advisable at this stage to weld the corrective weight in place in the form of a neat single weight so as to consolidate the improvement that has been achieved. The residual vibration is small and must be countered with a small weight.

It is noticed that the fan exhibits the characteristics of a one-plane rotor. After welding, the vibration has increased to more than twice what it had been. This is not unusual, and the available facts are seldom adequate to explain the discrepancy. The rotor is now viewed as a new problem with an original vibration of 7 at phase 3:00. Previous knowledge from runs 1 and 5 suggest that blade 17 is the best place for the next weight. Because, if an original phase of light spot of 7:30 can be corrected by a weight on blade 8 with a pickup at 9:00, and the index mark is in phase with blade No. 24, then blade 8 is 2.5 hours from the pickup when the strobe flashes. Likewise, if the strobe flashes at 3:00 for the index, blade 17 is 2.5 hours from the pickup and is probably the light spot. It is to be noted that runs 1 and 5 are not unduly reliable in this connection because the phase in run 5 should be 7:30 and not 6:00 or 5:00; but runs 1 and 5 are the best available information. Nevertheless, blade 17 turns out to be a useful choice and after run No. 7 a phase shift of the weight, based on the vector diagram, makes the balance almost perfect. At this stage the weight is welded on and a final check is made. Weights less than about 2 oz may be deposited entirely in the form of welding rod, allowing for the weight of the flux. Larger weights require both a properly shaped piece of metal and some welding rod.

For the type of fan described in the case history, the balancing requires about three hours. It might be done in thirty minutes if the greatest haste were necessary. Usually a variety of preparations is required, such as perhaps bypassing the fan to cool it, the arranging of dampers, the removal of access doors, the checking of rotor and coupling and the checking of alignment. Fastening and unfastening of access doors is heavy work, and the fumes and heat, as well as dust, are often more than unpleasant.

## 4. SCOPE OF PHASE MEASURING METHOD

For the phase measuring method and for any other method of in-place balancing it is necessary to see some part of the rotor (preferably a shaft end) and to be able to attach trial and final weights or make drillings or equivalent corrections. Small electric motors are not accessible and so must be balanced in the shop. Centrifugal pumps are partially accessible but are seldom balanced in-place because of the difficulty of attaching trial weights. These rotors are often balanced on knife edges in the shop. Experience shows this to be adequate although a balancing machine will make the correction faster.

**Forced and Induced-Draft Fans and Centrifuges.** Forced-draft boiler fans appear to hold balance indefinitely and seldom require field correction. Induced-draft fans usually stay within vibration limits for a year or more, after balancing. Centrifuges with horizontal shafts are easily balanced, although welding at the desired place is often not allowed so that the basket must be taken to the shop. Link-suspended and shaft-suspended centrifuges usually operate above the critical speed with considerable stability. Balancing seldom requires more than five runs and holds almost indefinitely. The load distribution during operation often causes large vibrations, but this cannot be corrected by balancing the basket.

**Overhung Rotors.** A perennial nuisance is the overhung fan or rotor, especially those requiring two-plane correction. These exhibit every caprice of which a machine is capable, especially if in bad mechanical condition or mounted on loose or weak supports. This author has made as many as thirty runs on some of these, taking perhaps nine hours to achieve an acceptable correction. When the job was complete, it was evident that the procedure was little more than educated guesswork, and the phase measuring contributed nothing at all except perhaps false information. The phase measuring method is not at fault itself. Other balancing methods would be no more successful. The difficulty lies in the presence of factors which lie outside the assumptions on which balancing theory rests. Despite some persistence the writer has never succeeded, on the whole, in identifying these factors and combating them. It is highly doubtful if a successful resolution of the factors is generally possible, as they probably include one or more resonances, some looseness and, perhaps, a tendency to behave like a "flexible" rotor. With experience and luck, some difficult rotors can be balanced but others must simply be rebuilt before balancing is possible. Some success has been achieved with overhung rotors and with rotors that lie between the bearings, using two trial weights at once. [1] The inboard bearing is corrected using weights on the inboard shroud of the wheel or the inboard end of the blades. The outboard bearing is then corrected using two weights at once, one of which is placed on the outboard shroud and one on the inboard. If one weight is changed in amount so also is the other. The two weights are always equal and always located 180 degrees away from each other. The phase angle is taken to be that of the outboard weight and balancing proceeds as if only one weight were used.

Fans, when difficult, are perhaps the most difficult of all rotating items to balance because of wear, contamination, erosion, looseness, cracks and other faults. Machines of almost any kind, with good bearings, foundations and fastening, are usually easy to balance. The problems become more amenable to the use of theory as they ascend the scale of precision finally arriving at the turbines or other high-speed rotors. Turbines are generally best corrected by experts because drilling or grinding at a wrong location could lead to dangerous stresses. Data that is desirable as the basis for a balancing job is probably available only to the manufacturer.

## 5. PRINCIPLES

The purpose of balancing may be said to be the reduction of the vibration at a bearing to an acceptable level. A rough bearing usually vibrates in a sine curve manner when seen on an oscilloscope. As the correction progresses and the amplitude of the fundamental sine curve decreases, other vibrations are usually discovered. These may be harmonics or externally generated vibrations, and it is often difficult to push the level of balance below the level of these other vibrations. Sometimes the filter will separate them and allow the balance to proceed; sometimes it will not. When it will not, the strobe signals become erratic or meaningless. This is not always to be regretted because it seems profitless to correct a rotor below the local level of vibration even when possible. And in such cases as it is in fact profitable, for example in balancing one fan in the presence of the vibrations of an adjacent fan, the filter system is usually of no assistance. In brief, the aim is suppression of a selected sine wave (the fundamental) in the presence of other vibration; suppression to zero is sometimes neither possible nor desirable.

The defining of balance in terms of bearing vibration is usually adequate. The definition may be taken a step further into terms of shaft vibration measured with a shaft stick on the pickup. This is likely to give more stable measurement and more refined balance, if there is no objection to the fatigue of holding the stick. Even this definition is merely fragmentary. In fact, there is probably no universally applicable definition. Good balance means different things in different applications and an understanding of the various facets of this fact is quite helpful in avoiding trouble.

Apart from other trios of principal axes, a rigid body has one trio passing through the center of gravity. It is this trio that is of interest in the theory of balance, and a rotor may be said to be in balance if the axis of rotation coincides with an axis of the trio. Such a principal axis of inertia through the center of gravity is called a principal axis, for the sake of brevity. The coincidence reduces external centrifugal force and centrifugal couple, to zero. For example, a thin, rigid rod is in balance if rotated about its long axis. It is, however, unstable because if the long axis departs slightly from the axis of rotation, it does not want to return; rather it tends to the configuration wherein the long axis is perpendicular to the rotational axis.

If the shaft distortion within the range of operational speeds, caused by centrifugal force, is negligible it may be called a stiff shaft or stiff rotor, as the case may be. Any two planes may be used for correction. Suppose the mass distribution is such that flexural distortion at one speed destroys the balance that has been induced at another. Correction in two planes is then generally insufficient, and the rotor is called *flexible*. Thus for rigid shafts, they may be said to be balanced if the axis of rotation coincides with a principal axis; and for flexible shafts the additional condition required is that flexural distortion and, perhaps, shear forces within the operating speed range be negligible. Beyond this note of caution no more is said here about flexible rotors, since they are mostly beyond the scope of this chapter.

**Rigid Rotors—Two Problems.** Returning to rigid rotors, the problems may be divided into two families that are distinguished by various descriptions which are helpfully summarized as follows.

|  | KINETIC PROBLEMS | DYNAMIC PROBLEMS |
|---|---|---|
| Otherwise known as: | One-plane | Two-plane |
| Axis of rotation: | Does not contain the center of gravity, but it is parallel to the principal axis | May or may not contain the center of gravity and is not parallel to the principal axis through the center of gravity |
| Unbalance amenable to correction by: | A single force | A couple, with or without a force |
| Example: | A circular disc having an eccentric hole, and having a shaft that forms the polar axis | Two circular discs, some distance apart with eccentric holes not in phase and having a shaft that coincides with the polar axes |

Ideally, three runs, with trial weights in two different planes for the second two, are sufficient for the complete computation of the phase and amount of the final weights. In practice this approach is only valid for shop balancing machines and some precision machinery. It is not recommended for field use. The method involves about forty-five minutes of concentrated computation with innumerable chances of error in manipulating the vector algebra and *complex* quantities (except when a computer is used). The burden of this is too much to remember for those who use it infrequently and too much to put into practice under the typical field conditions of noise and fatigue. In the field the computer results are almost invariably worthless for most large fans and similar machinery. Worse than this, they are often positively dangerous. In some careful runs on fans of about 200 hp, the results called for weights and positions that experience could foretell would wreck the fan and perhaps injure personnel.

The most valid and generally useful principle is contained in the vector diagram of Fig. 16.3, and even this must be used with some prudent mistrust.

## 6. RECOMMENDED PROCEDURES

Some procedures are given as guidelines. Because it is difficult to generalize on a subject offering so many variations, the recommendations should be taken as tentative and general unless it is obvious that they are intended categorically.

The use of small weights is desirable. Even in such instances which permit an initial exact computation of the weight that is apparently needed it is best to use about one-quarter of the computed amount, working up slowly.

A flawlessly methodical approach is desirable. Blades or rotors should be clearly marked with some phase-numbering system. The sense of sequence of the clock face numbers should be the same for rotor and external clock face; and to make matters simple, the rotating index should correspond with zero phase on the rotor markings. Without this approach, costly confusion may result and communication with those who are fixing weights is difficult.

Trial weights should be carefully attached and access doors replaced where possible. As an example of the potential danger: at 1800 rpm and 18-in radius, the centrifugal acceleration is of the order of 1500-g so that one pound becomes about three-fourths of a ton.

Caution is far more important than meticulous measurement and calculation. A working accuracy of 10 or 20 percent is about adequate. When a correction of 50 or 75 percent has been achieved it is generally profitable to fasten the weights permanently and proceed with a new problem. It is most undesirable to leave screw-on weights or temporary weights of any kind on a finally balanced rotor. These can upset the airflow or other functions of the machine, and they can become unfastened at some inopportune time causing much inconvenience or perhaps danger. Examples of satisfactory final weights are shown in Figs. 16.4 and 16.5. An unsatisfactory example is shown in Fig. 16.6. It is not recommended that the new problem should proceed before final fixation because this fixation usually causes a significant change in vibration and will harm the later and more refined balance, if postponed.

Fig. 16.4 A view of the final-welded balancing weight on a fan wheel (described in the case history). The ridges are Stellite weld deposit to limit erosion.

Before balancing starts (preferably the day before) the machine should be checked for maladjustment of any kind. Looseness of hold-down bolts is a common nuisance. It is not unusual to decrease the vibration at a fan bearing pedestal from say 15 down to 2 mils by merely tightening the pedestal bolts. Cracks in fan wheels or loose stays are insidious troubles. A thorough hammering usually discovers a crack if it is present. Couplings and alignment should also be checked. These can cause subtle changes during balancing runs. Because of friction, misalignment and torque, some couplings will lock in a crank-like configuration leading to a whirling of the driven shaft around the driver. This is not unbalance but it reads as such on the vibration meter. When the rotor is stopped and started and a correction has been made, the phase location of the crank-like configuration will have changed so that fruitless hours may be spent without knowing the cause of the spurious signals.

An oscilloscope is most useful for discovering trouble either before balancing or during the runs. Any departure from an approximate sine curve is indicative of trouble or of the fact that further refinement of balance is not possible.

A most difficult and unwelcome problem is that of the rotor that is so rough as to be positively dangerous before balancing starts. This author has worked on fans that showed 75 mils at 1800 rpm. Without prior knowledge there is no way of computing the proper position for the first trial weight, and the rotor is danger-

Fig. 16.5 Two, final-welded correction weights on the same wheel as shown in Fig. 16.6

ous enough as it is without the additional aggravation of a wrongly placed trial weight. Some such fans are so located that it is not possible to make phase readings without being quite close to them. One then suffers the danger and hopes for the best, or preferably he retires to a safe distance and makes amplitude measurements only until the trial weight has sufficiently moderated the rotor vibration.

## 7. ADVANTAGES

Because in-place balancing is a corrective technique of the greatest economic importance, it is worthy of study by persons responsible for machine performance. It is a technique that shows obvious and immediate results; one that gives tangible satisfaction to machine operating personnel. Besides this, the procedure usually gives a marked feeling of personal achievement to engineers, mechanics and helpers. The writer gladly recalls with affectionate appreciation the help he has so often received from welders, riggers, machinists and operators, at night and by day, in heat, in cold, in dust and dirt, in safety and in danger. He regrets that convention does not permit the recording of names, because they are surely among the best workmen in the world.

Fig. 16.6   Trial weights on the wheel of a 250 hp, 900 rpm, squirrel cage induced-draft fan

# REFERENCE

1. Ellis, R. "Balancing Overhung Rotors." *I.R.D. News* 2, I.R.D. Corp., Worthington, Ohio (March 1966): 1.

# BIBLIOGRAPHY

Langlois, Armond B., and Rosecky, E. J. *Field Balancing*. Milwaukee: Allis Chalmers Mfg. Co., 1968.

# 17

## IN-PLACE BALANCING
## OF COOLING TOWER FANS*

*Michael P. Blake*

Cooling tower fans can be balanced in-place at considerable cost savings. The costs that result from unbalanced fans are usually considerable: the tower structure may be damaged rapidly or the fan or running gear may be damaged or may fall apart and damage adjacent property. Many typical towers, with two cells, are operated at a slow speed because of known vibration troubles. Others are operated at a dangerously high speed because they are situated on top of other structures, so that the vibration is not noticed. Among about twenty towers measured by the author from time to time, approximately 75 percent exhibited unacceptably high vibration.

**Shop versus In-Place Balancing.** Sometimes fans are dismantled and taken to the shop where they are statically balanced. This knife-edge balancing is quite satisfactory and, except for the great expense and downtime, it leads to worthwhile saving. When a fan is balanced in-place, roughly another $500 per fan may be saved, apart from tower downtime cost which is eliminated. In this chapter, methods are described that require about two hours per fan per year to maintain a satisfactory vibration level. Usually, an operator, a mechanic and an observer make up the best working group for balancing towers where the fan controls are remotely located. Thus, about five man–hours per year per fan has been found by experience to be the effort required to keep average fans at the optimum economic vibration level.

## 1. PROCEDURE

A simple vibration meter pickup is clamped horizontally to a rigid member on the tower deck. A dial indicator cannot be used because there is no base for attachment, as the entire tower vibrates. For the less common forced-draft fans, a dial indicator could be used.

---

*Reprinted with permission from *Hydrocarbon Processing*, June, 1967.

Four trial runs are made: one with the fan as it is and three with trial weights in various positions. Figure 17.1 shows five 2-oz weights attached to a 13-ft diameter induced-draft fan during a trial run. Using a graphical method, the position and amount of the necessary final correction weight is then computed.

Vibration is commonly measured in mils, Peak-to-Peak (PP). One mil is simply 0.001 in. PP signifies that the *total* amplitude of the vibration is measured, just as TIR signifies the *total* runout when a dial indicator is used to measure an out-of-round movement.

Safety is the paramount consideration. Where motor controls are not on the cooling tower deck, a well-understood means of communicating with the operator should be established so that a fan will not be operated when a man is standing between its blades and so that it may be stopped immediately in any emergency.

Preliminary runs are made at the lowest speed. If the vibration seems dangerously high, it should be measured using a meter with an extension cable on the pickup so that the observer and other personnel may make readings at ground level or from a safe location several levels below the deck. Trial weights should be firmly fixed. The centrifugal force is high. For example, at 300 rpm (or say 30 radians per second) and a radius of 6 ft, the acceleration is 6(900/32) which is 170-g, so that a one-pound balancing weight becomes virtually 170 lb, when considered as a projectile. Many fan rings are too light to stop a flying weight and there is no place of refuge on the deck.

The entire tower is shaken during the trial runs, and where possible, the deck vibration should be kept less than 120 mils PP. This sounds like a very modest vibration, but in practice it is sufficient to make most people seek the stairs.

Fig. 17.1   Five weights, each two ounces, are shown attached to blade
of 13 ft in. diameter induced-draft fan, during trial run

Stairways and railings are often in a weak condition so that any happening that might lead to a quick exit from the deck is to be carefully avoided.

The gearbox, motor, couplings and shaft should be in reasonably good condition before an effort is made to balance. The box fastening bolts should be tight. The pitch of all blades on the same fan should be made equal by setting each blade at the same angle. Hollow blades should be checked for holes, water or foreign matter.

For multiple-cell towers that are longer in plan than they are wide, the vibration should be measured parallel to the shorter side and in line with the cell that is being measured. For square towers, the vibration should be measured horizontally, parallel to either edge.

To avoid fatigue and to yield or obtain consistent readings, the vibration meter pickup is clamped to some solid member of the tower at deck level near the edge of the deck. At this location gearbox chatter is often less than near the fan ring and so also is the bumping vibration from the blades, both of which upset some meters. Sometimes it is advantageous to clamp the pickup to the lube pipe that is often run from outside the fan ring to the gearbox. Because this pipe always moves more than the tower, the result is that a more refined balance is often achieved than might be achieved if the pickup were clamped to the deck. The lube pipe will follow the vibration of the gearbox.

Two important balancing or measurement methods are: the stroboscope method, requiring only two runs per correction, and the four-run method, carried out without a stroboscope, using almost any kind of vibration meter. Both methods are satisfactory; however the stroboscope method is the more common. This author prefers the latter, because the stroboscope light is difficult to see in glaring sunlight. Many stroboscope meters do not work consistently at the relatively low speed of tower fans, and they are generally less available than the simpler meters. For these reasons, this chapter is written around the four-run procedure, although it is readily paraphrased in terms of using the stroboscope procedure by those who prefer it.

If adjacent fans are not too rough, a fan may often be balanced with the other fans in the same tower running. A brief survey will locate the roughest fan and that should be corrected before the others, then the next roughest and so on. It may be necessary to return to the first fan when the others have been improved, because a fan cannot always be balanced to a satisfactory level if the adjacent fan is very rough. The fan to be corrected is now run at top speed if its vibration does not exceed about 60 mils PP. Otherwise it is run at a lower speed. Three blades are marked 1, 2 and 3 (preferably blades set 120 degrees apart) although the graphics are valid for any spacing (see Fig. 17.2). The vibration is measured with no trial weight and then with a trial weight successively on blade 1, blade 2 and blade 3, always at the same radius and running at the same speed. The graphic construction is then worked out as shown in Figs. 17.3 and 17.4. The final corrective weight is then fixed at the correct phase angle and at the desired radius.

It will be noticed that each graphic construction among the case histories leads to a vector T that indicates the proper phase location of the final corrective weight. The correct amount of the final weight is $E/T$ times that of the trial

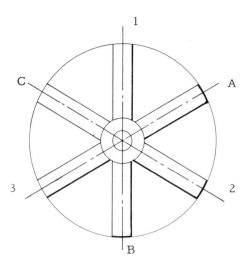

Fig. 17.2   Three blades are marked 1, 2, 3 spaced 120° apart

weight, where E is the original vibration of the fan and where the final weight is placed at the trial radius. If it is placed at another radius, the correct amount is then E/T times the trial weight amount multiplied by the trial radius and divided by the final radius. Whereas the best place for trial weights is at the blade tip, the best place for final weights is often near the center of the fan, where it is usually possible to fix weights at any angle. This ease of setting weights at any phase angle is an advantage when the desired phase lies between two blades, as

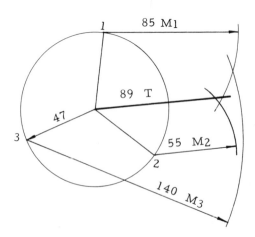

Fig. 17.3   Graphic construction of trial runs on Fig. 17.2 fan

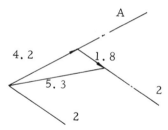

Fig. 17.4   Graphic construction for final corrective weights

it avoids the extra work of dividing the weights between two blades. Placing the weights near the center avoids interfering with the airflow at the blade tip or damaging the tip by welding or other methods of fastening. In the majority of cases, the required phase of the final weight coincides with the position of a blade.

## 2. BALANCING EQUIPMENT

All that is required for the four-run method is a simple vibration meter that will measure down to about 10 mils PP at 300 cpm and up to about 120 mils PP true vibration. Many meters read very erroneously (low) below about 300 cpm. The most commonly used meters have velocity-type pickups. Two of these that were checked were found to read only one-third of the true vibration at 300 cpm and one-fifth at 200 cpm. This is of no consequence as long as the readings are consistent, as far as the balancing procedure is concerned. But it is of concern when tower performance is reviewed against standards because standards are necessarily based on true vibration. For this reason a rough calibration of the meter down to 200 or even to 300 rpm is very desirable. It is indeed a rare person, although there are such, who can consistently judge the acceptability of a fan by sensory observation without appeal to a true meter reading.

Besides a meter, about one dozen 2-oz weights are required. When the final corrective weight is known, it is usual to consolidate it into one custom weight, specially made, neatly and rigidly attached at a convenient radius. No other equipment is required.

## 3. CASE HISTORIES

Three case histories are given.

**Case History 1.** The first is concerned with a typical two-cell tower supported by a building about 60 ft tall. Although the building and tower vibration was about 15 mils with idle fans, no difficulty was experienced in achieving moderately good balance.

Data. The fan has six blades, is made of solid aluminum and is 12 ft in diameter. It operates at 300, 200 and 150 rpm.

The motors are rated 25, 11 and 6.2 hp and operate at 1760, 1170 and 880 rpm.

Procedure. Make four runs: at 300 rpm with no weights—then with 10 oz successively at positions 1, 2 and 3 and at a 5.5-ft radius.

### TABLE 25
### RESULTS OF CASE HISTORY 1

| RUN NO. | VIB. AT DECK MILS, PP | VIB. SYMBOL | WEIGHT AND LOCATION |
|---------|-----------------------|-------------|---------------------|
| 1 | 47 | E | None |
| 2 | 85 | M1 | 10 oz (Blade 1) |
| 3 | 55 | M2 | 10 oz (Blade 2) |
| 4 | 140 | M3 | 10 oz (Blade 3) |
| 5 | 18 | E1 | 4.2 at A (2 at 2) |

Instrumentation. A vibration meter set to read displacement was used. The battery-powered meter reads displacement without a filter or with a filter that cuts out displacements below 20 cps. Here it was used without a filter. The meter also reads velocity and acceleration and is accurate at 300 rpm. A strip chart recorder was used to check the tower vibration for unusual faults.

Graphics. Draw a circle of radius 47 to scale (see Fig. 17.3), and arcs 85, 55 and 140. The correct phase of the final corrective weight is at T and the correct value is E/T times 10 oz, which is 5.3 oz. Draw a small vector diagram, Fig. 17.4, with a line 5.3 units long parallel to T and the other two lines parallel to blades A and 2.

Note: At 140 mils tower vibration, it is quite impossible to write. Because the tower vibration with both fans idle was 15 mils, no attempt was made to balance better than 18. The four runs required forty minutes.

**Case History 2.** Figures 17.5 and 17.6 relate to a very small single-cell tower with a four-bladed fan.

Data. The fan has four blades, is solid aluminum and is 6 ft in diameter. The high and low speeds are about 300 and 450 rpm.

Procedure. Four runs were made at top speed, first with no trial weights— then successively with two ounces at a 3-ft radius on blades marked 1, 2 and 3. The observations made are shown in Table 26.

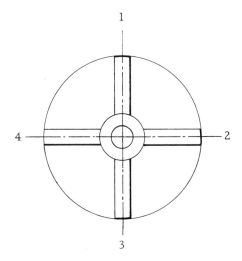

Fig. 17.5   A single-cell tower with a four-bladed fan

**T A B L E   2 6**
**RESULTS OF CASE HISTORY 2**

| RUN NO. | VIB. AT DECK, MILS, PP | VIB. SYMBOL | WEIGHT AND LOCATION |
|---------|------------------------|-------------|---------------------|
| 1 | 7 | E | No wt. |
| 2 | 8 | M1 | 2  oz  (Blade 1) |
| 3 | 10 | M2 | 2  oz  (Blade 2) |
| 4 | 8.5 | M3 | 2  oz  (Blade 3) |
| 5 | 2.5 | E1 | 4  oz  (Blade 4) |
| 6 | 1.5 | E2 | 5  oz  (Blade 4) |
| 7 | 1.0 | E3 | 5.5 oz  (Blade 4) |

Note: In runs 6 and 7, slightly heavier weights were added without further graphics to make the fan run acceptably smooth.

Instrumentation. A small, portable, battery-powered meter was used with a velocity-type pickup.

Note: Readings given for runs 1 through 7 are actual meter readings. For true vibration values, these readings should be multiplied by about two to allow for meter error. True readings are not necessary for balancing, but they are necessary when comparing tower vibration with national standards.

Graphics. Draw a circle with radius E (see Fig. 17.6) and locate points 1, 2,

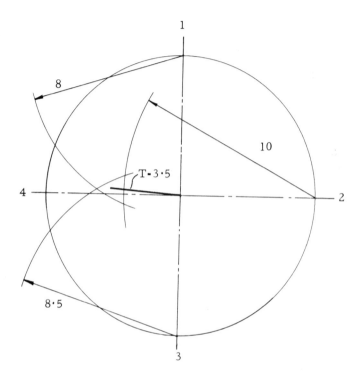

Fig. 17.6   Graphic construction of trial runs on Fig. 17.5 fan

3 and 4 in the same position as shown on the fan diagram, Fig. 17.5. With 1 as a center and a radius of 8, draw an arc. Draw additional arcs of radius 10 with 2 as a center and radius 8.5 with 3 as a center. Draw line T from the center of the circle to the common intersection. T is the correct position for the final corrective weight. The position of T is nearly always in line with a blade rather than between blades. The amount of the final weight is E/T times 2.0 oz.

In the graphic construction, the three arcs should intersect at one point but there is almost always some error so that the length of T must be estimated, to an extent. In the given case history, when a 4-oz weight on blade 4 did not give quite the good balance desired, the weight was changed to 5.5 oz by trial and error. This produced a very smooth balance.

**Case History 3.** This case relates to balancing the fans of a five-cell tower of typical size, with all fans operating. It is seen that corrections of 50 percent and 80 percent were achieved with four runs, which is about as much as can be expected from any balancing method. A sketch of the tower, the gearbox and lubrication pipe arrangement is shown on Fig. 17.7. The graphics are shown on Figs. 17.8 and 17.9. A seismic vibration meter, rather like a miniature seis-

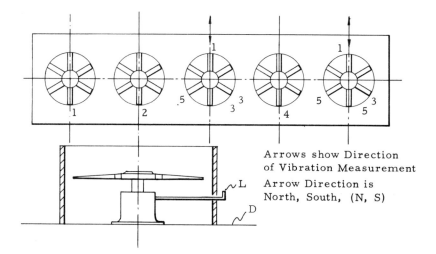

Arrows show Direction
of Vibration Measurement
Arrow Direction is
North, South, (N, S)

Fig. 17.7  Five-cell tower with gearbox and lube pipe shown below

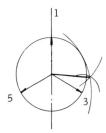

Fig. 17.8  Graphic construction for Fig. 17.7 vibration measured on
tower deck

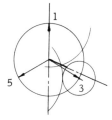

Fig. 17.9  Graphic construction for Fig. 17.7 vibration measured on
lube pipe

mograph pendulum, was specially designed for this test. It required no battery, weighed about 5 lb and proved to be very satisfactory.

Data. The tower had five cells, all in a line, with 13-ft-diameter fans operating at 270 rpm. The tower was 18 ft tall and located at ground level. Balancing was done in place with all fans operating. The results of the tests are shown in Table 27.

Note: The graphic procedure is shown on Figs. 17.8 and 17.9. It was found that the percent correction achieved was greater when the vibration was measured on the lube pipe L than when measured on the deck D. The instrument used was a mechanical seismoscope designed by the author for cooling tower use.

The fans were made with a disc at the center 3.5 ft in diameter. So, the final corrective weights were easily placed at the desired angular position. Usually the desired position for the final corrective weights coincides with a fan blade, if the blades are set at the same pitch angle and the tower is in good overall condition.

## 4. VIBRATION STANDARDS

Proposed standards are given in Fig. 17.10.

The purpose of these standards is to provide some indication of the economic and safety factors connected with fan balancing. Making a fan safe is almost a mandatory matter, leaving little choice. Beyond that, considerable savings may be obtained by balancing further. Beyond that, money is simply wasted by more refined balancing. So there exists a dangerous area indicated by AA, an uneconomic area indicated by A and B and finally an economical area. For greatest savings, the vibration should be corrected so as to lie in the C region.

Because cooling towers, like all equipment, are so variable in their construction and performance, the reader is urged to devise his own service factors, based on his own experience and his own situation. Fig. 17.10 and the given factors can be no more than a rough guide when better data is not available.

## 5. NATURE OF VIBRATIONS

What has been written so far is quite adequate for typical balancing work. But the reader who may wish to take a deeper interest in the subject must go somewhat further and take into account the nature of the vibration. Simple experiments in the field are most instructive. An oscilloscope is of little help. Although it can be photographed, it cannot be visually appreciated because at the low speeds of tower fans most of the vibrations will appear in the form of a star which bounces leisurely on the screen. The ideal apparatus is a meter which gives an output in the form of displacement, velocity and acceleration.

Displacement. For a rough fan, most of the displacement comes from the fan in the form of an approximate sine curve of frequency corresponding to the fan rotation. As the fan balance is improved, the displacement is often dominated by a harmonic that corresponds to the number of blades on the fan. This is because fans are usually near their drive shafts and each blade makes a distinct

**TABLE 27**
**RESULTS OF CASE HISTORY 3**

| RUN NO. | SEISMOSCOPE LOCATION | READING MILS, PP | TRIAL WT. OZ | TRIAL WT. POSITION | TRIAL WT. RADIUS | CELL NO. | CORRECTION |
|---|---|---|---|---|---|---|---|
| 1 | Tower D, NS | 2 | — | — | — | 5 | |
| 2 | " | 3 | 2 | 1 | 6.5 | 5 | |
| 3 | " | 1 | 2 | 3 | 6.5 | 5 | |
| 4 | " | 4 | 2 | 5 | 6.5 | 5 | |
| 5 | " | 1 | 6.2 | 2.5 | 1.85 | 5 | 50 % (Fig. 8) |
| 1 | Lube pipe, NS | 20 | — | — | — | 3 | |
| 2 | " | 24 | 2 | 1 | 6.5 | 3 | |
| 3 | " | 10 | 2 | 3 | 6.5 | 3 | |
| 4 | " | 26 | 2 | 5 | 6.5 | 3 | |
| 5 | " | 4 | 14 | 2.8 | 1.85 | 3 | 80 % (Fig. 9) |

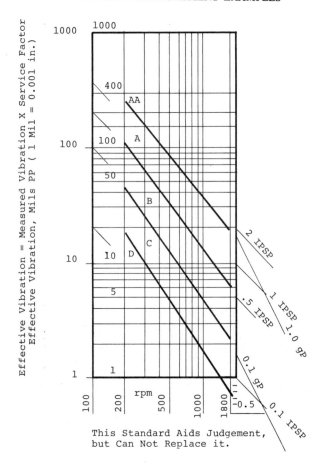

Fig. 17.10   Cooling towers—proposed vibration standards

*Service factors:*

Typical towers, measured on deck; tower sitting on industrial building: 1.0. Typical towers, on ground measured on deck: 1.5. Typical gearboxes or box attachments, multiply above tower factors by .75 (induced draft fans) to derive factor for box. Forced draft fans, measured horizontally, radially: 1.0. When using a vibration meter of unknown calibration, use these additional factors: 3 300/500 rpm.; 2 500/700; 1.5 700/900; 1. 900 and upwards. Example: the deck of a tower that is mounted on ground measures truly 55 mils, at 375 rpm.: what is the vibration classification:? 55 × 1.5 = 82.5, which is classification A.

ACTION

| | | | |
|---|---|---|---|
| AA | Shut down now | to avoid failure | Unsafe |
| A | Correct within 48 hrs. | to avoid failure | Very rough |
| B | Correct within 14 days | to save maintenance $ | Rough |
| C | No action required | Action wastes $ | Smooth |
| D | No action required | Typical new equipment | Faultless |

Note: velocity is indicated by IPSP (inches per second, peak); and acceleration is indicated by gP (max. acceleration, g units)

bump as it passes the shaft. This, and the presence of aerodynamic forces, explains why fans cannot be balanced to perfection or, to state it more accurately, why vibration cannot be reduced to zero. A paper strip recorder (DC recorder) connected to the vibration meter output will illustrate these points.

**Velocity.** The velocity form of the vibration is of considerable interest in monitoring the vibration of a tower from day to day, using a remote reading meter or alarm. Velocity is usually dominated by blade bumping, except when fans are very rough, in which case the fundamental sine wave exhibits well-marked ripples corresponding to the number of blades or corresponding to an unbalanced motor or shaft. For a 300 rpm fan with six blades, the blade frequency and motor frequency coincide. Such a fan is shown in Fig. 17.11.

**Acceleration.** The acceleration form of vibration is of most interest in determining the condition of the gearbox. Vibration in the frequency range between 300 cps and about 3000 cps will tend to dominate in the case of a gearbox in good condition. Severe gear trouble might be expected to lower the top frequency of the dominating vibration.

Fig. 17.11   Trial run: 50 hp fan, 18 ft diameter, induced draft

# 18

# A VELOCITY
# MEASURING PROGRAM

*Steve Maten*

## 1. INTRODUCTION

Most of today's maintenance records of machine vibration are in displacement values, rather than velocity. However, as an aid to maintenance, simple measurement of overall vibration velocity at bearing housings will tell quickly how smoothly a machine is running, *regardless of its operating speed*, type of construction, type of bearings or its environmental conditions. The energy of heat and wear generated in running machinery due to its vibration forces are proportional to the square of the measured velocity. Consequently, measurements taken in terms of either displacement or acceleration generally mean something only if the corresponding frequencies are also known. For quick field measurements and continuous vibration monitoring, the overall velocity measurement is rapidly proving to be a most valuable tool in assessing machinery condition.

A program of periodic vibration velocity measurements will reap the following benefits.

1. It allows evaluation of the initial installation of new equipment, thereby often eliminating costly and time-consuming hot-alignment checks.

2. It reduces the high cost of usually complex and cumbersome scheduled preventive maintenance programs.

3. It detects malfunctioning of equipment before major damages occur, resulting in a marked reduction in parts consumption.

4. It allows equipment to run continuously until actual trouble occurs. This increases the on-stream time of major nonspared equipment.

5. It reduces the number of emergency work orders from plant operators usually involving considerable overtime at premium cost.

6. It permits the maintenance crew to schedule their work more evenly with sufficient lead time to assure parts availability when a repair is scheduled. This improves maintenance performance and reduces the costs of lost time and overtime labor.

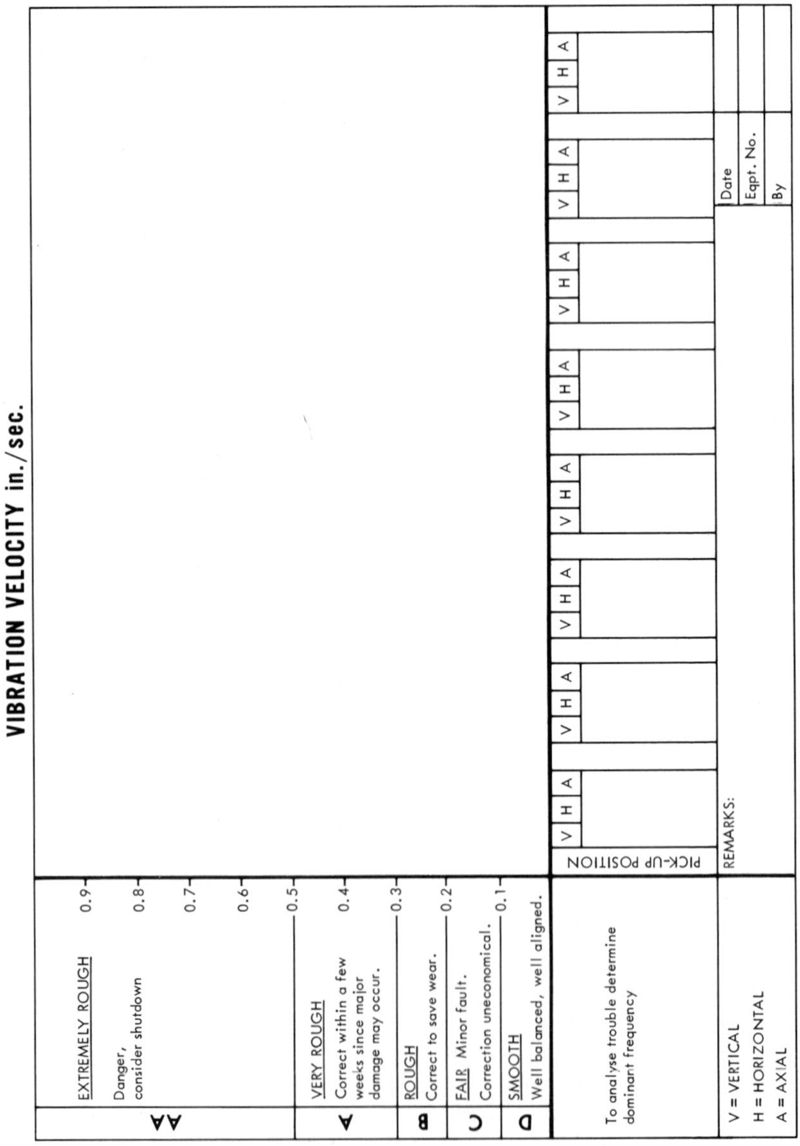


Fig. 18.1 Vibration velocity—analysis work sheet

## 2. MEASURING TECHNIQUES AND PROGRAM

For the maintenance mechanic's use, one of the simplest instruments is an IRD model 306 battery-operated vibration meter which now has both a velocity and a displacement output. With this or any other suitable instrument a mechanic

| FREQUENCY IN TERMS OF rpm | MOST LIKELY CAUSES | OTHER POSSIBLE CAUSES AND REMARKS |
|---|---|---|
| 1 × rpm | Imbalance | Eccentric journals, misalignment, bent shaft, bad belts if rpm of belt |
| 2 × rpm | Mechanical Looseness | Misalignment, rubbing, reciprocating forces as in auto engines, bad belts if 2 × rpm of belt |
| | Pulsation | Lack of pulse absorption in piping of reciprocating equipment |
| 3 × rpm | | Rare—Usually a combination of misalignment and looseness, sometimes bad anti-friction bearings |
| Less than ½ × rpm | Oil Whip or Whirl | Occurs only on high-pressure lubricated machines with plain bearings |
| Synch. | Electrical | Synchronous frequency is 60 cps or 3600 cpm for most AC power sources |
| 2 × Synch. Many times rpm | Torque Pulses Bad Bearings | For single-phase motors this is 7200 cpm. Refers to roll and ball bearings This is normally very high frequency; may be unsteady readings |
| | Gear Noise Aerodynamic Forces | Gear teeth times rpm of bad gear Fan blades times rpm |
| | Hydraulic Forces Misalignment | Impeller blades or lobes times rpm When axial is greater than one-half of either vertical or horizontal |
| Fractions of rpm | Beats, Forced Vibrations | Undamped support structures Natural frequencies of subsoil Lack of isolation between different machinery |

Fig. 18.1a

measures the vibration velocity of drivers and other equipment in three directions on each bearing housing, using the guide as shown in Fig. 18.1.

If all readings are in the D zone, velocity less than 0.1 in/sec for a particular installation, no record is made. The equipment is merely checked OK on a check-list. If any of the readings are in the C zone, 0.1 to 0.2 in/sec velocity, a complete set of velocity measurements is recorded as in Fig. 18.1. The report is then kept in the equipment file for future reference and comparison. If any of the readings are in the B zone, 0.2 to 0.3 in/sec velocity, a complete set of velocity measurements is taken. By taking amplitude readings as well, the mechanic can determine the dominant frequencies in the vibration (see Fig. 18.2). With the guide given in Fig. 18.1a, he analyzes the fault and writes a low-priority work order to correct the installation. In case of difficulty in interpreting the measurements, he can call on the plant engineering department which should be equipped with a frequency analyzer. An IRD model 320 is suggested. If any of the readings are in the A zone, 0.3 to 0.5 in/sec velocity, a similar procedure is followed as described for B-zone readings, except that the work order is given a high priority. If any of the readings are in the AA zone, velocity greater than 0.5 in/sec, shutdown of spared equipment is immediately effected upon analysis of the faults. In the case of major nonspared equipment, readings in the AA zone are immediately reported to the plant engineering department for analysis and a decision on plant shutdown. Any mechanical work performed as a result of the vibration measurements is recorded on a repair report card (see Fig. 18.3).

Upon completion of the work, a set of velocity readings is recorded on the original vibration report for comparison of results before and after repairs. The repair report is then attached to the vibration report and both are kept in the equipment record file. To be most effective, the mechanic's measurement program should be carried out on a biweekly basis. Figure 18.4 and associated notes represent a typical vibration analysis performed by a plant engineering department on the drive train of a major centrifugal compressor installation.

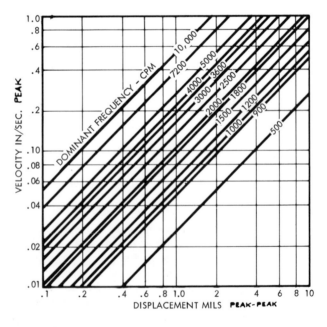

Fig. 18.2   A graph of velocity versus displacement

EQUIPMENT # _____          DATE _____

**REPAIR REPORT**                    **MATERIAL REPLACED**

_____     _____

_____     _____

_____     _____

_____     _____

_____     _____

_____     _____

_____     _____

_____     _____

_____     _____

_____     _____

_____     _____

_____     _____

_____     _____

Fig. 18.3   Repair report card

## 3. START-UP INSPECTION

Due to the effective feedback of information resulting from a set of running vibration readings and vibration analyses, it is possible to have installation contractors completely responsible for installation and precommissioning inspection of mechanical and rotating equipment. This results in less interference with the contractor's work and an overall improved morale.

The client's rotating equipment engineer can limit himself to a final check-out inspection and running check-out which includes a set of vibration readings. It is

## VIBRATION REPORT

JULY 11/67

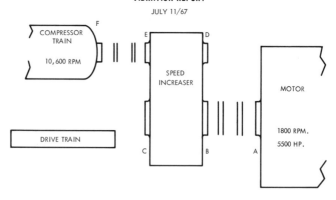

| PICK-UP POSITION | FILTER OUT | | FILTER IN | | | | | | | |
|---|---|---|---|---|---|---|---|---|---|---|
| | AMPL. MILS | VEL. IN/SEC. | MILS | CPM | MILS | CPM | IN/SEC. | CPM | IN/SEC. | CPM |
| **A** H | 0.3 | 0.09 | 0.4 | 1800 | | | | | 0.1 | 27,000 |
| V | 0.2 | 0.05 | 0.2 | 1800 | | | | | 0.025 | 32,000 |
| A | 0.25 | 0.09 | 0.1 | 1800 | | | | | 0.03 | 27,000 |
| **B** H | 0.2 | 0.12 | 0.1 | 1800 | | | 0.07 | 32,000 | 0.08 | 27,000 |
| V | 0.14 | 0.07 | 0.07 | 5500 | | | | | 0.06 | 10,600 |
| A | 0.60 | 0.22 | 0.55 | 5500 | | | 0.18 | 5500 | 0.10 | 10,600 |
| **C** H | 0.25 | 0.15 | 0.085 | 10,600 | 0.065 | 32,000 | 0.015 | 10,600 | 0.15 | 32,000 |
| V | 0.2 | 0.08 | 0.07 | 10,600 | | | | | 0.06 | 32,000 |
| A | 0.45 | 0.22 | 0.4 | 5500 | 0.1 | 1800 | 0.11 | 5500 | 0.10 / 0.09 | 27,000 / 32,000 |
| **D** H | 0.25 | 0.15 | 0.11 | 5500 | 0.08 | 22,000 | 0.11 | 32,000 | 0.10 | 22,000 |
| V | 0.2 | 0.15 | 0.11 | 5500 | 0.08 | 22,000 | 0.08 | 32,000 | 0.09 | 22,000 |
| A | 0.35 | 0.14 | 0.35 | 5500 | 0.13 | 10,600 | 0.10 | 5500 | 0.10 | 22,000 |
| **E** H | 0.22 | 0.14 | 0.08 | 22,000 | | | 0.09 | 27,000 | 0.09 | 22,000* |
| V | 0.17 | 0.12 | 0.10 | 5500 | | | | | 0.08 | 32,000 |
| A | 0.23 | 0.16 | 0.22 | 5500 | | | | | 0.16 | 27,000 |
| **F** H | 0.65 | 0.23 | 0.65 | 5500 | ← | | 0.16 | 5500 | 0.13 | 27,000 |
| V | 0.55 | 0.22 | 0.46 | 5500 | ← | | 0.13 | 5500 | 0.10 | 22,000 |
| A | 0.40 | 0.16 | 0.3 | 1800 | | | | | 0.14 | 27,000 |

H = HORIZONTAL
V = VERTICAL
A = AXIAL

Fig. 18.4   Vibration report: 1. Readings taken at the compressor's driver end F indicate oil-whip. The radial displacement at this bearing is about 0.55 mils at one-half shaft speed. This particular vibration has shown up on previous analyses as far back as 2½ years ago at about the same amplitude. 2. The axial readings taken at the gearbox bearings indicate that the box walls are in resonance at the disturbing frequency of 5,500 cpm. The highest amplitudes are measured in the wall centers. 3. Gearbox radial vibrations at bearings are low, particularly at shaft speeds; the same goes for the motor-vibrations. Axial vibrations at shaft speeds on motor, gear and compressor are practically non-existent; hence, it is concluded that alignment of equipment is acceptable. 4. Almost all velocity readings respond to high-frequency harmonics (4th, 5th and 6th harmonic) of the disturbing frequency except in the radial direction of the compressor and the axial direction of the gearbox where response is obtained at 5,500 cpm as well as the higher frequencies.

during this initial running check that one can indeed verify that the alignment is correct and that the installation has been properly set up. Experience has shown that one can find the faults, including pipe strain, in an installation more effectively in this manner, rather than, as has been a popular practice, by having a field inspector closely following the contractor's rotating equipment installation work.

**Scope:**  The inspection includes all mechanical and rotating equipment, their drivers and immediately associated controls.

**Phases:**  1. *Data cards are to be completed* by client personnel for purposes of familiarization with the new equipment and advance completion of maintenance records.

       Note:  One card is to be completed for each component in an equipment train, i.e. compressor casings, pumps, mixers, turbines, gears, etc.

2. *A lubrication list is to be completed* by client personnel. This list shall be handed to the contractor in time to lubricate all mechanical equipment components and to facilitate completion of the precommissioning check-out.

3. *A precommissioning check-out is to be performed* by the contractor on all mechanical equipment to insure a sound and reliable installation. The manufacturer's instruction manuals shall be followed and, where needed, service representatives shall perform the check-out.

    The contractor will record the precommissioning check-out on a control sheet using one sheet for each major item. There are two types of check-out control sheets: one for compressor/pump and motor driver combinations and one for compressor/pump, gear and turbine driver combinations (see Figs. 18.5 and 18.6). The sheets can also be used for fans, mixers or blowers, etc. The contractor will complete the precommissioning section of the form and sign and fill in the date at the bottom of the sheet under precommissioning check-out accepted. The client's representative will sign below the contractor for verification purposes and acceptance of alignment readings. The contractor will also record a set of final alignment readings on the reverse side of the check-out control sheet.

4. *A running check-out is to be performed* by the client's rotating equipment engineers on all mechanical equipment after commissioning. Manufacturers' instructions and start-up procedures shall be followed.

    The running check-out and operating values shall be recorded on the lower portion of the check-out control sheet for each item of equipment.

    A vibration velocity record card (Fig. 18.1) shall be completed for each installation. Equipment having vibration velocities above 0.1 in/sec shall be further analyzed and corrected until a smooth running classification is achieved. The client's representative will sign for acceptance of the running check and the contractor will countersign same for verification purposes.

| | | Check out OK | | | | | |
|---|---|---|---|---|---|---|---|
| | | Not Applicable | | | | | |
| | | Needs Further Action | | | | | |

**COMPR./PUMP AND MOTOR DRIVER**
**INSPECTION CHECK LIST**

Unit: _____

Eqpt. No: _____

| PHASE | TYPE | ENTRY | ACTION | 1st Check | | 2nd Check | | REMARKS |
|---|---|---|---|---|---|---|---|---|
| | | | | Eqpt. | Motor | Eqpt. | Motor | |
| PRECOMMISSIONING CHECK-OUT | Mechanical | 1 | Grouting (Hammer Base) | | | | | |
| | | 2 | Appearance of Equipment | | | | | |
| | | 3 | Strainer Direction | | | | | |
| | | 4 | Check-Valve Direction | | | | | |
| | | 5 | Proper Vents & Drains | | | | | |
| | | 6 | Seal Flush Properly Connected | | | | | |
| | | 7 | Wearing Flush Connected | | | | | |
| | | 8 | Cooling Water Connected | | | | | |
| | | 9 | Piping Strain-Free | | | | | |
| | | 10 | Piping Flushed & Tested | | | | | |
| | | 11 | Pipe Supports Adequate | | | | | |
| | | 12 | Packing or Seal Correct | | | | | |
| | | 13 | Lubrication (Couplings & Bearings) | | | | | |
| | | 14 | Alignment (Record Data) | | | | | |
| | Controls | 15 | Pressure Gauges Installed | | | | | |
| | | 16 | Safety Valves Installed | | | | | |
| | | 17 | Tripping Devices | | | | | |
| | | 18 | Other Protective Devices | | | | | |
| | | 19 | Electrical Check-out Compl. | | | | | |
| RUNNING CHECK | All Types | 20 | Running Amps. (Rec. Full Load) | | | | | |
| | | 21 | Eqpt. Bearing Temperatures | | | | | |
| | | 22 | Motor Bearing Temperatures | | | | | |
| | | 23 | Controls Functioning | | | | | |
| | | 24 | Noise Level | | | | | |
| | | 25 | Vibration Velocity (Rec. Data) | | | | | |
| | | 26 | Capacity Normal | | | | | |
| | | 27 | Pulsation or Cavitation Susp. | | | | | |
| | | 28 | Any Leaks | | | | | |
| | | 29 | Seal Functioning Properly | | | | | |
| | | 30 | Data Cards Completed | | | | | |

| Precommissioning Check-out Accepted by: | Running Check-out Accepted by: |
|---|---|
| Contractor: | Client: |
| Client: | Contractor: |
| Date: | Date: |

Fig. 18.5   Inspection check-list—compressor/pump and motor driver

| | | | | | | COMPR./PUMP, GEAR AND TURBINE INSPECTION CHECK-LIST | | | | |

**COMPR./PUMP, GEAR AND TURBINE**
**INSPECTION CHECK-LIST**

✓ Check out OK
✗ Not Applicable
Needs Correction

Unit: _____
Eqpt. No: _____

| PHASE | TYPE | ENTRY | ACTION | 1st Check | | 2nd Check | | REMARKS |
|---|---|---|---|---|---|---|---|---|
| | | | | Eqpt. | Driver | Eqpt. | Driver | |
| PRECOMMISSIONING CHECK-OUT | Equipment | 1 | Grouting (Hammer Base) | | | | | |
| | | 2 | Appearance of Equipment | | | | | |
| | | 3 | Strainer Direction | | | | | |
| | | 4 | Check-Valve Direction | | | | | |
| | | 5 | Proper Vents & Drains | | | | | |
| | | 6 | Seal Flush Properly Connected | | | | | |
| | | 7 | Wearing Flush Connected | | | | | |
| | | 8 | Cooling Water Connected | | | | | |
| | | 9 | Piping Strain-Free | | | | | |
| | | 10 | Piping Flushed & Tested | | | | | |
| | | 11 | Pipe Supports Adequate | | | | | |
| | | 12 | Packing or Seal Correct | | | | | |
| | | 13 | Lubrication (Couplings & Bearings) | | | | | |
| | | 14 | Lubrication (Gear) | | | | | |
| | | 15 | Meshing of Gear Teeth | | | | | |
| | | 16 | Alignment (Record Data) | | | | | |
| | Controls | 17 | Pressure Gauges Installed | | | | | |
| | | 18 | Safety Valves Installed | | | | | |
| | | 19 | Tripping Device Set | | | | | |
| | | 20 | Governor Set | | | | | |
| | | 21 | Hydraulic System Functioning | | | | | |
| RUNNING CHECK | All Components | 22 | Equipment Bearing Temperatures | | | | | |
| | | 23 | Gear Bearing Temperatures | | | | | |
| | | 24 | Controls Functioning | | | | | |
| | | 25 | Noise Level | | | | | |
| | | 26 | Vibration Velocity (Record Data) | | | | | |
| | | 27 | Capacity Normal | | | | | |
| | | 28 | Pulsation or Cavitation Suspected | | | | | |
| | | 29 | Any Leaks | | | | | |
| | | 30 | Seals Functioning Properly | | | | | |
| | | 31 | Data Cards Completed | | | | | |

| Precommissioning Check-out Accepted by: | Running Check-out Accepted by: |
|---|---|
| Contractor: | Client: |
| Client: | Contractor: |
| Date: | Date: |

Fig. 18.6   Inspection check-list—compressor/pump, gear and turbine

## FINAL SHAFT ALIGNMENT READINGS

TAKEN BY: _____

APPROVED BY: _____

DATE: _____

NOTES:

1. All readings to be taken with dial indicator graduated in thousands of an inch and recorded below.

2. All readings to be taken when viewed in the same direction starting from the outboard end of the driver.

3. Maximum parallel misalignment across flexible couplings for most installations is 0·003" and angular misalignment across coupling faces maximum 0·001"

4. For some larger installations where a long span across the coupling is involved, higher misalignment values are permitted. Consult Manufacturer's installation instructions and confirm same with clients Plant Engineering Dept. (Figure 8)

5. Equipment subject to higher operating temperatures must be aligned so as to allow for the thermal deformation. Consult Manufacturer's installation instructions for expected deformation values.

ROUGH PICTORIAL SKETCH OF INSTALLATION:

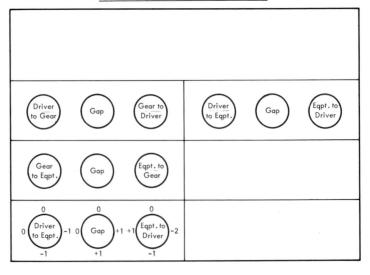

Fig. 18.7   Check-out control sheet for final shaft alignment readings

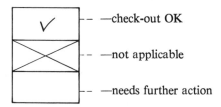

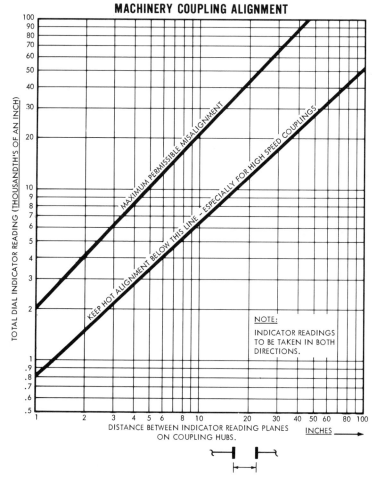

Fig. 18.8   Machinery coupling alignment for check-out control sheet

**Notes:**  a) It is important that the nomenclature on the inspection checklist is followed consistently, i.e.:
In the case of actual readings, the values are entered instead of the abovementioned check notes.

b) The inspection checklist allows for the recording of a second check in case the first check calls for further action. If the first check is OK the block for the second check can be left blank.

c) For the running check, vibration velocity measurements are taken in three directions (vertical, horizontal and axial) on all bearing housings. The readings are recorded on the vibration record sheet (Fig. 18.1) as a vertical bar for each pickup position. The pickup positions shall also be recorded.

## 4. MONITORING VIBRATION

For application in the industrial plant, the writer advocates the use of velocity monitors rather than the conventional displacement or acceleration types.

A monitor proportional to velocity will detect both low- and high-frequency vibration. Consequently, the signal will produce an alarm for any malcondition in rotating machinery. An economical installation, if more than one point is to be monitored, would be one monitor fed via a scanner from a series of velocity pickups. The monitor would provide an alarm in the case of excessive vibration at one point, and the scanner would show a light for the point reporting the trouble. Pressing the test button of this particular point would show the velocity value on the monitor. This then provides a direct clue to the degree of malfunction of the machine involved. A further analysis of the vibration can subsequently be made with a portable vibration analyzer.

Displacement monitors generally provide protection for low-frequency vibrations, such as shaft unbalance, misalignment or mechanical looseness. But what about a bad anti-friction bearing? If we measure a one-half-mil displacement with a vibration displacement meter on a machine operating at 1800 rpm, then the monitored vibration is indicative of the machine condition so long as the frequency of the disturbance is equivalent to 1800 rpm or 30 cps. But the displacement meter gives no information about frequency. The disturbing frequency could be equivalent to 18,000 rpm if the tenth harmonic were in fact the dominant frequency; and under such conditions of displacement and frequency failure might be imminent. This would be the case for a bearing that is ready to seize up. The corresponding velocity would be close to 0.5 in/sec so that a velocity monitor would provide an alarm.

Acceleration monitors are usually applied for higher-frequency vibrations (high-speed machinery, gears and anti-friction bearings). But what about the high-frequency resonance which usually occurs in high-speed machinery? Such resonance can cause g-switches to be set higher than is desirable to protect equipment against such phenomena as shaft whip occurring at lower frequency. Another, even more striking example, is the cooling tower which would fall down long before a switch, set at one-g protecting the gear, would be actuated by a low-frequency sway of the tower. At 100 cpm the tower movement would be about 7½ in if a switch set at one-g is to be actuated, whereas a velocity monitor set at 0.5 in/sec would sound the alarm at only 100 mils of movement.

Acceleration and displacement values mean very little to the plant operator who is unfamiliar with vibration charts. And the fact that different allowable values are used for different machines only adds to the confusion. Some of the disadvantages of acceleration and displacement monitors may be overcome by low- and high-pass filters; but, at the same time these limit the range of protection. Recognizing this problem, one can also resort to frequency selective monitors and now the poor plant operator is completely in trouble, unless he takes a course in vibration theory. Furthermore, multi-point monitors with a value output in terms of displacement or acceleration require a separate monitor installation for each point instead of the economical single monitor plus scanner system made possible if the value output is in terms of overall velocity.

# 19

# A VIBRATION
# MEASURING PROGRAM

## Michael P. Blake

The establishment of a vibration measuring program, for plants having about 100 men devoted to machine maintenance, is discussed here from the point of view of justification, organization, surveying, measuring, monitoring, records, checklists, safety, types of problems, instrumentation, correction and feedback. The purpose of this chapter is to provide help in the starting of new programs or in the improvement of existing ones.

## 1. JUSTIFICATION

The justification of a program is both economics-related and safety-related. A continuing program helps significantly to control cost and safety in operating most machines. The purpose of a machine is to provide a needed function at the lowest cost per unit of performance. This cost includes both the capital cost and the cost of ownership, the latter including such items as power, depreciation and maintenance. The concept of safety may be envisaged as being the extreme end of the spectrum of economics. Apart from ethical considerations, it is invariably a matter of good economy to operate a machine so as to avoid danger to it or to persons affected by it.

Often, the basic function of a machine is not clearly known to the designer or, for that matter, to the operating personnel: Does it mix or comminute or pump, or does it do all of these? Moreover, reasons for installing specific equipment can change with time. The function of each machine should be clearly defined as a prelude to achieving economy. For example, during World War II a pump that had been used in England for a century to control the local level of drainage water stopped for lack of coal. After a period of time it was found that the drainage water rose by only a negligible amount. The pump was no longer a necessity. An error existed as regards function.

A machine in good condition is usually safe and economical to operate, whereas one in bad condition is usually unsafe and uneconomical. A spectrum for the condition classification of a machine may be envisaged as including the following classes or their equivalents.

D—Faultless

C—Minor fault; correction wastes money

B—Significant fault; correction saves money

A—Acute fault; correction mandatory

AA—Dangerous; stoppage mandatory

These five classes have been found to be adequate by this writer. Other workers use, perhaps, ten or more classes to cover the same span. Early workers tended to use classes such as rough and smooth, without any indication of the economic or safety implications of the classification. The five given classes may be reduced to three: economical, uneconomical and unsafe, or to two: safe and unsafe. The level interval between classes is an empirical fact of life and turns out to be about 3:1 per class. For example, a dangerous fault shows a vibration amplitude of about three times that of an acute fault. In practice, the writer has advocated the use of a level interval of three which is approximately the square root of ten, and which corresponds conveniently with ten decibels, when measurements are expressed in that way.

In the foregoing terms the justification of a measuring program may be put as follows.

1. It determines the classification.

2. It diagnoses the fault.

3. It verifies the correction.

4. It predicts probable dates of required correction.

5. It minimizes maintenance and operating costs through numerically expressed measurements and analyses.

6. It lowers future costs by influencing such persons as purchasing agents and designers through a feedback of accurate information to them.

A large installation stands to gain more than a small one where each machine is intimately known to its operator. The measurement is invariably better than the opinion of the operator because even a valid opinion is seldom number-related. But the measurement cannot replace human judgment and human judgment cannot replace the measurement. Judgment is most powerful when applied to measurements, and least valuable when applied to mere physiological or psychological reactions.

In monetary terms a measuring program is usually expected to save about 20 percent of machine-related maintenance in the first year in such industries as refining and chemical processing. Savings of up to 25 percent in the first three months have been reported to this writer by others. On the other hand, a measuring program related, for example, to engine lathes might not justify itself. The program is only valuable when it provides valuable information as a basis for the making of decisions. The resulting decisions are in no way mysterious or novel, but are of the everyday kind such as *shutdown* or *do not stop* or *inspect this machine.*

## 2. ORGANIZATION

The discussion here is in terms of plants having upward of twenty men, and typically about one hundred, devoted to machine maintenance. In smaller plants, the person doing the measuring may also engage in operation and analysis and even maintenance so that there is less need for work organization. In every case good records should be kept.

Assuming that vibration-measuring experience is not available within the plant, much grief and money can be saved by working at first with a consultant. After an initial period of evaluation and training, it is to be expected that one engineer can attend to all analysis and organization, using perhaps 50 percent of his time; or it may be preferred to have all measuring done by an outside group. After the initial period, perhaps only one case in 100 will require the services of a consultant. In larger companies with corporate engineering facilities for analysis, the more difficult problems may be handled on a corporate basis. In such companies normal annual experience should be shared for the benefit of all. The primary aim is to establish and maintain a program that brings the most benefits. The cost of the program must be subtracted from the gross benefits. About half the work time of one mechanic is required for surveying and other work. It is desirable to encourage the cooperation of the instrumentation department as regards monitors and other devices that may come under its care for maintenance, alteration or installation. A summary of *organization* is given in Figs. 19.1 and 19.2.

| TECHNIQUES | RELATED SOURCES |
|---|---|
| **CLASSIFY** | |
| By survey, field request, monitoring, sighting-in | A stick held in the hand |
| | Portable vibration meter |
| AA—dangerous: stop now | Classification standards |
| A—acutely faulty; correct 48 hr. | Experience |
| B—faulty; correct within 21 days | Local knowledge |
| C—normal; correction wastes $ | Comparison with similar local machines |
| D—faultless | |
| **IDENTIFY** | |
| Measure | Portable vibration meter (50% of cases) |
| Analyze | Simple vibration analysis (25% of cases) |
| Use traditional methods | Auxiliary equipment: |
| | Tape recorder |
| | Wave analyzer |
| | Level recorder |
| | Oscilloscope |
| | Panoramic analyzer |
| | Shaker table for calibration and simulation |
| | Camera |
| | Frequency time analyzer |

Fig. 19.1   Vibration techniques as an aid to performance

| TECHNIQUES | RELATED SOURCES |
|---|---|
| **CORRECT** | |
| Mass-unbalance (40%) | Traditional corrective technique. |
| Alignment (30%) | Portable balancing equipment usually |
| Couplings (20%) | designed for use in simple analysis |
| Looseness, bearings; | Shop balancing equipment (shop balanc- |
| Resonance, gear teeth | ing usually bypasses classification and |
| Foundation cracks | identification, if the machine is down |
| Wear. Electromagnetic, | for repair) |
| hydraulic, aerodynamic effects | Isolation |
| Chatter, drive chains, belts | Redesign (revision) |
| | Foundation revision |
| | Tightening hold-down bolts |
| | Better belts or chain |
| | Correct pitching of blades |
| | Elimination of equipment |
| | Replacement |
| | Installation of monitor |
| **FEEDBACK** | |
| To Purchasing Dept. | Maintenance Supervisor |
| Vendor, repairer; | Maintenance standards |
| Design, Maintenance Dept.; | Maintenance equipment records |
| Safety, Standards Dept. | Safety Dept. |
| Equipment records | Purchasing Dept. |
| New specifications for: | Inquiries |
| Balance, isolation | Specification |
| Foundations | Selection |
| Chronic faults | Modification |
| Unsuitable equipment | Elimination |
| Undesirable equipment | Design Dept. |
| | Resonance, foundations, |
| | Isolation, . . . |

Fig. 19.1—Continued

Adequate classification leads to a decision as to whether or not to correct. Adequate fault identification, via analysis, leads to elimination of unnecessary maintenance when a correction is planned. After correction, the original source of trouble may be mitigated in future instances by feeding back pertinent information to such persons as designers and purchasing agents.

## 3. SURVEYING

This is the periodical measurement of all significant pieces of equipment, with a view to classification only. This writer's method is to carry the smallest velocity meter and a piece of stick about a yard long and about ⅜-in in diameter. This is used to touch every significant machine point, in passing. The meter is used if the feeling of the stick is suggestive of trouble. This saves time because most

| TRADITIONAL STRATEGY | PROPOSED STRATEGY |
|---|---|
| **DISCOVER**<br>Advanced acute faults or seeming<br>faults<br>Breakdown<br><br>**CLASSIFY**<br>By intuition, experience, or not at all | **CLASSIFY**<br>1. Survey or monitor<br>2. Vibration information<br>3. Fault or not<br>4. Classification standards<br>5. Classified case<br>Fault or not |
| **IDENTIFY**<br>By experience or not at all;<br>Implied criteria; vagueness<br><br>Delay, argument, emotion<br>Loss of money | **IDENTIFY**<br>1. Classified case<br>2. Measure and analyze or not<br>3. Identified fault, or<br>case closed |
| **CORRECT**<br>Use traditional methods;<br>Vague, blanket-type repair orders,<br>such as "Repair pump";<br><br>Doing more than needed;<br>Doing it too late | **CORRECT**<br>1. Identified fault<br>2. More precise repair order such as<br>"Replace bearing and balance"<br>3. Fault elimination<br>Number-related data |
| **FEEDBACK**<br>Little or none, mainly because of lack<br>of a number-related description of<br>fault severity or of frequency of<br>fault occurrence | **FEEDBACK**<br>1. Number-related data<br>2. Feedback resources<br>3. Decrease in need for<br>future corrective effort |

Fig. 19.2   Vibration techniques for machine-performance improvement

meters are quite inconvenient in use due to their cables and heavy pickups and the need for range stepping. However, the use of a stick is not recommended to beginners, and even in the hands of the experienced it may lead to missing some trouble. The stick is not, of course, used on critical machinery. Meter readings are classified as shown in the top left column of Fig. 19.1, using local standards. Such a survey is made at least twice annually for all equipment and perhaps more often for critical equipment. The survey may be expected to require from one to four weeks at eight hours per day in a plant having 100 men devoted to machine maintenance. Oil leaks, hot spots and other matters that are not vibration-related may also be reported to the maintenance supervisor.

## 4. MEASURING

Measuring implies a more searching review than that connected with surveying. A frequency analysis may be made or the vibration may be measured in both acceleration units and velocity units, for example, in order to evaluate its severity and diagnose the fault, if any. If the fault is obvious, the results of surveying are passed directly to the supervisor who is to correct it.

## 5. MONITORING

This connotes a continuing survey of some kind, usually linked with annunciation devices. The objective is the safeguarding of the operation of critical machinery as well as the prediction of the probable date when correction may be required, together with the elimination of all unnecessary inspection and correction. Although newer critical equipment is adequately monitored at the time of installation, it is given as an opinion here that more monitoring could be profitably used in connection with older and less critical equipment.

## 6. RECORDS

The best method of keeping records varies from plant to plant. In every situation care must be taken that every piece of equipment is identifiable, preferably by number and building and the level in that building. It is very difficult to communicate if machines have no numbers or if different descriptions of the same machine are used by different people. The record should indicate the points that are measured and the spatial direction of measurement, such as axial. For the less critical machine it is usual to keep no record if the classification is C or D. For the more critical machine it is usual to preserve records of all measurements, irrespective of classification. This forms a basis for trend prediction. As far as possible, records should be intelligible to persons other than the maker at a later time. Magnetic-tape records should give date, machine and so on, and particularly the level at which the recording was made and a taped calibration of some standard signal, such as an equivalent of a velocity 1 in/sec, peak. A key must be provided so as to show where on the tape the record is to be sought. In general it is preferable to keep the original tape intact and make loops which are used for purposes of analysis and review. As an example of record keeping, the reader is referred to Chapter 18 of this volume.

### Checklist for Measuring
1. Machine number, date, time, building, level, horsepower and speed
2. Location and direction of measurement
3. Instrument and operator, units of measurement
4. Local structural vibration, process-related vibration and machine-related vibration
5. Is measuring scheduled or requested?
6. When tape recording, are calibration and other necessary information included on the tape? Is it desirable to take advantage of the ease of recording local noise at the same time? Is the microphone adequately calibrated?
7. Is the predominant vibration frequency below about 2000 cps and, therefore, amenable to use of a velocity pickup, or is it of higher frequency, requiring an accelerometer?
8. Are all pickups and meters adequately maintained and calibrated?

9. Are pickups or meters affected by severe environments such as local heat or magnetic fields?

10. Is the operator in danger, and what form is failure likely to take if it occurs?

11. Are operating personnel aware that measurements are being made?

12. Is there a hazard due to using meters in an explosive environment?

13. Have all safety aspects of measuring received consideration?

14. Is the program considered to be partial or optimum in terms of discovery, classification and feedback?

15. Are all existing monitors reliable? What do operators expect from them? Are they behaving as expected?

16. Verification of the effect of correction: Has classification changed to an acceptable level?

17. What savings and benefits are due to the program?


## 7. SAFETY

The person making a survey of measurements must, of necessity, encounter a wide variety of equipment, and he can hardly be familiar with it all. It is therefore advisable that he gain an understanding of the working of the machines in each building from operating personnel. Also, machines with large vibration are usually prone to danger from failure of one kind or another. An advantage of tape recording is that less than a minute is all that is required at the site of the machine. This lowers the risk which is undertaken when, perhaps, an hour or more is spent at the site making analyses, besides relieving operating personnel of the need to operate the machine in a given manner or in a dangerous manner for an extended period.

It is advisable to keep out-of-line with couplings and other exposed components when measurements are made, to take care that the transducer cable does not wrap around a shaft, to keep out-of-line with temporary balancing weights, to be alert for surprise start-up of remotely controlled machines and, in general, to use imagination so as to predict such hazards as are latent in each situation. When the axis of rotation is vertical instead of horizontal, it is often not possible to keep out-of-line of danger. For example, the fans of cooling towers often rotate with the fan in the horizontal plane, surrounded by a flimsy fan ring that offers no protection to a man on the deck of the tower and who must spend some time there measuring or balancing. Some comminuters (similar to a fan wheel) have vertical axes. The casings are usually strong. However, it is to be borne in mind that there is no place of refuge from flying components or goods if an access door or casing is removed.


## 8. TYPES OF PROBLEMS

Mass unbalance and misalignment together are likely to exceed 50 percent of all problems. The larger machine attracts more attention because of the drama connected with it and the high cost of operation and ownership. The main drift of the problems varies from plant to plant. Some programs are almost exclusively devoted to rotary compressors and turbines while others are devoted to other specialties or to a miscellaneous group of machines. The task is easiest when several machines of the same type are surveyed. Then it is usually an easy matter to determine the classification levels, because when one machine is in trouble its normal counterpart may be used as a basis for establishing normal levels. Small machines are often neglected, either because they are considered not important or because the maintenance department is short of men and, therefore, must ignore the small machine almost until the moment of failure if it is noncritical. Some types of problems necessitate a solution and information, whereas others may be included or rejected by the measuring program. The situation must be approached on its overall merits. It is usually wise to survey all machines of every kind at least once at the beginning of the program in order to evaluate the scope of the operation.

## 9. INSTRUMENTATION

The basic measuring instrument is a small portable velocity meter of the most convenient kind (bearing in mind that this may be used hour after hour under all kinds of conditions). It is an empirically grounded opinion, held by persons familiar with measuring, that velocity is the most useful overall indication of the severity classification of a machine. The velocity-type pickup has been used, perhaps, more than others for survey work and it requires only a simple meter to read out the signal. Meters are being improved constantly. Traditionally, their response has been what is called average, although the meter may be calibrated in peak or rms. A meter with a response that corresponds with true rms or quasi-rms is better than one having a mere averaging response.

The second most essential instrument is the vibration analyzer, as it is often called. For field analysis of the frequency-amplitude relation of the signal from the pickup or for in-place balancing, this analyzer, along with its stroboscopic facility, is most useful. Such instrumentation is available, for example, from International Research and Development, Inc., Reliance Electric Co. or Carl Schenck. For signal analysis in the office, a somewhat more discriminating analyzer is often used that has no pickups or stroboscopic facility. Constant-percent and fixed-bandwidth types are common, and are offered by several makers, such as General Radio Co., Bruel and Kjaer Co. and Spectral Dynamics Corp. Such instruments are usually called wave analyzers.

A small, accurate, versatile oscilloscope is very helpful and comes high on the list of desirable instrumentation, followed perhaps by a wave analyzer, an acceleration measuring meter and, possibly, a tape recorder. It is to be borne in mind that the common audio recorder is most useful but cannot record frequencies that are significantly below about 50 cps. An automatic printout for the wave analyzer gives the frequency-amplitude spectrum which is of great value and requires considerable labor if manually produced. A combination that is used to process a signal is that of a sweep generator driving an analyzer through the frequency range, leading to a DC analog output of the signal level at each frequency, which is finally plotted automatically by an X-Y plotter.

The vast majority of problems is managed with the survey meter and vibration analyzer. Other instruments that are mentioned are needed for more intractable problems. Apart from the mentioned instruments, a phase measuring instrument other than a stroboscope is useful.

## 10. CORRECTION

The corrective measures applied to machines as a result of a measuring program are usually the same measures as would have been applied if the fault were discovered in a way other than vibration measuring. In perhaps half of the corrections the object is to eliminate the fault that was diagnosed by vibration measuring, rather than to eliminate the vibration. The vibration is eliminated as a by-product of correction, although it may not be particularly harmful of itself. In the other half of typical cases, the vibration itself is the fault, most usually caused by mass unbalance.

Except in those instances where the mechanic may diagnose the fault more cheaply than the vibration analyst, every effort should be made to pinpoint the fault before correction is undertaken. Classification alone saves considerable amounts of money, and so does monitoring. But, respectively, in both efforts it is very worthwhile to take the second steps of identification and trend prediction. The larger and more complex the machine the more valuable the identification, which is the diagnosis of the root of the trouble so as to avoid unnecessary work of correction.

Consider two actual examples of identification. A forced-draft boiler fan of 75 hp exhibited about 10 mils peak-to-peak displacement of vibration at its rotational frequency with a significant amount of second harmonic, perhaps 20 percent, as indicated by the portable vibration analyzer. Several hours were spent examining the coupling and other features until it was discovered that one of the foundation holding bolts, concealed under the sub-base, had no nut. Replacement of the nut eliminated the fault. In this instance the expected correction, according to tradition, is to dismantle the coupling and balance the fan either in-place or in the shop. Correct diagnosis saved this cost and the downtime that goes with it.

In another instance a pair of induced-draft fans of about 200 to 400 hp each provided a variable speed service in the range of 0 to 1100 rpm on the fan wheels. Normal operating speed was perhaps 800 rpm, and the vertical vibration in the floor was large at about 18 mils peak-to-peak at operating frequency. The fans were remotely operated, and some thousands of dollars were devoted to balancing them. For some reason the apparent balance remained poor so that the fans were balanced once more to about 1 mil peak-to-peak at 500 rpm and then run through the speed range, disclosing the fact of structural resonance at about 900 rpm with floor vibration of 2 mils at 1100 rpm and 1 mil at 500. The fans were remotely operated, manually. By keeping at least 100 rpm away from 900 rpm, up or down, the vibration was kept within acceptable limits. The cost of mechanical alterations to the structure, to raise or lower the critical frequency, was considered unjustified. Here the problem was more than unbalance. The added factor of resonance led to the temptation to balance frequently, until it was pointed out that no amount of balancing could insure safe operation at 900 rpm due to occasional buildup of debris on the blades and the residual errors in all balancing work.

## 11. FEEDBACK

Feedback is the effort to influence such persons as purchasing agents and designers to act so as to avoid repetitions of faulty machinery by choosing equipment that is shown to be best by vibration measuring programs and by avoiding machinery that is known to be notoriously troublesome. The objective of feedback is to profit by experience. The experience accumulated in a measuring program is considerable and runs far beyond the vibration-related aspects. Machine histories begin to show trends and bring to light costly experiences and unsuitable designs. In the case of existing equipment, it is seldom possible to justify either its replacement on the ground of chronic trouble or a radical overhaul and rebuilding. On the other hand, similar or analogous equipment is purchased or built from time to time. At the right moment the right person may be influenced to act in a way that experience has shown to be profitable, if this person will heed the promptings that arise from the measuring program. A brief list follows, illustrating opportunities for feedback.

1. Control noise level by selecting the quietest new machinery.

2. Take great care in the acceptance of reciprocating machines that may alter the soil vibration in the plant or neighborhood. Much of this machinery, including impactive hammers and the like, cannot be satisfactorily balanced.

3. Consider structural resonance in relation to new installations on the upper levels of buildings.

4. Consider the environmental effect of specific vibrating machines, such as sifters and feeders, that sometimes cannot be balanced and often lead to vast expense in order to satisfy neighbor complaints or the needs of delicate laboratory equipment that may be adversely affected.

5. Measure local environmental noise and vibration before the erection of new plants, laboratories or other facilities.

6. In general, take advantage of adverse experience by using it to avoid its own repetition.

# 20

## MONITORING AND SIGHTING-IN

*Michael P. Blake*

### 1. SCOPE

An effort is made here to cover the most important aspects of monitoring. Particular attention is given to the matter of designing suitable objectives and insuring that these are met, since many monitoring systems have deluded their operators in the past into a false sense of security of supposed protection. Also discussed are numerical standards, annunciation, monitor types, instrumentation and trends.

### 2. INTRODUCTION

A monitored machine in the present connotation is one in which the significant vibrations are continually classified, corrective action is implemented quickly and the safeguards against catastrophe are total or nearly so. A vibration is classified when it is compared with a standard, leading to such classifications as economic, in need of correction or dangerous. Thus, in a monitored system either the operator or some automatic element of the system is made aware of classification changes. When the danger class is reached inadvertently, the machine is usually shut down automatically. But, it is undesirable that the danger class be reached without warning. With a system that is adequately designed and operated, the sighting-in predictions that arise from a review of the monitoring trends are aimed at avoidance of dangerous levels and the consequent tripping of automatic shutdowns that are usually both costly and troublesome.

By way of comparison, a system that is merely measured is one wherein it is simply hoped that the machine will not change by more than about one class between the occasional measurements, and that there will probably be no automatic shutdown. If the measuring is done so frequently that the need of correction is always known, or nearly always known, before the danger level arises, the work of manual measuring then approaches toward the efficiency of a monitored operation. Trend extrapolation is an activity that makes the most of the monitoring or measuring commitment by facilitating maintenance scheduling and avoiding superfluous maintenance besides bringing to light any trends that are process-related, such as the accumulation of material on rotors and the like. This is commonly called sighting-in.

## 3. THE APPROACH

Among operators, monitoring has often fallen into the worst disrepute because the monitor has not done what the operator expected in terms of what he was told to expect. For the simpler machines he will often tend to rely on his own observations. But he must accept monitoring help as the number of machines increases, together with their value, power, speed and complexity. It then becomes the task of the designer to provide monitoring designs that perform as they should, not only to restore the faith of operators, but because there is now no second line of defense by way of old-fashioned personal observation. Many a turbine of 30 megawatts was monitored by watching the ripples on the surface of the operator's cup of coffee.

The value or number or criterion that is in question is that of vibration expressed in suitable terms, such as acceleration, velocity, displacement, phase or frequency or a combination of these, perhaps sometimes combined with a logic system such that C cannot fire unless A and B both trip. Synonyms such as fire and trip are often used to denote the activation of a relay or trip or like element. Some of the essentials of an acceptable monitoring system are:

1. Adequate numerical classification standards
2. Adequate and timely information
3. Human or mechanical decision triggered by call-out, read-out or a shutdown signal.

We define a monitored point as that from which vibration information is continuously available. The primary interpretation or action resulting from the information may be called event monitoring. This can protect a cooling tower fan that is unbalanced with ice or a centrifuge that has become unbalanced because of some process irregularity. The vibration itself may be process-related or maintenance-related. Although what has been called event monitoring is valuable and necessary as a second line of defense, ideally all monitoring should be in terms of trend monitoring, with its resulting predictions that can so greatly lower the cost of maintenance and process corrections.

**Single Point Monitoring.** An inherent difficulty that affects almost every monitoring effort may be stated as follows. It is a delusion to suppose that a pickup or other device intended to monitor a particular vibration at a particular point is not affected by other vibrations or impacts, or transient or accidental events that may each or all lead to corrective action or concern. For example, excessive vibration during start-up may exceed the preset level; and it is simply wrong to increase the preset level because this leads to a situation wherein there is no protection, although the operator believes that there is. This is worse than no monitor, and the monitor should be protected by a time delay or other suitable device.

**Summary.** Three intensities of monitoring emerge in the previous discussion.

1. Manual measurements are made at fixed intervals, the frequency being such that a significant change in vibration level does not occur between measurements. By recording the measurements, trend monitoring becomes possible. This, in turn, can lead to increasing the interval between measurements, which of course saves money, when possible.

2. Measurements are made continually and reviewed at intervals, providing complete effectiveness of event monitoring and moderately effective trend monitoring. This intensity of monitoring is perhaps typical of most monitored systems.

3. Continual measurements are continually reviewed, furnishing complete effectiveness of event and trend monitoring.

It is to be borne in mind that trend monitoring is the first line of defense and that the coming into play of the second line, by way of unscheduled corrections, should only occur when the actual trend is far different from normal, informed expectation.

## 4. NUMERICAL STANDARDS

When considering the values of measured vibration for a given point, it is necessary to classify the measurements in accordance with classes, bands or orders of magnitude, all based on experience. Some guidelines are offered in Chapter 33 of this volume. For example, consider four possible and typical classes:

C—Correction unprofitable
B—Correction profitable, but not urgent
A—Correction urgent
AA—Correction immediately imperative to avoid danger.

The classes may be process-related, as for example, a centrifuge may vibrate greatly due to the process material or method of operation. But, in the majority of cases the classes are maintenance-related. It becomes clear that the second line of defense in monitoring is the avoidance of class AA. Such monitoring is only partially effective in that although danger is avoided, great inconvenience and expense may result from unscheduled maintenance, long downtime, spoiled goods, disrupted production and so on. In this way of thinking, monitoring achieves its optimum effectiveness when corrective action is taken in the region of class B or A. Thus, safety is the ultimate concern and at that level the economic factors must be accepted as they come because it is too late to control them. Considered in these terms, the two kinds of monitoring that were previously christened event and trend monitoring may now be envisaged as safety and economic monitoring.

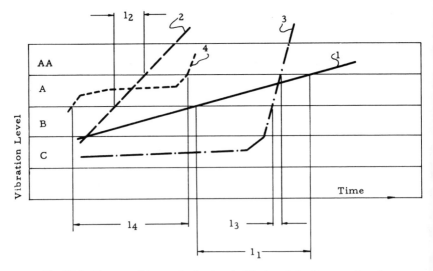

Fig. 20.1   Four possible monitoring trends. The trends for four monitored vibrations are shown. Trend extrapolation and forecasting becomes easier when the interval I required for one class change, such as A to AA, is several weeks or months, as shown for trend No. 1. Forecasting becomes increasingly difficult as the interval becomes small, as, for example, in trend No. 3.

**Summary**

1. Safety monitoring is a second line of defense and may be made totally effective through a continuous collection of information and automatic correction.

2. Economic monitoring is optimum monitoring, but it can only be made totally effective if the trend is such that it does not climb by more than about one class in the interval between review of measurements.

Review is usually a matter of human judgment, and this writer is not acquainted with any automated procedure. It is, of course, of little advantage to be aware of a trend, even if discoverable, that crosses one class in say an hour, because an hour's warning is generally useless as regards profitability. Each case must be considered on its own merits.

It is noticed that the word level, rather than value, is used in Fig. 20.1. This is because machines differ greatly so that the class is determined, not by the absolute vibration, but by that quantity multiplied by a factor of criticality which this author has called a service factor in previous writings. Thus, it is what the writer has called effective vibration that determines the classification. Furthermore, different parameters are used in different cases, for example, acceleration, phase, frequency and so on.

## 5. ANNUNCIATION

Information that is made available through call-out, read-out or direct action (such as shutdown) may be grouped under the heading of annunciation. Annunciation facilities that fall short of the safety level may be considered to be related to economics. Annunciation may be further classed as follows.

1. Trend annunciation: read-out, call-out, record, others short of shutdown
2. Safety annunciation: usually direct action without human agency

Annunciation arrangements may be further considered in the following way.

1. Read-out via meter dial, colored lights or printed chart. The classes may be represented by colors such as green, yellow, red and black.

2. Call-out via horn, bell or whistle. The passing from one class to the next may be announced by call-out.

3. Direct action finds little favor with operators, particularly if they have experienced expensive shutdowns through monitoring error. But direct action must be provided on all critical machines, with the intention that it comes into play only if class AA is reached before manual correction can be made. Direct action is envisaged only in relation with abnormal trends and is simply an emergency provision, not a normal annunciation.

The annunciation policy depends on a variety of factors, such as local circumstances, the type of machine, its speed, power, criticality and value, the value of the product material and the hazards connected with it, the skill of the operators and whether or not the process is automated.

## 6. BENEFITS OF MONITORING

The benefits are mainly economic, although the motivation may stem from safety consideration. Poor safety is invariably uneconomic, to state it in crudest and lowest terms.

Slow trends, such as are illustrated in Fig. 20.1 (curves 1 and 2), are typical of the mechanical aspects of machinery and make trend monitoring meaningful because the trends are slow. In this way the economic benefits become available. Taking explicit examples of economic benefits, the following comes to mind. A centrifugal blower is inspected annually at a cost of perhaps $8000. As a result of trend monitoring, the inspection may be postponed three years. In general, trend monitoring leads to the least amount of maintenance at the most convenient time, besides bringing into effect such process changes as may favor the machine and its maintenance. Process-related trends are often more rapid than mechanical trends; and process-related vibration standards often differ from maintenance-related standards, although the two may be the same. For example, the process-related classification level for the basket-end bearing of a centrifuge may be 20 mils PP, whereas the maintenance-related (unbalance) classification level for the empty machine may be 2 mils PP, both corresponding to the C classification.

Among explicit examples of the benefits of safety monitoring, the following come to mind. Large cooling towers with fans of about 20-ft diameter are saved from disaster due to icing of blades or the loss of one blade. A comminuter operating at 5000 rpm is saved from the destructive vibration that results from an unbalanced distribution of product material. Other destructive vibrations are "nipped in the bud," for example, when a turbine loses a blade, when a centrifuge load is unbalanced, when troubles arise in washing or knifing, or when a cross-head bearing becomes loose in a reciprocating compressor. Monitoring almost invariably avoids hazard to personnel, although it is not always possible to prevent some damage to the machine.

## 7. MONITOR TYPES

**Inertia Switches.** Among the oldest devices are the inertia switches that are acceleration-sensitive and are usually nothing more than safety monitors. These may be magnetic or purely mechanical. Simple arrangements of balls, cups, hammers, pendulums, magnets and springs are so arranged that a relay or microswitch is activated or a flag signal is given if a predetermined acceleration is exceeded. For example, Fig. 20.2 illustrates the scheme of a magnetic switch. By adjusting the screw, 2, the spring tension is increased to the point where,

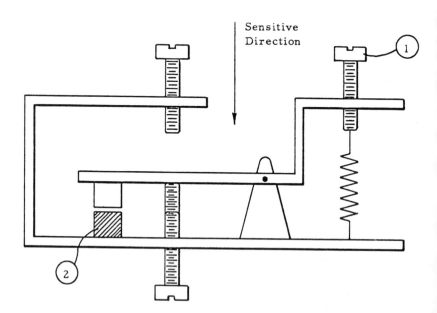

Fig. 20.2  Magnetic type acceleration relay. The magnet 2 holds its armature against the tension of the spring adjustment 1. When the spring tension is increased, the switch fires at a lower value of acceleration. When the armature separates from the magnet, it flies back and hits a microswitch or other relay for purposes of annunciation.

perhaps, one- or two-g or the desired value is sufficient to cause separation of magnet and armature, followed by the activition of a switch. A valuable feature of a magnet is this irreversible tripping, together with reliability of normal holding and adequate sensitivity to rather brief impulses, properties which are not readily duplicated in a mechanical way.

The inertial device will respond to merely one impulse, which is sometimes an advantage. Typical electronic systems may ignore impulses of short duration. Thus, the inertial switch has found favor with operators of reciprocating machines. But it does not usually annunciate any classification short of a single preset level so that it is not usually a basis for trend monitoring. It is also limited as regards its ability to respond to high-frequency vibration, although this author has not measured the limits nor has he seen any quoted by makers. Casual observation of magnetically held transducers and magnets, in general, suggests that they do not detach at the higher frequencies in the same way as at the lower. If the acceleration force of a long impulse exceeds the holding power, the armature and magnet separate irreversibly. On the other hand, the inertia, moment of inertia and the spring force of the moving arm may be so great that in the event of a short impulse the arm begins to move away and is, so to speak, recaptured by the returning magnet; or the transient or impulse is lost due to the isolating properties of the mechanical elements at high frequency. On the other hand, electronic devices respond well to high frequencies. In short, the high-frequency performance of inertial devices depends on a variety of factors, such as depth of latching, gradient of the magnetic field, inertia and fulcrum position.

**Velocity Pickup.** This came later than the inertial type. It affords complete information at any level and has an upper limit for frequency response of about 2000 cps, which is adequate for most application except, for example, some involving rolling bearings or the vibrations caused by dry friction. The cost is somewhere above $200, which is about 50 to 100 percent higher than the cost of the inertial equivalents. A read-out device and amplifier of some kind are normally required. It annunciates all of the classification spectrum and has the further advantage that velocity readings are more indicative of mechanical condition, in general, than are acceleration or displacement readings. The lower frequency limit is about 10 to 15 cps, which is rather high for cooling towers. The latter usually have a fundamental below 5 cps. Low-frequency velocity pickups are available but the price reaches, perhaps, $1000 when the frequency of free vibration is as low as about $\frac{1}{2}$ cps. A disadvantage is that the velocity pickup may be affected by magnetic fields, and some former users are bitterly aware of this.

**Proximity Pickup.** The use of this pickup has increased in the past decade. It is a relative type, meaning that it measures the relative vibration of, say, shaft and bearing housing or shaft and earth. The velocity and inertial types are seismic, meaning they are self-contained and require no anchor point. When displacement is the criterion of interest, this pickup is attractive, provided that the application is not such as a cooling tower deck whereon there is no earth point to fix the pickup. It monitors shaft displacements and alignment, running or idle, in many applications. It is not self-generating in the way that the velocity pickup performs by generating its own signal, but requires external excitation from a power source. The excitation is modulated by variations of proximity without contact with the running shaft, and the frequency of the variations (vibration) may be as low as zero. An electronic power supply and read-out device is required. The signal may be conditioned so as to read in terms of velocity or acceleration, although this adds a complexity; electronic differentiation conditioning is, in general, more troublesome than integration conditioning. The proximity pickup is likely to become increasingly popular for monitoring the shafts of high-speed, large, powerful drives, such as turbines, reading directly in displacement which, in this application, is usually the parameter of most interest.

**Displacement Pickup.** Besides the proximity pickup the linear variable differential transformer (LVDT) is sometimes used. Again, this device is relative (or driven) and is not seismic, although a seismic design is obviously possible if required. A coil with three windings is excited by an AC signal from a power source, and it gives an output signal proportional to the position within the coil of a cylindrical armature which can move axially. Again, a DC (reading to zero frequency) response is typical and it is, of course, most useful in some applications. The LVDT is usually used at zero frequency, rather than for monitoring running shafts, because it has no special advantage when running since it must contact the shaft. Its use as a velocity pickup can hardly be justified. It is not likely to displace the velocity pickup or proximity pickup. However, because of its configuration, which is more self-contained than that of the proximity pickup and because it does not depend on the contour or chemistry of the measured piece, the LVDT is likely to offer advantages in some applications. Displacement-range from about 10 to 300 mils PP is available with electrical excitation frequency from about 60 to 10,000 cps and an output signal of the order 1 millivolt per volt excitation per mil displacement. Resolution to about 1 microinch is possible without any steps in the signal output, although resolution to a fraction of one mil is more usual.

**Acceleration Pickup.** A growing tendency among those who measure vibration favors the acceleration pickup. Almost all of the traditional objections have been circumvented, and it is probably true to say that it is now the most versatile and universally attractive pickup. Some features now available are:

1. Frequency response to about 1 cps. Frequency response to 0 cps is also possible without significant sophistication,
2. Freedom from effects of magnetic fields,
3. Size and weight of the order of 25 mm and 25 grams, respectively,
4. High-frequency response to 10,000 cps or higher,
5. Integration for velocity and displacement is not troublesome:
6. Almost no limit to cable length,
7. Displacement to zero cps cannot be measured.

An accelerometer made by the Kistler Company, for example, uses a DC-powered amplifier to eliminate the difficulties connected with high-impedance matching requirements, with long cables which give rise to added capacitance and noise, and with low-frequency response. This accelerometer is readily capable of 2 to 5000 cps operation, and the range may be extended without significant trouble or expense.

## 8. FIDELITY

Having decided what vibrational effects are to be monitored, it is of paramount importance to insure that these effects are, in fact, monitored. This may sound like exaggerating a pitfall. However, many monitoring efforts have fallen into disrepute because of this pitfall, which is discussed here by grouping the difficulties under the term fidelity.

A simple example will illustrate what is involved. Consider a cooling tower vibrating at 5 cps with a gearbox having a predominant frequency of 400 cps and a motor vibrating at 30 cps, together with the drive shaft which, in turn, induces 30 cps bumps in a six-bladed fan. First, the safety of the tower takes precedence. Second, a strong 30 cps vibration may indicate the possibility of a fire. Third, it is desirable to protect the gearbox for economic reasons only. Some monitoring schemes may be summarized as follows.

### Case A

1. Tower is safety monitored for a single preset level of 120 mils PP.
2. Magnetic acceleration switch monitors the tower deck.
3. Switch is set experimentally or otherwise to operate at 0.155-gP. It is likely that the preset level of the switch must be raised to about 0.9-gP in the first few hours of operation in order to prevent tripping due to normal gearbox acceleration. Then no protection whatever is available for the tower structure, although operating personnel have no means of knowing this and will tend to assume that the preset level was raised to accommodate the tower structure.

The gearbox in normal operation may be expected to generate approximately 0.9-gP at about 400 cps. If the switch is not so arranged as to be isolated, it will trip until the preset level exceeds 0.9-gP. The motor, the shaft and the blade bumping may be expected to generate 0.1-gP at about 30 cps; and the switch may be expected to pick up most of this because 30 cps is not readily isolated, except by rather resilient intervening mechanical members. Thus, if the gearbox were absent so as to make matters most favorable, the switch would be tripped if the 30 cps vibration exceeded the set level by about 50 percent. Again, the preset level must be raised to a value at which the tower is unprotected. An array of typical figures is given in Table 28 since such arrays must be appreciated as a preliminary to monitoring design.

The upper three rows of figures give data for typical danger levels, classified as AA earlier in this chapter. The economic or normal level is in each instance about 10 percent of this. The lower three rows show the levels of velocity and

**TABLE 28**
**TYPICAL VIBRATION VALUES FOR CASE A**

| | | LEVEL FOR DANGER | | |
| ITEM | FREQ. CPS | DISPLACE- MENT MILS PP | VELOCITY IPSP | ACCELERA- TION gP |
| --- | --- | --- | --- | --- |
| Tower structure | 5 | 120 | 1.91 | 0.155 |
| Motor and shaft and blade bumping | 30 | 21 | 2.0 | 0.98 |
| Gearbox | 400 | 1.19 | 1.5 | 9.8 |
| Tower structure | 5 | 7600 | 120 | 9.8 |
| Motor and shaft and blade bumping | 30 | 220 | 21 | 9.8 |
| Gearbox | 400 | 1.19 | 1.5 | 9.8 |

displacement that must be achieved by the 5 and 30 cps vibration in order to trip a switch that is set for 9.8-gP. For example, the tower must achieve 7.6 in PP, which means that it does not have any protection.

The solution in this case is to monitor in terms of velocity because the classification levels for all the frequencies and sources are about the same, namely, 1.5 to 2 in/sec when expressed in terms of velocity. In many velocity pickups, the seismic coil hits the end stops at displacement amplitudes less than 120 mils PP. For the best solution the natural period of the pickup should be not much less than 1/5 of one second.

**Case B.** A centrifuge operates at 1800 rpm, and it is decided to monitor for basket-end bearing displacement and bearing vibration. The aim is to provide shutdown for a danger level of unbalance or a danger level of bearing fault. Taking typical values again, let it be supposed that the stress on the machine and the piping is the safety criterion and that the displacement at 30 cps must not exceed 20 mils PP. Again, let it be supposed that it is estimated or known that severe bearing trouble corresponds with about 20-gP at 1500 cps. The data may be arrayed as in Table 29.

**Solution 1.** Use only one velocity pickup on the bearing housing. Set the safety level to 2 IPSP, allowing the bearing vibration to reach that level before shutdown alarm. This corresponds with about 50-gP which is far higher than the stipulated 20, but probably not unduly risky. On the other hand, if a displacement of 50 mils were permitted instead of the stipulated 20 mils PP, a severe risk would likely exist because stress is about proportional to displacement amplitude. The velocity monitor performs well at 30 cps. However, at 1500 cps it is approaching its upper-frequency limit and the displacement amplitude is very small, being of the order of one-tenth of one mil. In this frequency region,

TABLE 29
TYPICAL VIBRATION LEVELS FOR CASE B *

| ITEM | FREQUENCY CPS | LEVEL FOR DANGER, AA | | |
| --- | --- | --- | --- | --- |
| | | DISPLACE-MENT MILS PP | VELOCITY IPSP | ACCELERA-TION gP |
| Process unbalance | **30** | **20** | 1.88 | 0.92 |
| Bearing vibration | **1500** | 0.00017 | 0.82 | **20** |
| | | LEVEL FOR ACUTE FAULT, A | | |
| Bearing vibration | **1500** | 0.000043 | 0.205 | **5** |

*The bold face data is stipulated and the remainder is computed.

the velocity pickup may not perform well, either because of its own sluggishness or because of the isolation property of its mounting. As pointed out later, because the 1500 cps vibration shall appear as a harmonic, it is necessary to insure that the signal detection circuit have an rms type of response; or the 1500 cps signal may be totally ignored by the annunciator. It is easier to mount accelerometers with great relative rigidity and little isolation because they weigh perhaps 25 grams or less, whereas velocity pickups range from about 200 to 1000 grams. The pickup is mounted with its sensitive axis in the direction of interest, which is usually radial horizontal.

**Solution 2.** A preferred solution is to use a velocity pickup and an accelerometer with respective safety settings of 1.88 IPSP and 20-gP. Or a single accelerometer may be used, annunciating acceleration directly and velocity by integration. But this requires a scanning device. It is, of course, hopeless to try monitoring via acceleration alone because when the monitor is preset at 2 to 3-gP to avoid tripping due to bearing vibration at normal levels, it then is set three times as high as the danger level for process unbalance; and there is no protection of that aspect of operation.

**Solution 3.** It is not uncommon to monitor unbalance with a simple driven (relative) microswitch. This is often mounted on the building structure about 10 mils from the basket housing or bearing housing, allowing a movement of 20 mil PP before tripping. This is usually most unsatisfactory because misalignment and structural movements both combine to vary the amount of displacement required for tripping. Nuisance tripping then leads to setting up of the tripping distance so that the machine is, in fact, unprotected and the switch only trips after a catastrophic displacement. Bearings are not usually monitored for acceleration as a means of avoiding unheralded failures, leading to shaft seizure or other damage. However, in view of the time and effort that is devoted to the measuring of critical bearings, it seems easy to justify perhaps $500 for an acceleration monitor.

**Summary**

1. Decide what points are to be monitored, and in what direction, to achieve desired annunciation.

2. Decide the minimum series of effects to be monitored and estimate the classification levels and frequencies.

3. Design a monitoring system that truly monitors the chosen effects.

4. Consider the possibilities of transient effects or of steady local vibrations or other circumstances that may require the use of time delays or other signal conditioners.

## 9. TYPICAL INSTRUMENTATION

Many eminent companies in the United States and abroad offer a variety of vibration measuring and monitoring equipment. The very few that are given here are quoted at random so that the reader may have a few immediate references to sources of supply.

| ITEM | EXAMPLE OF SUPPLIER |
|---|---|
| 1. Pickup, proximity type | Bently-Nevada |
| 2. Pickup, accelerometer | Endevco, Statham, Kistler |
| 3. Pickup, velocity | IRD, CEC, Reliance, MB Electronics, Schenck. |
| 4. Pickup, LVDT | Daytronic Corporation |
| 5. Acceleration switch, magnetic | Robertshaw Fulton |
| 6. Shock switch | F. W. Murphy, Inc. |
| 7. Impact recorder | Impact-o-graph Corporation |
| 8. Shock indicator and recorder | Inertia Switch, Inc. |
| 9. Monitoring systems | Reliance Electric Co., IRD Corp., Bell and Howell |

## 10. TRENDS

Monitoring seems to offer profitable opportunities so that its applications are likely to increase. However, the majority of new applications are likely to appear in new installations because monitoring is often imperative, due to automation, complexity or criticality. Nevertheless, the effort to review existing installations in terms of monitoring possibilities is worthwhile. Future installations are likely to perform more satisfactorily than the older systems, particularly since the supplier of the monitoring system will probably design it, rather than the user who usually has no experience. Phase measuring for monitoring may receive more attention as it is an important criterion of mechanical continuity which may be used, for example, in the detection of foundation cracks.

Meter circuits that respond to the average signal amplitude may be unaffected by the onset of significant harmonics. Thus, they may not respond to a possible indication of severe trouble. For this reason the true rms meter is being considered more than formerly because its response is a somewhat more faithful indication of mechanical condition. Pressure, temperature and other measurements are likely to find use in large installations, as well as vibration measurements. Instead of comparing an all-pass read-out with a single-valued standard, it seems probable that actual amplitude-frequency spectra shall be compared with standard spectra. The vast increase in the complexity, value and criticality of new machinery is likely to force a technology improvement that will benefit the monitoring of less critical machinery. The scope of monitoring may tend to include the relatively small machine if cheap and reliable devices are forthcoming.

The monitoring field is largely untapped and misunderstood. Although many well-conceived installations are working today, a measure of the backwardness of the situation may be gained from the fact that this author can recall seeing few systems that are more than makeshift in his experience with fans, turbines and processing machinery, in general.

# BIBLIOGRAPHY

Foster, G. B. "System Engineering Aspects of Machine Vibration Protection." Paper P-7005, IEEE Pet. Ind. Tech. Conf., October 1967.

Foster, G. B. "Recent Developments in Machine Vibration Monitoring." Paper P-7006, IEEE Pet. Ind. Tech. Conf., September 1966.

# 21

# SHAFT COUPLINGS

*Michael P. Blake*

## 1. SCOPE

A brief overall review of the more common shaft couplings is given, under the headings of types, operating characteristics, design considerations and applica - tion. Because very little has been written on couplings, in general, the discussion goes far beyond the topic of vibration in order to make the discussion clear to the reader who may have no other sources of information.

## 2. TYPES

The variety of coupling types reflects the effort to satisfy a multitude of func- tional requirements at lowest cost. Just as each condition of application is char- acterized by features that are peculiar to themselves, each of the many different couplings tends also to be unique in some respects and, therefore, particularly suited to some application or group of applications. However, there does not exist a coupling that is universally versatile, whatever its cost or complexity, although some couplings are far more versatile than others.

**Function.** Because the function explains the various types and because an under- standing of function is necessary in order to select the best coupling, a typical list of possible functions follows.

Operate safely
Fail safely
Provide a warning of failure
Transmit power via a solid shaft coupling
Transmit power via a flexible shaft coupling
Tolerate angular or parallel misalignment, or both
Tolerate extremes of speed, temperature, torque or vibration
Permit rapid original and subsequent alignment checks
Permit coupling disengagement without disturbing shafts

Permit spacer-type or radial-type disassembly
Provide lowest weight and easiest assembly
Eliminate lubrication and other maintenance
Provide aesthetic appearance or sanitary features
Operate in an environment such as hot oil
Disengage or slip when a specified torque is exceeded
Limit the stresses caused by misalignment
Limit the lateral and torsional vibration caused by misalignment
Limit the transmission of torsional vibration across the coupling
Limit torsional amplitude near resonance
Act as a damper for torsional vibration
Provide accurate stepping or indexing
Provide constant velocity across the coupling
Operate in a confined space
Transmit torque while supporting the shaft laterally
Permit axial freedom
Permit variations in torsional stiffness
Endure extremes of acceleration or of torsional vibration

**Typical Examples.** Six couplings are illustrated diagramatically and are referred to as 1, 2 and so on. These typify some of the important couplings which are currently in everyday use. Number 1 is extremely simple, having a jaw configuration and driving through a rubber spider. Thousands of this kind find application in the smaller horsepower applications, while a smaller number are

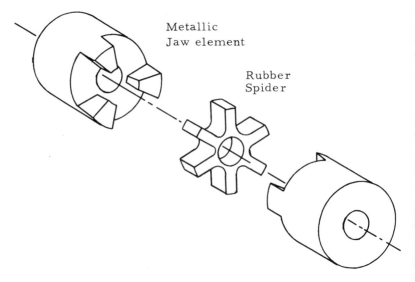

Metallic
Jaw element

Rubber
Spider

Fig. 21.1   Jaw-type coupling, single engagement

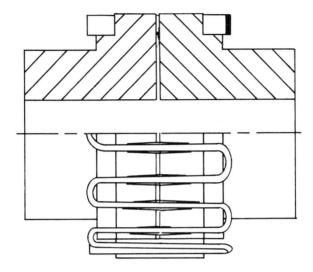

Fig. 21.2    Metallic-ribbon-type coupling, single engagement

used for torques as high as say 100 hp per 100 rpm. It is to be noted that couplings are often rated in this way, where 1 hp per 100 rpm corresponds to about 630 in-lb of torque. Number 2 is more sophisticated and is lubricated, having a shroud that is not shown; and it has all metallic elements. Torque is transmitted through a ribbon spring that is confined between jaws so shaped that the coupling becomes stiffer as it winds up. The torsional stiffness is not constant. Moreover, the torsional stress-strain relation is not a linear relation. This is often an advantage because natural frequency in the case of a nonlinear stiffness element is a function of amplitude such that severe resonance is often avoided. However, the dynamic behavior of nonlinear springs must be carefully considered in order to avoid misconceptions that may lead to unacceptable performance.

Number 3 comprises a rubber element subjected to shear, rather than to compression as in No. 1, which is held in the metal hubs by jaws or serrations without a bond. Number 4 is similar to No. 3 except that the former has a clamped rubber element which avoids all backlash. Also, disassembly is slower than for No. 3. Coupling No. 5 illustrates the elements of the classic gear coupling having two geared hubs connected by a spool that has internal gear teeth on both ends and is filled with oil. The elements are all metallic, and the load-carrying capacity is high in terms of external dimensions. This coupling is used to a large extent on critical equipment such as large turbine drives. Coupling type No. 6 drives through two stacks of thin steel discs which provide flexure without rubbing friction and without lubrication. An outstanding feature of this coupling is the absence of backlash and the relatively high torsional stiffness.

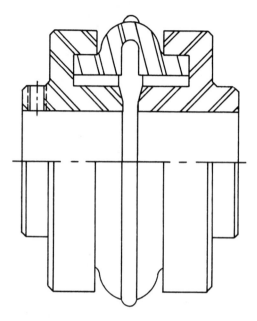

Fig. 21.3  Coupling comprising unclamped rubber doughnut between toothed metal hubs

The six styles may be briefly compared as follows.

### TABLE 30
### CHARACTERISTICS OF COUPLINGS

| | FIGURE NUMBER | | | | | |
|---|---|---|---|---|---|---|
| COUPLING | 1 | 2 | 3 | 4 | 5 | 6 |
| Backlash | yes | some | yes | no | yes | no |
| Single engagement | yes | yes | yes | yes | no | no |
| Double engagement | no | no | no | no | yes | yes |
| Lubricated | no | yes | no | no | yes | no |
| Torsional stiffness | med | med | low | low | high | high |
| Spring rate | var* | var | var | var | lin* | lin |
| Lateral stiffness | high | high | med | med | low | low |

* var . . . variable, lin . . . linear

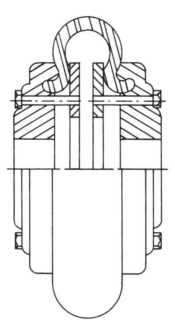

Fig. 21.4  Coupling comprising clamped rubber doughnut between metal hubs

Gear teeth                                    Gear Teeth

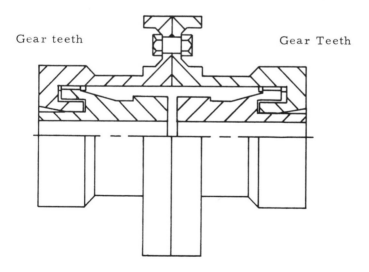

Fig. 21.5  Gear-type coupling, double engagement

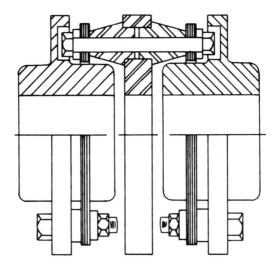

Fig. 21.6   Flexible metallic disc, connecting metal hubs with double engagement

The foregoing and all flexible couplings may be divided into two radically differing families: the single engagement and the double engagement. Couplings Nos. 3 and 4 are in fact hybrids. Lateral misalignment induces an eccentricity in the flexible elements of the single type, whereas the flexible element of the double type maintains its concentricity and takes up the lateral deflection by means of an angular displacement of the spool element.

## 3. OPERATING CHARACTERISTICS

**Vibration.** Generally speaking, flexible couplings are not among the most usual sources of vibration. Double-engagement couplings generate almost no vibration, except when unbalanced. For example, a gear coupling may generate vibration in the frequency range of 1000 cps and upward if the lubrication fails and fretting commences. Single-engagement couplings do generate some vibration when misaligned or unbalanced. It is not at all possible to appreciate the dynamic behavior of couplings without an understanding of the elementary kinetic considerations typical of jaws, teeth, pins and other driving elements, all of which may usually be reduced to a diagram of the kind shown in Fig. 21.7. This figure merits considerable attention, particularly since space does not permit more than the briefest introduction to kinetics in this article.

We shall try to enumerate the salient points, referring to Fig. 21.7.

1. When parallel misalignment exists, the output velocity is variable for a constant input velocity.

2. Torque is transmitted by moment and not by couple; hence a lateral force of an unbalanced kind is generated.

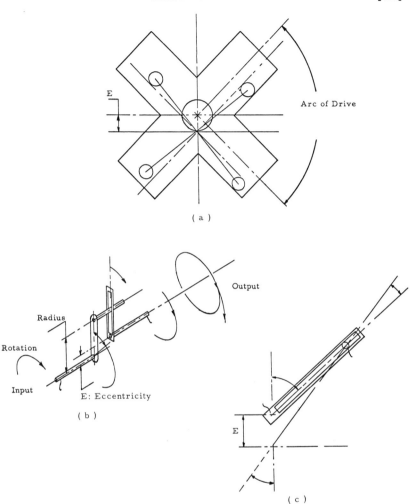

( a )

( b )

Output

Radius

Rotation

Input

E: Eccentricity

( c )

E

Arc of Drive

E

Fig. 21.7   The internal cross drives the external four jaw element. Driving
contact takes place only on one of four driving elements at any instant.
Then the next element takes over the drive. Rotation is clockwise. For a
constant input angular velocity, the output experiences a fourth harmonic
acceleration. A torsional vibration of four times the frequency of rotation
is induced, together with a lateral vibration of the same frequency. Drive
is by moment and not by couple. Single pin coupling element in Fig. 21.7b
and 21.7c illustrates variable output velocity for constant input velocity.

3.  The lateral force varies in magnitude and direction. For one pin or jaw it
rotates 360 degrees. For two, three and four pins, it rotates 180, 120 and 90
degrees, respectively, increasing and then decreasing as it rotates, after which it
resumes its first phase and value and starts to rotate and increase once more.

4. The frequency of the lateral force and of the torsional pulsation is the same as that of the jaws or pins on the driving or driven element, i.e. $f = N \cdot$ rpm.

5. The driver and driven elements continually tend to self-center, mutually. The lateral force is a measure of the self-centering tendency.

6. The previous remarks apply to rigid elements. The introduction of flexible cushions and the like does not eliminate the unwanted forces, it merely mitigates them.

7. The misalignment in a single-engagement coupling may be parallel or angular or both. In a double-engagement coupling each driving element of the pair is constantly self-centered so that double-engagement couplings experience parallel misalignment as angular misalignment.

8. Typical single-engagement couplings are capable of about ¾-of-one-degree angular misalignment. Because the tangent of this angle is about 0.012, it follows that a double engagement can endure about 12 mils of parallel offset of the shafts for every inch of length of spool. For typical speeds and loads, gear couplings will not endure more than about 10 percent of this amount.

9. Angular displacement of the element of a single or double coupling leads to an axial shear motion at the points of engagement, having a frequency the same as that of rotation and a velocity proportional to the angle, the rpm and the radius of the point measured from the axis.

10. The foregoing velocity causes rubbing in gear couplings, a shear-like action in cushions and a flexure in elements such as those in Fig. 21.6. Ideally, the cushions are so sized that static friction prevents their slipping along the jaw surfaces so that there is no rubbing.

11. Single-engagement elements often use rubber in shear instead of in compression and have no jaws. The purpose is generally to increase torsional flexibility. These elements generate no vibration when subjected to parallel or angular misalignment, although lateral stiffness may be very high for the single-engagement element.

Thus, the only vibration of any significance usually arises from single-engagement couplings, when misaligned. The hybrids, which refers here to Nos. 3 and 4, show in their displacement a fundamental frequency response and almost no harmonics. The single-engagement kind often shows a second harmonic, laterally and axially, greater in displacement amplitude than the fundamental. Although the third harmonic or fourth and so on, depending on the number of jaws or slots, might be expected to be very pronounced, it is usually weak as compared with the displacement amplitude fundamental and second harmonics. A second harmonic, especially in the axial direction, is often interpreted as a sign of misalignment.

Figure 21.8 shows what can happen to a pin coupling after a few weeks at 3600 rpm when applied to a situation of inherently poor alignment, together with local vibration and a tendency to resonance. When replaced with a coupling of the style of Fig. 21.4, the problem was eliminated. In many applications the flexible coupling is used and relied upon to control or damp such vibrations as arise from internal combustion engines, compressors and other sources.

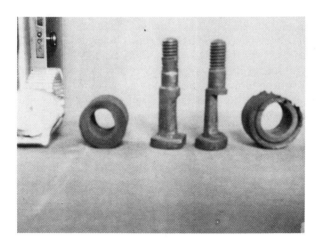

Fig. 21.8 Coupling wear; The effects of chronic misalignment and local
vibration. The coupling pins are about ½-in. in diameter and have worn
as shown in just a few weeks time

**Lateral Stiffness.** The load upon the adjacent bearings depends on the dynamic
lateral stiffness. This usually depends on both torque and speed. For example, the
coupling of Fig. 21.1 exhibits a small lateral stiffness if equipped with a very
loose spider and when driving into a very small load. As explained, it drives
through an arc of 120 degrees, and the lateral force is proportional to the output
torque and the moment of inertia of the connected machinery. As the torque
increases, the lateral force increases because it is this force that generates the
moment that drives the power train.

In general the lateral stiffness is high for single-engagement couplings and low
for double-engagement couplings. The dynamic stiffness bears no relation at all
to the static stiffness, either for double or single engagement. A loose-fit, gear
configuration, single-engagement coupling may appear to have a zero lateral
stiffness. However, a consideration of Fig. 21.7 makes it clear that lateral force is
porportional to torque. This force may be diminished considerably by increasing
the driving radius, but this is seldom practicable. For double-engagement couplings the lateral stiffness is usually low and depends on rubbing friction, shearing forces, coupling geometry, speed and other factors. In general, it is not possible to estimate lateral stiffness, except by actual measurement on a special rig
that can measure lateral force under actual operating conditions. When a coupling is in good alignment, the lateral force does not change when the power is
turned on or off, providing a useful means of alignment checking for test
purposes.

The effective lateral force in a coupling is generally less than that computed on
the basis of shaft eccentricity and known stiffness. Likewise, the observed stiffness is usually the combined stiffness of two shafts and the coupling, all in series.

The restoring force that tends to self-center the two halves of a single-engagement coupling is usually a function of shaft eccentricity. When the machine starts, the shafts deflect until the lateral forces in the shafts and coupling are the same. At this point, the internal lateral deflection in the coupling is less than the statically observed value.

The effective lateral stiffness of most couplings increases significantly with speed as well as with torque. There is almost no published data relating to these phenomena and very little unpublished research has come to this writer's notice.

Some examples of test data are given below. Couplings 1 and 3 are those illustrated in Figs. 21.1 and 21.3; 1D is more or less a double-engagement version of Fig. 21.1.

### TABLE 31
### EFFECTIVE LATERAL STIFFNESS OF COUPLINGS

| COUPLING STYLE | EFFECTIVE LATERAL STIFFNESS K LBS/MIL | TEST HP | TEST RPM |
|:---:|:---:|:---:|:---:|
| 1D | 0.34 | 2 | 1800 |
| 3 | 0.72 | 2 | 1800 |
| 1D | 0.47 | 5 | 1800 |
| 3 | 0.64 | 5 | 1800 |
| 1D | 0.66 | 10 | 3600 |
| 3 | 1.62 | 10 | 3600 |

Average vibration data in the range of up to 62 mils for 1D and 3 couplings, and up to 32 mils for No. 1 (lateral static misalignment), are given in Table 32.

### TABLE 32
### LATERAL VIBRATION OF COUPLINGS

| COUPLING STYLE | LATERAL VIBRATION MILS PP | TEST HP | TEST RPM |
|:---:|:---:|:---:|:---:|
| 1D | 0.28 | 2 | 1800 |
| 3 | 0.82 | 2 | 1800 |
| 1D | 0.43 | 5 | 1800 |
| 3 | 0.53 | 5 | 1800 |
| 1 | 1.1 | 5 | 1800 |
| 1D | 0.5 | 10 | 3600 |
| 3 | 2.8 | 10 | 3600 |

Table 33 lists the torsional vibration data due to lateral misalignment; average values through 62 mils eccentricity for 1D and 3 and 35 mils for No. 1.

**TABLE 33**
**TORSIONAL VIBRATION OF COUPLINGS**

| COUPLING STYLE | TORSIONAL VIBRATION ALL-PASS DEGREES PP | TEST HP | TEST RPM |
|---|---|---|---|
| 1D | 0.055 | 2 | 1800 |
| 3 | 0.080 | 2 | 1800 |
| 1D | 0.115 | 5 | 1800 |
| 3 | 0.07 | 5 | 1800 |
| 1 | 0.2 | 5 | 1800 |
| 1D | 0.16 | 10 | 3600 |
| 3 | 0.35 | 10 | 3600 |

The torsional vibration that is generated by the coupling itself, as distinguished from that generated by pulsating sources outside the coupling, is almost invariably of no consequence.

Typical data for torsional vibration arising out of a total shaft misalignment of 5 degrees are given in Table 34. None of the listed couplings are capable of this misalignment except for a period of a few hours. All of the foregoing test data and that below is taken from M. Blake, Monograph No. 21.47, Lovejoy, Inc., April 1965.

**TABLE 34**
**ALL-PASS VIBRATION OF COUPLINGS**

| COUPLING STYLE | ALL-PASS TORSIONAL VIBRATION DEGREES PP | ALL-PASS LATERAL VIBRATION MILS, PP | TEST HP | TEST RPM |
|---|---|---|---|---|
| 1D | 0.095 | 0.315 | 2 | 1800 |
| 3 | 0.175 | 0.17 | 2 | 1800 |
| 1D | 0.15 | 0.46 | 5 | 1800 |
| 3 | 0.13 | 0.30 | 5 | 1800 |
| 1D | 0.30 | 0.27 | 10 | 1800 |
| 3 | 0.18 | 0.65 | 10 | 1800 |
| 1D | 0.15 | 0.37 | 10 | 3600 |
| 3 | 0.11 | 0.55 | 10 | 3600 |

**Torsional Stiffness.** For the small miscellaneous applications the torsional stiffness is of no consequence. For the larger, and in applications wherein the coupling plays the role of a system component, specifically computed and chosen, torsional stiffness is important. Sometimes high stiffness is required and sometimes low. Couplings Nos. 5 and 6 have high torsional stiffness, whereas Nos. 1 and 2 have an intermediate stiffness and Nos. 3 and 4 have the lowest stiffness. In general, the all-metallic coupling is the stiffest. When rubber is used, the shear-type transmission is less stiff than the compression type. For this reason alone, rubber is often used in shear. The static torsional stiffness of Nos. 1 and 2 increases with amplitude, whereas that of Nos. 5 and 6 is about constant; that of Nos. 3 and 4 may show a softening at first and then a hardening with increasing amplitude, depending on the type of rubber. The torsional stiffness of Nos. 2, 5 and 6 is more or less independent of frequency, whereas that of Nos. 1, 3 and 4 is significantly dependent on the frequency of torsional vibration.

**Backlash.** If torque is plotted as a function of angle for positive and negative values, backlash is said to exist if the curve dwells at zero torque for an appreciable angle instead of passing through zero at or near zero-degrees. Backlash is, again, of no consequence in most applications. It exists in couplings Nos. 1, 2, 3 and 5 and not in Nos. 4 and 6. However, for applications involving torque reversal, a coupling with no backlash is normally required. For example, the gear coupling No. 5 might fail in a few hours on some diesel engine drives. If the coupling is used to transmit information rather than power, exact positioning may be extremely important. Then backlash must be zero, and torsional stiffness high.

## 4. DESIGN CONSIDERATIONS

Most couplings are either all metal or combinations of metal and rubber. Plastics have found little application because there is no plastic that even approaches rubber in the matter of enduring fluctuating loads without fretting and failing. Large critical couplings are usually made of steel. Perhaps cast iron is the most frequently used material, gradually being overtaken by aluminum. Die-cast zinc and magnesium are also used as well as some other nonferrous metals. Sintered iron is currently used for couplings to about 2-in diameter.

**Geometry.** The torque capacity of a coupling or shaft is about proportional to the cube of the diameter. One hp per 100 rpm corresponds to about 630 in-lb of torque (this is a typical measure of couplings). Some shafts are less stressed than others, but there is no universal practice, so the optimum maximum bore for a given coupling rating is often discussed without clear agreement. Forgetting theory and observing what is actually purchased in the United States for large numbers of applications up to about 5 hp per 100 rpm, the conclusion emerges that the optimum maximum bore in inches is about 1.3 times the cube root of the hp per 100 rpm. However, the designer must take careful note of the motor shaft sizes because most couplings are driven by motors. For example, reviewing the highest motor torque that is associated with a 1-in shaft in the NEMA standards, let it be supposed that this is 0.15 hp per 100 rpm. Then it is about optimum to design the coupling having a 1-in maximum bore for 0.3 hp per 100 rpm because the average service factor will be about 2. For higher service factors the user must purchase a higher-torque-rated coupling of potentially greater maximum bore and then underbore it. This is a fictitious example to illustrate an important point. If exceedingly large shafts are used for low torques, it is often more economical to turn down the shaft diameter rather than purchase a very bulky overrated coupling in order to provide the excessive maximum bore.

## 5. APPLICATION

Traditionally, the flexible coupling was chosen in terms of its static characteristics without much reference to the dynamic characteristics of the system. In the past few decades, with the advent of adequate torsional and other measurements, many of the larger couplings have been chosen on the basis of being a mass-elastic component of a dynamic system. Stiffness, backlash and other features are increasingly taken into account in making a choice.

**Service Factors.** Despite some variety of treatment, the service factor is the concept that is most usually used in the matter of determining the strength of a coupling required for a given application. Drivers are rated from about 1 to 3, turbines being about 1 and some diesel engines being about 3. Engines and compressors having 3 or less cylinders are usually associated with high-service factors. Driven items are also rated, usually into light, medium and heavy categories. The product of the driver and driven factor is the service factor. Supposing it is three, a 10 hp coupling at 1800 rpm becomes a 3.3 hp coupling at the same rpm; or, conversely, a 10 hp load becomes a 30 hp load. The service factor represents the difference between the dynamic operational situation and the mere static torque or average torque that is associated with it.

**Safety.** Safety is a paramount consideration. It is important to envisage the mode of failure of the coupling so as to provide against injury to personnel or equipment. In the smaller sizes failure is of little consequence. In the larger sizes failure must be avoided by careful design and maintenance. For example, a coupling that supports one end of a generator that has only one bearing cannot be allowed to fail; neither can a coupling that takes power from a large turbine, or that drives some critical piece of equipment, either on land, sea or in the air. The question must also be asked if failure of the elastomer means failure of the drive. Some couplings will drive after the elastomer has partially or totally failed. It is desirable to have some warning of failure, even for small couplings.

**Maintenance.** The couplings of Figs. 21.1, 21.3, 21.4 and 21.6 require no lubrication. Other than lubrication, couplings require little maintenance. However, the driven machinery must be aligned occasionally, particularly at the time of overhaul. At that time a complicated coupling or one that is inconvenient to manipulate or align can add considerably to the cost of maintenance. For the smaller drives the simplest nonlubricated couplings are often chosen. For large drives styles 2 and 5, or perhaps 6, are more typical. These are all metallic and do require some care. At speeds in excess of about 5000 rpm and for the larger horsepowers, for example 5000 hp and upward, it is usually necessary to make rather elaborate hot alignment checks besides monitoring the performance of the couplings almost continually. The advent of extreme speeds, such as 15,000 or 20,000 rpm, together with horsepowers in excess of 10,000, have led to the need for greater maintenance attention, more sophisticated methods and more sophisticated coupling design. Often the coupling is enclosed in a sealed space at elevated temperature, in an atmosphere of oil or gas.

**Alignment.** Irrespective of the capacity of the coupling to endure misalignment, it should be well aligned at the time of installation in order to preserve the greatest insurance against the hazard of later accidental misalignment. The coupling shown in Fig. 21.8 failed in a few weeks as illustrated because of unstable alignment and local resonance. The alignment procedure should be as simple as possible. For example, for coupling No. 5 the shroud must be removed in order to effect alignment or check it later. Shroud removal takes time and releases oil, and when the shroud is free the alignment procedure is not at all convenient. By comparison, coupling No. 1 may be aligned or checked very quickly and may be observed as regards alignment when actually operating.

**Lateral Flexibility.** Shafts are always misaligned to some extent. The load on adjacent bearings is determined by the amount of misalignment, by the stiffness of the shafts in flexure, by the coupling design and by the torque and rpm. For machines wherein the misalignment is normally below a static amount of say 15 mils eccentricity at speeds to about 3600 rpm, the single-engagement coupling is usually adequate in that it performs well and does not impose an unacceptable load on the bearings. It is, of course, difficult to give precise rules. But to gain some notion of the orders of magnitude, consider an electric motor of about 10 hp at 1800 rpm which corresponds with about 350 in-lb average torque or 350-lb lateral pull on a chain sprocket of 2 in diameter. Thus, the motor bearing may be supposed to be adequate for a lateral load of about 350 lb. If this load is not to be exceeded at 15 mils shaft eccentricity, then the dynamic lateral stiffness of the coupling must not exceed about 23 lb per lateral mil. A single-engagement coupling will fulfill this requirement as a rule, and the flexure of the shafts tends to reduce the lateral load below the statically expected value.

For the higher loads and speeds, couplings such as those illustrated in Figs. 21.5 and 21.6 are usually chosen. The double-engagement design reduces lateral load and lateral and torsional vibration to a minimum. Although these couplings are capable of more than 15 mils lateral eccentricity at low speeds, the capability decreases with load and speed so that as the speed increases beyond perhaps 3600 rpm the lateral eccentricity must be held to a static hot value of 10 mils and sometimes much less. For slow speeds the lateral eccentricity may be very high if the spool is long enough, and many double-engagement couplings can endure about 12 mils (lateral) per inch length of spool at slow speeds. Some of the rubber designs are capable of as much as 8 degrees or say 140 mils per inch of spool length.

Figure 21.9 illustrates a special rig used to measure torsional and lateral vibration as well as dynamic lateral stiffness of small couplings up to about 1 hp per 100 rpm and to about 5000 rpm.

Fig. 21.9  Test Apparatus. Test rig for lateral stiffness with variable load and speed; capable of measuring lateral and torsional vibration

# BIBLIOGRAPHY

*Standard Nomenclature for Flexible Couplings.* American Gear Manufacturers Association, AGMA 510.01, 1965.

Jackson, C. "Shaft Alignment, Using Proximity Transducers." ASME, 68–PET–25. New York, 1968.

Lohr, F. W. *Kupplungs-Atlas.* Ludwigsburg: A.G.T. Verlag Georg Thum, 1961. A comprehensive listing of couplings from the simplest to the most sophisticated types.

Nestorides, E. J., ed. *A Handbook on Torsional Vibration.* Cambridge, England: Cambridge University Press, 1958.

# 22

# ROLLING BEARINGS

## Michael P. Blake

## 1. SCOPE

The topic of estimating the condition of installed rolling bearings is discussed for speeds to about 3600 rpm and for bearings from about 2 to 12 in inner diameter, as found in typical machinery. Techniques for measurement and analysis are given that are likely to detect 90 percent of all faulty bearings before a failing bearing causes damage to other machine components. The frequencies of interest lie in the range of about 200 to 10,000 cps, but mainly in the region of 2500 cps. In this range acceleration measurements are most appropriate, whereas other machinery measurements are likely to be made in terms of velocity. The technique of estimating the condition of rolling bearings is not as well developed as that pertaining to other machine faults on the whole. Very little data has been published in this area, so this chapter, while not conclusive, may be found to be valuable.

## 2. BACKGROUND

It is desirable to be able to determine if a given bearing has a fault and to estimate its probable future life. Safety and economics are the usual background considerations. A measuring program is likely to eliminate unheralded failure and so take care of safety and economic requirements.

The unfiltered oscillograph of Fig. 22.1 is typical of many bearings that are approaching the end of life with a gradual increase in wear and fatigue. The vibration is seen to be narrowband random in a frequency range about 2.5 kcps. As the bearing condition becomes worse, the lower-frequency amplitudes in the spectrum become more pronounced, while the dominant frequencies also tend to exhibit greater amplitudes. Sometimes, occasional or even a continuous series of blips are seen in the oscillograph. Figure 22.2 illustrates the occasional blips. Bearings that are in good condition exhibit much the same kind of oscillograph, with the difference being that the amplitude is less than perhaps 4-gP or thereabouts. Bearings in the best condition tend to exhibit no dominating frequency; or, as it is said, the spectrum becomes broadband.

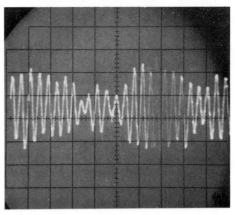

Fig. 22.1 Tape playback of accelerometer measurement on bearing housing of 25 hp, 720 rpm shaft-suspended centrifuge; Ball bearing about 4-in inner diameter. One division horizontal is 1 millisecond, and one vertical division is 2-g. As shown, maximum gP is about 4, with unshown blips to about 8-gP. Condition fair to bad.

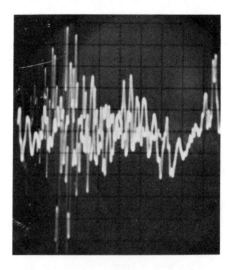

Fig. 22.2 Oscillograph for a water pump bearing, 7.5 hp, 1750 rpm, about 2-in inner diameter, vertical scale, 0.4-g per division. Meter reads about 1-gP, whereas blips extend to about 4-gP. Bearing considered almost bad and in need of replacement.

**Vibration Characteristics.** This chapter is based on some hundreds of measurements obtained mainly in the years 1963 and 1964 on bearings from 2- to 12-in internal diameter, at temperatures to about 300°F and in a corrosive, adverse environment, in general. The basic mechanism of the vibration, or noise as it is sometimes called, was not determined. It is often stated, and might be expected, that a knowledge of bearing geometry and of bearing defect is sufficient to enable the prediction of the frequency of the fault-induced vibration. But field observation shows that the matter is not so simple. For example, ball bearings have perhaps 8 to 12 balls, and roll-type bearings may have about the same. With one defect in the bearing race or one in a ball, a frequency of perhaps from three to five times that of rotation is generated. The frequency formulae for defects are as follows.

**Spot (Defect) on the Outer Race.** Blips per relative revolution of outer and inner race is

$$A = \frac{n\, S_2\, d_1}{d_1 + d_2}.$$

**Spot on the Inner Race.** Blips per relative revolution of outer and inner race is

$$B = \frac{n\, S_1\, d_2}{d_1 + d_2}.$$

**Spot on the Rolling Element.** Blips per relative revolution of outer and inner race is

$$C = \frac{4s\, d_1\, d_2}{(d_2 + d_1)(d_2 - d_1)}$$

assuming spot hits both bearing races, where n is the number of rolling elements (balls or rollers), s is the number of spotted rolling elements, $S_2$ the number of spots on the outer race, $S_1$ the number of spots on the inner race, $d_2$ the track diameter of the outer race and $d_1$ the track diameter of the inner race. The three expressions give the greatest frequencies that can occur. It is possible that a spot on a rolling ball will not contact the race at times or that a race defect may be relatively or totally free of load, for example, if the inner race is fixed and if the load is in a direction that bears on the unspotted side of the inner race.

Actual experiment shows that bearing components behave more or less like planetary gears, and the effective track diameter for balls is about the same as the greatest and least diameters of the track grooves. At very high speeds and low loads, the centrifugal effect can make the balls more or less cling to the outer race so that the cage rotates faster than expected on a planetary basis. Nevertheless, the expressions A, B and C are adequate as an indication of expected frequencies.

Taking for a typical race n equals 8, $d_1$ equals 38 and $d_2$ equals 62, the expressions give the following values.

$$A = 3\, S_2 \qquad B = 5\, S_1 \qquad C = 4\, s$$

Thus, if the inner or outer race is fixed, the bearing makes N revolutions per minute and the outer race has three spots, the generated frequency due to spots is not more than 9 · N. If the inner race has two spots, the generated frequency is not more than 10 · N. If there are two spotted balls, the generated frequency is not more than 8 · N cycles per minute (blips per minute).

The greatest value of generated blip frequency is likely to be that corresponding to about 4 · s or 5 · $S_1$ because all balls could be spotted and the number of spots on a race is not likely to exceed the number of balls. The number of spots would be equal to the number of balls if the race were brinelled; that is, it would have latent or actual defects corresponding to the ball positions due to vibration when in transit or at rest.

Thus, the greatest frequency for rotation at 1800 rpm is likely to be about 40 times 30, which is 1200 cps; and the least and, perhaps, the most probable is that due to one defect, which is say 5 times 30 or 150 cps. Figures 22.3 and 22.4 suggest that typical observed frequencies are from 1000 to 3000 cps.

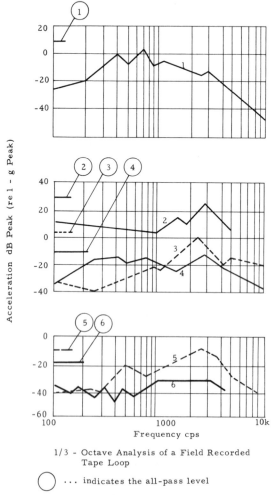

1/3 - Octave Analysis of a Field Recorded Tape Loop

◯ ... indicates the all-pass level

Fig. 22.3   Rolling bearings, diagnosis of condition from bearing housing measurement

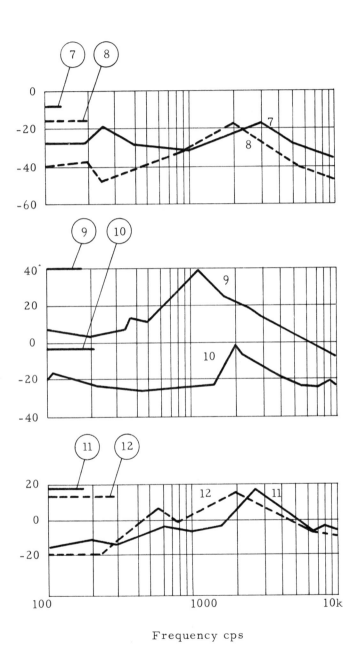

Fig. 22.4   Rolling bearings, diagnosis of condition from bearing housing
measurement

The point of interest here is whether the observed frequencies for defective bearings are those corresponding to expressions A, B and C above or whether what is observed is a decaying natural vibration, excited by occasional impacts at frequencies corresponding to A, B and C. This writer takes the latter view because the observed frequencies are, as a rule, greater than expected in terms of A, B and C; because, and this is important, the observed frequency does not appear to increase with an increase of rotational speed and because the natural frequencies of bearing elements and adjacent machine elements do fall in the frequency range observed for defective bearings.

| CLASS* | CURVE | ITEM |
|--------|-------|------|
| fair | 1 | Gearbox of cooling tower fan, about 30 hp, 1800/300 rpm |
| bad | 2 | Centrifuge, 30 hp, stiff shaft, 1800 rpm, 27 in dia., roller bearings with grind-like marks |
| good | 3 | Centrifuge, same as No. 2, with bearings replaced, 1800 rpm |
| good | 4 | Centrifuge, same as No. 3, but now at 1500 rpm, bearings about 5.5 in I.D., barrel roller, oil lubricated |
| good | 5 | New electric motor, fan end, 20 hp, 1800 rpm, mounted on rubber pads |
| good | 6 | New electric motor, fan end, 2 hp, 1750 rpm, mounted on rubber pads |

* Condition of bearing verified by physical inspection.

| CLASS | CURVE | ITEM |
|-------|-------|------|
| good | 7 | Centrifuge, 1100 rpm, Baker-Perkins, stiff shaft, 36-in dia., 30 hp; bearings 3 years old |
| good | 8 | Centrifuge, 900 rpm, shaft suspended, about 50 hp, 48-in dia |
| bad | 9 | Electric motor, 1800 rpm, 75 hp, driving centrifugal pump. See Fig. 22.7, same bearing |
| good | 10 | Electric motor, 1800 rpm, similar to No. 9 above and similar duty |
| bad | 11 | Mill, Williams, 15 hp, 3600 rpm |
| bad | 12 | Electric motor, 10 hp, 1800 rpm, driving agitator gear reducer |
| bad | 13 | Spectrum not shown: centrifuge, stiff-shaft, 24 in, 1100 rpm, 25 hp, measuring about 1-gP with oscilloscope blips to 8-gP |

Thus, the writer envisages the observed vibration as being generated by impacts due to defects, the impacts being of lower frequency than the local natural frequencies. These local natural frequencies are brought to life at each impact and play the part of free damped vibration for, perhaps, from 3 to 15 cycles before the next impact arrives. If a bell of 200 cps natural frequency were struck several times per second, the corresponding oscillograph would tend to exhibit 200 cps and ignore the striking frequency.

Actual measurement of the natural frequencies of an outer race of about 7-in outer diameter showed both 1500 and 10,000 cps response frequencies when the race was suspended on a piece of string with its polar axis horizontal and struck with a piece of metal. For smaller bearings the natural frequencies may be expected to be higher. Again, it may be remarked that sufficient work was not done to lead to firm conclusions. But the facts are given so as to offset the often accepted theory that the observed frequencies correspond with A, B and C, which has not at all been proved, so far as this writer is aware.

**Mode of Failure.** Three phases of degradation are sometimes discussed and are easy to envisage: A new bearing exhibits noise of low amplitude with a broad band of frequencies, analogous to surf or wind noise. As degradation proceeds, the emergence of one or more dominant bands or frequencies tends to make the noise narrow in band and higher in level. This may be called the noise phase. If the bearing survives it may exhibit relatively large displacements at low frequencies. Large-size defects now exist and the bearing "rumbles" rather than "hums." This may be called the vibratory phase. If the bearing survives, it may enter a final convulsive phase wherein it can become as hot as dull red. This may be called the thermal phase. But the above is an oversimplification, and an attempt is now made to clarify what has been observed.

**The Noise Phase.** Using a stethoscope or stick to lead the noise to the ear, a faultless bearing sounds like sea surf. The development of a small fault sounds like the occasional dropping of a few stones into the surf. When the bearing is near the end of its life, it sounds as if a continual load of stones or rocks is being dropped into the surf.

The bearing itself changes from a polished condition to a condition wherein macro defects begin to appear. These may cause small damage to small, noncritical machines and large damage to large, critical machines. Thus, it is the objective of design and maintenance to replace the bearing before the development of macro defects. Its physical life should be one of slow, orderly fatigue and wear without spalling or cracking. The end of useful life is indicated by a bearing housing acceleration of about 4-gP.

Sometimes in the presence of corrosion and very slow wear without roughness the acceleration may not increase when the bearing has constructively reached the end of its life. In such an instance, the life of the bearing may exceed the duration of the maintenance cycle so that the bearing is replaced at the time of overhaul. Nevertheless, it is advisable to monitor critical bearings, by either indicator test for radial play or shaft stick or proximity pickup measurements in order to detect gross wear which may not reveal its presence by an increase in housing acceleration.

Significant wear without much increase in roughness and acceleration has been observed in several centrifuge roll bearings, having internal diameters of about 5 in with double rows of spherical rollers in bronze cages. This kind of wear has not been observed in ball bearings. It is to be expected, and seems confirmed by observation, that rollers are more stable than balls, not being subject to the unavoidable sliding and gyrating motions inseparable from ball bearings.

**The Vibratory Phase.** The geometrical defects are such that the dominant feature is a somewhat more ordered, lower-frequency vibration, rather than a noise. There is no sharp transition point, and this second class is simply an approximate way of describing a condition of more or less gross defects. Ideally, all bearings should be replaced before this phase becomes severe. Perhaps nine out of ten bearings are, in fact, replaced in the noise stage. If they survive and do not become slack, the vibratory phase may be induced by an increasing fatigue rate, leading to large defects in the race or rolling elements. The housing acceleration increases up to perhaps 50-gP, and the probable life of the bearing is now merely weeks or even hours. Note that 50-gP at 2000 cps requires a displacement of only about 0.24 mils PP. But, if a single defect induces a frequency of perhaps 150 cps, about 5 mils PP displacement leads to only 5-gP. A defect of 5 mils is very large, and the corresponding housing movement is likely to be far less, so that perhaps 1-gP may be observed, depending on the mass of the housing, rotor and other features.

If the bearing survives, it may exhibit a relatively high displacement together with a relatively low acceleration. This is the vibratory phase in its full maturity. Such a bearing sounds to the ear like the dumping of a load of large stones. For example, a bearing in a 5-hp, 1800-rpm motor driving a small gear reducer was observed for about one year. The risk to person or property was negligible, although the bearing was obviously excessively bad. The acceleration never exceeded 4-gP, and the displacement was less than 5 mils PP, giving the computed frequency as about 130 cps. This bearing refused to fail. Bearings in such situations sometimes reach a favorable accommodation with adjacent bearings and then may last indefinitely for the simple reason that they no longer perform any function that is really useful.

**The Thermal Phase.** Although bearings can last indefinitely after the onset of the vibratory phase, it is more usual, especially in the larger bearings at moderate to high speeds, that the vibratory phase leads rapidly to a thermal phase. The convulsions of the bearing become more violent, leading to such a great amount of friction that the rolling elements may turn blue, while rolling elements and races may also be greatly deformed. An example is shown in Fig. 22.5. If this bearing had managed to relieve itself in the vibratory phase, it might have lasted indefinitely if not replaced.

The severe vibratory or thermal phase is to be avoided, especially in the case of critical bearings. For example, a high-energy machine, such as a 24-in centri-

Fig. 22.5  Bearing from a 125 hp, 1800 and 1200 rpm motor for a single-stage centrifugal pump. Balls and inner race have turned blue. Left hand ball is distorted into a lemon shape, and flanges of inner race are pushed outwards from the center of the track, axially. The bearing is about 3-in inner diameter.

fuge with perhaps half a million foot-pounds of energy, may do great damage if the bearings fail, even though the input may be less than 40 hp. In general the higher the horsepower and speed, the greater the care that is taken in monitoring bearings.

The question of why some bearings remain more or less in the noise stage and some do not may be answered to a limited extent in terms of load and speed. However, most bearings are correctly chosen so that it is perhaps very likely that brinelling, corrosion or the entrance of dirt before installation or incorrect installation are the probable causes. It is generally accepted in maintenance experience that more than half the unscheduled failures of rolling bearings are due to dirt inclusion at the time of installation or to corrosion or damage at that time.

**Lubrication.** Oil, or perhaps an oil mist, is usually used for higher speeds. It requires careful attention, provides little emergency reserve and sealing against ingress of dirt or corrosive elements and tends to be easily contaminated. It is usually easy to check, and damage by the person who replaces it is unlikely to occur. Grease is usual for the lower speeds. It provides considerable emergency reserve and so requires less monitoring. It provides some seal against foreign matter and is not as easily contaminated, although it cannot be filtered. It is impossible to check, and damage may result at the time of replacement if the operator does not follow correct procedures such as removal of flushing plugs, etc.

With either oil or grease, many bearings larger than about 4 in internal diameter operating above about 1200 rpm tend to run hot. For example, 4.5 in internal diameter, double, spherical, roller bearings at 1500 rpm in a machine at room temperature have been observed to operate at about 180°F. The machine was a stiff-shaft centrifuge with the reservoir of oil at about the level of the center of the lowest roller and using a bearing internal fit of one fit slacker than the maker's normal. The following rule is useful. In general, up to 175°F on the bearing housing may be considered normal. Up to 212°F (where it will fry a drop of water on the housing) is possible if monitored. Above 212°F is to be regarded as abnormal, although possible.

## 3. MEASUREMENTS

As always, the policy may be stated: to achieve the most of the machine function, however that may be defined, for the least expenditure, while not risking the safety of personnel or machine. For small machines this means that unscheduled failures are curbed to the point where more measuring leads to fewer failures but higher overall cost and vice versa. For large and critical machines, bearing failures must be eliminated, whatever the cost of measuring or monitoring.

**Measurement Practice.** Noncritical bearings are surveyed, measuring the velocity on the housings. Velocities exceeding about 0.2 IPSP are given to engineering personnel for further analysis. (Please see Standards and Tables, Chapter 33 of this volume.) Critical bearings merit more careful measuring.

It is desirable to classify all bearings in a given installation when regular measuring is undertaken. The speed, power, value, energy and function of the machine together with danger to personnel or process are some of the determining factors. At least the all-pass (unfiltered) acceleration on the bearing housing must be measured at regular intervals.

### A History of Measurement Possible Examples

1. Noncritical bearings are surveyed with a velocity meter, leading to the discovery of perhaps 70 percent of the defective bearings. If a simple acceleration meter is available, it should be used for bearing surveys, although not for surveying of miscellaneous machine components. The response limit of velocity meters is about 2000 cps; the velocity parameter is not, apart from this, as useful as the acceleration. Acceleration measuring may uncover up to about 80 or 90 percent of defective bearings. Decisions regarding corrective action are based on a single all-pass measurement, and no analysis is made.

2. A critical bearing is measured at the time of installation, at least in terms of all-pass acceleration and as often in the next few days as may be considered necessary to insure that it has settled down to a steady condition at an acceptable level. Measuring techniques of increasing refinement are given in Table 35.

**TABLE 35**
**MEASURING TECHNIQUES AND RELATED FAILURES**

| TECHNIQUE | PROBABLE UNSCHEDULED FAILURES, PERCENT |
|---|---|
| A. All-pass velocity of bearing housing, when acceleration measuring is not available | 30 |
| B. All-pass acceleration | 20 |
| C. Oscillograph display, acceleration (with technique B) | 10 |
| D. Velocity or displacement, with shaft stick or proximity pickup (with technique C) | 5 |
| E. Amplitude-frequency spectra of all-pass acceleration (with technique D) | 2 |
| F. Continuous monitoring of B and D above | 1 |
| G. Audio tape record of B above | |

The tape record is useful both as a comparison and a great storehouse of information. The purpose of the oscillograph is simply to uncover blips or other peculiarities that may not be revealed by the all-pass measurement or the spectrum analysis. An example of a typical record used when analyzing magnetic tape data is given in Fig. 22.6. A similar record is used in the field.

At intervals of about one month, after the bearing trend has become steady, all-pass acceleration is recorded. A sharp upward trend indicates the need for measuring more frequently. A slower trend calls for no action. In both instances an attempt is made to predict when the acceleration level is likely to reach say 4-gP or whatever level is considered to correspond with the end of bearing life. The estimate may call for unscheduled overhaul, or it may show that scheduled overhaul will occur before the end of life. The bearing is then replaced.

If the bearing shows a zero trend, interest becomes focused on total hours recommended by the maker or on an increase in radial slack. If the increase in radial slack is small or negligible, this writer prefers not to change the bearing at the time of machine overhaul, except when it is smaller than about 2 in internal diameter.

An occasional velocity measurement, or spectrum analysis, which is more searching, may be desirable. For example, an all-pass acceleration reading may show 5-gP which is not generally dangerous (see Chapter 33, Standards and Tables; also Figs. 22.3 and 22.4.) However, if the frequency is merely 100 cps or thereabouts due to the development of some major low-frequency defect, then the condition is dangerous. If no better information is available, Fig. 33.1 may be used as an envelope for acceleration spectra.

| | A | B | D | S | F | G |
|---|---|---|---|---|---|---|
| PLAYBACK SETTINGS | 0 | 4 | | 3.75 | | 2.1 |

| CALIBRATION SETTINGS | FREQ. CPS | INPUT mVPP | OUTPUT mVPP | LEVEL CONTROL | TAPE SPEED |
|---|---|---|---|---|---|
| | 1000 | 138 | 25 | −2 | 3.75 |
| | 1000 | 138 | 80 | 0 | 3.75 |
| | 1000 | 138 | 380 | 2 | 3.75 |
| | 1000 | 138 | 800 | 3 | 3.75 |

| | C REC'D LEVEL | mVPP FOR 1-gP | mVPP STEADY | mVPP BLIPS | ACCN. gP STEADY | ACCN. gP BLIPS | CLASS A/B/C/D | COMMENT: freq./pitched/unpitched/%/remaining life/ fault/photo/strip/freq. analysis |
|---|---|---|---|---|---|---|---|---|
| 1 | 2 | 380 | 240/100 | 400 | .63/.26 | 1.04 | | Motor, repaired, 1760, 25 hp, Q. Plant |
| 2 | | | | | | | | Motor, repaired, faulty tape |
| 3 | 2 | 380 | 50/80 | | .13/.21 | | | Motor new, 1735, 2 hp |
| 4 | 3 | 800 | 500/800 | | .63/.63 | | | Motor new, 1755, 20 hp |
| 5 | 3 | 800 | 360/500 | | .46/.63 | | | Motor, new, 1755, 20 hp |

| | | | | | | | |
|---|---|---|---|---|---|---|---|
| 6 | Pump, tower, NNN bldg. | | 2 | 380 | 300 | 0.80 | D |
| | | | | 380 | 100 | 0.26 | D |
| | | | | 380 | 200 | 0.53 | D |
| | | | | 380 | 200 | 0.53 | D |
| 7 | Pump, tower distrib. #2 NNN bldg. | 2 KC | −2 | 25 | 1400 | 56.0 | A |
| | | 1 KC | | 25 | 2100 | 84.0 | A |
| | | 2 KC | | 25 | 700 | 28.0 | A |
| | | 1 KC | | 25 | 800 | 32.0 | A |
| 8 | IR, close coupl. pump, NNN bldg. | 15 KC | 0 | 80 | 300 | 3.7 | B |
| | | 3 KC | | 80 | 300 | 3.7 | B |

Comments: Items Nos. 6 and 7, where four measurements are given, are listed:

—motor, outboard
—motor inboard
—inboard of driven machine
—outboard of driven machine

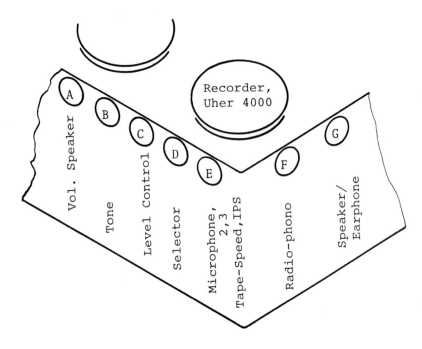

Fig. 22.6    Magnetic tape playback data work-sheet

**Instrumentation.** Various types of instruments are required for measuring the vibration response of bearings. A partial list is given below.

1. Vibration meter reading acceleration and velocity; for example, a General Radio meter type 1533, using a pickup suitable for frequencies to 10,000 cps

2. Any standard oscilloscope, such as Hewlett Packard type 120 B or Tektronix type 422

3. Shaft stick (see Fig. 22.9)

4. Proximity measuring system for shaft displacement measurement, such as made by Bently-Nevada

5. Wave analyzer with a 1/3 and 1/10 octave range, for example, as made by General Radio or Bruel and Kjaer

6. Tape recorder as made by above makers

7. Level recorder as made by above makers, compatible with wave analyzer for automatic plotting of spectra

## 4. CASE HISTORIES

The following list is typical. All were recorded on a portable Uher No. 4000 tape recorder directly from a Bruel and Kjaer model No. 4328 accelerometer with a sensitivity of 69 mils per g, and flat to about 10 kcps. The recordings exhibited less flatness than is desirable because of the inadequate input impedance, but they were of great help as a means of preliminary investigation in this field. No effort was made to correct the final figures. Tape sections of about 30 sec duration at 3.75 in/sec were identified by microphone, using identification numbers giving the date and item. The sections were made into loops of about ten seconds each on a Roberts 997 recorder. The Roberts recorder was played back into an oscilloscope and, at the same time, into a General Radio type 1564 A vibration analyzer. The system was calibrated by recording a signal of 1-gP at 1000 cps.

Referring to Figs. 22.3 and 22.4, Item 1 includes gear and bearing vibration. The input rpm is 1800, and a pinion of about thirteen teeth may be expected to give a frequency of 390 cps. The peaks at 400 and 700 cps are probably gear vibration. The spectrum peak is 2 dB, and the all-pass level is 6 dB (dB level is referenced to 1-g.)

Item 2 was measured on the housing of a bearing of about 5-in internal diameter with spherical rollers and oil reservoir, operating at 1800 rpm. The spectrum peak is about 25 dB at 2700 cps, and the all-pass level is about 29 dB. In terms of Section 4 in Chapter 33, this represents an acute fault. The bearing exhibited an unexplained pattern, as if from a process of rough grinding. It was suggested that the local corrosive environment may have caused this pattern. The writer knows of no reason why corrosion should exhibit itself in this way. The reader is asked to note that the decibels of Figs. 22.3 and 22.4 may be made compatible with those of the above-mentioned Section 4, by adding 40 dB to the former.

Item 3 is the same as Item 2 after bearing replacement. The spectrum peak and the all-pass levels are both about 2 dB. It is not known why the frequency at the spectrum peak is lower than in Item 2.

Item 4 is the same as Item 3, but now running at 1500 rpm. Again, it is not clear why the frequency at the spectrum peak has returned to the original value of 2700 cps. The peak and all-pass levels have fallen to about −12 dB, which is a decrease of 14 dB due to lowering the speed from 1800 to 1500 rpm. It was observed that overall vibration and the risk of rapid degradation always increased greatly in this bearing when the speed was increased above 1500 rpm. The bearing has two rows of rollers with a bronze cage and sixteen rollers in each row.

The life of a bearing is inversely proportional to speed and about inversely proportional to the cube of the load.[1] If the speed is doubled, the life is halved; but at the same time, the centrifugal unbalance increases as the square of the speed, and the radial load due to this varies from batch to batch when the machine in question is a centrifuge receiving successive batches of material. Suppose it is 10 percent of the static load before speed increase, as a conservative estimate. Then the bearing load is 1.1 before speed increase and 1.4 thereafter because doubling the speed multiplies the force of mass-unbalance by four. This load increase decreases bearing life by about two to one, while the decrease due to speed increase alone is two to one, giving a total decrease of four to one. Practical observation of the subject bearing suggested that its life expectancy is about inversely proportional to the cube of the operating speed when this speed is about 1500 rpm.

Item 5 is a new electric motor rated at 1800 rpm and 20 hp, with peak and all-pass levels both of −8 dB. Item 6 is a new electric motor rated at 2 hp and 1800 rpm, with peak and all-pass levels of −29 and −18 dB, respectively. Both motors were supported on rubber pads, as indicated in Fig. 33.3.

Item 7 is a bearing of about 5-in inner diameter with double-row spherical rollers that has passed 30,000 hours of operation. When new it was probably one fit looser than normal, which is the usual fit chosen for this kind of centrifuge. It is, in fact, worn out by a slow process of fatigue corrosion and wear; and the radial looseness is of the order 0.005-in greater than when installed (total side play of the order 10 mils greater than when installed). Bearings in this condition often perform better than new ones. The spectrum peak level is −16 dB and the all-pass level is −7 dB.

Item 8 has a peak level of −17 dB and an all-pass level of −14 dB. This bearing is about 6-in inner diameter. Again, the absence of significant vibration is not a guarantee that the bearing is not worn.

Item 9 corresponds with Figs. 22.7 and 22.8. The peak level is 38 dB and the all-pass level is 38 dB. The dominant frequency is 1300 cps. Taking the acceleration to be 80-gP, the velocity is about 4 IPSP and the displacement is about 1.0 mil PP, according to the nomograph of Fig. 33.11. Because the frequency is well within the response limit for velocity pickups (which is about 2000 cps), this defect is discoverable by means of velocity measuring. However, the sensation induced in the fingers is almost negligible, despite the very high acceleration; this is because of the high frequency, as illustrated in Fig. 33.4.

Item 10 is an electric motor similar to Item 9, but of 100 hp instead of 75. The peak level is −3 dB and the all-pass level is also −3 dB. Comparative spectra

0    1 0    2 0    3 0    4 0    5 0    6 0    7 0
Centimeters

Fig. 22.7  Balls from a loading-slot race bearing, 75 hp, electric motor, 1800 rpm driving a centrifugal pump for a cooling tower. Item No. 4. Measurement on bearing housing exceeded 80-gP, although vibration was barely detected by fingers

ROLLING BEARINGS [461]

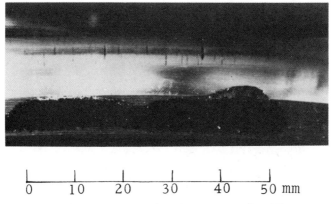

|____|_____|_____|_____|_____|
0    10      20      30      40      50 mm

Fig. 22.8   Outer race of bearing; same as Fig. 22.7

such as these for similar machines, when one is known to be in good condition, are of the greatest help in the formulation of service factors and of policies and decision for standards and maintenance.

Items 11 and 12 illustrate bad bearings, having internal diameters 2 to 3 in with respective peak and all-pass levels of 13 and 18 dB, and 12 and 16 dB. The dominant frequency in each is 1900 and 2700 cps so that the frequency in the 3600 rpm bearing is not at all twice that in the 1800 rpm bearing.

Item 13 illustrates how acceleration measuring may reveal obscure faults. Two similar stiff-shafted centrifuges, having a 24-in bowl diameter and rated at 25 hp and 1100 rpm, operate side by side. One was suspected because of its sound. Both were measured. The suspected bearings showed a steady display of about 1-gP. The unsuspected showed the same with blips to about 8-gP. Experience suggests that a blip of amplitude 2 is about equivalent, as regards severity of bearing condition, to a steady display on the oscilloscope of amplitude 1. Thus, the blip of 8-gP is regarded as equivalent to 4-gP steady vibration. Replacement of bearings was recommended. However, when they were dismantled they looked as good as new. So they were sent to the maker to resolve the puzzle in the interest of advancing field-measuring techniques. The maker's superficial examination revealed merely slight corrosion and a turning of the outer race in its housing. A further examination was made in view of the 8-gP measurement. The maker stated, "Several rollers from each bearing show grinding-chatter patterns, and these bearings should never have left our plant in their present condition." The replaced bearings exhibited 1 to 4-gP in the first week of operation. Acceleration often increases and then falls in the first weeks of operation. By way of illustration, Fig. 22.2 shows a display wherein the blips extend far beyond the, more or less, steady level.

**Summary of Measurements.** The all-pass and peak measurements differ but little on the whole. The frequency band of interest is about 200 to 10,000 cps. Wear may not always reveal itself in acceleration measuring.

Fig. 22.9    The use of a shaft-stick

## 5. CONCLUSION

1. Acceleration measuring usually indicates bearing condition.

2. An all-pass velocity reading on the housing is likely to uncover say 70 percent of defective bearings, and an all-pass acceleration reading 80 percent. An oscillograph display is likely to reveal about 90 percent. The number of unrevealed faults may be reduced to zero by adding further measurements, such as shaft movement, spectra and monitoring.

3. It appears probable that velocity meters and acceleration meters having a true rms response, as compared with common meters which have an average response but may be calibrated to rms, are likely each to reveal perhaps 10 percent more faults. Thus, an rms velocity or acceleration meter is likely to uncover 80 or 90 percent, respectively, in a single all-pass reading. Meters commonly used for vibration rectify the amplified signal in a simple way, and it is read-out on a d'Arsonval-type meter which indicates an average. For a simple sine signal the ratio of average to rms level is constant. However, if a signal with a tenth harmonic of 20 percent fundamental amplitude superimposed on the fundamental is fed to such a meter, it reads merely the fundamental; it is not significantly affected by the harmonic, which may itself indicate a severe fault. Thus, for the purposes of this study, averaging meters, even when calibrated in terms of supposed rms, are less desirable than true rms meters. (The meters may be the same, but the response is determined by the conditioning of the signal by an appropriate circuit.) A special peak-response circuit of some kind is likely to give the best results for bearing measurement.

4. This chapter is simply an introduction to a field which has received little attention. The root mechanism of the generation of vibration has not been determined; neither has the vibration been adequately classified and described except in terms of simple manual spectra, giving the maximum meter readings of the analyzer at the more significant peaks and valleys of a 1/3 octave analysis.

# REFERENCE

1. "Engineering Data." In *General Catalog*, no. 425. SKF Co., 1960.

# SECTION V

# MISCELLANEOUS MEASUREMENT EXAMPLES

# SECTION V CONTENTS

# SYMBOLS

| | | |
|---|---|---|
| S | Static deflection | in |
| F | Force | lbs |
| msec | Millisecond (one msec = 0.001 seconds) | —— |
| $X_r$ | Peak residual (steady state) displacement response due to pulse | —— |

# 23

# SHOCK AND IMPACT
# IN MACHINERY

*Michael P. Blake*

## 1. SCOPE

This chapter briefly considers the measurement and aims connected with such machine impacts as are not amenable to correction. These impacts arise in the normal operation of hammers, presses and even large broaching machines and are transmitted away from the machine through its support or foundation. Impact control amounts to control of transmission, since generation cannot be controlled. The task of control is considered as regards sources of impact, measurement, instrumentation and theoretical considerations, with a view to providing an introduction for the nonspecialist who may be faced with this task.

## 2. INTRODUCTION

Impact transmission is often controlled by the machine owner at its point of generation so as to avoid physical or physiological nuisance in his own interest or that of his neighbor. This is the control that is considered here. The converse control, that of controlling reception, arises when it is decided that the impact environment is too severe for delicate machines, or persons, and that it is either impossible or uneconomical to control the environment. From the point of view of reception, all impact is, to an extent, a nuisance. From the point of view of transmission, the impact is usually seen as being beneficial, although some nuisance impact, such as arises from looseness or maladjustment, may exist and should be corrected. Thus, the discussion is confined to outward transmission. The topic of reception is considered to an extent in Chapters 11 and 24 of this volume.

Many machine processes that are executed by impact could be done by static force. But static-force machines cost about twice as much as impact machines. A cost and impact-control compromise may be achieved by using counterblow hammers so that the impact generated in the machine frame approaches zero. The hammer is quick, which is important in forging. A hand hammer used by a skilled smith, miner, rivetter or carpenter achieves results that either cannot be achieved by static force or would lead to prohibitive cost. The existence of impact is due then, to an extent, to its being a poor man's force generator.

The available means for impact control are quite analogous to those for vibration isolation. Those nuisance items that can be corrected, such as looseness, are corrected after measurement. If the beneficial (functional) impact from the hammer or the vibration from the vibratory conveyor is then a source of transmission troubles, it is isolated by altering the machine-mass or stiffness of the machine supports. In controlling vibration transmission, significant progress is usually possible if the response curves with their resonance and decay features are understood. The response of simple systems depends only on the ratio of the disturbing frequency to the natural frequency. The same is true for impact control, except that the important ratio is now that of impulse duration to the natural period of the system. The shock spectrum takes the place of the response spectrum; however, it is used in a somewhat different way.

In order to make measurements that are profitable, it is usually necessary to know beforehand what is to be done with the measurements. They become profitable only when subjected to some analytical process. For this reason some analytical considerations are introduced here that properly belong under the heading of analysis.

If the impact and vibration implications of machines are considered at the design stage, much subsequent expense is often avoided at little cost. If they are not considered, subsequent control may prove uneconomical or impossible.

Measuring methods and instrumentation for research purposes are expensive, and much work remains to be done. Not too much is known about what occurs in the head of a hammer or about why an anvil, costing perhaps $30,000, lasts only eighteen months before failing by cracking. On the other hand, for day-to-day purposes of control, useful progress can result from the simpler measurements and simpler analytical considerations.

## 3. IMPACT SOURCES

In this chapter we consider impact and shock as the same thing and loosely define them as being a transient transmission of mechanical energy of brief duration, into or out of a mechanical system. The word brief usually connotes not exceeding about one natural period of the system. However, this is too confining. Impact is a classification concept with a gray boundary (i.e. it is not precisely defined). Impulse is defined exactly as being the time integral of the applied force.

The nuisance impacts arise from loose bearings, loose foundation bolts and other maladjustments. The functional impacts (beneficial) arise, for example, from hammers, mills, punches and explosive operations. The fact that crude hammers operated by boards or straps are being manufactured today bears testimony to the low cost of such machines for the purpose of generating large forces.

## 4. PRINCIPLES OF IMPACT CONTROL

Having eliminated the nuisance impacts by conventional mechanical corrections, the task is then to isolate the functional impacts. Profitable measurement depends, to a large extent, upon understanding the principles of this task. A brief description follows. In general, when the forcing frequency exceeds the natural frequency of the system, isolation is improved as the natural period T of the machine on its support is increased. This implies an increase in static deflection S, but it may not always be possible to increase S sufficiently to achieve a significant improvement of isolation. Through measurement it is possible to estimate the desirable correction and decide whether this correction is feasible. The relation between T and S is simple. The expression

$$T = 0.32\sqrt{S}$$

gives the natural period of a simple system in seconds. The system is shown in Fig. 23.1. In the above, S = W/K, the weight in pounds divided by stiffness in lbs/in; S is expressed here in inches, which is, in effect, the static displacement of the system.

If the form of the impulse is known (see Fig. 23.2) it is possible to determine the response of the system using suitable analytical methods. Of course the response may be found by measurement. The value of the analytical method is that it predicts the change in system response, due to a change of T, in the properties of the system. The method may tell us that any feasible change will lead to no significant improvement, or it may make matters worse. This is also true in vibration isolation, wherein it is not always proper to add more concrete in order to improve vibration isolation. In many instances this addition makes matters worse.

The phase-plane method is discussed in Chapters 10 and 11 of this volume. It is a graphical method for determining response to a given impulse. Figure 23.3 gives the phase-plane derivation of the shock spectra for a rectangular step excitation. During the impact and afterward there occurs, due to the impact, a greatest initial (transient) and a greatest residual (steady) displacement. These are discussed in terms of peak (semi) displacement excursions. Figure 23.3 shows that the possibilities of isolating the system cannot be estimated until F, K and $T_p/T$ are known. T is known if W and K are known.

In the actual example of Fig. 23.3 the value of $T_p/T$ is about 0.31. It is evident from the illustration that significant reductions of $X_i$ and $X_r$ (the maximum initial and residual peak displacements) are possible if $T_p/T$ can be reduced to say 50 percent of 0.31. In fact, $X_i$ is reduced in the ratio of about 1.4/4.8 and $X_r$ in the ratio of about 3.3/5.5. Since the impact duration $T_p$ is usually a fixed quantity, the value of T must be increased in order to diminish $T_p/T$. But in order to double T, because it is proportional to root W, it is necessary to multiply W by four or to divide K by four or, in effect, to multiply the static deflection by four. This may not be acceptable in practice.

In any event, the shock spectra do afford an insight into what is possible, what is involved in achieving it, what is the effect of the separate operable variables

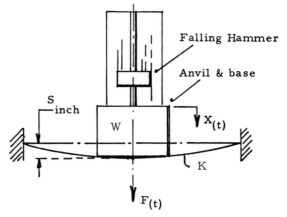

Simple Machine

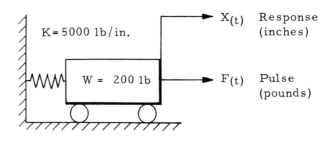

Idealized Analogy

$$S = W/K \qquad\qquad T = 2\Pi \sqrt{\frac{S}{12g}}$$

$$= 0.04 \text{ in} \qquad\qquad = 0.064 \text{ sec}$$

Fig. 23.1  Expression of the natural period T of a system subjected to impact by means of an idealized analogy, where T is the natural period, the time for one complete vibration of the machine on its supports when free of external constraint

such as T and what, above all, it is desirable to measure. Figure 23.3 shows the gist of what the task is; the topic is amplified in Chapters 10 and 11.

## 5. IMPACT MEASUREMENT

The objective of measurement is to classify and, if the classification is unac-

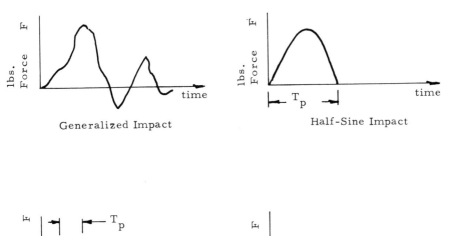

Fig. 23.2   Form of the impact. (Typical duration $T_p$ is 0.010 sec, and
the range is from 0.001 to 0.10.)

ceptable, to correct. A brief explanation of corrective principles as given above
paves the way to an understanding of profitable measurement. It is noticed in
Fig. 23.3 that there are several zeros of residual amplitude, occurring at $T_p/T$ =
1, 2, 3 and so on and, in the imaginary case, where $T_p/T$ is large. In general,
for symmetrical pulses like the half-sine, the versed sine and the rectangle,
this family of zeros does occur for some series of values of $T_p/T$. But for some
pulses wherein the rise or decay time is zero, as in the vertical-flank sawtooth,
these zeros do not exist, nor do they exist for some other asymmetrical pulses.
However, although $X_r$ cannot always be reduced to zero it can often be reduced significantly by simply increasing the value of $T_p/T$ to find a minimum
point in the undulating spectrum of $X_r$. On the other hand, the undulations are
not great for some pulse forms, so that nothing is to be gained by varying $T_p/T$
except by diminishing it below about 0.75, which is the value below which the
sine-like pulses exhibit a diminution in response toward zero. It is to be remembered that if $T_p/T$ is increased instead of diminished, at least one cycle of $X_i$
must be accepted, and this may not be acceptable for stress or other reasons.
   This finally leads us to the simplest measurement, that of the motion of the

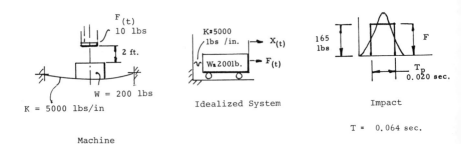

Machine          Idealized System          Impact

T = 0.064 sec.

Weight of 10 lb drops 2 ft;
Velocity, V=√(2gh) = 11.2 ft/sec;
Momentum,MV= $\frac{10(11.2)}{32.2}$ = 3.3 lb-sec

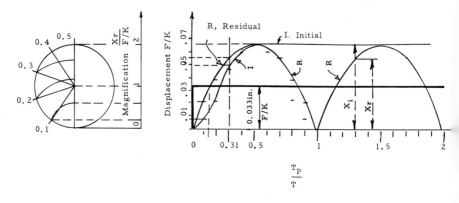

Fig. 23.3 Shock spectra for a rectangular step where F/K = 0.033. If the actual impact as shown has a duration of 0.035 sec, the equivalent rectangular duration $T_p$ is about 0.020 sec. Thus 0.020 F = 3.3, giving F = 165 lbs and F/K = 0.033 in. Since, $T_p/T$ = 0.31 the greatest initial response deflection is about 0.049 and the greatest residual is about 0.055, as shown. The spectra R and I give the greatest values of transient and steady displacement response $X_i$ and $X_r$ for a rectangular pulse of any relative duration $T_p/T$. For example if $T_p/T$ = 1.3, then $X_r$ and $X_i$ have the respective values of 1.65 F/K and 2 F/K as shown. The greatest response for any value of F/K is obtained from the dimensionless scale giving the ratio of maximum displacement to F/K. For rough computation any simple pulse may be replaced by a rectangle. Using several rectangles, any desired accuracy is obtained.

machine. A dial indicator may give an adequate reading and even separate the transient from steady displacement. If an alteration is then made in the weight or stiffness of the supported machine, it becomes apparent when an improvement is being effected. Common displacement, velocity or acceleration meters may be used for the same purpose, bearing in mind that, whereas the readings are more likely to be apparent rather than true, a diminution in reading is likely to indicate an improvement.

However, the cost of machine alteration is seldom so small that ill-informed alterations are permissible. So, it is generally necessary to measure further and determine the natural period T, the stiffness K and the form of the impact. A quick eye can often detect the value of T from an oscilloscope trace. However, if a storage-type oscilloscope or a pen oscillograph is available, the measurement is improved. Knowing the period, the static deflection is easily derived from the equation $T = 0.32\sqrt{S}$. It may be possible to measure K by adding weights to the machine and measuring the corresponding deflection, particularly if the spring is concentrated between the machine and the floor. This in turn gives W.

The form of the impact and its value are determined by attaching an accelerometer to the impact source. The hammer-like element may be hot or inaccessible, or it may have such a great drop as to make measuring too expensive or impossible. If this is so, an educated guess is made regarding the appropriate correction, based on the partial facts yielded by measuring the machine motion; that is to say, the amplitude F, the duration $T_p$ and the form of the impact must be estimated. The assumption that the form of the impact is nearly that of a half-sine is valid in many cases. From this, if one of the pair of F, $T_p$ is known, the other can be calculated. For example, as is shown in Fig. 23.3, the value of MV is 3.3 lb/sec from simple calculation and the duration of the impact $T_p$ is assumed to be 0.035 sec. The area under the half-sine curve is 3.3 because this 3.3 is the momentum change. This change must be equal to the impulse, which in turn is the area under the half-sine curve. Further, the area under a half-sine curve is about 2/3 that of the enclosing rectangle, so that $0.035 \times F \times 2/3 = 3.3$, which gives the peak force F as 142 lb. In some instances, by reversal of the phase-plane procedure, for example, it may be possible to estimate the pulse from the response curve. But to do this would take us beyond the scope of this introductory chapter.

In a forging operation wherein a die is closed to the point of generating flash, having compacted the forged metal, it may be expected that the peak force F of the impact increases, whereas the time duration $T_p$ decreases with each succeeding impact. In such an instance it is probable that the first impacts, because they last longer, may be the more troublesome. On the other hand, because a succeeding impact may, in fact, start at an instant of maximum velocity in the residual vibration of the preceding impact (velocity being in the same direction as the new impact force), the succeeding impacts may be equally troublesome, particularly as the peak force increases. Therefore, because the impact duration is variable, it is not possible to hold the residual vibration at a zero point, since $T_p/T$ must be constant to achieve this. Because it is not possible to achieve a satisfactory phase relation between the new impact and the old residual vibration, the only available approach to better isolation is to have $T_p/T$ significantly less than about 0.75, which is to say $T/T_p$ must be significantly greater than 1.3.

A consideration of the phase plane shows that phase relation is all important. The displacement of the machine due to one or more impacts depends greatly on impact duration and on the phase position of the system when impact begins. To achieve ideal phasing would require timing of the blow to within a few milliseconds, which timing must accord with variations in impact duration, all of which is, in practice, impossible. Or, to be more precise, phase control can hardly be justified as an economic possibility.

## 6. INSTRUMENTATION

The more refined the measurement, the more expensive the instrumentation and labor. Often, the investigator is restricted to very simple, unsuitable instrumentation. He may decide to make no measurement, and therefore achieves no result. However simple the instrumentation, it is usually possible to obtain valuable measurements if ingenuity is combined with a study of the instrumentation, the problem and the objectives. In any measuring effort the instrumentation errors must be assessed, usually in consultation with the maker.

/ The most suitable instrumentation is a combination of accelerometer system and high-speed oscillograph or storage oscilloscope. A tape recorder may prove valuable if its errors, such as those related to transient amplitude and phase, are known to some extent. Beyond these, much useful information may be obtained from a common accelerometer connected to a suitable oscilloscope. The accelerations typical of the impact (the hammer) are far greater in amplitude than those of the response (the machine). Impact measurement, in general, is more difficult than response measurement. Table 36 gives some of the features of suitable accelerometers.

Typical errors inherent in accelerometer systems are overshoot and undershoot (reading higher or lower than true), phase distortion and the superpositioning of the natural vibration of the accelerometer on the true waveform. In some instances correction charts, supplied by makers, must be used to cope with these errors. /

The nonspecialist, becoming aware of the limitations of instrumentation, may be inclined to *give up* and do nothing, or he may make no study of the measuring system at all and assume that all measurements are ideal. The best attitude is compromise. The significance of acceleration measuring is that if a body is accelerated by a mechanical force, then this force is directly proportional to the acceleration according to the formula:

$$\text{Force (lbs)} = \frac{\text{Weight of body (lbs)}}{32.2} \times \text{acceleration (ft/sec}^2)$$

TABLE 36
COMMERCIAL ACCELEROMETER CHARACTERISTICS

| PURPOSE | SHOCK | SHOCK | VIBRATION |
|---|---|---|---|
| Accelerometer type* | 4333 | 4336 | 4328 |
| Sensitivity, millivolts/g | 14–20 | 4–6 | 35–70 |
| Principal resonance, kcps | 45 | 125 | 18–25 |
| Maximum g | 10,000 | 14,000 | 500 |
| Frequency range, cps (10 %) | 0.5–15,000 | 0.5–40,000 | 10–10,000 |
| Weight, grams | 13 | 2 | 33 |

* Accelerometers made by Bruel & Kjaer, Cleveland, Ohio. Comparable accelerometers are made by Kistler Corp. and others.

## 7. IMPACT-CONTROL EXAMPLES

An example is given in Table 37.

Thus, while the dynamic deflection (peak) has risen from 55 to 124 mils, the transmitted dynamic load (peak) has fallen from 275 to 155 lbs. Transient loading is of no consequence because it is less than the residual when angle $\alpha$ is less than 180 degrees.

A further example (Table 38) shows the effect of increasing W to increase T. Thus, the dynamic load ratio of the revised/original transmitted force is still 0.57. While the dynamic deflection (peak) has fallen from 55 mils to 31 mils, the transmitted dynamic load (peak) has fallen from 275 to 155 lbs. The conclusion in this example is that the same dynamic load reduction is achieved whether $T_p/T$ be reduced either by decreasing K or by increasing W by equal ratios. However, when K is decreased the dynamic deflection is increased; whereas when W is increased, the dynamic deflection is decreased.

## 8. SOME CONCLUSIONS

The unwelcome transmitted vibration resulting from functional impact may be controlled in three ways: the use of statically generated force, the use of counterblow devices or the isolation of the machine. The cost of isolation must be weighed against the extra cost of counterblow machines. Static force generation is, in general, too costly or inappropriate as a solution (i.e., it may be too slow for forging).

**TABLE 37**
**RESULT OF INCREASING SYSTEM STIFFNESS**

|  | ORIGINAL | REVISED |  |
|---|---|---|---|
| Weight of machine, W | 200 | 200 | lbs |
| Rectangular impact force, F | 165 | 165 | lbs |
| Support stiffness, K (this is the revision) | 5000 | 1250 | lbs/in |
| Natural period, T | 64 | 128 | msec |
| F/K, displacement for F, statically applied | .033 | .132 | in |
| Impulse duration, absolute $T_p$ | 20 | 20 | msec |
| Impulse duration, relative, $T_p/T$ | .31 | .155 | ratio |
| Angular duration, impulse, $\alpha = 360\,T_p/T$ | 112 | 56 | degrees |
| Peak residual displacement $X_r = F/K\sqrt{2(1 - \cos \alpha)}$ | .055 | .124 | in |
| Magnification factor, $\sqrt{2(1 - \cos \alpha)}$ | 1.66 | .94 |  |
| Static load | 200 | 200 | lbs |
| Additional peak dynamic load, $KX_r$ | 275 | 155 | lbs |

The ratio of the revised/original transmitted force is 0.57

The master key for isolation is the ratio $T_p/T$, the ratio of pulse duration to natural period of the machine. In general this must be reduced significantly below about 0.75 for pulses approaching the half-sine form. A tabular example shows that reduction is often best achieved by increasing the machine mass rather than by decreasing the machine-support stiffness. And, conversely, if a decrease in natural period T were desired, for like reasons it may be better to increase stiffness rather than decrease mass. The phase (timing) of successive impulses has a large effect on the vibration that the impulses cause. But it is not practicable to control this phase relation as a means of nuisance control.

## TABLE 38
### RESULT OF INCREASING SYSTEM WEIGHT

|  | ORIGINAL | REVISED |  |
| --- | --- | --- | --- |
| Weight of machine, W | 200 | 800 | lbs |
| Rectangular impact force, F | 165 | 165 | lbs |
| Support stiffness, K | 5000 | 5000 | lbs |
| Natural period, T | 64 | 128 | msec |
| F/K, displacement for F, statically applied | 033 | .033 | in |
| Impulse duration, absolute, $T_p$ | 20 | 20 | msec |
| Impulse duration, relative, $T_p/T$ | .31 | .155 | ratio |
| Angular duration of impulse $\alpha$ | 112 | 56 | degrees |
| Peak residual displacement amplitude $X_r$ | .055 | 031 | in |
| Static load | 200 | 800 | lbs |
| Additional peak dynamic load, $KX_r$ | 275 | 155 | lbs |

# REFERENCES

Harris, C. and Crede, C. *Shock and Vibration Handbook*, 1st. ed. New York: McGraw-Hill Book Co., 1961.

Jacobsen, L. S. and Ayre, R. S. *Engineering Vibrations*. New York: McGraw-Hill Book Co., 1958.

Kittelsen, K. E. *Tech. Rev. 3*. Cleveland: Bruel and Kjaer Co. Inc., 1966.

# SYMBOLS

| t | Time | seconds |
|---|---|---|
| $X(t)$ | Displacement as a function of time | feet |
| $\ddot{X}$ | Acceleration | ft/sec² |
| $\ddot{X}_m$ | Maximum acceleration during a pulse<br>Note, If actual acceleration shows a maximum of 70 ft/sec², then $\ddot{X}_m = 2.17$ | g units |
| g | The acceleration due to gravity, 32.2 | ft/sec² |
| F | Force | lbs |
| $F(t)$ | A force which is a function of time | lbs |
| V | Velocity | ft/sec |
| K | Stiffness of spring | lbs/in |

# 24

# SHOCK AND IMPACT
# IN TRANSIT

*Michael P. Blake*

## 1. SCOPE

Here we briefly consider shocks that arise in handling and during the ride—on truck, aircraft, rail and ship. The treatment is from the point of view of the owner of the freight or transportation who seeks to control damage, rather than from the point of view of overall research. Some shock sources and levels are given, and simple mechanical instrumentation is described. Measurement units, categories, objectives and impact control are discussed.

## 2. INTRODUCTION

Loss-and-damage costs associated with transit amounted to $140 million in the year 1960, and they show an increasing trend.[1] Of the total damage en route, more than 60 percent related to packaged items and more than 15 percent to unpackaged items. Other causes of damage include improper handling, fire and accident.

Although much research has been devoted to measuring roads, rides and packages with the grand objective of improving the economics of freight transportation, success has been only partial because of the endless factors and uncertainties involved. Speaking statistically, research data can give useful information regarding the lowering of total cost of packaging and freight, where choices exist concerning different routes or vehicle types. The statistical approach satisfies the needs of a vast volume of shipping. Packaging and freight costs and routes are such that the sum total of these costs, including damage, is least. Some damage is then expected. It is, in fact, not possible to eliminate the possibility of damage.

However, many freight owners experience a need to improve the economics of their particular operation by refining their own statistics. This significant improvement is often the result of using simple shock measuring devices, whereas overall research requires the use of expensive transducers, recorders and analytical methods. Measurement is the basis of improvement. Simple instruments and simple methods can lead to most valuable results.

Measurement, for example, helps carriers select suitable trucks and routes, thereby achieving an optimum compromise between damage claims and cost of operation; it can help in evaluating suspensions, in pinpointing tire bubbles or unbalance as well as defective alignment, shock absorbers, springs or grabbing brakes. It can reveal the time and date of freight movement and delay as well as which carrier may be responsible for particular damage. Measurement is a great help in gaining control of the various related categories such as vehicle maintenance, allocation of responsibility, vehicle and route selection, package design and testing, criteria and standards and unknown causes of damage.

As noted, overall research is expensive. It labors under the burden that its results must be highly communicable. For this reason the measurements must reflect the phenomenon being measured with a known degree of accuracy. To achieve this, the cost of instrumentation and labor must often be increased far above what might otherwise be necessary. On the other hand, communicability is quite secondary when measurement is devoted to improving a particular operation. The instruments may be poorly calibrated and it may not be possible to tell whether velocity or acceleration is being measured or whether the measurement is average or peak. The main question is: Are the measurements profitable? With very simple instruments, they can be highly profitable. The applications are almost endless and increasing, and it is our purpose here to awaken the reader to some of the possibilities of measuring and to indicate how it is done.

## 3. IMPACT SOURCES

Impacts arise from the collision of rail cars, road surface, dropping of packages during handling and other sources, as illustrated in Fig. 24.1. The collision of cars during switching may be expected to be more damaging than the ride. Switching practice expects to limit the impact speeds to, say, six miles per hour. However, when perhaps 3000 cars are switched in a classification yard in one day, impact speeds up to 10 mph and greater do occur. Typical collision impacts lie in the range of 1 to 8-g in the longitudinal direction, and the vertical acceleration may be expected to be of the same order of magnitude. Several factors influence the impact felt in the car, such as the type of draft gear, the speed of impact and the load on the car. The impact felt in the package is, in turn, influenced by factors such as solid blocking or flexible bracing.

Grabbing of brakes or the taking up of slack between cars when a train starts also cause impacts. The severest impact is likely to result from dropping the package. If the package is quite light, there is a risk of its being thrown or dropped several feet. If it is beyond the strength of one or two men, it is probable that impact due to drop shall be minimal. On the other hand, impacts resulting from very small drops tend to be far greater than impact in transit due to the ride or switching of cars. An adequate estimate of drop impact may be made by assuming that the impulse is sinusoidal in form. The maximum acceleration $\ddot{X}_m$ expressed in units of g, the rise time $t_r$ in seconds and the drop height h in inches are then related as follows.

$$(\ddot{X}_m t_r)^2 = 0.013 \ h$$

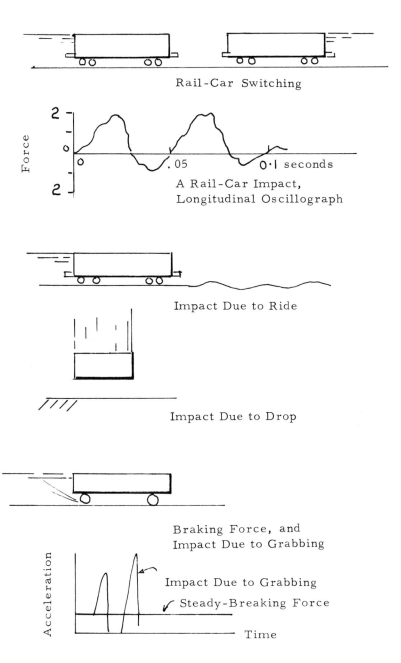

Rail-Car Switching

A Rail-Car Impact,
Longitudinal Oscillograph

Impact Due to Ride

Impact Due to Drop

Braking Force, and
Impact Due to Grabbing

Impact Due to Grabbing

Steady-Breaking Force

Time

Fig. 24.1   Impact sources

For typical packages, the rise time ranges from about 0.002 sec for an unpackaged container to 0.015 sec for a container cushioned with three inches of fibrous material. Taking a drop height of six inches and a rise time of 0.010 seconds, the maximum acceleration is about 25-g for these rather favorable conditions.

Prolonged braking and prolonged or very low-frequency accelerations, such as experienced in ships and aircraft, induce forces that are relatively small in comparison with the impactive forces discussed above. Because of the long duration, it is not possible to isolate the package content, which experiences the full force corresponding to the acceleration of the package. Forces due to braking may be reduced by increasing the braking time, but the slow forces in ships and aircraft cannot be reduced at all.

During the ride the package experiences impacts and vibration, and it is not possible to make a clear distinction between the two. Impact connotes a transient or occasional condition. The truck is likely to give the roughest ride, typically about 1-g and occasionally rising to about 4-g. Rail and aircraft figures are about one-half these values, and ships about one-quarter. These are illustrated in Fig. 24.2. As can be appreciated, braking decelerations are likely to be greater than traction accelerations and upward accelerations are likely to be greater than downward counterparts.

## 4. THE PACKAGE

Items that are called fragile are usually packaged. The presence of a container may invite careless handling and thus offset some of its value. Fragility may be expressed in terms of the maximum acceleration that the contents can endure. If the contents comprise delicate components that may resonate, this may have to be taken into account. Sometimes the fragility cannot be estimated, for example, if the item is so expensive that destructive testing cannot be justified. For this reason and because of unknown factors, package design is to an extent an art, guided by considerable scientific literature.

An idealized package is shown in Fig. 24.3c. Actual cushioning materials, such as latex hair, exhibit an increase in stiffness with increasing compression, and the permanent deformation is less than about 10 percent of the cushion thickness. Such a cushion is called elastic, whereas other cushions perform in a crushable (nonelastic) manner and must provide sufficient cushion thickness to absorb successive impacts. Analytical data is available giving the properties of cushion materials, making possible the analytical design of a package to meet specified requirements when these requirements are known. When they are not known the design may have to depend on trial and error. A requirement that the acceleration of the contents must not exceed a specified value for a specified drop is a useful criterion, leading readily to an acceptable cushion design by experiment or calculation. An impact measuring device attached to the contents during a drop test compares various cushions effectively. The drop-test criterion is an effort to simulate an impact severity unlikely to be surpassed during all phases of transit and handling.

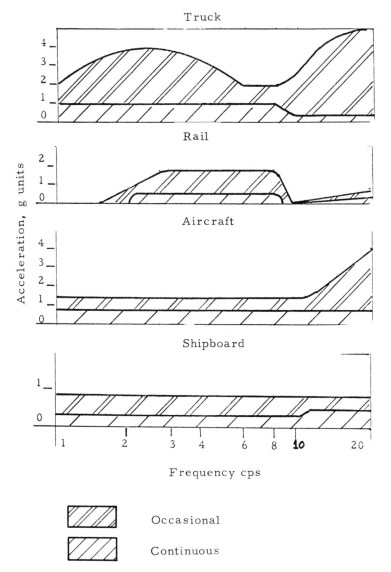

Fig. 24.2   Vertical vibration in transit (after L. C. Simmons and R. H. Shackson [1])

## 5. INSTRUMENTATION

For research work instrumentation must be chosen with care and knowledge in terms of the impact durations and amplitudes to be measured. It is necessary to know the probable errors and the possibilities of misinterpreting the observed measurements. The transducer is usually an accelerometer of suitable range,

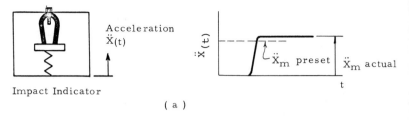

Impact Indicator

( a )

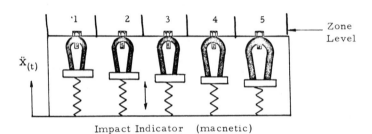

Impact Indicator   (macnetic)

Impact Indicator   (mechanical)

( b )

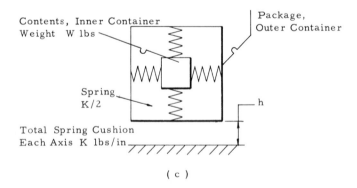

( c )

Fig. 24.3   Idealized package and impact indicator elements

Fig. 24.3a   Impact indicator reveals when the acceleration has exceeded
a given present level.

Fig. 24.3b   Magnetic indicator establishes a zone for maximum acceleration; mechanical indicator produces an absolute value.

Fig. 24.3c   Idealized package subject to impact

damping and response flatness. This is connected with an amplifier, filter and recorder or oscilloscope. The form of the impact is read-out, for example, on a storage-type oscilloscope that is triggered by the onset of the impact and then retains its display for one-half hour or more.

For the improvement of a particular operation, the task is less sophisticated. Any oscillograph with any accelerometer is likely to give measurements useful for comparison. Simple precautions, such as doubling the drop height to determine if the read-out increases proportionately, help to avoid gross misinterpretation. Some electronic measuring instruments are self-contained and read out on a low-cost meter a number that is proportional to peak acceleration. However, in

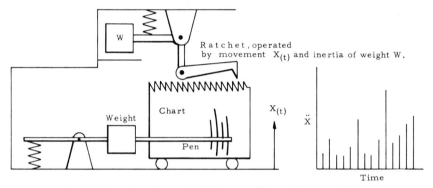

Ratchet Recorder: Records All Impacts Above a Minimum Value

(a)

Clock Recorder: Records All Impacts and the Time of Occurrence of Each

(b)

Fig. 24.4   Impact recorders, mechanical type
Fig. 24.4a   Ratchet recorder records all impacts above a minumum value.
Fig. 24.4b   Clock recorder records all impacts and the time of occurrence of each.

measuring shock in transit, the mechanical-type instrument is, perhaps, used more than any other.

Figure 24.3a shows a magnet and armature which separate when a particular acceleration is exceeded. Figure 24.3b shows a series of magnets of increasing strength and a mass-spring stylus writing on a stationary chart. In Fig. 24.4a the chart is moved by a ratchet actuated by impact, and in Fig. 24.4b the chart is moved by a clock that runs about one month. Comparison of the four types of instrument is given in Table 39.

## TABLE 39
## INSTRUMENT INDICATION

| INSTRUMENT STYLE AND FUNCTION | QUESTION ANSWERED BY INSTRUMENT |
| --- | --- |
| Fig. 24.3a Indicate, classify | Did at least one impact exceed a particular value? |
| Fig. 24.3b Measure, record | What was the greatest impact? |
| Fig. 24.4a Measure, record | What was the number of impacts above a particular value and the value of each impact? |
| Fig. 24.4b Measure, record | Same as Fig. 24.4a together with the time of occurrence of each impact. |

Type 3a is useful for simple monitoring of shipments, and 3b is useful for simple laboratory experiments. For adequate monitoring of packages in transit, Type 4b is used. Because the impact direction cannot be predicted, a three-directional version of Type 4b is necessary in order to give the shock components in each of three mutually perpendicular directions. Corresponding components are added vectorially by taking the square root of the sum of the squares. This is illustrated by Fig. 24.5 which shows the angular position of the line of action of the impact. The direction of the impact along this line is determined by noting the directions of the three components on the chart.

## 6. UNITS OF MEASUREMENT

Ideally, an impact is described by giving the time history of the acceleration. In practice, the maximum acceleration alone is often measured. This conveys much useful information and often proves to be hundreds of times less expensive than ideal measurement. The usual unit is the gravity unit, g. The mechanical instruments are considered to be acceleration meters. It is to be borne in mind, however, that they measure acceleration for the longer impacts and may tend to read in terms of velocity or displacement for very rapid impacts. This does not greatly impair their utility because the user is more interested in comparative readings than in absolute values. Readings are recorded in terms of apparent g.

Rail car impacts are sometimes discussed in terms of zones which is a tradi-

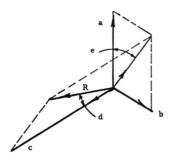

a, b, c, ... Mutually Perpendicular Components of an Impact

$$R = \sqrt{a^2 + b^2 + c^2}$$

Angular Directions

$$e = \tan^{-1} b/a$$

$$d = \tan^{-1}\left[\frac{\sqrt{a^2 + b^2}}{a}\right]$$

Fig. 24.5a The resultant R of an actual impact, from three mutually perpendicular components of an impact given by a three-directional impact recorder

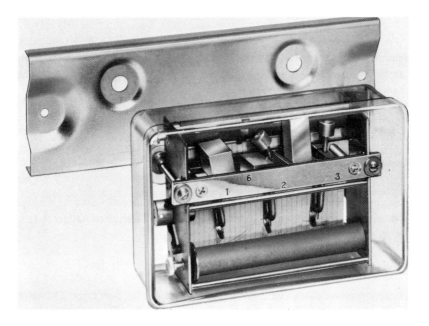

Fig. 24.5b Three-directional Impact-o-Graph

tional measure, although not as useful as true or apparent acceleration. Zones range from 1 to 5, the first two being considered normal and the remaining three rough. The corresponding speeds of impact range from about 4 mph to 14 mph.

## 7. MEASUREMENT OBJECTIVES

We consider some of the many possible objectives.

1. Allocation of responsibility

A refrigerator manufacturer experienced a situation of gross damage. The use of an impact recorder, Fig. 24.4b, revealed that the damage occurred in his own shipping department.

A record chart for a truck showed that the driver had taken an extended rest period and had then speeded to make up for the lost time, leading to severe damage of the cargo.

A record chart for a package alleged to have been shipped on Wednesday showed that it was not moved until Friday. Another chart showed that one carrier had ignored the direction, "This side up," and had placed the package on its side.

### 2. Maintenance

An impact meter attached to a truck can often indicate bubbles on tires, tire unbalance, grabbing brakes, faulty shock absorbers and other faults.

### 3. Classification of the ride

Prior to shipment of fragile packages, an impact recorder may be used to monitor vehicle or packages over several trips to determine the sources, value and number of impacts typical of the route. In this way better routes may be chosen or more care may be devoted to packaging.

### 4. Monitoring

It is often desirable to monitor valuable shipments, such as glass-lined tanks or other plant machinery. This is easily done by attaching an impact recorder to the item of interest.

### 5. Research

Recorders attached to steel pipes during transcontinental shipment revealed that each pipe rotated about fifteen times and was subjected to a serious swaying action. Damage elimination was facilitated.

## 8. IMPACT CONTROL

Impact definition states that it is brief, but it does not say how brief. The longer transient accelerations in vehicles and aircraft cannot be controlled with respect to isolating passengers or freight. The criterion that ultimately determines control of brief impacts is the ratio of impact duration to natural period. The affected system, such as a package or vehicle, comprises spring and mass. The mass may sometimes be increased with advantage, but it is more usual to soften the spring. Everyone knows that a package wrapped in a mattress is not as subject to damage as an unwrapped package. Some related analytical data are given in Chapters 10 and 11. These data, together with that given below, throw some light on the simpler analytical considerations. The data in these chapters relate to the impacts in transit, wherein the time-displacement relation may be fixed as regards the container. Data given below is concerned with the fixed impulse delivered to a container due to a drop.

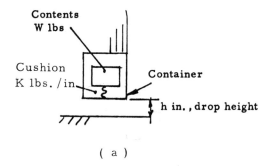

( a )

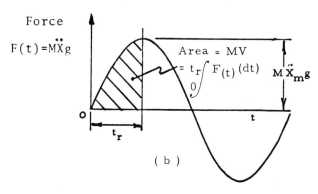

( b )

Fig. 24.6   Container subject to impact

Example.: Container falls from height h(inches) and strikes the floor with velocity V(ft/sec). Weight of contents is W lbs and mass M is W/32.2. Contents come to rest instantaneously at time $t_r$ after container strikes the floor, contents then vibrate sinusoidally, if undamped and linearly cushioned, as shown by the sine curve. Hence

$$t_r = \frac{\pi}{2}\sqrt{\frac{W}{386\ K}}\ \text{sec} \qquad (\ddot{X}_m t_r)^2 = 0.013\ h$$

$\ddot{X}_m$ is the acceleration in units of g. If $\ddot{X}_m$ is 1 then acceleration is 1-g or 32.2 ft/sec$^2$
Force, F lbs $= Mg\ddot{X}_m = W\ddot{X}_m$; F is the force, in lbs, exerted on the contents by the cushion.

The contents of a package dropped flat from a fixed height experiences a fixed impulse. By softening the stiffness of the system, the impulse duration is increased. The overall rise-time of the impact $t_r$ is defined here in terms of a sinusoid as one-quarter of the natural period T so that $T_p/T$ has the fixed value of ½. Because for the drop the impulse is fixed and not the maximum force, the maximum force may be diminished by increasing T, which also increases $T_p$. The impulse is fixed because the velocity of impact is determined by the drop height. The momentum destroyed during the rise-time is MV, where M is W/32.2. The change in momentum is equal to the impulse, where the latter is defined as being the time integral of the force acting on W. Thus

$$\int_1^2 M\ddot{X}\, dt = M(V_2 - V_1).$$

For a sine pulse the area under the curve is $2/\pi$ times the area of the circumscribing rectangle of duration $t_r$, leading to the relation

$$\int_0^{t_r} M\ddot{X}\, dt = 0.64\, M\ddot{X}_{max} t_r = MV.$$

$V = (2gh/12)^{1/2}$, where h is the drop height in inches. Expressing acceleration in g units, we have

$$0.64\, \ddot{X}_m\, g\, t_r = \frac{(gh)^{1/2}}{6}$$

$$\text{or} \quad (\ddot{X}_m t_r)^2 = \frac{gh}{6(0.64g)^2}$$

$$\text{or} \quad (\ddot{X}_m t_r)^2 = 0.013\, h$$

which is to say the maximum acceleration in g units multiplied by the sine rise time in seconds, all squared, is 0.013 times the drop height in inches. The value of $t_r$ may be derived from measurement if instrumentation is available or from the literature. It may be computed assuming a sine pulse and remembering that $t_r$ is one-quarter of the natural period T. Thus

$$t_r = \frac{\pi}{2}\sqrt{\frac{W}{386K}}$$

where $t_r$ is the rise-time in seconds, W is the weight of the package contents in lbs and K is the spring stiffness in lbs per inch deflection.

Supposing K is divided by four to make a more resilient cushion, then $t_r^2$ is multiplied by four; $\ddot{X}_m$ (the maximum acceleration in units of g, which determines the maximum force $W\ddot{X}_m$) is divided by two. In effect when K alone is varied for h constant:

$$\ddot{X}_m = C(K)^{1/2} \text{ where C is a constant.}$$

# REFERENCE

1. Kennedy, R. Bulletin 31, Section 2. In *Proceedings 13st Symposium for Shock Vibration and Associated Environments*. (See also L. C. Simmons and R. H. Shackson, Bulletin 31, Section 2.)

# BIBLIOGRAPHY

Harris, C. and Crede, C. *Handbook of Shock and Vibration*, 1st ed. New York: McGraw-Hill Book Co., 1961.
Kittelsen, K. E. *Tech. Rev. 3.* Cleveland: Bruel and Kjaer Co., 1966.

# SYMBOLS

P signifies the peak (semi-amplitude) value of a vibration

PP signifies the peak-to-peak (complete amplitude) of a vibration

gP peak acceleration in units of 1-g, i.e. units of 32.2 ft/sec$^2$

IPSP velocity, inches per second peak

IPS$^2$P acceleration, inches per second per second peak

# 25

# SOIL AND FOUNDATIONS

*Michael P. Blake*

## 1. SCOPE

An extremely condensed review is given of a few facets of this large and growing subject. The discussion is written under the headings of sources, concepts, purposes, general principles, isolation principles, instrumentation and case histories. Most of the discussion is devoted to nuisance-type steady vibrations in the soil.

## 2. INTRODUCTION

The soil and other components of the earth behave more or less elastically and have the capacity to transmit vibration and impact from such sources as earthquakes, explosions, the breaking of sea waves, traffic, reciprocating compressors, internal combusion engines, hammers, punches, vibratory conveyors and other machinery. Another class of vibration arises when devices such as radar or telescope towers are subjected to their own more or less sudden forces generated by tracking action and so oscillate and reflect a nuisance upon themselves, rather than transmit a significant vibration to other receivers. Of course, these towers can be affected as the receivers of vibration from outside sources. In order to design such towers, to design foundations that will not deteriorate, to control the level of transmission and reception, to avoid soil deterioration, to design buildings that will withstand earthquakes and explosions it is necessary to devise theories of soil behavior and to make measurements. Considerable attention has been devoted to theory and measurement in recent years and, also, to the dynamic behavior of structures.

   The earthquake is the most significant of the natural vibrations. It is not an enduring vibration; rather, it is a group of transient events. Despite seemingly small peak accelerations of 0.5-g, it may devastate all that lies in its path. Professor R. E. Richart[1] states that the response spectra are quite often used to indicate limiting values of displacement, velocity and acceleration generated within a structure from earthquake excitation; and that quite often the velocity criteria are more significant than the acceleration. On the other hand, the greatest steady vibrations in the soil are not as severe as those measured in machines, although the average is about the same as regards displacement and, perhaps, less as regards acceleration. This is because frequencies in the soil are typically less than those generated by cavitating liquids or rolling bearings. For the latter, accelerations may amount to perhaps 25-gP at 4000 cps. Usually such machine vibration is not especially noticeable, whereas the relatively low accelerations of earthquakes literally shake the world.

All of the usual concepts of vibration are used in the consideration of soil, together with some that are not, perhaps, significant in mechanical vibration, such as the three types of wave transmission: the Rayleigh wave, the shear wave and the P wave. Internal damping capacity and shear modulus are of particular interest.

The purpose of measurement is to gain control, through either design or later modification, of the soil or foundation. In general the vibration source is accepted as being beyond modification.

Although a well-developed body of theory exists, its application is, to an extent, frustrated by the endless variability: of soils, of their state of compaction, of rock structure, of voids and faults of various kinds.

Electronic instrumentation has proved to be an incomparable aid in assessing specific cases and in providing the kind of confirmation that theoretical hypotheses require for their refinement and further advance. The civil engineer in this field, like his brothers in acoustics, mechanical vibration or even in parts of medicine, must acquire a working knowledge of electronics that seems surprisingly broad. Whereas a small part of the instrumentation, such as that used for determining the dynamic or complex shear modulus, is unique, the instrumentation, in general, is the same as that used commonly for other purposes.

Measurement on the whole presents no difficulty, particularly on the surface, although measurements of subsurface vibration or of earthquakes, where the apparatus is kept in readiness for a score of years, is expensive rather than difficult.

The task of isolation arises continually, either as regards isolating the source or the receiver. Some modern devices, such as the telescope, microscope or zone refiner, for example, may be so sensitive that isolation and choice of site are part of the original design. Despite all advances, isolation of existing installations may provide as many failures as successes.

## 3. SOURCES

These include earthquakes, sea waves, explosions, traffic, reciprocating and impactive machinery, functionally vibratory machines such as the vibratory conveyor, and the self-generated vibration of the radar, caused by transient accelerations. An example of an earthquake is given in Fig. 25.1, together with an example of soil vibration transmitted by a large horizontal compressor of the slow-speed reciprocating type. An earthquake is propagated over thousands of miles, and the oscillograph may be expected to show less of the higher-frequency

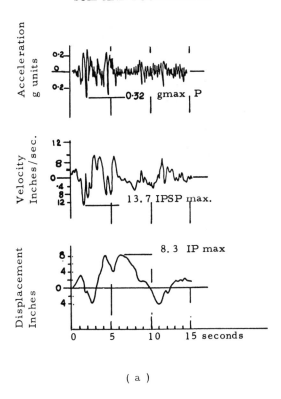

( a )

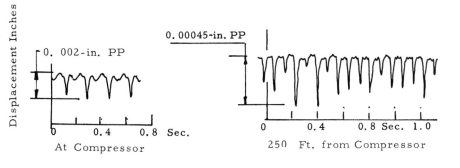

( b )

Fig. 25.1   Soil motions due to earthquake and compressor

Fig. 25.1a   An example of a strong motion seismograph, horizontal
motion redrawn from R. V. Whitman [2] (see also N. M. Newmark [3])

Fig. 25.1b   Horizontal soil displacement at the surface due to 327 rpm,
2000 hp, horizontal reciprocating, 5-cylinder compressor

components than of the lower as the distance increases. The duration ranges from the order of, perhaps, one second to several minutes.

## 4. CONCEPTS

Concept is distinguished from principle in the same rough way that philosophy is distinguished from science. A prime concept is that of classification: it is desirable to say whether a condition is dangerous or severe or harmless. To classify, a measurement is compared with some kind of standard. The measured condition may be considered against the broad classes: functional, beneficial, accidental, symptomatic, normal or nuisance. Respective examples are the vibratory conveyor, the soil compaction machine, the reciprocating compressor, the change in phase or harmonic content noticed at a crack in a foundation, the omnipresent slight earth vibrations and the vibration that leads to physical, physiological or psychological nuisance.

The standards arise from physical considerations, such as soil stress or the malfunction of a particular, sensitive machine. Measurement and experience are our guides. Standards also arise from the complaints of persons. This is a far more difficult category because in practice the reaction of the person is psychological rather than physiological and, therefore, unpredictable. The eminent Scottish engineer, Nasmyth, made engines in the same building in which wineglasses were made. Nasmyth, as the transmitter, felt no nuisance; but his neighbor did. The severity of the vibration then depends on the point of view. The receiver, unless he identifies himself with the interests of the transmitter, may be supersensitive and beyond placation, particularly as he will tend to expect and seek out and tune his receptive apparatus to its utmost, at which it may easily detect motions in the range 0.00005 in peak-to-peak, depending on frequency. The receiver tends to be more tolerant and to identify himself with the transmitter if he considers the source activity to be justified rather than arbitrary.

The concepts illustrate the various ramifications of soil vibration and are more valuable during the design activity than later. For example, measurements before installation of machinery and periodical later measurements may help in exonerating a suspected source or in focusing the blame. Another conventional concept is concerned with the definition of an order of magnitude of severity. There is a general agreement that one condition is the same as another except when separated by a factor of about three to one. Inasmuch as some choice exists here, it is preferable to adopt a convenient number. Existing preferred numbers appear to offer no convenience, and this writer prefers and suggests the use of the square root of ten, approximating an rms power-like change of ten decibels. Besides the empirical standard, there is the arbitrary or conventional standard that determines whether measurements should be given in microns or mils, or decibels, or rms or peak, etc.

## 5. PURPOSES

Basic measurements connected with research provide data for design through the general evaluation of elastic constants, modes of propagation and other data. Applied measurements are used for classification purposes, for identification of type and source, for analysis or for the monitoring of a given situation. They may be used to evaluate a change in condition or an intentional correction of some kind or for purposes of geological exploration.

## 6. GENERAL PRINCIPLES

The notions of mass, elasticity and damping arise in the vibration of soil and foundations just as they do in mechanical vibrations. However, soil vibration is distinguished in that it connotes great distances, endless variations in the medium, several wave paths and modes of transmission as well as a comparatively delicate medium having a complex mathematical model, even when considered homogeneous. The principles are, therefore, concerned with a variety of detail and uncertainty, going far beyond the relatively simple dashpots and springs that typify mechanical vibration systems.

Three paths of propagation are important and to some extent in the following order. The surface (Rayleigh) wave, often carrying most of the energy along a cylindrical front with a particle motion that is a retrograde ellipse at the surface, is first. The body waves which include the shear (S) wave and the compression (P) wave, next. The body waves have a spherical front. The particle motions are transverse for the S wave and push-pull or longitudinal for the P wave. All waves experience geometrical and hysteretic attenuation along the path. The geometrical attenuation results from encountering more mass. It is analogous to the attenuation of illumination as the distance increases and is called, perhaps inaccurately, geometrical damping. The hysteretic attenuation (called material damping) is a true damping effect, resulting from the internal energy losses. Although the material damping is apparently viscous when considered in terms of its measured effects, it is not considered to be the result of viscous behavior. Several features of the real earth distinguish it from an ideal medium and are revealed by measurement, such as curvature, layering and voids.

As an example of measuring surface waves for steady vibration, Fig. 25.2 illustrates some of the principles involved. The task here is to measure wavelength and frequency and so deduce velocity from the relation $V_r = fL_r$, where $V_r$ indicates velocity in ft/sec, f indicates cps and $L_r$ indicates wavelength. It is

**TABLE 40**
**SILTY FINE SAND, SURFACE WAVE VELOCITY**

| FREQUENCY f CPS | WAVELENGTH $L_r$ FT | PHASE VELOCITY $V_r$ FT/SEC |
|---|---|---|
| 500 | 0.7 | 350 |
| 350 | 1.1 | 385 |
| 250 | 1.7 | 420 |
| 100 | 5.4 | 535 |

to be borne in mind that velocity can be frequency dependent for a given formation. This is exemplified by experimental data of R. D. Woods,[2] Table 40.

The longitudinal wave-propagation velocity or longitudinal phase velocity,
denoted by $V_r$, is analogous to the apparent velocity of ripples on a lake as they
move across it. The particle velocity is reflected to an extent by the bobbing of a
floating cork. For a particular frequency, phase velocity depends only on the
medium, whereas particle velocity depends also on the lateral amplitude (stress).
This is illustrated in Fig. 25.3.

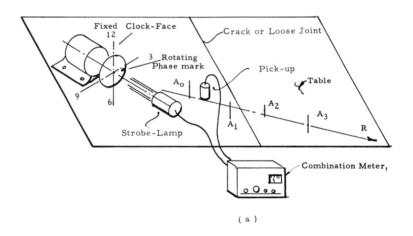

( a )

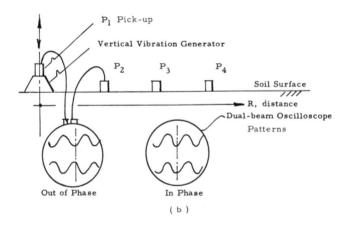

( b )

Fig. 25.2   Phase measuring surface waves

Fig. 25.2a   Small, unbalanced, 3600 rpm motor generates vertical
vibration. Strobeflash is synchronized with vibration and fires when
displacement is zero (maximum velocity).

Fig. 25.2b   Illustrating use of the oscilloscope

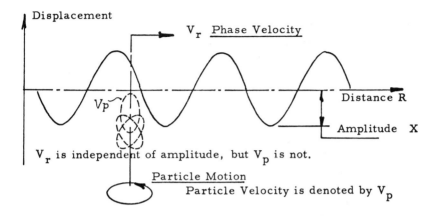

Fig. 25.3    Phase and particle velocity for surface wave

Reverting to Fig. 25.2, an instructive experiment is illustrated which may be arranged on a tabletop as well as on the soil in the field. This experiment reveals some of the principles and measuring techniques of interest. A tabletop, being more or less homogeneous, behaves only somewhat like soil but very much like a foundation. A small motor, 1/15 hp with perhaps thirty centimeter-grams of unbalance rotating at 3600 rpm, is placed on the table to induce a vibration that is mainly vertical. A phase mark is made on the unbalanced wheel, and a fixed clock face is provided for reference. A common vibration meter-analyzer-stroboscope combination instrument is used to measure vertical vibration along a radius R, by moving the pickup. At any location $A_o$ the meter can give three measurements: vibration amplitude, harmonic content and phase. For example, at $A_o$ the strobe light may illumine the phase mark at three o'clock. As the pickup is moved to $A_1$ and $A_2$ at a larger radius, the phase mark will rotate. Finally, if the table is big enough and the frequency high enough, the mark will return to three o'clock. The distance between this pickup location and $A_o$ is one wavelength. Since $V_r = f L_r$, $V_r$ is easily deduced, which is the phase velocity 60 $L_r$. For a pure sine wave, the vertical component of particle velocity may be estimated from the vibration displacement amplitude. The peak value of this velocity may be estimated from the relation $V_p$ (inches per second) is equal to displacement amplitude (inches peak) multiplied by frequency in radians/sec. Furthermore, a crack or loose joint in the table is likely to alter the pure sine form of the wave by the addition of second or third harmonics of the order of 10 to 40 percent of the fundamental amplitude or by the addition of a spike in the waveform, as measured vertically. An abrupt phase change may be expected at the crack. These observations are useful when considering soil and foundations.

Other phase measuring instruments are also used and must be used where there is no easily seen index for the stroboscope flash. For example, in the lower part of Fig. 25.2, a dual-trace oscilloscope is used. Common oscilloscopes are adequate for frequencies down to about 20 cps, and below that a storage oscilloscope or two-channel oscillograph is desirable to enable the eye to estimate phase.

The foregoing notes are indicative of practical measurement and the tabletop is an instructive, convenient kind of experiment. It is to be borne in mind however, as Professor Richart remarks, that the wavelengths measured on a tabletop may well represent the structural vibrations of the tabletop and thus would depend on its geometry, rather than representing only the Rayleigh-type wave which propagates within the surface of a mass.

The shear modulus and other properties have been successfully measured by exciting a column of soil, perhaps 4 cm in diameter and 25 cm long, encased in a membrane, either torsionally or longitudinally. This method is called the resonant column test. The free vibration decay curve is shown in Fig. 25.4. The logarithmic decrement varies widely, depending on material and test condi-

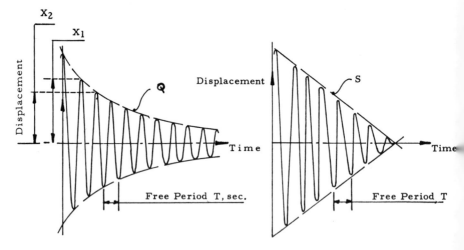

a. Free Vibration Decay of Soil, Resonant-Column Test. The Curved Asymptotic Envelope, Q, is Typical of a Viscous Effect.
Friction Force = $C\dot{X}$
Logaritmic Decrement:

$= \mathrm{Log}_e\ (X_1/X_2)$

b. Free Vibration Decay with Coulomb Damping. The Straight Envelope, S, is Typical of Dry-Friction-Response, $= +-\ C$

Fig. 25.4   Comparison of viscous effect (typical of soil) and dry friction effect in vibration

tion, and values of the order of 1 to 20 percent may be expected. The specific damping capacity is defined as the ratio of energy absorbed per cycle to the potential energy at greatest displacement in that cycle. The coefficient of attenuation, indicating the attenuation due to internal energy loss, is defined as being proportional to logarithmic decrement by frequency, divided by phase velocity.

Poisson's ratio exhibits a wide variation and affects or is affected by other parameters. When its value lies in the range 0.25 to 0.5, the phase velocities of Rayleigh and shear waves may be expected to exhibit a respective ratio exceeding 90 percent, increasing as Poisson's ratio increases.

## 7. ISOLATION PRINCIPLES

In general, the Rayleigh wave may be expected to attenuate less than the body waves as the distance from the source increases. In the following the Rayleigh wave, being the more likely to be significant, is in mind. Isolation procedures may be considered under three classes and under two aspects of each class, as follows.

### Classes

1. Wherein advantage is taken of attenuation with distance from a source (distance attenuation).

2. Wherein the receiver is most favorably located at a fixed distance, by use of a deeper foundation or by resting a foundation on the most favorable geological formation (location attenuation).

3. Wherein a barrier, such as a trench or spring system, is interposed between the source and receiver (barrier attenuation).

### Aspects

1. Active attenuation (source attenuation, transmitter attenuation, total environment attenuation) wherein the means of attenuation is located at the source.

2. Passive attenuation (receiver attenuation, fractional environment, or selective environment attenuation) wherein the means of attenuation is located at the receiver.

A variety of synonyms and near-synonyms is used to take account of the fact that the terminology is not well settled and that some of what is accepted is questionable. For example, geometrical damping, which is well accepted, is not damping in the usual sense of the word. The terminology used in this article is chosen for its clarity and precision.

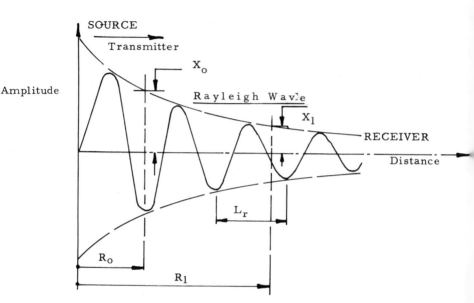

Fig. 25.5 Isolation principles, distance attenuation. Total attenuation $X_1/X_o$, is due to the combined effects of geometrical and hysteretic attenuation. Geometrical attenuation (geometrical damping) is given as: $X_{1g}/X_{0g}$ $= (R_o/R_1)^{\frac{1}{2}}$. This somewhat typifies a fixed frequency condition f and fixed soil condition s. For variable conditions, total attenuation is determined by measurement and may be written as, $X_1/X_o = f\,(R_o/R_1, R,f,s)$, where $f$ indicates *function of.* In practice geometrical attenuation is greater than hysteretic.

The three classes are illustrated in Figs. 25.5, 25.6 and 25.7. For distance attenuation existing theory seems inadequate. There are two contributing components: geometrical and hysteretic. The geometrical effect is greater than the hysteretic and is given as being the inverse square root of the ratio of the distances from the source (see Fig. 25.5) according to the equation

$$\frac{X_{1g}}{X_{0g}} = \left(\frac{R_o}{R_1}\right)^{1/2} \text{ (for Rayleigh wave)}$$

This formula is only a close approximation since it takes no account of the wave frequency and it implies an increase of attenuation without limit as $R_o$ approaches zero. A combined expression for geometrical and hysteretic attenuation, after Bornitz,[5] gives

$$\frac{X_{1g}}{X_{0g}} = \sqrt{R_0/R_1 \cdot e^{-(R_1-R_0)\alpha}}$$

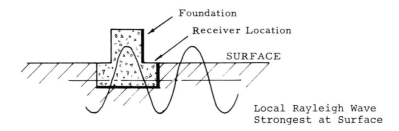

a.  Foundation On Surface
    Elevation View

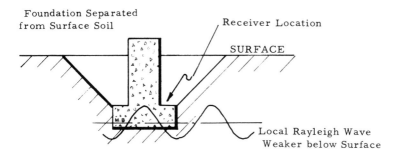

b. Foundation Below Surface, Same Geological Formation
   Elevation View

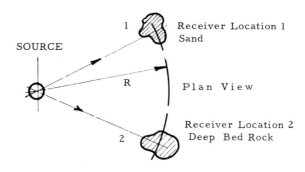

c. Isolation Improved by Moving Location from 1 to 2

Three Diagrams drawn from point of view of  Passive Isolation,
(Receiver Attenuation): They may also be considered from point
of view of Active Isolation (Transmitter Attenuation).

Fig. 25.6  Isolation of a Rayleigh wave by relocation of the receiver at
the same radius

where $\alpha$ is a coefficient having the dimension of 1/meters.

The combined attenuation is readily measured by moving the vibration pickup farther from the source, as in Fig. 25.2. Field measurements in fine silty sand by R. D. Woods,[6] for values up to $R_1 = 6 L_r$ (where $L_r$ is the Rayleigh wavelength), show a total attenuation in this range increasing from about $(R_o/R_1)^{0.5}$ to about $(R_o/R_1)^{0.8}$ as $R_1$ increases from about $L_r$ to 6 $L_r$, where $L_r$ is 2.25 ft and frequency is 200 cps.

The terms hysteretic damping and material damping appear to be accepted. The latter is perhaps the better, inasmuch as the friction is inter-particle, rather than inter-molecular.

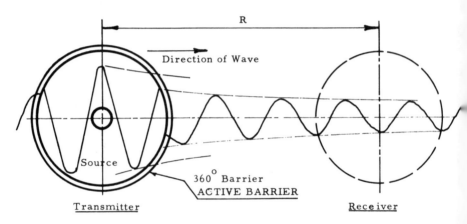

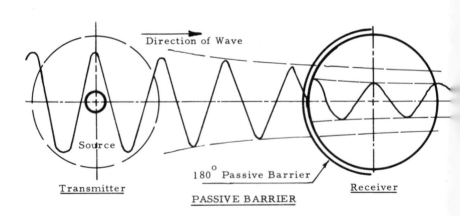

Fig. 25.7 Barrier isolation of a Rayleigh wave for a fixed distance R. Active barriers may be circular or straight, usually void trenches, subtending an angle from 90 to 360 degrees at the source. The 360 degree active barrier improves all of the environment. Passive barriers may be circular or straight, usually void trenches, subtending not more than 180 degrees at the receiver. They improve a small part of the environment.

In some, and perhaps in most practical cases of troublesome soil vibration, the frequency lies in the range of 1 to 10 cps. The author has measured not less than four cases wherein these frequencies caused trouble to householders and office workers and office machines at distances up to about 1500 ft. Distance isolation is not usually an acceptable solution after design is complete. In the cited cases the evidence suggested that neither distance, location nor barrier attenuation could be affected at acceptable cost and that the source itself had to be modified. In one instance attenuation over a distance of 250 feet was observed to be zero.

Regarding location attenuation, the technique is evident from Fig. 25.6. However, for the mentioned low frequencies, below 10 cps, there appears to be little data available to indicate the order of magnitude of the attenuation that may be achieved.

Regarding barrier attenuation, the same remarks apply. This attenuation is illustrated in Figs. 25.7 and 25.8. Woods[6] has exhaustively measured the effects of void-trench barriers in silty fine sand for frequencies about 200–350 cps and over distances of about six wavelengths. In this range his results suggest that effective active and passive attenuation may be achieved with trench depths not less than about 0.6 and 1.5, respectively, where given trench depths are expressed in terms of actual depth divided by Rayleigh wavelength. Sheet walls, such as aluminum sheets buried in a vertical plane, are not as effective as void trenches.

In mechanical engineering the frequencies above about 15 cps and impacts lasting less than perhaps 60 milliseconds may not be expected to offer too great a difficulty as regards isolation. But as the natural period of the vibration or the duration of the impact increases, isolation by common means becomes difficult or impossible due to the unstable jelly-like spring elements that are required. The same may be expected in the behavior of the isolation procedures for soil and foundation. The required distance for attenuation may be impossibly great and so also may be the trench depth. At the moment it seems that almost no data are available for low frequencies and long distances. About half of the reported cases of trench digging have been failures. In this connection it is to be remembered that failure must be defined, and this writer prefers to classify a barrier or other device as a failure if it does not achieve an average amplitude reduction factor (amplitude at the site with a trench divided by the amplitude at the same site before a trench is dug) of one order of magnitude. He further prefers to define one order of magnitude as being 10 dB, which is an arithmetical ratio of $\sqrt{10}$ which is 3.16 or nearly $\pi$.

Ward and Crawford[7,8] have studied the response of buildings from about ten to forty-five stories, and so have many others. But the response and dynamic characteristics of common brick and frame dwellings seem to be unknown and are often important. It may be stated in conclusion that there is much to be learned. In one case where a trench about 15 ft by 15 ft depth in cross section was dug, against the writer's advice, it provided no attenuation 250 ft from the source at frequency of 10 cps.

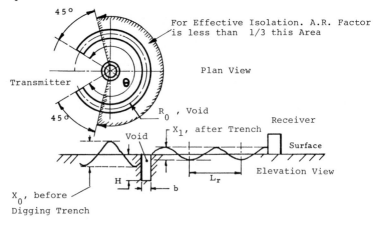

a. Transmitter Attenuation, Active

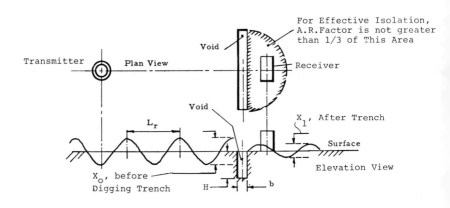

b. Receiver Attenuation, Passive

Fig. 25.8   Barrier attenuation of Rayleigh waves

## 8. INSTRUMENTATION

Typical instrumentation is listed in the following.

a) Oscilloscope, dual beam is the most useful

b) Oscilloscope, storage type that holds display for about an hour

c) Oscillograph, pen type, multi-channel from 0 to 125 cps, and galvanometer, light-beam type, to several thousand cps

d) Phase meters and phase marking devices

e) Stroboscope, tuned with a pickup signal

f) Common velocity pickups, natural frequency of 15 cps and sensitivity of about 500 mv peak per in/sec peak, with various degrees of damping

g) Common piezoelectric accelerometers for vibration and shock, to 10,000-g
h) Camera
i) Exciter (vibration generators), steady vibration, both mechanical and electro-mechanical
j) Exciter, impactive
k) Geophone velocity pickups
l) Low-frequency velocity pickups
m) Servo accelerometers
n) Other instrumentation

In the above list, all but the last five items are more or less common, whereas the last five are, to an extent, peculiar to seismic measurement.

**Exciter.** The electrical type is usually vertical and comprises an armature and coil which is excited, usually in a sinusoidal fashion, by a signal generator. It is

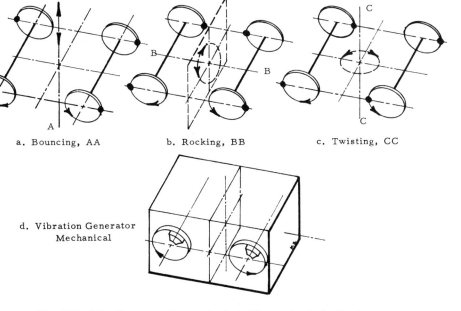

a. Bouncing, AA          b. Rocking, BB          c. Twisting, CC

d. Vibration Generator
   Mechanical

Fig. 25.9  Vibration-generator, mechanical. The mechanical vibration-generator has two rotors geared together so as to rotate in opposite directions. At each end of each rotor is an unbalanced weight, indicated by heavy dots in Figs. 25.9a,b,c. In Fig. 25.9d the weights are shown in the vertical position. The generator has a power source and tachometer generator. Its operation may be varied in three separate ways as follows:
1. *Mode.* 9a generates a force component AA, only. 9b generates a rocking about BB, only. 9c generates a torsion around CC, only.
2. *Speed.* Speed variation intensifies or mitigates the energy in each mode for a given unbalance.
3. *Unbalance.* By altering this quantity, the energy in each mode is varied for a given speed (rpm).

useful for frequencies of 10 to 1000 cps. It is like a large reversed velocity transducer and does not involve much labor in use. The mechanical type is used from about 30 cps to near zero; and it generates vertically, torsionally or in the rocking mode. The weight of these range from about twenty-five pounds to several tons. The usual type is shown in Fig. 25.9 and has been used to deduce the damping, spring and other constants for building structures and soils. Impactive exciters include drop hammers, explosive caps and explosive charges.

The mechanical vibration generator has two rotors, geared together so as to rotate in opposite directions. At each end of each rotor is an unbalanced weight, indicated by heavy dots in a, b, c. In d the weights are shown in the vertical position. The generator has a power source and tachometer generator. Its operation may be varied in three separate ways, as follows.

1. Mode. (a) generates a force component AA, only. (b) generates a Rocking about BB, only. (c) generates a torsion around CC, only.
2. Speed: speed variation intensifies or mitigates the energy in each mode, for a given unbalance.
3. Unbalance: by altering this, the energy in each mode is varied for a given speed (RPM).

**Geophone.** This word usually connotes a small, expendable velocity transducer of the kind that may be buried by the dozen for exploration work. Descriptive data are given in Figs. 25.10 and 25.11. The basic element varies as regards natural frequency and sensitivity. It is always encased in some suitable protective cover and costs about $15. Small geophones are also made so as to be pressure-sensitive.

**Low-frequency Velocity Pickup.** These are usually vertically-operable. The natural frequency ranges down to about $\frac{1}{2}$ cps with outputs up to about 10 volts per inch per second, and weight up to about 20 lbs and cost up to about $800. The price at 2 cps is about $350 and increases sharply thereafter. The devices are manufactured, for example, by the Geo Space Corp. (model HS-10, 2 cps, about $300) and Electro-Tech (model EV-17, 1 cps, about $400).

**Servo Accelerometer.** These have largely replaced the older piezoelectric and strain-gauge types, and they are usable for frequencies from 0 to 200 cps and capable of resolution to the order of one micro-g (one millionth-gP). The price ranges from $400 to $2000. An example of one is model 304, the Kistler Co. These transducers are new in the sense that they successfully combine known transducing elements and adequate circuitry to wipe out the position error of the seismic mass so that the acceleration is measured in terms of restoring force, which is in turn measured by its voltage or current analog. For example, see Fig. 25.12. This accelerometer is extremely sensitive and has the advantage of zero-frequency capability, which in turn means, among other things, that a reliable calibration-signal change of 2-gPP is always available when the accelerometer is positioned with an axis vertical and then turned upside down. Because the sensor does not read out, it need not be displacement linear. This allows

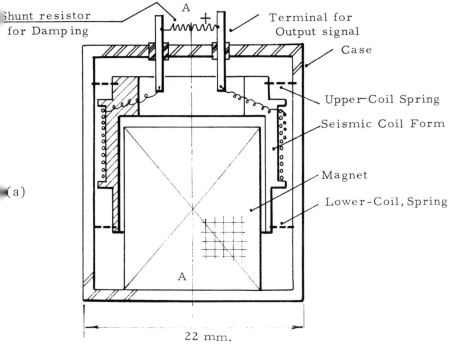

The geophone is a velocity-seismic, self-generating transducer.
Sensitive axis: AA.
Weight: say 55 grams.
Displacement limit: Bottoms out at about 0.030 in PP.
Operation: The coil is seismically suspended on flat springs, spiral, and is free to move on axis AA.
Natural frequency: about 15 to 30 cps.
When vibration is applied to AA an alternating voltage is generated at terminals, and is closely proportional to instantaneous velocity when vibration frequency exceeds natural frequency.

Damping: electrical, by means of Shunt Resistor.

Purpose: measures soil vibration, low-cost transducer.

(Based on Geospace Corporation, Type HS-J)

Fig. 25.10  Velocity transducer element (Geophone-Geo Space Corp. Type HS–J) (a) The Geophone—a velocity-seismic, self-generating transducer.

the use of elements, such as variable capacitors, which may be quite unsuitable for final read-out. The control signal is relatively large and easily read and must be force linear.

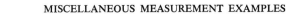

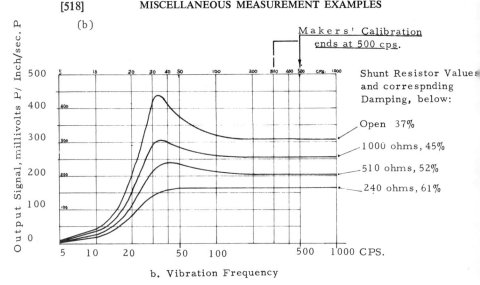

b. Vibration Frequency

See Fig. 25.10a for Physical Details.
Model K, Natural Frequency 28 cps., Coil Resistance 215 ohms
Open Circuit Damping 37%, Type HS-J, Seismic Detector
Re-drawn from Geo-Space Corp., data.

(See legend for Fig. 25.10(a))

Fig. 25.10   Velocity transducer element (Geophone-Geo Space Corp.
Type HS–J). (b) Velocity transducer, calibration data

**Other Instrumentation.** Several laboratory and field devices are in use. Notable among these are the machines that generate torsional or linear vibration for measuring laboratory samples. The resonant column machine takes a soil sample of perhaps 4 cm in diameter, sometimes hollow like a tube and perhaps 25 cm long, and enables the measurement of such parameters as damping and shear modulus. As ordinarily understood, the complex modulus which may also be measured is an operator of the form $Z\,e^{i\phi}$ which, when applied to the strain vector, yields the stress vector by modifying vector length and phase. It seems important to observe that in this form of the definition an instantaneous strain value multiplied by the complex modulus does not yield the corresponding instantaneous stress.

B. O. Hardin[9] has devised a practical torsional generator that yields valuable shear and damping measurements at low cost. These measurements form a useful, comparative and absolute key to probable soil performance and properties. A pad supporting a vertical shaft is placed on the ground, and the shaft supports a mass of large moment of inertia that can oscillate around the vertical axis. This mass is excited into torsional vibration by an electrical signal generator, which induces torsional shear in the soil surface, and its motion is measured by a pickup. In this way resonance may be detected and its frequency measured, together with the coasting (decay) pattern of the free vibration, leading to an

(a) Piezo Electric Accelerometer, 69 mVP/gP; Principal Resonance 24 KC with Flat Response to 10 Kcps; Mounted on Magnet (Bruel & Kjaer)

(b) Geophone Element, 150 mVP/IPSP for Horizontal Use; Amplitude Limit About 0.030 in. PP; Electronically Damped (Geo Space Corp.)

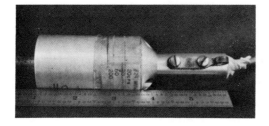

c. Geophone mounted in Case (Geo Space Corp.)

Fig. 25.11 Transducers. (a) Piezoelectric accelerometer, 69mVP/gP; principal resonance 24 kcps, with flat response to 10 kcps; mounted on a magnet (Bruel & Kjaer); (b) Geophone element, 150 mVP/IPSP, for horizontal use; amplitude limit about 0.030-in PP, electrically damped (Geo Space Corp.); (c) Geophone mounted in case (Geo Space Corp.)

evaluation of shear modulus and damping. This is a low-frequency torsional oscillator. An analogous low-frequency linear-vertical oscillator could be devised, although it is likely to be a clumsier machine.

The classical seismograph is comprised of a pendulum, weighing perhaps several tons, and a magnifying mechanical linkage that enabled a pen to draw a chart on clockwork-actuated paper. Two good examples (made by the amateur, O. Leary*) were seen working at Rathfarnham, Ireland, in the 1930s. Such instruments can record distant earthquakes, perhaps 4000 miles away, and often distinguish the arrival times of the P, S and R waves, which may differ by perhaps five minutes. The use of two pens makes possible the separation of mutually normal components. Such seismographs detect the occurrence of earthquakes and determine their location and magnitude, besides yielding much information about the interior of the earth from the behavior of the waves. They are, however,

---

* Fr. O. Leary, noted seismologist and Jesuit priest.

too delicate for recording the strong motions that occur near the earthquake location (epicenter).

The strong motions are of great interest to engineers and have only been measured in very recent years. Very recently the strong motion seismograph has come into such prominence that some building codes (e.g. Los Angeles) now require their installation in every major building to measure acceleration. It has often been humorously observed that when the seismograph is in working order no earthquakes occur. This is because in the past several ostensibly operating seismographs gave no record during major earthquakes. Because of the current proliferation of seismographs, this is unlikely to happen in the future.

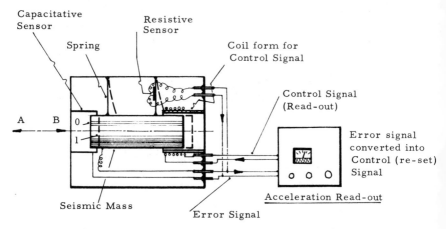

Fig. 25.12 Servo accelerometer element. When zero-frequency or frequency to about 200 cps acceleration is applied to the accelerometer in direction AB, the seismic mass tends to move from its neutral position 0 to position 1. The displacement is sensed, for example, by a variable capacitance or strain-gage. The sensing signal causes a control signal to be fed to a coil, the signal (control) being of such magnitude as to eliminate the displacement. Acceleration is read out in terms of the control signal, which is calibrated in terms of acceleration. Signal level is about 1000–5000 mVP/gP.

**Case Histories.** Five examples which tend to be failures rather than triumphs and which underline the gloomy aspects of practical problems connected with existing installations rather than with design, are taken from the writer's records. The examples relate mainly to work done some years ago. This explains why isolation is not stressed as a solution. In very recent years, the use of air mounts of various kinds has been very successful. The technology of air mounts is now so advanced that isolation is nearly always possible even in the very low-frequency cases such as those as low as 3 cps, for example.

Figure 25.13 illustrates a case classified as a physiological nuisance. Plates, loose windows and miscellaneous articles rattled in nearby homes, and persons lying on soft beds noticed a disconcerting horizontal movement of their bodies. Frequency is invariably the key. A survey of a local factory disclosed five vibratory conveyors, driven by separate induction motors operating at 375 cpm. When these were shut down, the vibration in the homes fell to zero. The oscillograph is from a photograph taken with a table-supported, model 110 Polaroid camera with one-second exposure to match the trace period of the oscilloscope. The vibration period appears to be 150 milliseconds which corresponds to about 390 cpm, and this is close enough. The conveyors were 15 ft long and carried variable loads of say 300 lbs of castings, just shaken from their molds. No corrective action was taken. Possible solutions are: the anti-phasing of conveyors in pairs, using synchronous motors and timing belt drives or the installation of balancing weights or rotors of the Lanchester style. The leaf-mounted type of conveyor oscillates at about 45 degrees to the horizontal so as to lighten the load or make it airborne in one direction and then increase the load and friction so as to make the goods move in the other.

A second case, classified as a physical nuisance to a machine operation, concerns self-generated vibrations in the machine. Slight vibration was observed by eye in the zone of a series of one dozen silicon zone refiners. The zone refiner, illustrated in Fig. 25.14, is a good example of a modern machine very sensitive to

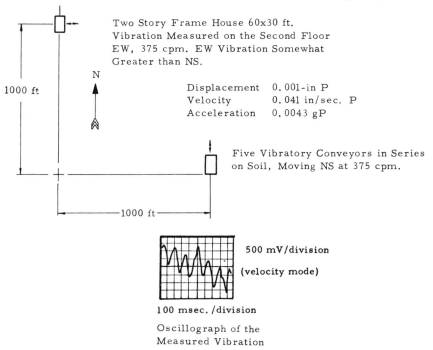

Two Story Frame House 60x30 ft.
Vibration Measured on the Second Floor
EW, 375 cpm. EW Vibration Somewhat
Greater than NS.

N

1000 ft

Displacement    0.001-in P
Velocity        0.041 in/sec. P
Acceleration    0.0043 gP

Five Vibratory Conveyors in Series
on Soil, Moving NS at 375 cpm.

1000 ft

500 mV/division

(velocity mode)

100 msec./division

Oscillograph of the
Measured Vibration

Fig. 25.13 Physiological nuisance vibratory conveyor. A General Radio meter 1553A with a Shure crystal accelerometer pickup was used to measure vibration. It was connected to a Tektronix 422 oscilloscope. The meter read directly 2 mil PP displacement, velocity 0.04 IPSP and acceleration 0.0052-gP. On the basis of 41 radians/sec and a displacement of 0.001-in P, the corresponding velocity and acceleration for a sine function are as given at the top of the figure.

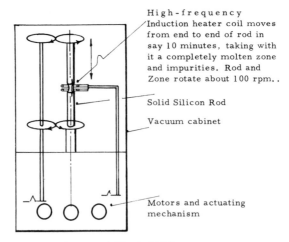

High-frequency Induction heater coil moves from end to end of rod in say 10 minutes, taking with it a completely molten zone and impurities. Rod and Zone rotate about 100 rpm..

Solid Silicon Rod

Vacuum cabinet

Motors and actuating mechanism

An Internally Generated Physical Nuisance

Fig. 25.14   Silicon zone-refiner schematic

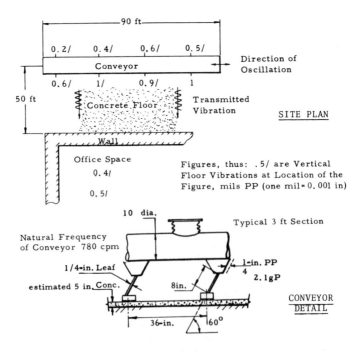

SITE PLAN

Figures, thus: .5/ are Vertical Floor Vibrations at Location of the Figure, mils PP (one mil = 0.001 in)

CONVEYOR DETAIL

Fig. 25.15   Physiological nuisance vibratory conveyor. 0.5 mil PP at 780 cpm is equivalent to 0.0042-gP. Human physiological threshold is about 0.006-gP at 780 cpm, but in this case the shaking of lampshades and furniture lowered the threshold.

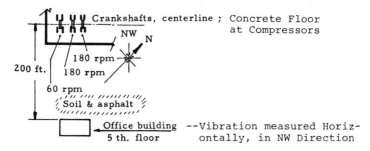

Plan-View of Site

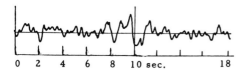

Oscillograph No. 15-Displacement in a NW Direction near
Compressors, operating respectively 60,118, and 118 rpm.
Maximum Displacement,2 to 14; Typical Value, 6 to 8 mils PP.

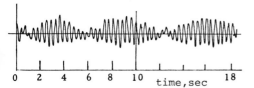

Oscillograph No. 9-Displacement on the 5th Floor of an Office
Building, 200 ft away. Compressor speeds of 60,164, and 170 rpm.
Displacement Range,4 to 26 ; Typical Value, 16 to 25 mils PP.

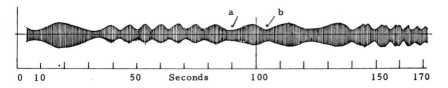

Typical Time Envelope of Oscillograph No. 9.  Sometimes the
Displacement falls to Zero at instants such as a and b.  Typical
values, 16 to 25 mils PP

Fig. 25.16 Physiological nuisance reciprocating compressors. Three
reciprocating, ammonia compressors of 800 hp each, steam driven with
top summer speeds of 60,180 and 180 rpm.

vibration. It prepares silicon for the electronic industries by reducing impurities to the order of one part per million. The zone is supported only by surface tension. The zone natural frequency for rods of about 20 mm diameter appears by eye to be about 5 cps. Operation was adversely affected by a vibration so slight that it was considered to arise from traffic tremors in the soil. An overall survey was made of acceleration with a General Radio 1553-A battery-powered, vibration meter. The point of interest appeared to be the lower end of the coil transporter. Observations ranging over several hours were taped, using a battery-powered Uher 4000 recorder. The tapes were looped and analyzed by playing back through a General Radio sound and vibration analyzer, model 1564 A with 1/3 and 1/10th octave filters. Three notable vibrations were unraveled from an unusually complex pattern: electrical-type vibrations showing clear 60, 120, 180 . . . cps patterns, steady mechanical vibration of about 0.0002-in PP at the coil transporter and impacts arising from electrical relays. An improving trend resulted from many recommendations relating to construction and maintenance standards.

A third case is classified as physiological nuisance in relation to office workers. Figure 25.15 shows a vibratory conveyor which shook the partitions, lighting fixtures and furniture in a plant office. The conveyor was 90 ft long and driven by a 5 hp electric motor. The vertical floor vibration was 0.5 mils PP in the office space, with 0.15 mils of horizontal vibration at the concrete walls and about 0.5 mils PP horizontally on the wood partitions. A recommendation was submitted to discuss balancing with the maker of the conveyor. As so often happens, this writer has no record of the outcome.

A fourth case is classified as a physiological nuisance to office workers. When measured, it was designated case 27 B. 45 (and is illustrated in Figs. 25.16 and 25.17). It was desired to ascertain if the vibrations on the fifth floor of an office building were too severe for the proper working of a proposed computer-type machine that was said to be sensitive to gyroscopic action resulting from vibration. Three steam-driven compressors producing the vibration were operated at

Fig. 25.17   Physiological nuisance reciprocating compressor, 60 rpm

speeds of 60, 100 to 180, and 100 to 180 rpm, respectively. Being ammonia compressors, they operated at top speeds during the summer.

Typical horizontal measurements of vibration for the fifth floor of the office building are given in Table 41.

TABLE 41
RESULTS OF VIBRATION MEASUREMENT,
CASE 27 B. 45

| ITEM | COMPRESSORS SLOW | COMPRESSORS FAST |
|---|---|---|
| Displacement, mils PP, max. | 11 | 26 |
| Velocity, IPSP, max. | 0.08 | 0.19 |
| Acceleration, IPS²P, max. | 4.3 | 5.9 |

Typical frequency is 150 cpm. This corresponds to about 15 radians per second so that 15 times displacement P should give velocity P, and it does so quite nearly. Also, 15 times velocity P should give acceleration P, which it does not, so that the vibration is not pure sinusoidal. Some high-frequency vibration was noted in the acceleration oscillographs.

The threshold of human sensitivity is about 0.005-gP at 150 cpm. The observed acceleration based on 15 times 0.19 divided by 386 gives about 0.0074-gP, which is above the threshold. But, the vibration was not generally physiologically perceived. Occasionally, when it was perceived, it led to a sickening sensation. There is a wide variation in human perception. This writer has observed 1/20 mil PP, 60 cps displacement sharply affecting a person sitting on a chair with the vibration vertical. This is about 0.009-gP, whereas typical published data gives about 0.05-gP as being the level between perception and unpleasantness at 60 cps.

A fifth and last case relates to a synchronous reciprocating compressor of 2000 hp leading to physiological nuisance in dwelling houses from about 300 to 1000 ft away. This is designated case 1019 and is illustrated in Figs. 25.18, 25.19 and 25.20. The data in this case and in other such cases, where there is no routine solution, usually amount to dozens of pages. Only a very brief review is given here. The typical waveforms are shown in Fig. 25.1. Some of the interesting general conclusions that arise in this case, taken in correlation with others, are as follows.

1. Persons in their homes can be upset for a psycho-physiological combination of reasons when the PP displacement at 650 cpm exceeds about 0.15 mils, measured in the basement. In typical frame houses the basement amplitude may be doubled for each floor above that, depending on the type of house.

2. Some compressor configurations give rise to large vibrations in the soil, whereas others with different cylinder and mechanical layout, but of comparable horsepower, appear to generate almost no vibration at all.

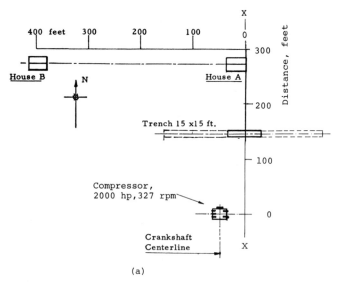

(a)

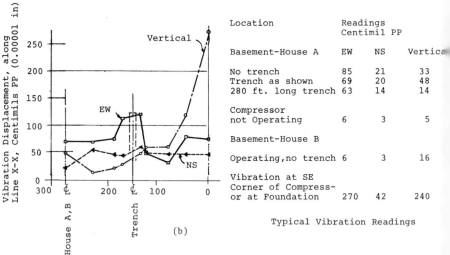

(b)

| Location | Readings Centimil PP | | |
|---|---|---|---|
| | EW | NS | Vertical |
| Basement-House A | | | |
| No trench | 85 | 21 | 33 |
| Trench as shown | 69 | 20 | 48 |
| 280 ft. long trench | 63 | 14 | 14 |
| Compressor not Operating | 6 | 3 | 5 |
| Basement-House B | | | |
| Operating, no trench | 6 | 3 | 16 |
| Vibration at SE Corner of Compressor at Foundation | 270 | 42 | 240 |

Typical Vibration Readings

Fig. 25.18  Physiological nuisance reciprocating compressor. The drawing shows a site plan of a compressor and the houses that are disturbed
by it. A graph of vibration measured on a 1.5 x 1.5-in. angle iron driven
into ground is given in centimils PP displacement for various points on
line XX (one centimil = 0.00001 in). The trench was later extended as
shown by dotted lines. Vibration measured at the first floor of house A
averaged 45 to 210 centimils without the trench, and the first floor of
house B, 10 to 50 centimils. In both houses rattling of plates and loose
articles was observed. Vibration frequency observed throughout was
650 cpm. Two other comparable compressors, at the same site with
different cylinder arrangements, showed 30 centimils longitudinal vibration. Waveform is shown in Fig. 25.1.

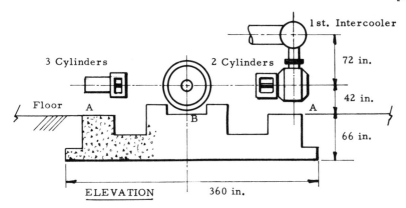

Typical vibration measured vertically at AA is 2. 7 mils PP, and
0. 75 mils PP at B, which demonstrates that foundation rocks.

Detail of 5 cylinder, 3000 PSIG. air compressor, 2000 HP. , 327 RPM. ,
Synchronous motor drive.    Looking North, Case 1019.

Fig. 25.19   Elevation of reciprocating compressor. Detail of a 5 cylinder,
2000 psig air compressor, 2000 hp, 327 rpm, with a synchronous motor
drive. Typical vibration measured vertically at level AA is 2.7 mils PP and
0.75 mils PP at B, which demonstrates that the foundation rocks

Fig. 25.20   Physiological nuisance reciprocating compressor

3. Based on the two cited compressor cases, there is reason to suppose that
the horizontal vibration is the most troublesome component and that if this
exceeds 0.5 mils PP at the compressor in the frequency range 2 to 10 cps the soil
environment shall exceed 0.15 mils PP, which appears to be the psychological
threshold. The stated 0.5 mils PP at the compressor refers to the greatest horizontal component measured on the foundation. The human being exhibits a

peak sensitivity at about 6 cps. It is to be noted that when neighbors talk, the least affected is convinced that he is over the threshold; and there is no end to the trouble when residential districts are in mind. A suggested criterion based on these cases is that the horizontal PP mils at the compressor foundation should not exceed 0.25 mils. In the cited cases, the compressors were reciprocating, in-line type which normally produce a minor vertical component of acceleration, especially the ammonia compressors. Being located in a structure, they would, therefore, excite horizontal as well as vertical modes of response. On the other hand, compressors with upright piston arrangements would excite modes of response that are more dominantly vertical.

4. When trouble exists, the environmental vibration must be reduced by an order of magnitude of at least three if the means used for reduction are to be considered successful. From the writer's experience combined with a few cases discussed by others, the probability of success using a trench of any dimension is about three to one against success.

5. All available knowledge must be brought to bear at the design stage, cooperating with the machine manufacturer. It is virtually impossible to rectify these cases without total reconstruction.

6. Attenuation does occur as the distance from the source increases; but over a given few hundred feet, no attenuation or even augmentation may often be observed.

7. Relatively little appears to be known about the behavior of surface waves in soil of frequency up to 10 cps. The writer has not had an opportunity to measure these wavelengths and velocities. The wavelength is likely to be of the order 100 ft and upward. Bearing in mind that the higher frequencies probably travel more slowly than the lower, the question of how a complex wave with several harmonics may travel is of interest. A seismologist remarked some years ago that the lower frequencies may carry the higher so that attenuation in the high frequencies may be less than expected. Lionel J. Lortie (vibration specialist)[10] has confirmed this, and points out that relatively high frequencies, that would normally attenuate and vanish, often exhibit a surprising persistence in the presence of low-frequency waves, by which they appear to be carried as if in a parasitic way.

# REFERENCES

1. Richart, R. E.; Hall, J. R.; and Woods, R. D. Lecture Course Notes: Vibration of Soils and Foundations, University of Michigan, Ann Arbor, 1968.
2. Whitman, R. V. Lecture Course Notes: Earthquake Engineering, University of Michigan, Ann Arbor, 1968.

3. Newmark, N. W. *Geotechnique* 15 (1965): 139–160.
4. Woods, Richard D. Lecture Course Notes: Elastic Waves in Layered Systems, University of Michigan, Ann Arbor, 1968.
5. Bornitz, G. *Uber Die Ausbreitung Der Von Groszkilbenmaschinen Erzugten Bodenschwingungen in Die Tiefe*. Berlin: Julius Springer, 1931.
6. Woods, Richard D. Lecture Course Notes: Isolation of Foundations, University of Michigan, Ann Arbor, 1968.
7. Ward, H. S. and Crawford, R. *Determination of the Natural Periods of Buildings*. National Research Council of Canada, 1964.
8. Ward, H. S. and Crawford, R. *Wind Induced Vibrations and Building Modes*. National Research Council of Canada, 1966.
9. Hardin, B. O. Personal communication, 1968.
10. Lortie, Lionel J. Personal communication, Montreal, 1970.

# BIBLIOGRAPHY

Barkan, D. D. *Dynamics of Bases and Foundations* (translated from the Russian.) New York: McGraw-Hill Book Co., 1962.

Leet, D. *Practical Seismology and Seismic Prospecting*. New York: Appleton-Century-Crofts, 1938.

Richart, R. E.; Hall, J. R.; and Woods, R. D. *Vibrations of Soils and Foundations*. Englewood Cliffs, N. J.: Prentice-Hall, Inc., 1970.

Richter, C. F. *Elementary Seismology*. San Francisco: W. H. Freeman and Co., Publishers, 1958.

# 26

# TELEMETRY

*Lionel J. Lortie*

## 1. INTRODUCTION

The word *telemetry* implies that measurement is done remotely or at a distance. To some it means a method of measurement with some loss of information or diminution in accuracy, as compared with direct methods. However, inaccuracy is the exception and not the rule. In typical applications of telemetry, the signal appearing at the point of display is an exact analog of what is generated by the original transducer at the measurement location. It is usual to arrange the measurement system so that the finally emergent analog signal has the same amplitude (1:1 ratio) as that of the original signal. This avoids read-out and analysis errors and the labor of applying factors.

Telemetry is then simply a means of transmitting information so that it may be recorded, displayed or analyzed at a location that is either remote from or in some way not directly accessible to a location where the signal is to be measured. Just as the tape recorder stores information, the telemetering system simply transmits it, but in a particular way.

Telemetry may be justified, for example, on the basis of human safety, on the basis of convenience and, perhaps most often, on the basis of making possible the measurement of signals from inaccessible transducers.

There are in common use today two basic methods of telemetering; namely, pulse-width amplitude modulation (PWAM) and frequency modulation to frequency modulation (FM-FM). However, for most of the ordinary problems that occur in vibration measuring, FM-FM systems are of far greater significance. Hence, all of the remarks that follow, except when qualified, apply to FM-FM systems and their application. Telemetry is also often accomplished via coaxial cable systems. However, cable transmission is not of particular significance for this article. It is implied here that telemetry is transmission via radio waves without wires between the pickup transmitter and the receiving system, with modulation and demodulation via FM techniques.

Typical examples of application include transducers mounted in submarines, road vehicles or aircraft or missile vehicles; also, a transducer-transmitter system mounted on an engine crankshaft, or a transducer-transmitter system for transmitting information about explosions or earthquakes. When the use of cable wire is technically possible, it is often avoided simply because its integrity cannot be guaranteed, for example, when it must traverse a few miles over ground in which it may be harmed by traffic, man, animal or other misadventure. So, radio-telemetry is also justified by a reliability of performance, that is, by its almost inviolable integrity.

Consider a large manufacturing or processing plant in which monitoring transducers are mounted at as many locations as seems desirable—on tanks, on engines and the like. At any chosen moment, a monitoring signal is available at a central station for display, analysis, measuring or recording. In this example, the keynote is that of convenience or of reliability rather than that of solving a problem of inaccessibility.

It should be noted that it is not possible to make a clear distinction between telemetry and common radio transmission. In general, the distance of transmission for telemetry ranges from a few feet to a few miles. Thus, it is mainly distinguished from radio transmission of the ordinary kind by the difference in the range of the physical distance of transmission. In addition to this, the common radio signal may carry one or perhaps two channels of information. By contrast, the radio signal used in telemetering may carry as many as twenty or more channels.

The outputs of the transducers shown in Fig. 26.1 are modulated, each by its own separate subcarrier oscillator. The oscillator outputs are then mixed and conveyed to a radio oscillator from which the signal is broadcast via an antenna. At the receiving station the radio signal is subjected to the usual processes of detection and rejection (demodulation). The resulting signal is fed to a series of discriminators, each having center frequencies corresponding to those of the subcarrier oscillators. Each discriminator processes its own particular channel of information and demodulates it so that the finally emergent analog is precisely that generated by the remote transducer.

It should be noted that each transducer-sender must have a source of power (usually a battery), an oscillator modulator and a transmitter. Sometimes the power source and the transmitter are common to several transducers, each having its own subcarrier oscillator. When the transducer is most inaccessible, as in the case where it is mounted on an engine crankshaft, the transducer, power supply, oscillator, modulator and transmitter are all combined into a single package, perhaps no larger than that of a large cigar. The unit cost of such packages may be of the order of $1000.

## EXAMPLES OF APPLICATION

**Example 1: Human Safety and Reliability of Transmission.** Suppose it is desired to subject explosive material to an impact and vibrational environment for the purpose of determining its dynamic limitations. Consider that the material is separated from the observer by a distance of perhaps two miles. The initial cost of cable and its installation as well as the hazards connected with its failure are sufficient to rule out its use. Instead, telemetry transmission is used and in two

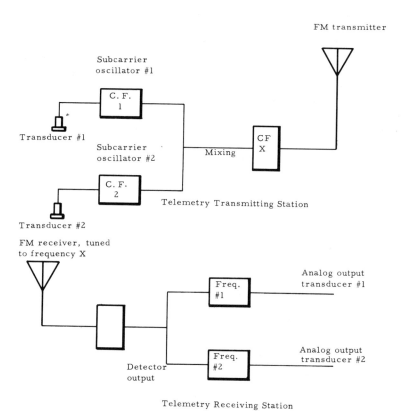

Fig. 26.1   A two-channel telemetry system

different ways: first, the dynamic shaker is controlled from the observation location via telemetry. Second, the induced vibration is picked up by a transducer
and its output is transmitted to the observation location via telemetry. If the
explosive material explodes, the information connected with the explosion, such
as the time of onset, signal waveform, amplitude and duration, are all transmitted
to the observation base. Moreover, human safety is guaranteed. It is to be
observed that typical test locations are usually in remote and difficult terrain. This
in itself enhances the attractiveness of radio telemetry.

**Example 2: Instrumentation Systems.** The elements of an **FM-FM** radio
telemetry system are suggested in Fig. 26.1. The center frequencies of the subcarrier oscillators depend to an extent on how many channels are to be used as
well as on the frequency response that is expected of the system. Factors to be

taken into account are: the frequency range of the components of each channel, considered in terms of the Fourier components that are involved in the signal waveform; the number of channels to be used; and the accuracy required in terms of the flatness of frequency response. The flatness of frequency response is considered ideal if after transmission and demodulation all frequency components in the frequency band are received on a one-to-one basis. It is to be noted that typical transmission systems are such that phase relationships between the various information channels are adequately preserved without difficulty. Depending again on the frequencies to be transmitted and the number of channels, the center frequencies of the subcarrier oscillators range from about 1 kcps to 200 kcps. Low-frequency channels of about 2 kcps are used when the signal frequency variation does not exceed about 20 percent or $\pm 200$ cps. Low-frequency channels are also used for synchronization, for calibration and for control signals of many kinds, such as when it is required to know how often a signal amplitude surpasses an assigned value.

**Example 3: Human Safety and Inaccessible Measurement Sites.** Consider the testing of electronic components at high levels of nearly constant uni-directional acceleration. For example, performance of a transistor, both electronically and mechanically, is often required at 5000-g peak. A test acceleration can be provided by means of an air-actuated centrifuge operating at say 35,000 rpm. Here, considerations of human safety and the feasibility of information transmission arise.

The usual testing method involves mounting the electronic components on the centrifuge and monitoring the centrifuge motion with an accelerometer. Batteries to power transistors and a transistor circuit are first potted in epoxy material. Then the assembly is mounted on the centrifuge and carefully balanced. Next, the monitoring accelerometer is locally mounted and it too is balanced. It should be noted that considerable acceleration flutter occurs in an air-actuated centrifuge so that computation of acceleration from the average rpm is not acceptable. In removing the test information from the centrifuge, the best slip rings, even those of mercury-wetted silver, are totally unacceptable due to the speed. Hence, telemetry is given prime consideration.

In a telemetering system, the accelerometer and its associated oscillator and transmitter might send mechanical information for one trace of an oscilloscope. The transistor circuits could send their information, again, via telemetry to other channels of the oscilloscope. In this way, the acceleration applied to the transistors is constantly correlated with their electronic performance. The simultaneous channels of information can also be recorded on tape. Conversely, the same system may be used for the calibration of DC accelerometers (those accelerometers capable of generating or providing an electrical analog of a fixed or very slowly varying acceleration) of unknown performance by comparing them with one of known performance. Known performance usually means knowing a value of signal output per unit of acceleration.

**Example 4: An Inaccessible Measurement Site.** Consider the testing of missile electronics in relation to their intended environment. By telemetry it is possible to measure and transmit all the vibration and shock information from an actual environment that is sought while a missile is carried, perhaps by its launching vehicle, while it is separating to take up its independent flight and while it is in flight. Thus, all of the necessary environmental information can be made available for analysis or for recording on tape for storage. The tape may be used later, if desired, for the purpose of simulating as closely as desired the working environment that has been previously measured. For example, a shaker device may be actuated by using the tape record that resulted from an actual environmental test.

As observed above, phase-distortion difficulties are not a characteristic of telemetering systems. Systems are usually so designed that the original phase relation between transducer outputs is preserved with great fidelity at the point of final read-out, analysis or tape recording.

## 3. CONCLUSION

The foregoing has briefly discussed the more important points connected with telemetry as applied to vibration measurement. Much more could be included here. But further considerations tend to be those connected with specialized studies of instrumentation systems and techniques which are outside the scope of this chapter.

# 27

# GAS PULSATION

*Lionel J. Lortie*

## 1. INTRODUCTION

Problems associated with gas pulsation phenomena often arise because they have escaped the consideration of the designer or they are beyond the state of the art of design. Consequently, design revisions in the field are often necessary. Such problems normally require somewhat different treatment from that which is typical of problems connected with rotating machinery. For the latter, amplitude and frequency data are usually sufficient; for gas pulsation-related problems and for the general problems associated with reciprocating machinery not only is the usual measuring instrumentation required, but the ability to measure phase and time intervals as well.

**Example.** The following example, a sketch of which is shown in Figs. 27.1 and 27.2, occurred in actual practice and typifies a large number of such cases, illustrating some of the measuring techniques and principles that are involved. A carbon dioxide compressor having an L-configuration with two cylinders, running at about 300 rpm and 150 hp, delivered gas to a 4-in diameter pipe at a nominal pressure of 1000 psig. Piping vibration was so violent that failure was expected before too long. For a solution, the piping was attached more securely to the building structure. The vibration of the piping was reduced dramatically, but the building structure vibration became violent.

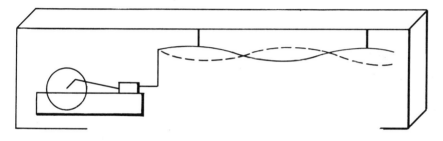

Fig. 27.1   Extreme piping vibration

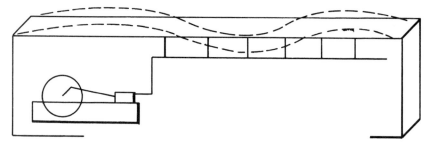

Fig. 27.2   When piping is more firmly attached to building, the piping settles down and the building vibrates violently.

A measurement instrumentation arrangement for dealing with this type of problem is shown in Fig. 27.3. More information about phase measuring instrumentation and techniques is given in Chapter 13. The particular purpose of the band-pass filters in this application is the elimination of the high-amplitude, high-frequency noise that results from the passage of gas in the piping and the elimination of those frequencies that correspond with the lateral motion of the piping as a whole. This latter motion is not of interest. The signal that is of interest is that which is the analog of the diametral piping strain engendered in the piping by the passage of gas pulsations, irrespective of whether or not the piping as a whole exhibits lateral vibration.

Filter tuning is accomplished by selecting a simple low-pass setting; the unwanted frequencies are all above that frequency, which is of principal interest. In this study, identical Krohn Hite filters were used. These are variable as to frequency and the width of the pass band. When set, the pass band is of constant width, rather than having a constant percent width; and they may be set to a bandwidth as narrow as 1.5 cps. This very fine discrimination is most useful when unwanted frequencies are very close to the frequency that is of primary interest.

In the present case, gas-actuated valves were used. These are more likely to generate pulsations or to permit the generation of pulsation than are cam-operated valves which, of late, are being used more as a consequence of the noted difficulty.

Two identical accelerometer transducers were attached to the outside of the piping. The following results were derived from the oscillograph of Fig. 27.3, which is illustrated enlarged in Fig. 27.4:

V—velocity of pulse transmission          30 in/sec
L—physical length of pulse                3 in

The next objective in measurement was that of discovering the distance along the pipe between adjacent pulses. A single trace of the oscillograph of Fig. 27.3 was observed, and is shown in Fig. 27.5.

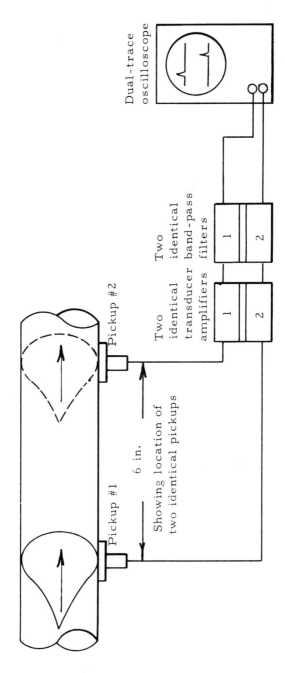

Fig. 27.3  Piping and instrumentation arrangement for measuring gas pulsation

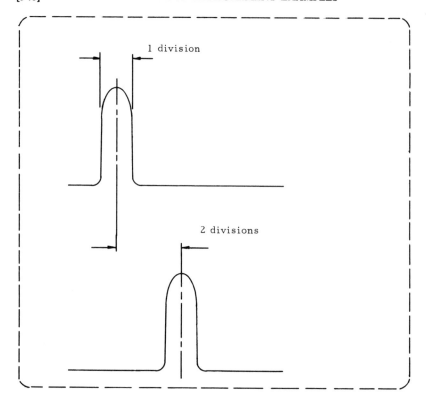

Fig. 27.4 Determination of pulse velocity, pulse-duration-of passage and physical length of pulse, based on oscillograph display of Fig. 27.3. One horizontal division is 100 milliseconds, in this example. Conclusions: each pulse takes 200 milliseconds to travel 6 in. Hence velocity of pulse transmission in piping is 30 in/sec. Since $V = 30$ in/sec, and since pulse duration as seen by the transducer is 100 milliseconds, it follows that the physical length of the pulse, measured along the pipe, is 30/10 or 3 in.

From Fig. 27.5, the following result was derived.

I—pulsation interspacing          15 in

Two design criteria were now developed in relation to the design revision that takes the form of an expansion chamber (stub branch).

$L_1$—the length of the chamber               L length of the pulse
$V_1$—velocity of pulse transmission across chamber

The design-revision solution is summarized in Fig. 27.6. The inlet to the expansion chamber may be angular (sharp). However, the outlet should make some

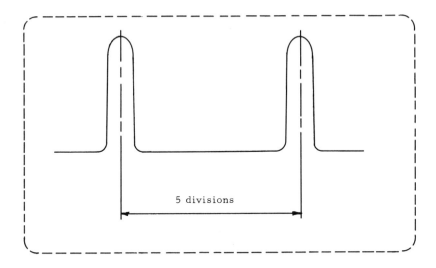

One horizontal division = 100 milliseconds

Fig. 27.5 Finding the pulse interspacing. In the above oscillograph is shown enough of trace 1 or of trace 2 of the oscillograph of Fig. 27.3 to exhibit two adjacent pulses in one trace. The objective is to find the physical distance between pulses, measured along the pipe. The time lapse between the arrival of adjacent pulses at one pickup is 500 milliseconds, and the previously computed velocity is 30 in/sec. It follows that the physical distance between adjacent pulses is 15 in.

attempt toward conforming with a rounded (or venturi) configuration such as tends to minimize turbulence.

Since the radial strain in the piping is implied by the pulse amplitude of the oscillograph and assuming strain variation is proportional to the pressure amplitude of the pulse measured above the local static gas pressure, it follows that the amplitude variation of the oscillograph trace is proportional to the pressure amplitude of the pulse. Because of this, an observation of the oscillograph that derives from a transducer placed at a position downstream of the expansion chamber provides for a comparison of the diminished pulse amplitude, as induced by the chamber, with the original pulse amplitude.

Because the expansion chamber exhibits a different relation of pressure to radial strain than does the piping, its internal pressure is not readily compared with that in the piping by simply referring to the vibration transducers. However, a pressure transducer will show that the pulse pressure in the chamber is about $a/A$ times that in the pipe.

For optimum correction, the length of the chamber is made the same as the length of the pulse, both measured in lineal units such as inches. If the chamber is longer, the emergent, diminished train of pulses will not be contiguous. In effect, a separation of pulses arises and this is not what is required. On the other hand, less than optimal smoothing results if the chamber is too short (shorter than a pulse length). When the expansion is too short, maximum pulse relaxation in the chamber and, therefore, maximum pulse diminution in the emergent piping are not achieved. The pulses tend to accumulate. For optimal smoothing the pulses should emerge from the chamber as a contiguous train, without intervening gaps and without accumulation (piling up of adjacent pulses). The foregoing comment may be expressed in the form of an equivalent criterion, namely: the time of passage of a pulse across the chamber must be made the same as the time interval between the arrival of two adjacent pulses at a given point in the piping, upstream of the expansion chamber.

Using the simple snubber (expansion chamber) illustrated in Fig. 27.7 which is made of standard piping components, an original pulse amplitude of 100 pressure units may usually be reduced to the order of ten or five units. If further smoothing is desired, a simple accumulator may be installed downstream of the expansion chamber. Any short length of larger diameter piping will provide some improvement. The most suitable diameter is about that of the snubber. The best length of accumulator is such that its volume is about ten times that of the snubber. Thus, the length of the accumulator is about ten times the net length of the snubber which excludes the anti-turbulence transition at the snubber outlet.

It is to be noted in this example that the velocity of propagation of the pressure pulse bears no relation at all to the local acoustic velocity. In the present example, the velocity of propagation is 30 in/sec, and this velocity is determined largely by

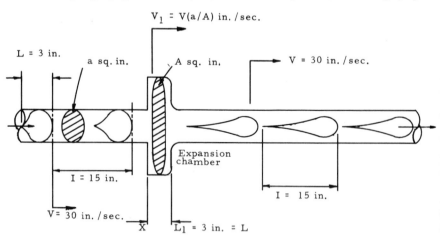

$$V_1 = V(a/A) \text{ in./sec.}$$

L = 3 in.

a sq. in.    A sq. in.    V = 30 in./sec.

Expansion chamber

I = 15 in.

I = 15 in.

V = 30 in./sec.

X   $L_1$ = 3 in. = L

Fig. 27.6   Length, $L_1$ and cross-sectional area, A, of expansion chamber. When a pulse enters expansion chamber at plane X, the velocity of transmission in the chamber must be such that the time taken by the pulse to move across the chamber is the same as the time taken for the front of the succeeding pulse to reach plane X. But the time lapse between successive pulse arrivals at X is 500 milliseconds, therefore the time of passage of the pulse across the chamber must be arranged to be 500 milliseconds. The velocity of transmission across the chamber must be made equal to 3 in per 500 milliseconds, or 6 in./sec. Whence: A/a = 5:1 (A/a = I/$L_1$; and $L_1$ = L).

Fig. 27.7   Pulse expansion chamber and ripple snubber on air compressor
discharge (courtesy of Gulf Oil Canada, Ltd., Shawinigan Chemicals Div.,
Varennes, Quebec, Canada)

the relation of peak-pulse pressure to local static pressure within the pipe. The
higher the peak-pulse pressure, the greater the velocity of propagation. It is also
noted that the pulse amplitude has no direct bearing on the computed solution
that is described in the foregoing discussion. Note in Fig. 27.6 that, after the
introduction of the appropriate snubber, the pulses which were of large amplitude
and separated from one another are now of small amplitude and contiguous.

The ratio of pulse pressure in the expansion chamber (pulse-peak amplitude)
to pulse pressure in the upstream piping is a/A, which in this example is 1/5.
This ratio is determined by the original parameters of the system so that there
is no choice as regards the subject ratio. However, the total diminution in pulse
amplitude that is achieved is far greater than suggested by the ratio 1/5. That is
to say, the ratio of pulse pressure downstream of the chamber to pulse pressure
upstream of the chamber is likely to be of the order 1/20.

# SECTION VI

# INSTRUMENTATION

# SECTION VI CONTENTS

# 28

## TRANSDUCERS AND METERS
## FOR MEASURING VIBRATION

*Michael P. Blake*

### 1. SCOPE

Velocity and acceleration pickups are described from the user's point of view, together with examples of survey meters and wave analyzers. Thus, it is the primary instrumentation that is discussed or, in other words, the instrumentation that is used at the machine site. The additional instrumentation that is often used to analyze the primary signal or reading is not discussed. The primary instrumentation is important because it is the source of the basic information upon which decisions are finally based and because many successful vibration measuring programs have resulted from the use of one or two meters alone without any additional facilities.

### 2. INTRODUCTION

Although transducers and meters are here separated on the whole from considerations of the vibration source and of such additional instrumentation as may be used, nevertheless, it is not possible to understand or make proper choices without some consideration of the compatibility of the transducer with its signal source and its read-out and of its read-out with other additional instrumentation. For large, massive systems where transducer weight is not critical, the velocity transducer is the preferred type for common vibration problems in the frequency range of about 10 to 1500 cps. The accelerometer is preferred for the frequency range say 1500 to 20,000 cps, in general. The simplest velocity-type survey meter, possibly with displacement read-out also, is preferred for routine survey of machine vibration, while the combined meter-stroboscope-wave analyzer is usually preferred for vibration analysis in the field and for field balancing. Some general requirements of instrumentation may be summarized as follows.

1. Calibration: some means must be available, if even a second meter. This avoids gross deception.

2. Units and detection mode: units must be clearly known. It is not possible to communicate with a person who cannot say, for example, if his units are peak or

peak-to-peak and so on. The detection mode is usually average, although the read-out may be in terms of peak or some other parameter. Root-mean-square is sometimes preferred as a mode of detection. Again, it is not possible to assign more than a partial meaning to measurements when the detection mode is unknown.

3. Response: the amplitude limitation must be known—for example, 0.100 in PP for bottoming of the velocity pickup. The frequency response must be known. There is no substitute for a graph showing, for example, the meter reading over the entire frequency range for a vibration of constant amplitude, whether velocity, acceleration or displacement. Many meters exhibit errors of say 60 percent at 300 cpm, and the user is totally unaware of this departure from linearity at the lower frequencies.

4. Compatibility: the meter performance and output must be compatible with envisaged instrumentation and measurement requirements. The analyzer-filter characteristic should be available in graphical form or in intelligible quantitative description. If the use of a tape recorder, pen oscillograph or oscilloscope is envisaged, the meter should have suitable outputs for these. For field balancing, at least two transducer input connectors are desirable.

5. Connectors: this topic must be considered at an early stage to avoid, as far as possible, the most troublesome proliferation of connectors that easily arises. Some proliferation is unavoidable because, for example, the connector for a 10 gram accelerometer cannot be the same as that used for the 110 volt power supply. On the other hand, it is simply indefensible to have two or three different kinds of connector on a single instrument when one kind will suffice. Here the user must assiduously defend himself until agreed standards come to his aid.

6. Modularity: as far as possible, instruments should not overlap in capability. The most suitable wave analyzers, for example, are separate items. If a wave analyzer is included in the vibration meter, it is redundant if a separate analyzer is available. In this way a suitable meter is simply a high-quality voltmeter. Combination-type instruments are often suitable for use by untrained personnel. They are, however, often lacking in the performance that is desirable for careful studies. Again, for example, an oscilloscope that offers interchangeable amplifiers and other variations may be more acceptable than one that is fixed as regards its capability.

7. Durability: instrumentation should be rugged and accurate. Portable instrumentation should be protected by covers or otherwise against all usual hazards of transportation and field use.

8. Weight: portable instrumentation should be as light as possible. It is hardly realistic to call an instrument portable if it weighs more than 45 lbs.

## 3. APPLICATION

Large-scale vibration measuring is little older than one decade. Tradition to date is such that the survey meter is used for condition classification, which is to say for judging the severity, such as normal or dangerous. If the cause of an abnormal condition is not obvious, then it is traditional to take a vibration analyzer to the machine and continue the measurements with a view to diagnosis.

If the machine malfunction is due to mass unbalance, it is good practice to take the analyzer to the machine because the analyzer is necessary for balancing correction. On the other hand, if the malfunction is not due to unbalance and the cause cannot be diagnosed from the survey meter reading, it is often more logical to make a tape record of the vibration and analyze the tape in the office, avoiding field work amounting to, perhaps, some hours in uncomfortable, dangerous locations. A satisfactory tape requires about thirty seconds of playback time. However, few plants possess suitable tape recorders so that the vibration analyzer is normally used in the field. The labor is considerable, and the accuracy sometimes suffers because the discrimination and overall performance of the vibration analyzer is not likely to be as good as that of a laboratory-type analyzer which usually offers printout facilities, oscillograph photographs and other valuable results. It is worth remembering that if a vibration analyzer is used in the field, an oscilloscope connected with it often provides very valuable and immediate supplemental information.

## 4. THE VELOCITY PICKUP

Because of its familiarity, its velocity analog signal and the comparatively high voltage of its signal, the velocity pickup has found much favor with the non-specialist. Some of its features are listed below.

**Favorable Features**

1. The signal is a voltage analog of velocity, and velocity is the most usually used criterion of vibration severity.

2. The signal is of relatively high voltage, being, in fact, in the approximate range of 150 to 5000 millivolts/in/sec (usually about 150 to 1000). Thus, the signal may be read directly on many oscilloscopes or voltmeters without additional preamplification. The signal may also be fed to instruments having comparatively low impedance.

3. The principle of operation, a moving coil in a magnetic field, is familiar to many users.

**Unfavorable Features**

1. Many engineers are incurably distrustful of the pickup because it can be seriously affected by magnetic fields.

2. It is relatively bulky and seldom weighs less than 2 lbs.

3. The displacement range is limited to about 0.1 in PP.

4. The frequency range is limited to the approximate band 10 to 1500 cps.

A velocity pickup weighing about 3 lbs is illustrated in Fig. 28.1. The sensitivity is 200 mV per millimeter per second, which is unusually high. Pickups of this general type are usually used for all routine field measuring and balancing. The illustrated pickup is shaken by the vibration, and the coil suspended inside tends to stay at rest so that the relative movement generates a signal proportional to the relative velocity. The relative movement, in general, is the same as the absolute movement of the casing. The exact relative response is discussed in Chapter 4 of this volume, under the heading, Pallograph Response. The arrange-

Fig. 28.1   Velocity pick-up (courtesy of Schenck-Trebel Corp.) Range: 2 to 1000 cps with 4 millimeter peak-to-peak maximum displacement and sensitivity of 200 millivolts/millimeter per second. Price about $250.00.

ment is called seismic. Another arrangement, called relative or driven, is such that the casing stays more or less at rest while the vibration shakes an internal element via a probe. The relative type is unusual in the United States. It has the advantage that it may be designed so that its appropriate natural frequency, the casing driven through a spring, may be made lower than the natural frequency of the seismic type so that the useful frequency range may be extended from the usual minimum of about 10 or 15 cps down to, perhaps, one-fifth of this. The human hand increases the effective mass and makes the natural frequency lower than it would otherwise be, and if the pickup casing is clamped, the frequency capability may be extended to near-zero.

For seismic work connected with earthquakes or oil exploration, expensive velocity pickups of very low natural frequency are often used, measuring to 1 cps or less. When measuring the earth itself, there is no other body absolutely at rest to which the casing may be clamped. Sometimes small velocity pickups, called geophones, are used when low-frequency response is not required and when it is likely that the pickup will be lost or damaged.

## 5. THE ACCELEROMETER

The accelerometer is a transducer (pickup) that generates a signal which is a voltage analog of the acceleration. The signal may be electronically integrated to give velocity or displacement, just as the velocity signal is often integrated to

(a)

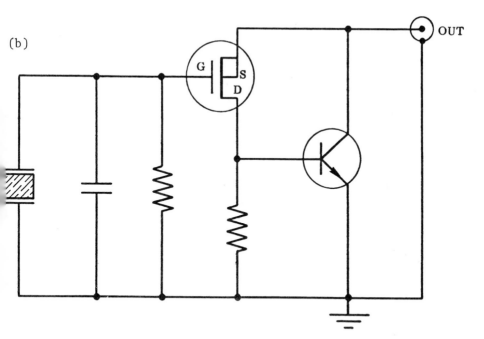

(b)

Fig. 28.2a   Accelerometer Piezotron Model 818 (for circuit, please refer to Fig. 28.2b) (courtesy of Kistler Instrument, Corp.)

Fig. 28.2b   Piezoelectric accelerometer with built-in impedance converter, showing the arrangement of the accelerometer illustrated in Fig. 28.2a (courtesy of Kistler Instrument Corporation)

give displacement. Most of the original disadvantages of the accelerometer have been mitigated so that it is likely to gain increased acceptance at the expense of the velocity pickup. Some advantages are listed below.

1. Typical accelerometers are less than one inch long and weigh less than one ounce.

2. Maximum displacement as such is limited only by the mechanical endurance of the cable.

3. A usable frequency range from about 2 to 10,000 cps is available in many accelerometers, and frequencies to 20,000 cps are easily measured.

4. The price (about $250) is about the same as that of the velocity pickup.

5. It is not affected, in general, by magnetic fields.

6. Sensitivity is almost unlimited. For example, one micro-g is readily measured.

7. Incomparable resistance to severe impact and large acceleration.

In the common arrangement the piezoelectric accelerometer comprises a crystal, say of quartz, which experiences a force when accelerated because of its own mass or of an attached mass. The common accelerometer operates far enough below its natural frequency to preserve linearity of frequency response by avoiding the flank of the peak response that is typical of resonating mechanical devices having elements similar to those of the accelerometer. Other types are available for special applications. For example, the servo type is often used in measuring soil vibration. Accelerometers may also be designed around bonded or unbonded strain gauges. These examples measure to zero frequency and may be calibrated by turning them upside down when a difference in signal level corresponding with 2-gP is observed.

A typical, versatile accelerometer is illustrated in Fig. 28.2, having the following approximate specifications.

| | |
|---|---|
| Weight | 0.7 ounce |
| Height without stud mount | 0.75 inch |
| Range | ± 250-gP, 1000-gP for shock |
| Sensitivity | 10 millivolts/g |
| Frequency response | 2 to 5000 cps ± 5 % |
| Resonant frequency | 30,000 cps |
| Amplitude linearity | ± 1 % from 1 to 250-gP |
| Output impedance | 100 ohms |
| Power requirement | 18 volts |
| Temperature range | − 65 to 250°F |
| Maker | Kistler Instrument Corp., Piezotron model 818 |
| Price | Approximately $280 |

It is noticed that this accelerometer system comprises an external impedance converter. Problems connected with impedance, losses in long lines, noise and other effects are solved in different ways by different makers. From the point of view of the user who measures typical machine vibrations, the great difference between the velocity pickup and the accelerometer is the fact that the velocity unit is limited to about 10 to 1500 cps, which indeed includes the majority of vibrations, whereas the frequency range of the accelerometer is, in effect, unlimited. For the high-frequency vibration associated with many bearing, cavitation, friction and other phenomena, there is no choice other than the accelerometer. For the majority of vibrations, however, in the more usual frequency range where the weight of the transducer does not alter system response, the user will use the velocity pickup simply because it is the one that is compatible with typical vibration meters.

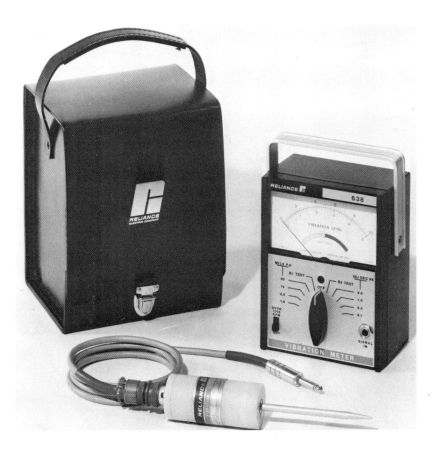

Fig. 28.3   Survey Meter Model No. 638, reading displacement and velocity (courtesy of Reliance Electric Co., Cleveland, Ohio)

## 6. SURVEY METERS

A survey meter is shown in Fig. 28.3, having the following approximate specifications.

| | |
|---|---|
| Weight (in carrying case) | 4.5 lbs |
| Displacement ranges | 0.1 to 50 mil PP |
| Velocity ranges | 0.1 to 1.5 IPSP |
| Power | Battery, 75-hour life |
| Make | Reliance Electric Co., model 638 |
| Price | $500 |

Another survey meter is shown in Fig. 28.4, having the following approximate specifications.

| | |
|---|---|
| Weight (in case) | 16 lbs |
| Velocity ranges (reads to 0.2 mm/sec) | 3.4 to 100 mm/sec |
| Power | Battery |
| Meter detection | rms |
| Make | Carl Schenck; Vibrometer |
| Price | $680 |

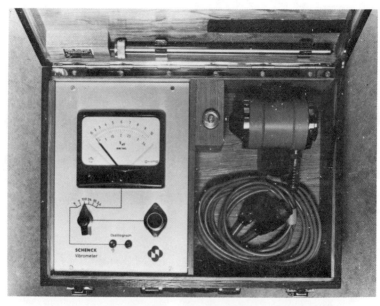

Fig. 28.4    Survey Meter (courtesy Schenck-Trebel Corp., N.Y.)

## 7. VIBRATION ANALYZERS

The survey meter is usually capable of only an all-pass or, as it is also called, a broadband reading. Although the nomenclature is not universally accepted, such meters as comprise a wave analyzer are often called vibration analyzers and usually comprise also a stroboscopic system that may be used as a field balancing instrument and as a tachometer. The analyzer often includes a frequency meter that reads the dominant frequency of vibration.

The wave analyzer, often called the filter, has two uses. It can analyze a signal, for example: 3 mil PP at 30 cps, 0.4 mil PP at 60 cps, and 0.1 mil PP at 90 cps. In this example the components are first, second and third harmonics. By knowing what sources are likely to generate these frequencies, the source of the vibration is usually discovered. Second, when using the analyzer for field balancing, the filter is often used when the balancing is nearly finished and the signal is weak, to stabilize the strobe flashing by excluding frequencies other than the fundamental and by excluding vibrations from neighboring sources. In this it usually gives

Fig. 28.5 Vibration Analyzer. Reading velocity, displacement and frequency and having a wave analyzer, and stroboscope actuated by an oscillator or an unbalance signal (courtesy of Reliance Electric Co.)

stability, but it may shift the phase in a misleading way so as to make its help quite useless. Makers of analyzers are at present making successful efforts to avoid this problem.

A vibration analyzer is shown in Fig. 28.5, having the following approximate specifications.

| | |
|---|---|
| Weight (analyzer alone) | 10 lbs |
| Range | 0.01 to 300 mils PP, also velocity |
| Power | Rechargeable battery |
| Read-out | Average, P and PP |
| Filter range | 2.8 to 5000 cps |
| Frequency meter to read dominant | Yes |
| Make | Reliance Electric Co., model 642 |
| Price | $2000 |

A second example of a vibration analyzer is shown in Fig. 28.6, having the following approximate specifications.

| | |
|---|---|
| Weight of analyzer | 15 lbs |
| Weight in case with strobe and pickup | 44 lbs |
| Power | AC, 110 volt |
| Capability | Vibration meter, stroboscopic flash wave analyzer, frequency meter |
| Make | Carl Schenck (Germany) |
| Price | $1800 |

## 8. VIBRATION METERS

There is no generally accepted nomenclature distinction between the connotation of survey meter and that of vibration meter. For the purposes of this chapter, a vibration meter is somewhat more versatile and accurate than a survey meter. The example shown in Fig. 28.7 is, in effect, a portable voltmeter with a wide amplitude range and extreme sensitivity. It offers a wide frequency range together with the capability of reading out in terms of displacement, velocity, acceleration or jerk. Jerk is the time derivative of acceleration, just as acceleration is the time derivative of velocity; it is sometimes used as a criterion of vibration severity. This meter is usually used with an accelerometer pickup. The typical field vibration analyzer cannot accept an accelerometer so that it cannot measure frequencies exceeding about 1500 cps.

Fig. 28.6  Vibration Analyzer. Reading velocity and frequency, and having a wave analyzer, and stroboscope actuated by an oscillator or unbalance signal (courtesy of Schenck-Trebel Corp.)

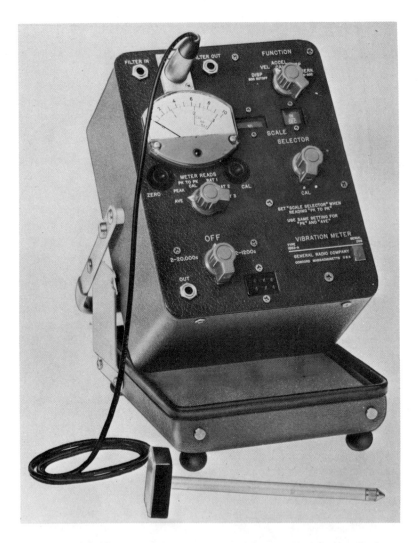

Fig. 28.7    Vibration meter and volt meter, using an accelerometer pickup, and reading Acceleration, Velocity, Displacement, and Jerk, either Average or Peak, or peak to peak (courtesy of General Radio Co., Inc.)

## 9. CONCLUSION

A wide variety of excellent instrumentation for use at the machine site is offered by many makers. The user's choice is affected by his intended program and by what instrumentation best fits with it. This chapter has provided an introductory hint of what is available and usual.

# 29

# THE OSCILLOSCOPE, CAMERA STROBOSCOPE AND SIGNAL GENERATOR

*Michael P. Blake*

## 1. THE OSCILLOSCOPE

The cathode ray oscilloscope is without doubt the most versatile and useful single instrument in all the range of instruments that can accept a signal and enable it to be measured. A bright spot on the screen has its Y coordinate determined by the instantaneous value of the signal voltage, and the X coordinate is usually a linear function of time. The response of the spot is so devoid of inertia that its trace on the screen is a true picture of the signal as a function of time. Some commercial oscilloscopes are capable of displaying signals having a frequency of the order of two thousand million cycles per second so that the display of audio frequencies of twenty thousand cycles per second is easily achieved on the simplest instruments. Whereas all vibration and sound meters and voltmeters read out some kind of conditioned result such as average, peak-to-peak or a weighted result, the oscilloscope gives a true display of the unconditioned, instantaneous signal.

The word oscilloscope is usually reserved for the cathode ray instrument. The word indicates that the result can be seen, whereas the word Oscillograph indicates that the result is drawn or graphically recorded (the word oscillogram often refers to the actual drawing). Pen recorders that record the signal form as a function of time and analogous optical oscillographs are all termed oscillographs. On the other hand, the instrument that generally records a meter reading without Following The Waveform and gives the average or rms value, for example, is called a graphic level recorder. It is helpful in distinguishing the oscilloscope and oscillograph from other instruments to consider the former as being a combination of a voltmeter and a clock that reads easily to a millionth of a second.

**Construction.** The basic construction of the instrument has changed little since its conception. The important improvements have been devoted to accuracy, stability and versatility. Thirty years ago the instrument was a laboratory device, giving useful qualitative pictures difficult to calibrate and tedious to use. Twenty years ago stability and calibration were so poor that this author considered the oscilloscope useless, for example, in the field testing of engine performance, and felt constrained to use such devices as the piston and pencil method or the Farnborough spark instrument in order to make indicator diagrams. In the past two decades, however, the oscilloscope achieved its great promise and is now as stable and reliable as a good wristwatch or radio. Voltage and frequency calibration requires a few seconds, and stability is almost perfect. For these reasons the oscilloscope has become the universal tool in the investigation not only of electronic devices but also of mechanical devices, the motions of which are sensed with some kind of transducer.

The basic elements are shown in Fig. 29.1. A beam of electrons is provided by a source such as a heated rare-earth element. The beam is accelerated to a desirable velocity along the tube axis and then passes between vertical and horizontal deflector plates. The vertical plate voltage usually follows the signal voltage. The horizontal plate voltage usually follows a uniformly increasing voltage provided by an internal circuit in the oscilloscope so that the horizontal movement of the bright spot is such that it traverses an equal number of screen divisions in any given time. If a fixed voltage is applied to either plate (DC voltage), the beam which is axial before entering the plate region undergoes a parabolic deflection, emerging from the plate region as a straight beam which is no longer axial. The deflection is quite analogous to that of a projectile which starts horizontally in a fixed gravitational field and continues with a more or less uniform horizontal velocity as it falls toward the earth. Thus it is possible, although it is not the purpose here, to make measurements relating to the charge and mass of the electron.

So as to avoid photography and, in particular, to display transients, the storage oscilloscopes are most useful. These are becoming more and more common. For general engineering purposes the storage type may be used first as a common oscilloscope in the ordinary way, and then to store any picture up to one hour or so, if required. The signal level that will cause the spot to sweep horizontally may be preset so that the spot does not move until the expected transient or steady wave induces a predetermined voltage in the vertical amplifier system.

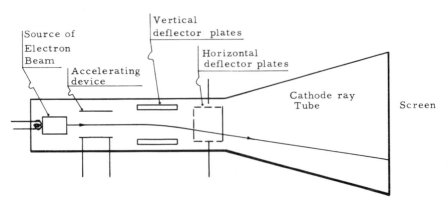

Fig. 29.1   Cathode ray tube, basic elements

The storage oscilloscope is of great value in the study of impacts and transients. The present cost of a suitable storage oscilloscope for mechanical and acoustic studies is of the order of $2000, which is about two or three times the cost of the corresponding simple oscilloscope.

In selecting an oscilloscope for vibration measuring, it is desirable to have sweep times and signal attenuation in terms of 2, 5 and 10 rather and 1 and 10, or 5, 10 and 50. If the 2, 5, 10 system is not available, the display is often too great for the screen or, conversely, too small to be measured when the particular knob is altered by a one-click position.

Recent oscilloscopes offer variable display persistence. The time for a trace to fade to one-tenth of its original brightness is called persistence, and this may be varied continuously in the range from about 1 to 50 sec. Some modern oscilloscopes are described as plug-in types, meaning that parts of the internal circuits may be removed and replaced with different modules so as to provide different capabilities such as panoramic analysis. For the extreme high frequencies of electronic engineering, a sampling circuit may be used. This relieves the deflection circuitry of having to operate unduly fast, by displaying each wave cycle as a series of say 100 spots. For example, when spot No. 10 has been displayed for one cycle, spot No. 11 may then be displayed for the next. The eye and the camera are, of course, content if the actual screen frequency exceeds some very low value such as 30 cps.

**Operation.** The oscilloscope is not easily harmed, and its operation should be understood to some extent by everyone who wishes to measure vibrational phenomena. Perhaps the best way to learn is to use a signal generator as a source, which provides typical waveforms, frequencies and amplitudes. In this way those unfamiliar with the oscilloscope can gain much knowledge and confidence in an hour or two. The basic operational aim is to be able to use the *voltmeter and the clock*.

The oscilloscope is turned on and warmed up. The appropriate knob is turned to the *calibration position*. An internal calibration signal then generates a screen display such as one volt PP at 100 cps. If the attenuation knob is set to 200 millivolts per division and the height of the display is not exactly five divisions as it should be, the screw marked *cal.* is turned to achieve five divisions. Another screw, labeled *balance*, is provided so as to correct any tendency of the trace to shift vertically as the signal attenuation knob is shifted. When the sweep-time is set to one millisecond per division, the 100 cps signal should occupy ten divisions for one wavelength.

A signal of say 100 cps and 70 millivolts PP is connected with the vertical amplifier. If there are ten vertical and ten horizontal divisions on the screen and if the input attenuator knob is set at ten millivolts per division, the vertical amplitude of the beam trace is seven divisions. If the horizontal sweep-time is set at 1/10 sec per division, the trace will move from left to right across the screen every second. What is observed is a vertical, bright line that moves from left to right once per second. If now the sweep-time is set at one millisecond per division, the width of the screen is swept every 1/100 of a second, and the trace appears as in Fig. 29.2.

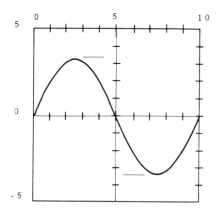

Oscilloscope setting:
Vertical sensitivity:     10 millivolts per division
Horizontal sweep time: 1 milliseconds per division
From this: -
Signal frequency is 100 cps.
Signal amplitude is about 70 millivolts PP.

Fig. 29.2    Oscillograph interpretation

When the signal frequency is lower than about 20 cps, the spot appears to bounce on the screen and the eye cannot perceive the form of the signal. To determine the signal form, a photographic exposure at least as long as one cycle must be made. For higher frequencies a photograph is also useful as a record and as an aid toward a more exact determination of frequency than is possible with the unaided eye. If the signal is periodic and of simple form, such as sinusoidal or triangular, a photograph is not usually necessary. The signal is then fully described by observing and noting its frequency and amplitude. In all cases the object is to determine frequency, amplitude and form. An actual field photograph is given in Fig. 29.3. This shows the output of a vibration meter.

An oscillograph such as is shown by Fig. 29.3 gives an excellent picture of what is happening and is sensitive to every change. For calibration of the picture, it is merely necessary to record a steady vibration of some item, such as a turbine, taking another photograph if necessary, leading to a result such as 1 mil PP is equivalent to 3.5 vertical divisions when the vibration meter attenuator and the vertical sensitivity of the oscilloscope are set at the same levels used in producing the original picture. If a tape recorder is used, it is desirable, in general, to record some known signal. Then when all other signals are played back and photographed on the oscilloscope, the amplitude calibration is known. Frequency calibration offers no difficulty.

Sine, square and triangular signals are fed to the oscilloscope, varying the frequency and voltage. After a little practice it is easy to identify the voltage and frequency. The Lissajou figure is a less common form of display. For this no time base is used. One signal is fed to the vertical amplifier and another signal to

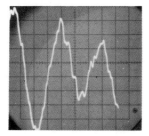

Volts per division: 1.0
Milliseconds per division: 10

Whence:
Periodic time is say: 30 milliseconds
Frequency is about 33 cps.
Amplitude is between say: 4 and 8 volts.

Fig. 29.3 Actual oscillograph interpretation. This oscillograph is a Polaroid photograph taken with a Polaroid 110A camera without an oscilloscope adaptor. The signal is an analog of the lateral vibrational displacement of the rear bearing of a horizontal, stiff-shafted centrifuge of about 30 hp at 2000 rpm. The signal is only approximately sinusoidal and contains a higher frequency about five times that of the fundamental.

the horizontal. For example, the signals might be the vertical and horizontal velocity of a turbine bearing. The display then gives an indication of the actual path in space of the bearing housing. It is possible from Lissajou displays to extract some information regarding the relative phase of the signals and the frequency ratio of one signal to another. (It is suggested that the reader refer to books that describe the detailed use of the oscilloscope.)

Useful information regarding phase is often available in other ways. For example, a sine-wave generator can be attached so as to display the rotation of a rotor as a sine curve. Sweep-rate is set to a convenient value, and there is no horizontal input. A periodic pickup signal displays the vibration displacement in a second display on the same screen. Some oscilloscopes have dual beams for this purpose. Others achieve a dual trace without dual beams by chopping or alternating a single beam so as to give the effect of two separate beams using two vertical amplifiers. When alternating, the beam first sweeps the screen width to display one signal and alternately to display the other. When chopping, the beam displays both signals in each sweep by moving between them at a rate so rapid as to be undetected by the eye. Suppose the crest of one trace is vertically in line with that of the other. Then the two are in phase, and the horizontal distance between crests is 360 degrees. Now, if a balance weight is applied to the rotor, the phase may exhibit a shift of 80 degrees. Using this information, the rotor may be balanced using the usual simple vector procedures.

As a final example of phase measuring, consider a sine generator giving two equal sine outputs. If these are phased at 90 degrees and fed respectively to vertical and horizontal amplifiers, the spot traces a circle on the screen. For such applications it is desirable to have identical vertical and horizontal amplifiers, which is not always the case. If the signal generator is a rotating machine, the trace frequency is that of shaft rotation. If, at the same time, matters are so arranged that a suitable voltage pulse comes from a vibration pickup at some fixed point, such as a maximum or zero crossing for the vibration signal, this pulse is connected to the Z-axis input, as it is called, and has the effect of blanking the trace momentarily. This Z-axis blanking causes a small dark gap in the circular trace. When a balance weight is applied to the rotor, the position of the gap is likely to change. It does change if the phase relation of rotation to unbalance changes. In this way phase may be measured with useful accuracy for various purposes.

**Specifications.** For vibration measuring, an oscilloscope of about the following specifications is adequate.

|  | MINIMUM CAPABILITY SPECIFICATION | MAXIMUM CAPABILITY SPECIFICATION |
| --- | --- | --- |
| Vertical sensitivity, volts per division (1, 2, 5 sequence) | 10 millivolts | 100 microvolts |
| Sweep-time per division (1, 2, 5 sequence) | 0.5 second to 0.5 microsecond | 5 seconds to 0.1 microsecond |
| Number of traces | 1 | 2 |
| Number of beams | 1 | 2 |
| Persistence | Fixed | Variable |
| Storage capability | No | Yes |
| Weight not exceeding | 45 lbs | 45 lbs |
| Price | $750 | $2500 |

A typical storage oscilloscope is illustrated in Fig. 29.4. The brief specifications are as follows.

| | |
| --- | --- |
| Model number | Hewlett-Packard type 1201 A |
| Number of traces | 2 |
| Sweep-time per division | 5 seconds to 0.1 microseconds |
| Vertical sensitivity, per division | 20 volts to 0.1 millivolts |

| | |
|---|---|
| Storage capability | Yes, with variable storage time |
| Variable persistence | Yes |
| Input | One horizontal, two vertical |
| Weight | Net 30 lbs |
| Price | $1800 |

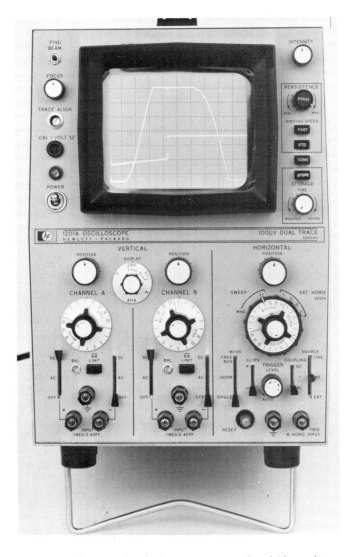

Fig. 29.4 Oscilloscope with dual trace, storage and variable persistence capabilities (courtesy of Hewlett-Packard)

As an example of a versatile lightweight oscilloscope, the following specifications are typical.

| | |
|---|---|
| Model number | Tektronix, Inc., type 422 |
| Number of traces | 2 |
| Sweep-time per division (1, 2, 5 sequence) | 0.5 seconds to 0.5 microseconds |
| Vertical sensitivity per division (1, 2, 5 sequence) | 20 volts to 10 millivolts |
| Storage and variable persistence capability | No |
| Input | One horizontal, two vertical |
| Weight | 20 lbs |
| Price | $1450 |

## 2. THE CAMERA

In the past decade the Polaroid camera has been used almost exclusively, both for oscilloscope work and for making other photographs in the field. The camera is a recording instrument, and may be grouped with the pen recorder and tape recorder because it makes data available for future reference. For some work a reflex camera is also desirable.

**Specifications.** Polaroid cameras constructed especially for oscilloscope use are offered with lenses of up to about f 1.9 aperture, together with additional features such as electrical remote control, synchronization signal and ultraviolet light for better contrast of the picture, as well as convenient viewing and mounting. Such cameras cost from about $400 to $550.

Some of the general-purpose Polaroid cameras without oscilloscope attachment, such as the model 110 A, 110 B and the 110 Pathfinder, are quite suitable for taking oscilloscope pictures. The 110 A and 110 B differ mainly in that the 110 B is provided with a parallax-corrected view finder. Prices range from about $150 to $250. The 110 A has an f 4.7 Rodenstock lens. Both the 110 A and 110 B have the EV system of correlating f number and exposure time so as to give a single EV reading on the shutter mount. This system is a great convenience, particularly as an aid to rapid oscilloscope photography in the field.

When laboratory work of a close-up kind arises, the use of a reflex camera is often unavoidable. The most versatile lens is that which covers the range infinity to about 4 in. Such lenses are offered with several cameras such as the Nikon and Alpa. Reflex Polaroid cameras are also available.

**Application.** Periodic or stored traces on the oscilloscope are still life so to speak and require no unusual technique. Sometimes, nonrepetitive traces are measured and if the exposure time exceeds the duration of one horizontal sweep, two or more superimposed and confusing traces appear on the photograph. In this case, exposure duration must be set to correspond with one sweep or less. Thus, the exposure time and the corresponding f number are determined and there is no choice. When in a given location and illumination it is determined, for example, that EV 11 is the most suitable exposure and the exposure time varies, the f number is readily selected by resetting the EV pointer to 11 each time the shutter speed is altered, without the bother of computing the required f number and without even knowing what it is.

Polaroid photography offers two great advantages: There can be no disappointment because of unacceptable pictures discovered too late. Also, standard film speed as high as 3000 ASA is available, which is 7.5 times faster (or three f stops faster) than fast ASA 400 panchromatic film. The cost per exposure is about $0.25 as compared with say $0.20 or less for pictures from 35 mm cameras. In general, all oscilloscope pictures and such field and laboratory pictures as may not be available for taking at a future time are best taken with a Polaroid camera if possible. Otherwise, and especially for close-up work and color slides, the 35 mm reflex camera is preferable.

The general-purpose Polaroid may be used without flash and the risk of explosion for the photographing of plant sites and equipment in the field. By using the close-up lens attachment and placing the camera on a table about ten inches from the oscilloscope screen, excellent oscilloscope photographs are taken without difficulty. The photograph of Fig. 29.3 was taken in this way. After the oscilloscope, the Polaroid camera is one of the greatest tools at the disposal of persons interested in the care of machinery via vibration measurement.

## 3. THE STROBOSCOPE

A stroboscope provides a periodically flashing light. In general, the shorter the flash duration and the higher its intensity and frequency range, the better the stroboscope. It is a well-developed, rugged, low-cost device that is much used in maintenance, research and development work. Usually the construction comprises a condenser that is charged between flashes and which may be triggered in a variety of ways, such as by using a high-voltage signal from an automotive ignition or a signal of perhaps a few millivolts connected with a suitable trigger circuit. The voltage in the flash circuit may be expected to be several hundred volts so that this safety consideration must be borne in mind when examining or repairing the instrument.

**Specifications.** Typical specifications are given below, corresponding with the instrument illustrated in Fig. 29.5.

| | |
|---|---|
| Description | Strobotac, type 1531-A, General Radio Co. |
| Flashing rate | 110 to 25,000 per minute |
| Accuracy | 1 % of flash-rate dial setting |
| Calibration | Against power line frequency |

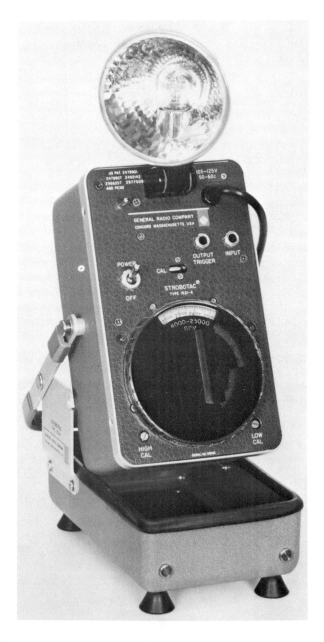

Fig. 29.5   Stroboscope, general purpose, Strobotac Type 1531–A (courtesy
of General Radio Co., Inc.)

| | |
|---|---|
| Flash duration at $\frac{1}{3}$ peak intensity | 0.8 to 3 microseconds |
| Peak light intensity | 0.21 to 4.2 million candlepower continuous, and 7 million for a single flash |
| Output trigger | 600 to 800 volts negative |
| Input trigger | 6 volts PP required |
| Power | 35 watts, 115 to 250 volts, at 30, 60 or 400 cps |
| Weight | 7¼ lbs |
| Price | $300 |

**Application.** The stroboscope is used universally as a tachometer to measure the speed of rotation without mechanical or electrical connections. However simple the procedure may appear, errors are often made. Ideally, a distinctive mark is made on the end of the shaft, to one side of the center. As the flash frequency is varied, the shaft will appear to stand still. By trial, a frequency is found that generates only one stationary image of the mark. Suppose the rotative frequency is $f_s$ and the flash frequency is $f_f$; the $f_f$ will have one of the following values when the one stationary image appears: $f_f = f_s/n$ where $n = 1, 2, 3, \ldots$ By decreasing the flash rate so as to find adjacent single stationary image frequencies, the flash frequency changes from $f_s$ to $f_s/2$, or from $f_s/2$ to $f_s/3$ and so on. The problem is how to determine $f_s$.

If the first flash frequency be denoted as $f_{f1} = f_s/n$, and if the second flash frequency be denoted by $f_{f2} = f_s/(n + 1)$, then from the above

$$f_s = \frac{(f_{f1})(f_{f2})}{|f_{f1} - f_{f2}|} .$$

This gives the following rule for the determination of shaft speed: discover a single stationary image on the shaft end. Increase or lower the flash frequency carefully so as to discover the least frequency change that will once more generate a single stationary image. It is more usual to lower the flash frequency, and this must be done if in the first instance it is equal to the shaft speed. Then multiply the two observed flash frequencies and divide the product by their absolute difference.

If the end of the shaft is not available, a difficulty arises when the shaft mark appears stationary. The flash rate may be twice that of the shaft speed, and the double image is not observed because it is on the far side of the shaft. This does not invalidate the formula for speed as given above, as long as the operator proceeds from any multiple of shaft speed to the next whole-number multiple.

To avoid confusion it is usual to set the stroboscope frequency at shaft frequency or a submultiple because this permits the measurement of shaft speeds that are as much as ten times higher in frequency than that of the flash. To insure that only one image is possible even if the shaft were illuminated all around its periphery, a chalk circle is drawn around the shaft. A gap is rubbed in the circle, and a short axial chalk line is drawn in the gap. A virtual double image is then detected by the disappearance of the gap as the shaft rotates. Conversely, an axial mark and a circumferential mark may be made opposite one another and on about the same circle. If the flash discloses a cross, it is two or more times faster than the shaft speed.

Steady lateral or rotative movement may be made to move very slowly instead of being held stationary. To determine the direction of shaft rotation, first make the single image stationary; then decrease the flash rate slightly. The apparent slow rotation of the shaft is now in the same direction as the actual rotation. By using this slow motion, the amplitude and dynamic behavior of a variety of mechanical vibrations may be examined at ease. Photographs of the dynamic performance, either still or in slow motion, are easy to make. The suggestion of absence of motion is so powerful that the safety precaution must be taken of remembering that motion, which may be dangerous to the person, is in fact taking place.

The stroboscope offers a unique phase measuring capability, which is used almost universally for the balancing of rotors in the field besides being often used in the shop. If the stroboscope is fired by a signal from a vibration pickup on a shaft bearing and the shaft carries an index mark, then phase shift between rotation and unbalance, when a balance weight is altered, is immediately and very simply disclosed by a shift in the index position. The shift is equal to the phase shift, although it is most usually in the opposite direction.

A variety of other phase measurements may be simply made with the least expense and complexity and with considerable accuracy. For example, a rotary shaker shakes the earth and carries an index mark. If a vibration pickup is moved over the earth while firing a stroboscope, the relative phase of the earth vibration at any position is readily measured. Again, for example, a pickup may be moved around the base and structure of a turbine as the stroboscope illuminates a shaft index. Sudden or gradual phase shifts in the structure vibration are immediately evident and are often a significant key in the matter of disclosing faults, such as foundation cracks. Signal generators are exemplified by the "sine generator" attached to shafts and by independent signal generators, which are in fact oscillators. The latter are described in the following section.

## 4. THE SIGNAL GENERATOR

The signal generator is usually an electronic oscillator capable of generating sinusoidal waveforms with little distortion and with continuously variable frequency. Instruments which provide waveforms other than sinusoidal are often called function generators, whereas those providing sinusoidal waveforms only are often called signal generators. Function generators and signal generators together with oscillators are usually grouped under the generic term, signal sources. The basic requirement is the generation of a periodic signal which is a replica or approximation of the sine function at various frequencies from about 1 to 20,000 cps. The instrument is vastly improved if it generates perfect waveforms such as the sine, triangle and square and if the output voltage remains steady as the frequency is altered, even in the presence of a slight output current.

In most signal generators the frequency is steplessly variable. For work related to mechanical vibration, a useful frequency range is from about 1 to 30,000 cps.

**Specifications.** Generators supplied in kit form and costing perhaps $50 are extremely useful. The more professional generators cost about $350. In procuring a generator it is to be remembered that those commonly used for radio and electronic work are often of little use in vibration work because they are designed mainly for high-frequency applications.

A typical, useful signal generator is illustrated in Fig. 29.6, and the brief specifications are as follows.

| | |
|---|---|
| Description | Function generator, Wavetek model 110 |
| Dynamic frequency | 0.005 to $10^6$ cps (cycle duration: from 3.3 minutes to one microsecond) |
| Output | Six, simultaneous |
| Waveform | Sine, triangular, square. Also synchronous pulse |
| Power | 105 to 125 volt, 200 to 250 volt, 50 to 400 cps, less than 10 watts; or rechargeable battery (110 B) |
| Weight | 7 lbs |
| Price | About $445 |

Fig. 29.6   Function Generator Model 110 (courtesy of Wavetek)

**Applications.** The applications are very numerous and include, for example, the simulation of mechanical vibration using an amplifier and shaker, the determination of response of transducers and seismographs, audio measurements and circuit testing. Since all of the measuring instrumentation used in vibration studies deals with electrical signals, it is of the greatest advantage to be able to simulate, more or less, such signals. The simulated signal may be used to test the instrumentation response, to check circuits for faults, or to provide an artificial vibrational environment for the testing of mechanical components.

When testing circuits for continuity and performance, the capability of varying frequency gives an assurance that the observed signal at some point in the circuit is, in fact, due to the input test signal. When testing instrumentation response, consider the use of a variable-frequency input signal, such as in the case of common direct-type tape recorders. Suppose at a fixed voltage of say 50 millivolts which is monitored by an oscilloscope at the tape recorder input, various frequencies are recorded such as 25, 50, 100, 1000 and 10,000 cps. The recording is then played back, using an oscilloscope to measure the output voltage and to identify the frequencies. In this way a graph of frequency response is obtained. Against a base of frequency, the playback voltage is plotted at each chosen frequency. The output attenuator of the tape recorder may be set to some arbitrary level such as 100 millivolts at 1000 cps. It is then left unchanged while the output voltages at other frequencies are measured. An ideal tape recorder gives a horizontal straight line when output voltage is plotted against frequency. The same kind of test may be applied to an FM tape recorder which nominally has an input/output voltage relation that is a fixed ratio, such as 1:1. The final graph of output voltage is then not only a criterion of flatness of response, but it is also a measure of the constancy of the ratio. The oscilloscope detects any departure of playback frequency from signal frequency.

As an example of using the generator to simulate an acoustic or vibrational environment, suppose that various acoustic frequencies and levels are required for architectural or physiological testing. It is merely necessary to connect the signal generator to a speaker via an amplifier, while adjusting the signal frequency to the desired ranges and adjusting the amplifier or generator so as to give the desired levels. If a shaker is used instead of the speaker, the shaker then simulates any desired harmonic amplitude and frequency, and it may be used to test components that are intended for use in such an environment. By using more advanced function generators, many kinds of complex vibration may be simulated. Sometimes it is simpler and more acceptable in the case of complex vibrations to make a tape recording of an existing analogous environmental vibration and to use the playback as a signal source to control the shaker.

# 30

# WAVE ANALYZERS

*Michael P. Blake*

## 1. SCOPE

Three distinct types of analyzers are described: the wave analyzer that reads the signal amplitude that corresponds to a chosen, tuned frequency band; the panoramic analyzer that exhibits the amplitude-frequency spectrum on an oscilloscope screen; the audio spectrograph that prints out the amplitude-frequency spectrum or the frequency-time-amplitude spectrum. Description, purpose, operation and specifications are discussed.

## 2. INTRODUCTION

A vast variety of signals are generated by connecting a microphone or motion pickup to a dynamic event, such as sound or vibration. An analog voltage is thus generated which follows the event, faithfully representing its frequency and amplitude as a voltage with the same frequency components and with an amplitude proportional to the amplitude of the event. In order to interpret the event, it is usually necessary to analyze the signal. A periodic signal with, perhaps, a fundamental and one harmonic component can often be analyzed by observing its oscilloscope display. The two frequencies and the corresponding amplitudes are readily evaluated, and that is all the analysis needed or possible, except for the phase relation, if the components are sinusoidal. But where several frequencies are present or the signal components are periodic but not sinusoidal, an analyzer is required. Any periodic signal may be regarded as a summation of a sufficient number of sinusoidal harmonics, the frequency and amplitude of each depending on the nature of the signal. The analyzer may be considered as a device that reads the amplitude of the harmonic to which it is tuned, for example, as illustrated in Fig. 30.1b. Sometimes the analyzer is capable of giving the phase relation of each harmonic to the fundamental. Amplitude and phase information as a function of frequency completes the analysis of a periodic signal for most purposes, although phase information is not usually required.

Another kind of signal is called noise. When all frequencies are present, analysis leads to a spectrum such as shown in Fig. 30.1a. A special kind of noise is

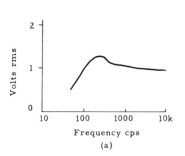

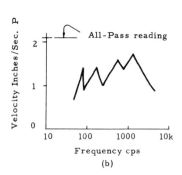

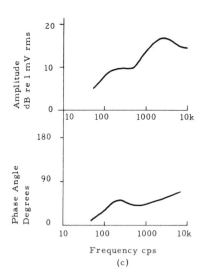

Fig. 30.1   Examples of spectra derived from wave analyzers. Figure 30.1a exhibits many frequency components and is suggestive of wide-band noise. Figure 30.1b exhibits a combination of approximately four pure tones. This type spectrum is most valuable in the identification of faults in machinery because the source of the fault may be determined from the frequency of the fault. The all-pass reading (without filter) indicates if some fault exists. Wave-analyzer spectra are plots or numerical displays of some amplitude as a function of frequency. Amplitude is plotted on a linear or log scale and may represent such parameters as sound pressure level, vibrational velocity, acceleration or displacement, power or voltage. Figure 30.1c shows a complex spectrum.

called random noise. It may contain a mere group of frequencies or all frequencies; the signal is not periodic. Thus, the spectral analysis is different at different instants of time. The noise is called *white noise* if the power in the signal is independent of frequency. *Pink noise* has constant energy per octave of bandwidth. The random signal can never be absolutely defined, and it is the most difficult to describe analytically. All that can be done is to give probability estimates in terms of parameters that are the most valuable in relation to a given study of the signal. Such parameters are the root-mean-square, mean-square power spectral density, cumulative distribution, probability density, autocorrelation function and Fourier integral transform of the power spectrum.

The mathematical aspects of random-vibration analysis are beyond the scope of this chapter. They are noted here in order to indicate the range of analytical methods and, therefore, to indicate the range of analytical instrumentation that lies outside that which is described later. The two constant-bandwidth analyzers that are described later are capable, with their related instrumentation, of evaluating many of the parameters of random signals.

Although some devices such as the wattmeter are at times used as filters, for example in instrumentation for rotor balancing, most analyzers contain electronic filters. The characteristics of an ideal and an actual filter are indicated in Fig. 30.2. Because of their overall electronic characteristics, all analyzers depart, more or less, from the ideal, so that the read-out often departs considerably from the true form of the spectrum. For example, two pure tones differing by perhaps 4 percent in frequency may appear as a single tone in the spectrum. A pure tone accompanied by much noise may not appear in the spectrum at all. In some cases, though, it may appear to be of greater amplitude or less amplitude than, in fact, it really is. For a vast array of measurements connected with the care of machinery, the departure of the analyzer from the ideal is of no consequence. On the other hand, when nonroutine signal analysis is undertaken, it is a part of wisdom to ask if the analyzer is giving readings true enough to be profitable, and if not, what corrective steps may be taken to insure the required accuracy and discrimination. Likewise, when communicating the results of signal analysis in nonroutine work, it is advisable to state the make and model number of the instrumentation used.

A signal having many harmonics is sometimes called a broadband signal and one having few is called a narrowband signal. The analyzer reading when the filtering system is inoperative is often called the broadband or all-pass reading. The analyzer behaves then as a voltmeter. The reading is indicative of the signal amplitude taken as a whole, and is therefore indicative of the amplitude or severity of the event of which the signal is the analog. In many actual signal analyses the all-pass reading differs little from the greatest reading in the corresponding spectrum, the all-pass reading being perhaps 5 dB above the greatest narrowband reading. Thus, the all-pass reading is indicative of the existence of a severity or energy level in a dynamic event, whereas the narrowband reading is indicative of (and relates to the indentification of) the significant mechanical faults or other components that contribute to the existence of the output level signal.

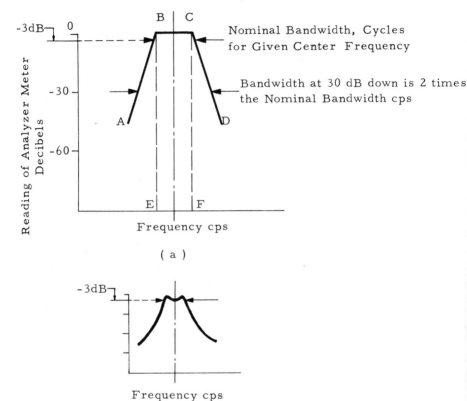

Fig. 30.2 Selectivity of filters. EBCF indicates ideal filter response. ABCD indicates response of the analyzer to a pure tone which is varied in frequency around a center frequency to which the analyzer is tuned. A detailed plot of ABCD for typical filters usually shows curved skirts and rounding and some ripple within the nominal passband as indicated in Fig. 30.2b. Figure 30.2a shows the typical selectivity of filters for the analyzer of Fig. 30.5. The shape factor here is 30 dB per octave bandwidth.

The output of the filter system is usually rectified and read out on a meter. Besides this, most analyzers have a sinusoidally alternating voltage output corresponding to the tuned frequency and filtered amplitude. Some analyzers have a DC analog output such that the DC voltage constantly reflects the rms or average value of the filtered output. This is necessary for automatic plotting of spectra on recorders that are not capable of converting the AC filtered output to its DC analog.

Manufacturers use various criteria to describe the selectivity of filtering. Selectivity is an index of the conformity of the filtering to ideal filtering as indicated in Fig. 30.2. The nominal bandwidth is often taken to be that which is measured 3 dB below the highest point in the response curve. It may be expressed in octaves, a ratio or in cycles. Selectivity may then be expressed in terms of the number of dB down at which the bandwidth is say twice the nominal bandwidth. But analyzers are not limited to instrumentation. The human ear with its related memory and perception faculties has remarkable analytic capability. It can detect, analyze, compare and recognize particular footfalls or voices almost instantaneously in a manner that a whole roomful of instrumentation cannot emulate.

## 3. THE WAVE ANALYZER

All of the instruments discussed in this article are wave analyzers, but the term is traditionally reserved for those simple instruments that amplify and filter the input signal and read out the amplitude for each frequency on a meter. A typical example is shown in Fig. 30.3.

**Constant-Percent-Bandwidth Analyzer.** This is the most common analyzer and the bandwidth is, for example, one octave or one-third or one-tenth octave. Thus, the actual width of the nominal pass band might be 1000 cycles at a center frequency of 2000 cps, and 10 cycles at a center frequency of 20 cps. This analyzer is sometimes called a proportional-bandwidth analyzer. The center frequencies of the internal filters are in accord with some set of preferred numbers. When the bandwidth is less than one-third octave, the analyzer is usually continuously tunable; that is, the center frequency may be set at any value within the range of the instrument.

The purpose of this analyzer is to determine with moderate precision the content of typical signals that are voltage analogs of mechanical or electrical events, usually periodic. It is the type most used for identifying faults in machinery and for field balancing. Machine faults are almost invariably identified in terms of frequency. For field balancing, particularly when the balancing is almost complete, it is usually desirable to filter out all frequencies except that of rotation. This analyzer is often used as a separate unit, and is sometimes found as part of a combination instrument such as a vibration or sound-level meter.

To operate the analyzer, the signal of interest is connected to the input. Care is taken to adjust the attenuator knob to the highest voltage range to avoid damage to the meter. The maker's special instructions must be followed. The meter-response knob is set to the slow (highly damped) position to avoid wild movements of the meter needle until the response behavior of the meter to the signal is known. The bandwidth knob is usually set to all-pass, where there is no filtering and the instrument is behaving simply as a voltmeter. By turning the attenuation knob, which is usually calibrated in decibels and volts, a midscale reading of the meter is obtained and noted as the all-pass reading. Then the pass band is set to the one or more bands offered by the instrument. It is customary to connect an oscilloscope to the analyzer input or output as a monitor. The tuning knob is then turned by hand from the lowest frequency range of interest to the highest. As the knob is slowly turned, the meter readings and

Fig. 30.3    Sound and Vibration Analyzer, Type 1564, 1/3 and 1/10th-
octave (courtesy General Radio Co., Inc.)

corresponding frequencies of interest are noted for later plotting. For acoustic
measurements the amplitude is invariably recorded in terms of decibels. The
decibel measure is sometimes used for acceleration, whereas vibrational velocity
and displacement are usually recorded in terms of volts, or inches per second and
inches, respectively. The analyzer, when in use, is in fact a frequency-selective
voltmeter, usually with an rms response. The type of response must be borne in
mind when presenting the resulting spectrum so as to make it clear as regards
what the amplitude units are on which it is based. The plotted spectrum might
be described as, for example, 'response of a true rms one-third-octave analyzer
to vibration signals from a blower fan, recorded directly on brand X tape recorder
and played into the analyzer. A calibration sine vibration of 10 mils PP at 100
cps gives a meter reading of 100 millivolts rms when the analyzer is tuned to
about 100 cps'.

    For routine use, the operator need not be too concerned with what the analyzer does to the signal. But if the results are communicated to other persons, great care is necessary to avoid misinterpretation. For example, many galvanometers average the input signal and are calibrated to read rms on the assumption that the signal is sinusoidal; and the reading has no readily ascertainable meaning when the signal is not simply sinusoidal. However, for many purposes the operator needs to know only the all-pass level and whether harmonics are present and what are their frequencies and approximate relative amplitudes. This information is readily obtained from the constant-percent-bandwidth analyzer.

It may be necessary to vary the meter attenuation when scanning the spectrum in search of very weak components. Some spectral plots may look like Fig. 30.1a, and enough readings are taken to outline the curve. Some plots may look like Fig. 30.1b, and it is usually adequate to find the consecutive maxima and minima and join them with straight lines.

Brief specifications for the analyzer illustrated in Fig. 30.3 are as follows.

| | |
|---|---|
| Tunability | Continuous |
| Frequency | 2.5 to 25,000 cps |
| Filters | At least 30 dB attenuation when signal is one-half or twice the selected frequency |
| Signal voltage | 0.3 millivolts to 30 volts full scale |
| Output | At least one volt open circuit when meter reads full scale |
| Calibration | Built in |
| Detection | rms, with three averaging times |
| Power | 105 to 125 or 210 to 230 volt, 50–60 cps; battery is also provided |
| Weight | 15.5 lbs |
| Price | $1500 |

The advantage of the constant-percent-bandwidth analyzer is that it yields the analytical information that is most usually required in vibration and acoustic studies conveniently, using moderate cost instrumentation. The limitation of this analyzer is that it does not usually discriminate too well between pure tones that have nearly the same frequency. It does not, in general, yield all of the information that may be required, for example, in the study of noise or random vibration wherein nominal filtering bandwidths as narrow as 5 or even 1 cycle are commonly used.

**The Constant-Bandwidth Analyzer.** In this analyzer the filter bandwidth is fixed at a value lying between about 1 cycle and 200 cycles, and it is independent of the center frequency. Two instruments are described below, both being continuously tunable by means of the heterodyne principle and a single highly discriminatory filter. For example, a filter centered at 50,000 cps and having a bandwidth of 10 cycles is used to analyze a signal. The signal is mixed with a signal from an oscillator, generating a sinusoidal signal of constant amplitude, and the oscillator frequency is varied as the analyzer is swept over its frequency range. The oscillator frequency is added to each frequency in the signal. If the oscillator is set at 49,000 cps, the signal components in the frequency range of 995 to 1005 cps will pass the filter. The tuning knob of the analyzer is arranged so as to indicate the frequency difference between the oscillator and the filter. In this way continuity is achieved and only one filter is required.

The purpose of the constant-bandwidth analyzer is to achieve more discrimination than is usual in the constant-percent type, to make detailed studies of noise and vibration, to analyze random vibration or, for example, to provide power spectral density data. To operate the analyzer, the signal is connected to the input after calibration and overall fitness have been checked. The attenuation knob is first set at its greatest voltage so as to avoid damage to the meter, and then turned to lesser voltage ranges until the signal gives a mid-scale meter reading without filtering. Usually a range of nominal bandwidths is provided, either by plug-in filters or by knob selection. A preliminary scan of the signal is often made with the widest bandwidth, turning the frequency knob and noting the meter reading at each frequency of interest. A narrow bandwidth scan takes far more time because, in general, the meter readings will swing more widely and more often.

Figure 30.4 illustrates a typical analyzer, and very brief specifications are given below.

### SPECIFICATIONS FOR ANALYZER, FIG. 30.4

| | |
|---|---|
| Bandwidths | 3, 10 and 50 cycle |
| Frequency | 20 to 54,000 cps |
| Selectivity | At least 30 dB down when signal frequency is plus or minus twice the nominal bandwidth away from center frequency |
| Voltage range | 30 micro volts to 300 volts full scale |
| Power | 105 to 125, 210 to 250 volts, 50 and 60 cps, 40 watts |
| Weight | 56 lbs |
| Price | $2650 |

The above analyzer features automatic frequency control to enable it to remain tuned to a slowly varying component that might otherwise drift out of the chosen bandwidth. The analyzer may also be used as a tracking generator.

Another typical analyzer is illustrated in Fig. 30.5. This constant-bandwidth analyzer is a general-purpose instrument besides being particularly adaptable to analysis of the vibrations of engines, for example, wherein the harmonics often bear a fixed ratio to the fundamental rotative speed; but the difficulty exists that the fundamental may drift unintentionally from an ostensibly fixed speed or it may be varied intentionally over a wide range during operation or test. These vibrations exemplify particular application needs, and these shall be taken up later.

Fig. 30.4    Type 1900, Wave Analyzer, constant bandwidth 3, 10 and 50
cycle (courtesy General Radio Co., Inc.)

Very brief specifications of the analyzer are given below.

SPECIFICATIONS FOR ANALYZER, FIG. 30.5

| | |
|---|---|
| Bandwidths | 1.5, 2, 5, 10, 20, 50, 100 and 200 cycle |
| Selectivity | 4:1 at 60 dB down |
| Frequency range | 2 to 50,000 cps |
| Input signal | 32 mV to 10 V rms, full scale, 10 dB steps |
| Power | 105 to 125, 210 to 250 volts, 50 or 60 cps, about 180 watts |
| Shipping weight | 22 lbs |
| Tuning | The analyzer is tuned to track the frequency of an independent sine-wave signal source |
| Price | $2900 |

Regarding the analytical needs of particular applications, a vast array of
problems connected with day-to-day care of machinery is amenable to solution

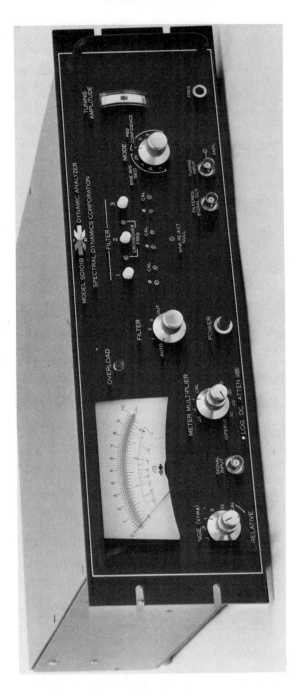

Fig. 30.5   Constant Bandwidth Analyzer Model SD 101 B   (courtesy of Spectral Dynamics Corp.)

using only a constant-percent-bandwidth analyzer of about one-third to one-tenth-octave bandwidth. All that is required are the approximate relative amplitudes of the various frequencies occurring in the spectrum. The analysis of torsional and lateral vibration for internal combustion engines sometimes requires that the analyzer be arranged so as to track the fluctuating shaft speed. The tracking filter may be locked to the shaft speed using a tachometer to generate a sine signal which tunes the analyzer always to the signal frequency. For example, as the engine is speeded up and down, it is possible to read out or print the vibration corresponding only to shaft speed, while excluding all other vibration. Another refinement of the tracking filter is the capability of printing an amplitude-frequency spectrum, not in terms of frequency but in terms of the ratio of the frequency of each harmonic component to the fundamental or to some chosen reference frequency. As the engine is operated at various speeds, the different harmonics, or *orders* as they are usually called, are readily recognizable on the special spectrum which is usually called a Signature Ratio. The harmonics 2, 3, 4 and so on appear at the same positions on the printed spectrum, irrespective of the operating speed.

For research and other purposes, far more exacting analysis than that described above is often necessary. For example, the phase relation of each harmonic component in relation, perhaps, to the fundamental may be of interest. A mathematical theorem states that any complex, periodic signal may be generated by adding together a sufficient number of harmonically related sinusoidal functions of time such as $A_1 \sin(\omega t + a_1)$ and $A_2 \sin(2\omega t + a_2)$, where $A_n$ is an amplitude, $a_n$ is the phase angle in radians and $\omega$ is the radian frequency of the fundamental. The general term is $A_n \sin(n\omega t + a_n)$. Here n is always a whole number such as 4 and never a number such as 4.2 (this is what is meant by *harmonically related*). The analysis of a time function into these harmonic components is called Fourier analysis. The signal of interest may be, for example, a sawtooth having no relation, as regards the vibration it represents, to a sinusoidal source. Nevertheless, for such a complex signal as well as for the static deflection curves of structural components, it is often found that the signal or geometrical curve may be better understood and more profitably manipulated in terms of its Fourier components even though these, as can be imagined, are partially fictitious. Beyond the Fourier analysis, it is often of interest to estimate power spectral density, cumulative distribution, autocorrelation and other functions. These are beyond the scope of this chapter. Perhaps the most important point to be kept in mind by the nonspecialist is that no matter what kind of analyzer is used, the reading almost never reflects the *true* spectrum. The deviation from truth is sometimes as much as 3 dB, which means that the observed amplitude is about one-half, greater or less than the actual. For these reasons, when absolute amplitudes become of interest instead of more or less comparative and approximate amplitudes that are so helpful in day-to-day care of machinery, the limitations and performance characteristics of the instrumentation must be understood. In this matter the reader is referred to standard texts and particularly to the bibliography at the end of this chapter.

The simplest analyzers have been discussed above. The more versatile combinations of these with their attachments include such capability as the generation of a sine signal that can be fed to a shaker and the ability to automatically tune the filter to that signal (or any chosen harmonic or to sweep automatically through the frequency spectrum) in order to read the response of some mechanical component on the shaker, which generates an analog response signal, via a pickup attached to it. A sweep oscillator is used for the generation of periodic signals that move through the frequency range at an adjustable preset linear or logarithmic rate. Combined sweep-generation and tracking capability greatly adds to the versatility of the instrumentation. The combination is called a tracking generator. Beyond these capabilities, many more are available; the reader is referred to the literature of the manufacturers, which is an excellent guide. Perhaps the most often used adjunct to the analyzer is the graphic-level recorder

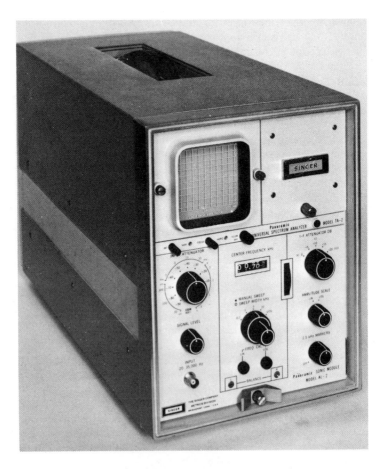

Fig. 30.6   Panoramic Universal Spectrum Analyzer, Model MF–2/AL–2
(courtesy of the Singer Company)

or a somewhat different instrument called an X-Y plotter, both of which print out the spectrum for permanent reference.

## 4. THE PANORAMIC ANALYZER

Using the simple wave analyzer, perhaps thirty minutes are required to make a typical manual scan of the signal of interest and to plot the results. If a graphic-level recorder is used, the time required to print one spectrum varies from about a minute to ten minutes. Thus, time is consumed because of manual operation or because of the mechanical or servo-electrical limitations of the recorder or plotter. The limitation is not in the analyzer because it may be tuned by tracking a sweep oscillator that can sweep the entire audio range (20 to 20,000 cps) in perhaps one second. If a heterodyne system is used with a filter having a center frequency of perhaps 100,000 cps, then the sweep oscillator would sweep from 99,985 to 80,000 cps in order to evaluate signal components in the range of 15 to 20,000 cps. If the sweep oscillator is connected so as to move an oscilloscope trace through one horizontal sweep as the oscillator sweeps the frequency band and if the output of the analyzer is connected to the vertical control of the oscilloscope, the spectrum of interest appears as an oscillograph, newly drawn at each sweep, perhaps once per second. Such an arrangement is called a panoramic analyzer when the sweep, tracking, filtering and oscilloscope facilities are all combined in one instrument. Panoramic-type displays may be achieved with analyzers, such as in Fig. 30.5, by using them in conjunction with suitable sweep oscillators, analog outputs and oscilloscopes.

The rapidity with which the spectrum is presented makes the panoramic analyzer desirable, for example, in presenting the varying amplitude-frequency spectrum of a gyroscope as it coasts to rest and in many mechanical, electronic and acoustic studies, such as the analysis of the human voice.

SPECIFICATIONS FOR PANORAMIC ANALYZER (FIG. 30.6)

| | |
|---|---|
| Frequency range | 20 to 35,000 cps with optional ultrasonic module, 100 cps to 700 kcps |
| Sweep widths | 200, 1000, 5000, 20,000 cycles |
| Scan rate | Once per second |
| Sensitivity | 30 microvolts full scale |
| Cathode ray tube | $3\frac{1}{2}$ inches square |
| Vertical and horizontal scale markings | Linear or log |
| Power | Internal battery, capable of four hours between charging. Line power; 95 to 130, 190 to 250 volts, 50 to 1000 cps |
| Weight | 40 lbs |
| Price | $3000 |

The resolution (ability to separate adjacent frequency components) depends on the instantaneous scan-rate and filter selectivity. For logarithmic scanning wherein the scan-rate varies continuously, optimum resolution is maintained by continuously varying the filter selectivity. The panoramic analyzer is offered in a great variety of capabilities, ranging in frequency from the acoustic through radio frequencies. It is capable of presenting a frequency-time spectrum which will be discussed later. Print-out capability for a permanent record is also available.

## 5. THE SOUND SPECTROGRAPH

The spectra for typical wave analyzers may be recorded as a set of numbers or on the graph chart of a level recorder or that of an X-Y plotter. The spectrum of the panoramic analyzer appears on a cathode ray display. It may also be

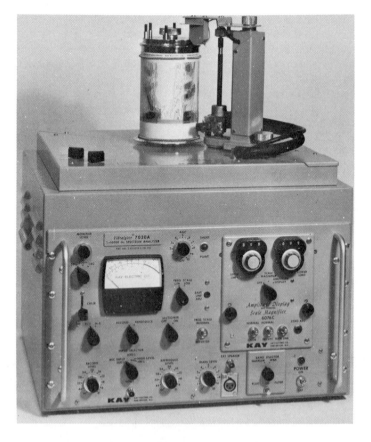

Fig. 30.7   Sound spectrograph, The Vibralyzer, 7030A (courtesy of Kay Electric Co.)

photographed or automatically printed on a chart. The typical sound spectro-graph records the spectrum on paper as a graphic display, usually written by a spark from a stylus. This instrument is based on work done at the Bell Telephone Laboratories with such objectives as a pictorial display of the human voice that might be interpreted by the deaf. Like the other analyzers, this one has its own peculiar capabilities and, therefore, its own most typical applications. Brief specifications for Fig. 30.7 are as follows.

SPECIFICATIONS FOR VIBRALYZER, FIG. 30.7

| | |
|---|---|
| Analysis time | 1.3 minutes |
| Recording medium | Nickel-cobalt-plated turntable |
| Microphone provided | Altec-Lansing 681 A, dynamic |
| Power supply | 117 volt, 50 to 60 cps |
| Dimensions | 25-in x 20-in x 18½-in |
| Price | $4445 |

Figure 30.7 illustrates the Vibralyzer of the Kay Electric Co., which has a frequency range of 1 to 16,000 cps. The signal to be analyzed is recorded so that

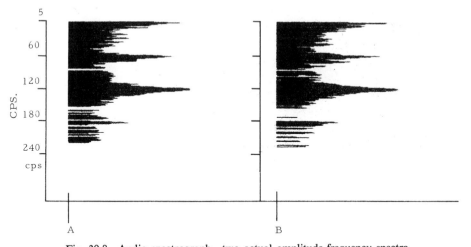

Fig. 30.8   Audio spectrograph—two actual amplitude-frequency spectra

Illustrating two amplitude-frequency spectra. Amplitude is plotted horizontally (around the drum) and frequency is plotted vertically (along the drum). The two spectra are derived from the same signal but one is taken at instant A and the other is taken at instant B. It is possible to select any instant during one drum rotation and the chart paper can contain about four spectra. Notice the difference between the two spectra: for example the amplitude of many of the components is different, showing that the amplitude of these components is varying with time. Notice also that some components appear and disappear. Close examination shows about fifty horizontal lines in the interval corresponding to 60 cps on the chart.

it plays back once each time the drum rotates. The description given here is general and does not necessarily correspond to Fig. 30.7. The specifications for Fig. 30.7 are given later. The signal is connected to an analyzer. While the analyzer sweeps across its frequency range, the spark stylus moves axially along the drum. The instrument may be operated to give a variety of graphic displays on a light-gray matte recording paper.

Figure 30.8 illustrates the amplitude-frequency display. At a selected instant during each rotation of the drum and for a very short time, the output of the analyzer filter is connected to the spark stylus in such a way that a line of constant blackness is drawn around the drum. The length of the line is proportional to the amplitude of the filter output. At the same instant one drum revolution later, the frequency of the filter has changed and a new line is drawn around the drum on the paper. The length of the line is again proportional to the amplitude of the filter output. The resulting spectrum is shown in Fig. 30.8. Different instants may be chosen so that four or more spectra may be drawn on one chart, in order to

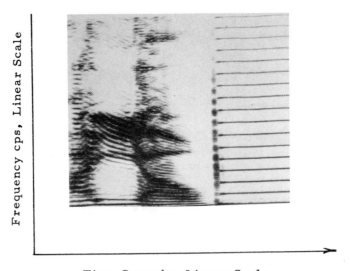

Time Seconds, Linear Scale

Fig. 30.9   Audio spectrograph, frequency-time spectrum (courtesy of Kay Electric Co.)

Illustrating a plot of frequency as a function of time for a selected frequency band. At any instant, if a vertical line is drawn through the point on the horizontal axis corresponding to this instant, then a particular frequency is present if the chart is darkened where the line corresponding to that frequency intersects the vertical line. For the greatest amplitudes the chart lines are full black and for lesser amplitudes the chart line is increasingly lighter. Thus the plot gives frequency amplitude information as a function of time. The horizontal lines are frequency markers.

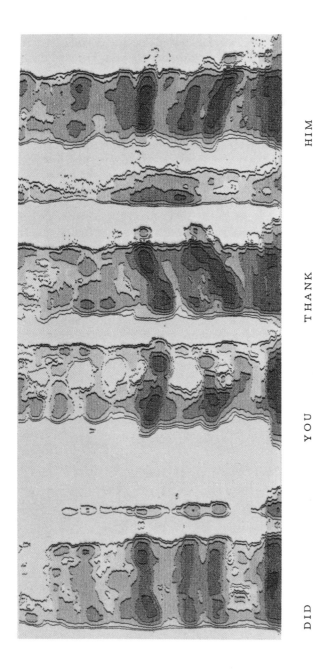

Fig. 30.10    Audio spectrograph contour plot (courtesy of Kay Electric Co.)

sample the recorded signal. This type of spectrum differs from that given by the ordinary wave analyzer in that the audio spectrograph display represents the frequencies and amplitudes existing in the signal at an instant, whereas the simple wave analyzer display covers a multitude of instants, as a rule.

Figure 30.9 illustrates the frequency-time display. For this the mechanical operation of the audio spectrograph is as before so that the frequency to which the filter is tuned is indicated by the axial position of the stylus along the drum. But now the spark intensity is proportional to amplitude of filter output. If every frequency is present in the signal, all of equal amplitude, the stylus draws a fine-pitch helix of constant blackness. If those frequencies that are near the center frequency of the filter during a given few instants are removed from the signal, the stylus draws no line during these instants. If the signal component corresponding to the tuned frequency varies in amplitude, the blackness of the line diminishes with decreasing amplitude. In this way it is possible to draw a picture of the human voice and of other signals and to recognize them even when attempts are made to disguise the signal.

What is termed a contour plot is illustrated in Fig. 30.10. In this remarkable mode of operation, the blackness of the display is arranged so as to jump in steps of intensity corresponding, perhaps, to 5 dB steps in filter output amplitude. The various displays that have been described and other variations have been used to analyze the performance of engines, human and animal sounds, and a host of other signals.

Besides the analyzers that have been described, there are many other electronic or electronic-optical instruments with prices ranging upward of $20,000. These are used for very rapid, accurate amplitude and phase analysis and for the derivations of such functions as power spectral density. Very rapid advances are being made in the development and application of fast analysers. The "real time" analyser is typical, and is likely to become a commonplace aid to machine maintenance. Its general performance is akin to that of the mentioned Panoramic analyser.

# BIBLIOGRAPHY

Beranek, L. L. *Noise Reduction*. New York: McGraw-Hill Book Co., 1960.

Beranek, L. L., *Acoustic Measurements*. New York: John Wiley & Sons, Inc., 1949.

Harris, C. M., ed. *Handbook of Noise Control*. New York: McGraw-Hill Book Co., 1957.

Keast, D. N. *Measurements in Mechanical Dynamics*. New York: McGraw-Hill Book Co., 1967.

Williamson, G. D. "Sonic Analysis." *Design News*, October 1967.

*Acoustical Terminology Including Vibration*. USA Standards Institute, S 1.1–1960.

# 31

# HOMEMADE INSTRUMENTATION

*Michael P. Blake*

## 1. SCOPE

The common loudspeaker, which can be used both as a shaker and as a velocity transducer, is discussed together with the photo-resistive cell which is used as a displacement transducer. Also discussed are the full-wave rectifier and some of the characteristics of meters that indicate in terms of average or rms.

## 2. INTRODUCTION

A familiarity with homemade instrumentation is most desirable and almost unavoidable for anyone who has more than just a superficial interest in vibration measuring. Familiarity is justified on the grounds of its great educational value as well as its versatility, suitability, low cost and ready availability. Homemade instrumentation is usually unsuitable for field work, wherein time must be conserved and there is often no chance for a second measurement if the first is incorrect. For that reason professional instrumentation is invariably used in the field, whereas homemade instrumentation is often used in the laboratory.

It is not possible to make significant progress in vibration measuring in the laboratory without at least an oscilloscope, a signal generator, an amplifier of 25 to 250 watts and a microammeter. The first three instruments can be obtained in kit form for a total of about $150. The last item can be had at small cost. A good example is the Calrad meter, reading either 1 milliamp or 50 microamps full scale, which sells for about $5 each. When these items are available many other items can be cheaply built to work with them. This chapter gives an indication of what can be done, and is intended to encourage initiative and learning through doing and testing.

## 3. LOUDSPEAKER AS A SHAKER

A shaker provides a simulated vibration environment by following the signal that drives it. Typical shakers are expensive. An ordinary 8- or 10-in speaker provides a useful substitute at very low cost. A speaker is shown, mounted in a rectangular box without sides on two faces, in Fig. 31.1. A washer about 1 in in diameter is soldered to a ⅛-in diameter rod. The washer is glued to the inside of the cone and the rod is led through the box to connect with a flexible bridge which, in turn, is mounted on pillars on the box. The bridge is preferably of wood or plastic, such as micarta, so that holes may be readily drilled in it for mounting test components. Metal is inclined to fret and is subject to buckling and magnetic troubles, besides being more difficult to procure and to make. The speaker costs about $8 and the entire apparatus costs little more.

**Operation.** The usual driving signal is sinusoidal, coming from an amplifier, which is in turn driven by a signal generator; or it may be driven by a recorded signal on magnetic tape. It is possible to obtain and sustain for periods of ten or fifteen minutes an amplitude of 1/10 in PP at 60 cps on the flexible beam. The useful frequency range is from about 10 to 2000 cps. The load on the speaker coil is greatly diminished if the bridge beam, the cone and the components mounted on the bridge are all arranged so as to have the same natural frequency as the driving frequency. This may be achieved by adding or subtracting restraints to and from the bridge, such as wedges or clamps, but it is not worthwhile when frequency is being changed continually.

As examples of utility, the apparatus may be used for calibration of pickups. A pickup of known performance may be mounted on the same axis as one of

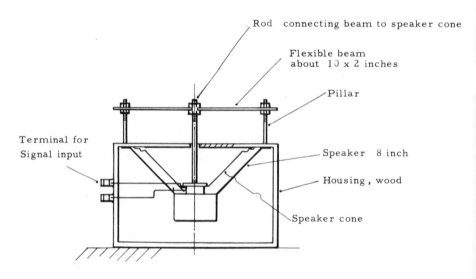

Fig. 31.1   A loudspeaker as a shaker

unknown performance. Calibration is obtained by shaking the pickups at various amplitudes and frequencies, while comparing their respective output signals. Components of up to about 8 or 10 oz may be shaken to determine weak points in design or to determine resonant frequencies. By driving a small secondary beam on which strain gauges are mounted, the performance of the gauge and its circuit may be evaluated. If a vibration meter is not available as a reference source, or if it is desired to check a meter response, a spherical ball of about 3/16 in in diameter is freely retained on top of an accelerometer by providing a small circular wall of modeling clay. As the magnitude of vibration acceleration is gradually increased by turning the amplifier output knob, the sine wave on the oscilloscope, driven by the accelerometer, increases in amplitude until it shows a small ripple or noise in the region of a crest. This indicates that the ball is partially airborne because the acceleration now exceeds 1-gP. By increasing and decreasing the amplitude at this point, a fairly accurate estimate of output for 1-gP is obtained. It should be repeated at various frequencies. When the frequency is halved, the amplitude for 1-gP should be four times what it was before. The hopping ball is a useful absolute test. But it must be used with much caution and comparative checking until the operator is assured that results are not unduly erroneous because the ball does often behave in unexpected ways. If its presence upsets the transducer reading, by virtue of axial pressure, it must be put at the side or on a small bracket overhead. An experimental arrangement is shown in Fig. 31.2.

The limiting factor of this apparatus is the current that the speaker will bear. This can be discovered only by careful experiment and failure. With an amplifier of 75 watts, it is possible to burn out a speaker in a minute or so if too much driving power is used. Occasionally, because the speaker is often driven far beyond its design amplitude of a few thousandths of an inch, the small leads from the cone to the coil will fail. This can usually be avoided by reinforcing the leads near the coil with the application of silicone rubber. This rubber is available in toothpaste-like form, made by General Electric Co., Dow Chemical Co. and

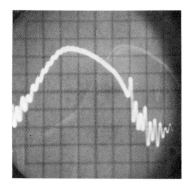

Fig. 31.2  Calibrating an accelerometer at 1-gP. Two accelerometers are shown mounted on the bridge of the shaker. The large accelerometer is 40 millimeters square and the balls are 8 millimeter diameter. The corresponding oscillograph for 100 cps shows a ripple in the sine curve due to bouncing of the balls. The sweep rate is 1 millisecond per division.

others. It sets to a rubber-like consistency in 3 to 24 hours and is a laboratory item of great value.

## 4. THE LOUDSPEAKER USED AS A VELOCITY TRANSDUCER

A common velocity transducer with an output on the order of 500 millivolts P/ in/sec/ P costs about $250. Much useful work can be done using a small speaker of about 4 in in diameter for the same purpose. The speaker costs perhaps $4. It is prepared by gluing a weight of about 2 oz at the apex of the cone and reinforcing the signal leads near the coil, both being accomplished with silicone rubber.

**Operation.** Velocity pickups are usually driven above their critical frequency. This frequency may be discovered by mounting the small speaker on a shaker and noting the output for various shaker frequencies. If the critical frequency is too high it may be lowered by adding more weight to the cone. The speaker as a velocity transducer is used like any other velocity pickup, by attaching it to the vibrating source so that its sensitive axis, its center line, is along the vibration axis of interest. In one case using a transducer of this kind, almost perfect linearity of response was obtained in the range of 0.2 to 3 IPSP at 20, 50 and 100 cps. However, the output varied with frequency and was 60 millivolts per IPSP at 20 and 50 cps and 100 millivolts per IPSP at 100 cps. Perfect linearity of response connotes that for a given frequency the millivolts per IPSP are independent of the amplitude. Sometimes a resistor is connected in shunt across the output leads of a velocity pickup in order to provide more damping with a view to minimizing the resonance hump and, therefore, extending the flat region of frequency response. When a shunt resistor was used here, the millivolt output per IPSP decreased; but it was then independent of frequency, on the whole.

This same small speaker may be used as a shaker, by attaching it to the point that it is desired to shake and then driving it from an amplifier and generator. Many useful experiments, for example, in physiological response may be made in this way.

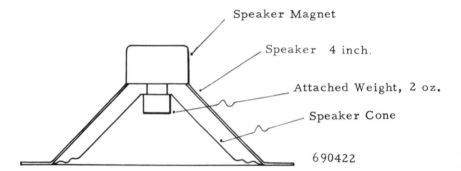

Fig. 31.3   Shaker or velocity transducer capability, using a 4 in. speaker

## 5. PHOTO-RESISTIVE CELL AS A DISPLACEMENT TRANSDUCER

Vibration displacement down to 0 cps may be measured with a linear variable differential transformer apparatus costing several hundred dollars. Down to a few cycles per second, it may be measured with an accelerometer or velocity pickup using electronic integration to obtain displacement. A transducer arrangement equivalent to that shown in Fig. 31.4 costs less than $10 and was used to measure displacement in the frequency range 20, 50 and 100 cps, from 0.06 to 100 mils PP. In this range, the graph of amplitude versus millivolts output for a fixed frequency was almost a perfectly straight line. The actual output was 100, 85 and 70 millivolts rms per mil PP at 100, 50 and 20 cps, respectively. These outputs were the same over the entire amplitude range for each frequency. At 200 and 500 cps the output signal did not reflect the vibrational input signal in any useful way. Supposedly, this was because of the frequency limitations of the particular photocell.

The same circuit can be used to generate a phase signal from a chalk mark on a rotating shaft or to measure rotative frequency, both responses being read on the oscilloscope. Using a dual-trace oscilloscope with the second trace being driven by a velocity pickup on the shaft bearing, the apparatus may be used to balance rotors.

There are two principal types of photocells; the photo-resistive and the photo-voltaic. The latter require no battery and generate a variable voltage that in-

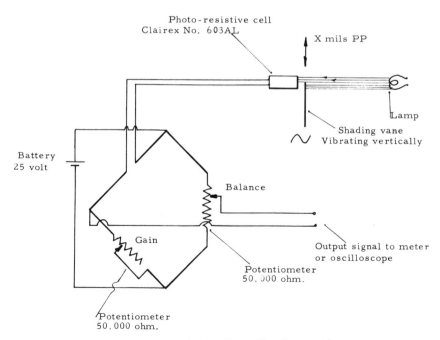

Fig. 31.4  Photo-resistive cell as a vibration transducer

creases with illumination although they may be used in conjunction with a battery bias. The photo-resistive type is characterized by an enormous variation of resistance as the illumination changes. The resistive change is not linear with illumination down to zero, and it exhibits its greatest variation near the zero end. The photo-resistive type is the most useful for vibration work and dynamic work, in general, since it usually provides far larger signals than the photo-voltaic type. The circuit in Fig. 31.4 is also useful for a variety of measurements such as detecting the movements of pendulums and the like. It has a DC response, which means that it will readily track very slow movements of the shading vane.

## 6. FULL-WAVE RECTIFIER AND METER CHARACTERISTICS

A rectifier to convert an AC signal into a pulsating DC signal may be made for about $1 with almost any of the common diodes or transistors. The transistor 2N 1303 and the diode 1N 34A are typical. The modern diode, with a diameter of perhaps $\frac{1}{8}$ in and a length of perhaps $\frac{3}{16}$ in, is one of the most valuable components in the gamut of electronic devices. It offers little resistance to current in one direction and great resistance in the other. A closely related but different diode, the Zener diode, has similar characteristics, with the difference being that the Zener diode avalanches (lets current flow freely) if the voltage in the direction of no flow exceeds a specified value. The transistor, which is not usually used as a rectifier, is often available and can replace the diode as a rectifier in experimental circuits. Some transistors, when connected as shown in Fig. 31.5, exhibit less resistance in the flow direction for small voltages than does the usual diode. The relation of through-put current to applied forward voltage in diodes and transistors is not linear down to zero. In fact, most diodes display a very high apparent forward resistance for applied voltages less than about 40 millivolts peak so that they are of little or no assistance in rectifying signals of less voltage than this. For this reason and others, the rectifier in a vibration meter is normally placed on the output side of the amplifier.

**Operation.** Alternating-current signals from about 100 millivolts P to 15 volts P or higher, depending on the rating of the diodes, are connected to the full-wave rectifier as illustrated in Fig. 31.5. The output of the rectifier exhibits about the same peak voltage as the input, pulsating in one direction only at twice the input frequency. When the output is connected to a DC (D'Arsonval-type) meter, the meter may be arranged to read the average output current or to read the output voltage. Because commonly used meters are of the DC type and read zero for alternating currents, rectification is a usual step in commercial meters and in laboratory circuits. For example, if it is desired to measure the output of the circuit of Fig. 31.4 using a meter, a rectifier is connected between the circuit output and the meter.

Whereas the oscilloscope always reads the true waveform, the meter merely reads some average value. If the oscilloscope reads 100 millivolts PP for a sine wave, the full-wave voltmeter reads 31.8, the half-wave meter reads 15.9 and the rms meter reads 35.4 millivolts. The half-wave meter has a rectifier with only

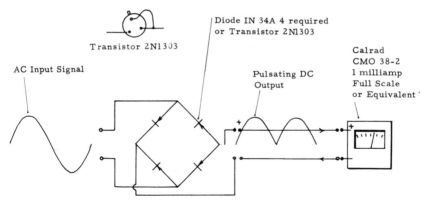

Fig. 31.5   Full-wave rectifier

two diodes, and the DC pulse frequency is the same as the AC frequency. If a fundamental sine signal is modulated with a harmonic that does not change its average value (a symmetrical ripple), the full-wave meter shows no change but the oscilloscope and rms meter do show a change. In vibration measuring the addition of the ripple connotes a greater severity of vibration, and for that reason the oscilloscope or rms meter is a more suitable monitor than the D'Arsonval meter with a simple rectifier. When a D'Arsonval meter is used for vibration measuring of signals other than fundamental sinusoids, a modification of the rectifier circuit is desirable to take note of such signals as can increase their rms value without increasing their average. Sometimes meters are labeled rms but they are simply averaging meters. The labeling is valid so long as there is a constant conversion factor from average to rms, as in the case of the simple fundamental sinusoidal signal. However, for complex signals, such meters are very likely to give a false reading. Due to its coil inductance, the rms meter of the electrodynamometer type is often limited to use below about 1000 cps.

# BIBLIOGRAPHY

Crowhurst, N. H. *Basic Audio*, vols. I, II, III. New York: Rider, Inc., 1959.

Mandl, M. *Directory of Electronic Circuits*. Englewood Cliffs: Prentice-Hall, Inc., 1966.

Schure, A. *Industrial Electronics Measurement*. New York: Rider, Inc., 1964.

# 32

# THE TAPE RECORDER, PEN OSCILLOGRAPH AND LEVEL RECORDER

*Michael P. Blake*

## 1. INTRODUCTION

While it is often possible to measure and analyze a vibration of interest without recourse to recording of any kind, recording techniques have, nonetheless, found their way into a surprising number of everyday, simple measurements. This has brought about the development of an endless variety of measuring and analytical procedures which would be quite impossible to put into effect without the aid of recording.

Of the various recorders, the tape recorder is perhaps the most versatile and useful of all, as it carries a vast amount of data in a small space. Frequency capability ranges from 50 to 10,000 cps for the typical commercial direct-recording type and down to 0 cps for the frequency-modulated type. The pen oscillograph gives immediate information regarding wave characteristics, but it is limited to a frequency range of from 0 to 125 cps. The level recorder gives a graphic presentation of a rectified signal level, while the pen oscillograph has a DC response and follows the instantaneous value of the signal. So also does the optical oscillograph in which several mirror galvanometers with moving elements that are only about ⅜ in in diameter respond satisfactorily at 2000 cps and often far higher. The storage oscilloscope is very useful as a recorder for storing a picture as long as an hour. It has the disadvantage that it stores information occurring only during a brief input period, usually far less than one second and, in fact, of a duration corresponding to the time taken for one or two horizontal trace excursions.

The tape recorder, pen oscillograph and optical oscillograph record functions of time. The storage oscilloscope, the graphic-level recorder and the X-Y plotter may be used to record functions of frequency or other parameters. For example, amplitude may be plotted against frequency to give an amplitude-frequency spectrum. This is a kind of secondary recording, a recording of an analytical process, whereas the primary recording is usually envisaged in terms of a function of time. In many cases the X-Y plotter has a fixed chart of approximately 15 in square, and the writing stylus takes up a position at any instant that corresponds to the DC voltage applied to the X and Y actuating mechanisms.

The responsiveness of the level recorder and the X-Y plotter is very small as compared with that of the pen oscillograph. Level recorders and X-Y plotters are generally servo-type instruments that find their appropriate stylus position by eliminating an unbalanced voltage in an internal bridge circuit in which the signal voltage and stylus response are compared. Although system response is slow, the power that is available for moving the stylus is very great.

The tape recorder is, in effect, the master instrument. Now that frequency-modulated versions are common, the slowest vibrations may be recorded with ease, and actually there is seldom any reason for using another kind of recorder for taking the original record. Afterward, when analysis is attempted, a variety of special playback procedures may be used.

## 2. APPLICATION EXAMPLES

**Cooling Tower Fans.** The keynote in most applications is that of frequency, and the frequency of rotation of a cooling tower fan is usually about 300 rpm maximum for typical towers. The motor and shaft frequency are nominally 30 cps and the gearbox exhibits various discrete frequencies ranging to a maximum of about 500 cps. When it is desired to balance the fan and determine beforehand if a significant looseness or other fault exists, it is most helpful to be able to see the signal waveform. The frequency of 5 cps is somewhat low for displaying on an ordinary oscilloscope. The storage oscilloscope is rather heavy and not too easily viewed when the sun is shining. The most compact and handiest instrument is the single-track pen oscillograph with the hot stylus, such as is made by the Sanborn Company, or the device of the American Optical Company with the cold stylus and carbon transfer paper. Both work well, but the carbon transfer unit may be easier to manage in the field because there is no problem connected with correct pen heat. The required information regarding the fan is made available immediately, and the work may proceed. In this application there is little attraction in using a tape recorder, except for permanent records or research, because too much time is required in order to obtain the playback in a graphical form. It is noted that the considerations which enter here are speed, weight, ease of viewing and frequency, all predicated around a frequency of about 5 cps.

**Slow-Speed Gears.** It is often necessary to analyze the action of slow-speed gearing on drum driers and similar machines. The frequency is often in the range of about 10 cps, and the machines are not usually too far from the ground, as are cooling towers with their unsteady ladders and flimsy handrails. For such gearing the pen oscillograph may be used if an immediate opinion is required. The mechanical, hand-held recorder can also be used because the frequency is now in its range. An example of the handheld mechanical strip recorder is the unit made by the Askania Company (Germany). Analysis information associated with this type of gearing is not usually required immediately, so that it is preferable to make a tape recording of the vibration signal, if a low-frequency recorder, such as an FM recorder, is available. The pickup is usually attached to the nearest bearing housing.

**Forced-Draft Fans.** Forced-draft fans for steam boilers usually operate at a nominal frequency of 30 cps, and they are often equipped with rolling bearings. Before balancing or overhauling such a fan it is desirable to analyze the vibrations, particularly if the characteristics of the fan or its history are unknown and if it is larger than about 50 hp. As always, the analysis may be done in the field without any recording. However, the ambient noise, heat and other discomforts, such as fly ash which plays havoc with some instrumentation, are very undesirable as regards spending perhaps an hour in examining the vibrations. The bearing vibrations can be tape recorded using about one minute of tape time. Since the fundamental frequency is only 30 cps, an FM recorder is required for the examination of this type of low-frequency vibration. Otherwise, it may be done in the field as it does not take long to decide with an oscilloscope whether all is well or not.

Bearings are among the slowest and most difficult items to analyze. Furthermore, it is desirable to retain records on recording tape so that it may be readily determined later if a bearing is deteriorating, even if the engineer cannot determine the nature of the fault through analysis.

**Induced-Draft Fans.** The frequency of rotation of induced-draft boiler fans seldom exceeds a nominal 30 cps, and the bearings are usually of the sleeve type so that little information is obtained from high-frequency analysis. Moreover, bearing problems are few if the fan receives adequate care. The usual requirement prior to balancing is a quick view of the fundamental waveform, in order to detect coupling performance, looseness or perhaps cracked stay rods, blades or shrouds. The wheel is usually hammered all over in order to detect faults through listening to the resulting ringing noises. Except for research projects, the common oscilloscope tells all that is required, and it is easily photographed when necessary.

## 3. SOME PRINCIPLES OF THE RECORDING TECHNIQUE

Although this chapter is written mainly in terms of machine maintenance, it applies to some extent to design work and prototype testing. In all instances the primary recorder is the tape recorder, with the pen oscillograph and level recorder being secondary devices. The tape recorder has opened up so many possibilities that some years may elapse before its full potential is realized. Admittedly, the low-frequency limitation of the acoustical-type or direct recorder that restricted it from applications having fundamental frequencies below about 50 cps has been responsible for a tardiness in taking advantage of the subject possibilities. Nevertheless, far more could have been done with the acoustical-type recorder than has been the case. Consider the following partial list of possibilities.

1. Seldom do field work investigations exceed one minute per point of interest.

2. When machines are specially loaded or adjusted for measurement purposes, the operation is upset only for a matter of minutes.

3. When the field location is dangerous or uncongenial, an excellent record is quickly made, thereby avoiding the temptation to make no measurement or to make one that is obviously inadequate.

4. The work of analysis may be done with comfort and complete lack of danger or distraction in the laboratory.

5. A tape loop of about one-minute duration is equivalent in almost every respect to having the subject machine in the laboratory as long as it is required for analysis and as long as it is required for permanent records.

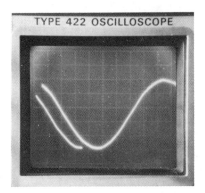

a. One Cycle per Second Recording, Playback
500 mV and 100 Milliseconds per Division

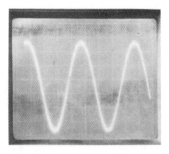

b. One Half Cycle per Second, Playback
500 mV and 500 Milliseconds per Division

Fig. 32.1   Recorded signal playback

6. After six months or a year, the old recording makes an excellent and informative basis for comparison and for research work if that is required.

7. Tape recording captures, so to speak, the whole story. Nothing is missed that might be required later. In the case of random vibrations, the tape recording captures more information than any other device.

8. Particularly as compared with optical or other strip recorders, the bulk and cost of the tape record is as near negligible as can be imagined. It is also very permanent and withstands much abuse in storage or handling.

## 4. THE TAPE RECORDER

**FM Recorder.** The low-frequency capability of a simple FM apparatus is illustrated in Fig. 32.1. The two signals were recorded and played back using the combination of signal conditioner and tape recorder shown in Fig. 32.2. The recorder is a commercial Aiwa recorder of the ordinary kind, costing less than $200. The signal conditioner was developed by Lovejoy, Inc. to enable users of ordinary direct-type recorders to record and play back low frequencies down to 0 cps. Although FM recorders having, for example, two to about ten channels have become common in recent years, they are not usually available for field applications, such as maintenance work. Lockheed, Honeywell, Hewlett-Packard, and Bruel and Kjaer are manufacturers of portable FM recorders.

The Lovejoy signal conditioner is shown more clearly in Fig. 32.3. An input switch selects either a zero signal, a shielded three-pin plug or phone-jack input. The signal goes to an integration stage where it is either passed without alteration or is integrated once or twice with respect to time. The signal is subsequently recorded, either directly or via a frequency-modulation circuit. When the signal is played back from the tape recorder, it passes to the output either directly or travels via a demodulation circuit. The FM conversion is used for signals up to about 800 cps, and the direct recording is used from about 60 to 10,000 cps.

The conditioner illustrated in Fig. 32.3 weighs about 7 lbs and is approximately 11-in × 8-in × 4-in. The FM response is almost without error up to 500

Fig. 32.2   Signal conditioner and audio recorder, capable of 1/10 cps
recording or less

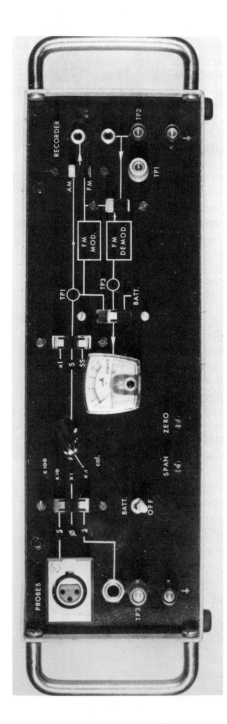

Fig. 32.3   Signal Conditioner for tape recording, FM and AM. For use in combination with common tape recorders for frequencies to zero CPS; with two stages of integration and playback demodulation (courtesy of Lovejoy, Inc.)

cps and is down about 3 dB at 800 cps. Signal inaccuracies are introduced by the mechanical imperfections of the recorder. The conditioner is designed to match the performance of the very best direct recorders. When it is used with recorders costing well under $200, then departures from accuracy do occur. However, these departures are not of great significance for maintenance work or for the evaluation of machine performance in general. Excellent records have been achieved with very modest recorders. Workers not having access to relatively expensive FM recorders have been able to record in the frequency band of 0 to 50 cps, which was hitherto impossible.

At present there are perhaps a dozen or more well-known FM recorders on the market. These are being improved and the weight and cost are being reduced. However, an FM recorder with two channels still costs about $4000. A tape recorder for the measurement of machinery must include the frequency range down to about 3 cps, if not to 0. It is to be expected, therefore, that the FM recorder will find more applications each year in the maintenance and design of machinery. ]

**Direct Recorder.** The direct or AM recorder is the usual acoustical-type recorder that is used in the home and office. Although a somewhat more refined version is desirable for vibration measurement, nevertheless, much useful work may be done with the home recorder, particularly now that accelerometers are available with very simple impedance converters not requiring expensive and heavy power supplies but merely requiring a battery device weighing a few ounces. With this type recorder, such measurements as those relating to rolling bearings may be carried out with satisfactory results if care is taken to account for the possible errors in the system.

To calibrate an unknown recorder, it is desirable to record a signal of 1-g peak at two frequencies, say 100 and 500 cps. The playback voltage should be fairly independent of frequency; or as it is more commonly expressed, the response should be flat. Thus, with particular settings of the input and output attenuators, it is possible to determine what output voltage corresponds with 1-g peak. In this way calibration is effected.

The ear plays a very small part in the analysis of vibration records, and the playback signal is almost invariably channeled to some form of graphic or analytical device. Usually a wave analyzer is the end point of the playback system, and an oscilloscope is invariably used to monitor the work. Other final read-out devices are: the panoramic analyzer, the real-time Fourier analyzer which gives very rapid analyses, the pen or optical oscillograph and the storage or common oscilloscope. The usual and most convenient read-out system is a combination of automatic wave analyzer and level recorder which prints out the amplitude-frequency spectrum.

An excellent example of a recorder that is suitable for acoustic and vibration studies is shown in Fig. 32.4. This is a more refined recorder than those normally used for speech or music, and it may be used with its built-in sound-level meter as a precision recorder in the frequency range of 15 to 16,000 cps. Brief specifications follow.

---

Model: Data Recorder, Type 1525-A, General Radio Co., Concord, Mass.

---

| | |
|---|---|
| Channels | Two with separate amplifiers |
| Tape speed | 7.5 and 15 in/sec |
| Frequency response | ± 2 dB, 50 to 15,000 cps, at 15 ips |
| Input level | 10 microvolt to 1 volt |
| Flutter and wow, recording | Below 0.2 percent rms |
| Tape | ¼ in; reel to 7-in diameter |
| Power | 105 to 125 volt, 60 cps |
| Accessories | Include guides for tape loop |
| Dimensions (portable version) | 21-in x 16-in high x 9-in |
| New weight (portable) | 53 lbs |
| Price | About $2650 |

---

## 5. THE PEN OSCILLOGRAPH

In this instrument the writing pen follows the instantaneous input signal wave and not a rectified average. One or two pens are usual, which may contain ink or

Fig. 32.4    Data Recorder, Type 1525 (courtesy of General Radio Co., Inc.)

operate through pressure on carbon paper or through heating of a waxed-paper chart. The motion may be circular or rectilinear. Circular motion, as if from a compass, is not as acceptable as that of the rectilinear stylus that moves on a straight line when the chart is idle. Rectilinear motion is obtained by mechanical linkages or by ingenious arrangements, such as, for example, in the American Optical Co. instrument which uses a sharp rectilinear ridge over which the paper is drawn while a stylus rod wipes to and fro across the ridge. The motive force for the stylus comes always from a powerful D'Arsonval galvanometer. Charts are about 50 millimeters wide per stylus and move at speeds from about 1 to 100 millimeters per second. Response is usually adequate up to 125 cps. The cost is about $500 for one channel. Optical recorders do about the same job, with the difference being that the upper frequency limit is several thousand cycles per second. This necessitates very high paper speeds.

**Application.** When an FM tape recorder is available, the pen oscillograph finds almost no application in the field except when very quick results are required. For example, it may be desired to make an immediate comparison of the vibration waveform at a compressor with one taken in a domestic dwelling to determine if the latter is caused by the former. Otherwise, the pen oscillograph is simply a very useful device for drawing waveforms that may be too fuzzy or too slow for display on the oscilloscope. The carbon-transfer oscillograph makes a chart of the greatest refinement with a clarity showing every little detail, thus permitting accurate comparisons of detail and phase where several pens are used at once.

There is no better instrument than the pen oscillograph for a preliminary investigation of unknown vibrations in cooling towers and low-speed machinery as well as in many cases of seismic vibration. It does not use much paper. At the other end of the scale comes the optical oscillograph, using reams of paper. Then the electronic oscilloscope which would fill the world with paper, if a chart could keep pace with it. The pen oscillograph tends to reveal form and phase, whereas the oscilloscope tends to reveal amplitude rather than form because, as it so happens, this is of more interest as the frequency increases.

## 6. THE GRAPHIC-LEVEL RECORDER

The graphic-level recorder plots the rectified average of a signal in some such nominal term as rms. It should be noted that it is not impossible to rectify, average and smooth the response of a pen oscillograph. This would, in effect, make it a level recorder and thus it would be more versatile. While the pen oscillograph is actuated by a galvanometer, the level recorder is invariably servo actuated, taking advantage of the permissibility of a very slow writing speed in order to achieve the forcefulness and accuracy of the servo system.

**Description.** The level recorder, as its name implies, is predicated upon level rather than upon form. In many periodic vibrations the level changes but little and slowly. When a complex, periodic vibration is broken into its frequency components by a wave analyzer, the level recorder easily writes fast enough to keep pace with the level variation that corresponds with the sweep rate of the analyzer. Writing speeds are, however, slow, but not too slow to permit a DC (unrectified) response on some recorders up to about 20 cps. In this way the level recorder plays the role of a very slow-speed pen oscillograph.

A level recorder is shown in Fig. 32.5, through the courtesy of the General Radio Co. It contains one stylus. Brief specifications are given below.

---

Model: Graphic-Level Recorder, Type 1521-B, General Radio Co., Concord, Mass.

---

| | |
|---|---|
| Range of levels | 20, 40, and 80 dB potentiometers |
| Paper speed | 75 in/min to 2.5 in/hr |
| Chart width | 4 in (writing width) |
| Power | 105–125 and 210–250 volt at 50 or 60 cps |
| Net weight | 50 lbs |
| Dimensions | 19-in x 9-in x 13.5-in |
| Price, approximate | $1325 |

---

**Application.** Usually the level recorder is used to plot the output of a wave analyzer. When the analyzer is mechanically linked to the recorder motor, it may be set to sweep at a rate proportional to the logarithm of the filter frequency. Thus, the abcissa of the chart becomes a frequency axis and the ordinate is proportional to filtered amplitude, giving the typical amplitude-frequency spectrum. Spectra are plotted in a few minutes with little labor. Sometimes it is desired to plot sound or some other level over a period of a day. Then the recorder is driven at a slow speed to produce a very useful record. The response of microphones and the decay times of auditoria are easily plotted automatically with a great saving in labor as well as a high order of accuracy. The level recorder is primarily predicated upon the plotting of integrals or averages and not forms.

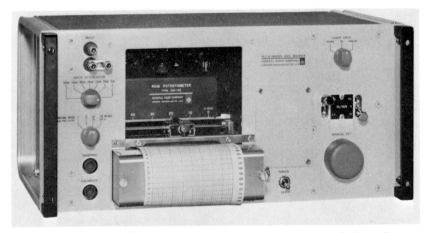

Fig. 32.5  Graphic Level Recorder, Type 1521 (courtesy of General Radio Co., Inc.)

# STANDARDS AND TABLES

# SECTION VII CONTENTS

# 33

# STANDARDS AND TABLES

*Michael P. Blake*

## 1. MISCELLANEOUS MACHINERY

National standards have not yet appeared. Those given in Fig. 33.1 are based on this writer's experience,[1] taking due note of the standards offered by Rathbone,[2] Nittinger,[3] Maten[4] and others. The offered standards are strictly experimental and any of the theories that are discussed in support of them come after the fact and are of secondary interest.

Measurements are usually taken on bearing housings or on the machine surface. Measurements made directly on shafts are usually about two or three times as great as the corresponding measurement on the bearing housing.

**Description.** The standards are presented in terms of acceleration, velocity and displacement. Each presentation is equivalent to the others. The form of the graphs is experimental. Whereas a vast amount of data was available to support the graphs up to about 1000 cps, it is regretfully admitted that higher-frequency data was meager. It is, therefore, hoped that future work will afford more assurance in the high-frequency range. The frequency of about 1 or 2 kcps is a dividing line between mechanical vibration and the higher frequencies of what might be called structure-borne sound or noise, or between discrete and continuous systems as a general rule. It is noticed that the interspacing of the graphs is a fixed ratio of about the square root of 10 up to 4 kcps. This is done in the interest of mathematical convenience, while it fits the experimental observations. Root ten is about 3.16, even though some other figure such as 2.8 or 3.5 is near enough to the facts. However, root ten is an internationally preferred number. It also corresponds with 10 dB on the basis of dB level of X re Y = 20 $\log_{10}$ (X/Y). In this way the given standards, if expressed in decibel notation, have a constant interspacing of 10 dB.

The most remarkable aspect of the graphs, which is also recognizable in the work of other authors, is that in the range from about 30 cps to 1 kcps constant severity connotes constant velocity. Again, the theoretical explanations of this striking feature came after the fact and are of little substance in this writer's opinion. The practical consequence is this: since, perhaps, 99 percent of all

[615]

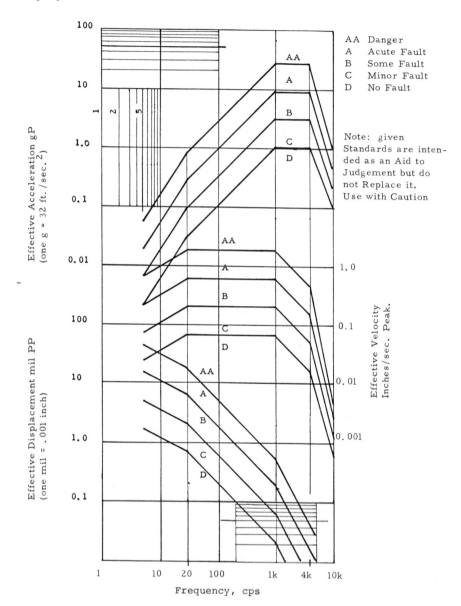

Fig. 33.1  Vibration classification standards for an extended frequency range of miscellaneous machinery. Effective vibration is measured vibration multiplied by the service factor. When service factor is unknown, use 1.0 for miscellaneous machines, 2 or 3 for critical machines and 0.5 for non-critical machines. Class C is the condition of optimum economy, i.e. optimum performance. The table shows effective vibration as a function of frequency for constant condition classification.

measurements lie in the frequency band of 30 to 1000 cps, the entire standards system may be committed to memory by taking 2 in/sec (peak) as the danger level, 0.63 as the acute level, 0.2 as the economy level and 0.063 as the faultless level.

The unfiltered vibration reading, the meter reading, is entered in the graphs, using an appropriate Service Factor, and at the fundamental frequency of the measured machine. Chapter 12 provides information regarding caution in the matter of using the graphs as limiting envelopes for measured Amplitude-Frequency vibration spectra, and regarding complex vibration standards in general.

**Utility.** The standards are intended to cover the greater part of miscellaneous machinery of all kinds. To avoid altering the standards to suit various machines, the concept of service factor is introduced. For example, a very critical machine might exhibit a vibration of 0.2 mils PP. This could be entered in the graphs as 1.2 mils by using a service factor of 6. A noncritical machine might use a service factor of 0.3. Examples of service factors are listed in Table 42. The reader is cautioned that these are a useful guide when no better knowledge exists, but that he should ideally decide his own service factors in the light of experience.

TABLE 42
SERVICE FACTORS FOR VARIOUS MACHINES

| TYPE OF MACHINE | SERVICE FACTOR |
| --- | --- |
| Electric motor or single-stage centrifugal pump or fan | 1 |
| Typical noncritical machinery | 1 |
| Turbine, generator, centrifugal compressor, multi-stage pump | 2 |
| Typical critical machinery | 2 |
| Centrifuge, stiff shaft measured on the basket housing | 2 |
| Centrifuge, shaft suspended or link suspended | 0.3 |

If machinery is tested while resting on a resilient support so that it operates above about three times the critical frequency of the supported system, use an additional service factor of 0.5 to allow for the fact that bolting of the machine to the foundation decreases free vibration usually by about 50 percent.

## 2. COOLING TOWER FANS

Cooling towers with fans up to about 25 ft in diameter operate at speeds generally below 350 rpm. Forced and induced-draft towers are mechanically aspirated and, thus, distinguished from the large concrete natural convection towers. Because the frequency is lower than the range of the standards already given for miscellaneous machinery, special standards are given here, based on experimental work and on a paper presented to the Cooling Tower Institute in 1964.[5]

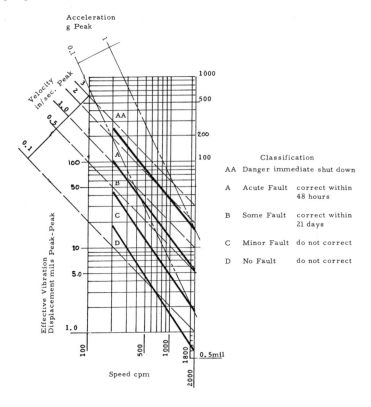

Fig. 33.2 Vibration classification standards for cooling tower fans. Given standards are primarily for 300 to 400 rpm fans with extended curves for higher speeds or for tower motors. Observed vibration × service factor gives effective vibration.

SERVICE FACTORS:

| | |
|---|---|
| Typical tower on a building, measured on deck | 1.0 |
| Typical tower on ground, measured on deck | 1.5 |
| (For induced-draft fans, gear boxes, or box attachments, multiply above factors by 0.75.) | |
| Forced-draft fan on a building, horizontal radial measure | 1.0 |
| Forced-draft fan on ground, horizontal, radial measure | 1.5 |

When using a vibration meter of unknown calibration, use these additional factors as multipliers (most velocity pick-ups show large errors below 900 cpm):

| | |
|---|---|
| 300 to 500 rpm | 3.0 |
| 500 to 700 rpm | 2.0 |
| 700 to 900 rpm | 1.5 |

EXAMPLE:

55 mil true vibration on deck of tower on ground, 375 rpm; effective vibration is 82.5 mils: Classification A.

Same as above with meter of unknown calibration; effective vibration is 247 mils: Classification AA.

**Description.** Figure 33.2 is based on true vibration multiplied by a service factor. The epithet, *true vibration*, is introduced to draw attention to the fact that most velocity meters read about 30 percent of true vibration at 300 rpm. If a velocity system is used for monitoring or measuring, its calibration at low frequency must be ascertained. Otherwise, Fig. 33.2 speaks for itself. The reader is also referred to the article on the balancing of cooling tower fans in Chapter 17.

## 3. ELECTRIC MOTORS, UNMOUNTED

The unmounted electric motor is an item of interest to repair shops and maintenance supervisors. Depending on whether the motor is new, repaired and balanced, or repaired without balancing, different standards are required in the interests of maximum overall economy. Standards are offered in Fig. 33.3, based on NEMA (National Electrical Manufacturers) standards and on this writer's observations of motors passing through the repair shop in the context of optimum maintenance practice.[6]

## 4. ROLLING BEARINGS, INSTALLED

A single all-pass (unfiltered) reading taken on the bearing housing with an accelerometer capable of at least a 5 kcps response is likely to give a correct classification of the bearing condition in seven cases out of ten. The velocity meter is somewhat less reliable. (See also Chapter 22.) Table 43 gives tentative standards which are based on observations of some hundreds of installed bearings in the range from about 1.5 to 8 in outside diameter.

**TABLE 43**
**TENTATIVE CLASSIFICATION STANDARDS FOR**
**INSTALLED BEARINGS**

| CLASS | gP | dB | BLIPS gP | VELOCITY IPSP | dB |
|---|---|---|---|---|---|
| Danger | 100 | 80 | 200 | 5.62 | 55 |
| Acute fault | 17.8 | 65 | 35 | 2.37 | 47.5 |
| Profitable to correct | 3.16 | 50 | 6 | 1.00 | 40 |
| Minor fault | 0.56 | 35 | 1 | 0.42 | 32.5 |
| No fault | 0.1–0.01 | 20–0 | 0.2 | 0.18–0.01 | 25–0 |

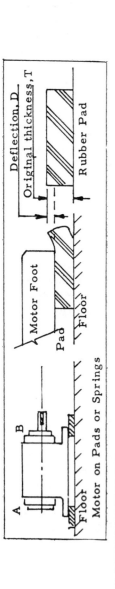

Motor on Pads or Springs

| FRAME SIZE | RPM | MILS PP NEW MOTOR | MILS PP REPAIRED NOT BALANCED AND BALANCED | MILS PP REPAIRED AND BALANCED | MILS PP, NEW MOTORS NEMA COMMERCIAL STANDARD MG1-12.06 | REVISED NEMA FRAMES |
|---|---|---|---|---|---|---|
| 180, 200, 210, 220 | below 3600 | 1 | 2 | 1 | 1 | H 143 T to 213 T |
| 180, 200, 210, 220 | 3600 | 1 | 1.5 | 1 | 1 | |
| 250, 280 | below 3600 | 1.5 | 2.5 | 1 | 1.5 | 215 T to 286 T |
| 250, 280 | 3600 | 1.5 | 1.5 | 1 | 1.5 | |
| 360, 400, 440, 500 | below 3600 | 1 | 2 | 1 | 2 | 324 T to 445 T |
| 360, 400, 440, 500 | 3600 | 1 | 2 | 1 | 2 | |
| Larger motors | through 3600 | 1 | balance | 1 | none | |

Minimum limits for deflection D: 3600 rpm, $\frac{1}{16}$ in; 1800 rpm, $\frac{1}{16}$ in; 1200 rpm, $\frac{1}{4}$ in; 900 rpm, $\frac{5}{16}$ in; 1 in.

Note 1: D must never exceed $\frac{1}{2}$ T for rubber; and springs, if used, must not bottom. Vibration is measured at A and B, in any direction at right angles to the shaft. A half key (flush with shaft) is taped to the shaft. Keep safety constantly in mind. Run AC motors at rated volts and frequency; DC motors at base speed; Series and Universal motors at operating speed. Typically observed vibration troubles when installed (approx.): resonance or loose mount: 40 percent; bearings and fans: 30 percent; unbalance: 30 percent.

Note 2: Vibration of bolted-down motors is less than half that which is measured when motor is on a resilient support. To measure vibration, rest motor on four resilient supports on firm floor. A motor can jump because of seizure, when on supports.

Note: NEMA MG1-12.06 and MG1-12.07 formerly MG1-4.23 and MG1-4.24.

Fig. 33.3   Classification standards for industrial electric motors, mass balance for motors not installed

## 5. SOIL AND FOUNDATIONS

The vibration level that causes worry to people in their homes is very small. Much greater amplitudes are acceptable in industry and to persons if the affected person is convinced of the necessity of the vibration. The least vibration between about 300 and 600 cpm that causes worry in a domestic dwelling is about 0.15 mil PP displacement. If the greatest horizontal vibration measured at a machine foundation exceeds about 0.5 mil PP in the range 2 to 10 cps, the horizontal component in the soil is likely to exceed 0.15 mil PP in the basements of homes within a radius of about 500 ft. In frame homes the horizontal vibration amplitude on the first and second floors is likely to be two and four times the value in the basement, respectively. For these reasons it is desirable to avoid unbalanced machinery altogether. If it cannot be avoided, it is then desirable to limit the maximum horizontal component in the foundation to 0.5 mil PP. (See Chapter 25.)

## 6. HUMAN PHYSIOLOGY, VIBRATION

When a particular group of persons is subject to various amplitudes of vibration at various frequencies, the results in terms of the average reaction appear somewhat as in Fig. 33.4. Because physiological sensitivity varies from person to person, Fig. 33.4 is merely an average. The perception or other response level for the most sensitive person may be half of that for the average, whereas the perception level for the least sensitive person may be twice that for the average. For this reason it is not at all possible to formulate standards for persons that are as representative as those for machines.

A physiological standard assumes that each person tested has about the same emotional attitude to the environment. In these terms Fig. 33.4 may be taken as the average physiological reaction for a uniformly neutral emotional attitude. The variability of the emotional attitude, which is psychological, further aggravates the difficulty of providing a standard. Depending on the psychological attitude or alignment of the affected person, the levels given in Fig. 33.4 may again be moved up or down by a factor of about three. Thus, a person who is making a profit from some product is likely to find his level for annoyance higher than given in Fig. 33.4, whereas another man, affected in his home by the vibratory sifter or compressors that are associated with the product, may find his level of annoyance below the perceptible level given in Fig. 33.4, except when he is favorably related to the manufacture of the product for some reason such as family or finance. In brief, the reactions of persons in given circumstances may not correspond even moderately well with any classification standard. Nevertheless, such standards are necessary.

**Description.** A significant amount of testing has been done in the frequency range of from 1 to 100 cps by workers such as Reiher and Meister, Steffens, Goldman and others. A useful reference is the graphical presentation given by Goldman,[7] based on the findings of several workers and giving average values and an indication of maximum deviation from average. From 100 to 1000 cps, the writer finds no help in the standard handbooks. Thus, he has drawn the curves in

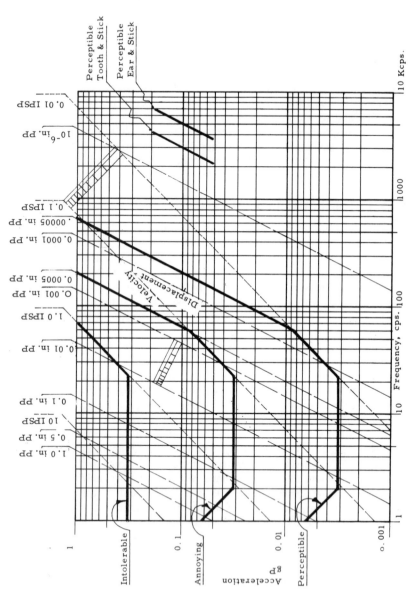

**Fig. 33.4** Human physiology. Vibration response of the fingers and body (except as noted). Displacement in inches PP, velocity in inches per second P and acceleration in g's P.

Fig. 33.4 for that range from his own experience, while consulting the data of the above mentioned workers and using his own measurements as a basis for the range 1 to 100 cps. Data from 100 to 1000 cps is based on fingertip sensations. The sensitivity of the fingertips, the spine and bone structure when the body receives vibration from a chair or of the body in general shows little difference in this writer's measurements.

Perception is a more easily understood criterion than annoyance or intolerance because it is more absolute. Measured reactions and the confirmation of measurements of low-frequency visceral resonances both suggest that the perception level is sharply decreased at about 5 cps and increases rapidly for lower or higher frequencies. This is not too well stressed in Fig. 33.4, wherein the curves are drawn as far as possible to coincide with convenient values while doing no essential violence to measured values.

A wooden stick held in the teeth increases the sensitivity of perception. A brief set of measurements made by the writer on himself showed that 1/10 of 1-gP is perceptible up to 2 or 3 kcps and that the same level is perceptible up to about 5 or 6 kcps if the stick is held to the ear. Thus, the fingers and the body, which are the usual detectors of ordinary vibration, do not notice the high-frequency vibrations, for example, of faulty rolling bearings, whereas the teeth and ear, when suitably connected, do notice these vibrations.

## 7. HUMAN PHYSIOLOGY, SOUND AND NOISE

Although the presentation of meaningful sound and noise standards is even more difficult than that of physiological vibration standards because of various factors such as the age of the affected person, more work has been done in noise than in vibration. Perhaps the topic of greatest interest is that of hearing loss through sound exposure. This is the subject of Fig. 33.5. The reader is particularly cautioned that Figs. 33.5 and 33.6 are included here merely for the purpose of giving some general indication of the classification levels, and may require significant modification and qualification when applied to specific cases. For further data, see Refs. 8, 9 and 10.

**Description.** Figure 33.5 gives an indication of the levels that lead to impairment of hearing and are called damage-risk levels. Figure 33.6 gives an indication of equal loudness. The decibel level at 1000 cps describes the loudness of each contour in terms of phons. A 60-phon contour has a level exceeding 100 dB at 20 cps. Figure 33.5 gives an indication of the perception level for young persons, the level of discomfort and the level of pain.

**Conclusion.** It is evident inasmuch as the subjective severity of vibration and noise varies enormously, depending on the variations of personal physiological sensitivity and psychological attitude, that it is quite impossible to relate any set of standards to a given person. This troublesome fact has made it almost impossible to stem the rising tide of noise and vibration by legislation with reference to standards. However, national and international standards may be expected in the not too distant future in simple terms of average human reactions.

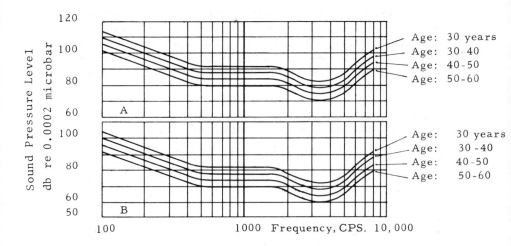

Fig. 33.5 Damage-risk, noise criteria. Proposed damage risk criteria for (a) wideband noise measured by octave, 8-hour continuous exposure; (b) pure tones or critical bands of noise. After Kryter, K. D., from *Noise Reduction*, ed. L. L. Beranek, McGraw-Hill, by permission, McGraw-Hill, New York, 1960

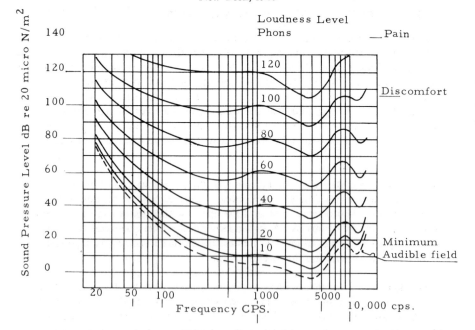

Fig. 33.6 Contours of equal loudness (after Robinson & Dadson, National Physical Laboratory, England). Pure tones with the observer facing the source (spherical propagation without obstacles, reflections or extraneous sound or noise)

To an extent, these standards for legislative purposes may be expected to be extremely simple. For example, it has been decided[11] that a continuous industrial-noise level exceeding 90 dB on an A-scale (re 0.0002 microbar) is not permissible, even though there are many observations that might prove this decision to be wrong in particular instances or very difficult to satisfy as regards compatibility with well-established industrial practice.

## 8. RESIDUAL MASS UNBALANCE OF MACHINE ROTORS

Traditionally, the balancing of rotors was an art without much orderly basis and largely without units of measurement that were meaningful in communication. Specifications were equally vague. Drawings often specified *good balance* or so many inch-ounces or neglected the subject completely. However, in the past decade two propositions have been universally accepted.

1. The measure of constant vibration severity for machines from 10 to 1000 cps, approximately, is a constant velocity of vibration.

2. The measure of constant severity of residual mass unbalance for machine rotors from 10 to 1000 cps is a constant velocity of vibration measured on the shaft surface.

Thus, when a rotor is balanced in-place, it is most usual to refine the balance until the vibration at the bearing housing corresponds to the provisions of Fig. 33.1. Except for the fact that it is unusual and therefore confusing, there is no objection to using Fig. 33.7 together with shaft-stick measurement made on the shaft in the field. The former practice is generally adequate; however, vibration at the bearing housing includes vibrations that may come from sources other than unbalance so that shaft-stick measurements are sometimes preferred. Furthermore, the bearing housing vibration depends to an extent on bearing tightness and it can vary for a fixed shaft-stick reading.

Figure 33.7 relates mainly to rotors that are balanced in balancing machines. It has been the usual practice to spin a rotor in a suspension arrangement of relatively low natural frequency and relatively low mass so that it could be assumed that the apparent eccentricity was the actual eccentricity. Usually, for horizontal machines the suspension is free in the horizontal direction only. Some of the vertical-axis machines have only one direction of freedom, and some are free to move in all horizontal directions. Omni-directional suspensions are sometimes called isotropic, and the others are called anisotropic suspensions. Suspensions and balancing methods for flexible rotors are different from those for stiff rotors. Of late, however, a tendency has appeared to use *hard* suspensions and it seems likely that this tendency may grow considerably. Thus, force instead of displacement becomes the actual measure of unbalance. But overlooking these details, the classifications shown in Fig. 33.7 are based on displacement. This is the sort of graph on which current American and European discussions of balancing standards are based. Later on, if the force measuring practice becomes more important, acceleration or force lines may be added to the primary standard.

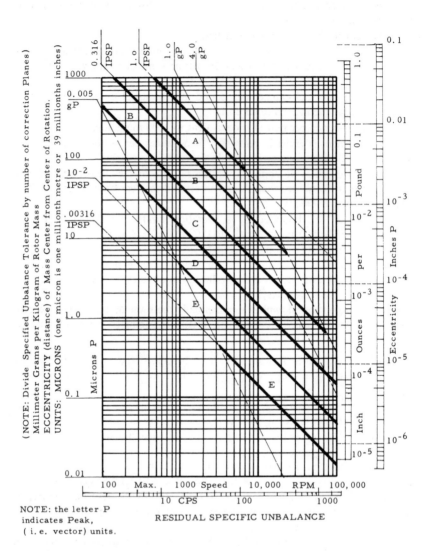

NOTE: the letter P
indicates Peak,
( i. e. vector) units.

RESIDUAL SPECIFIC UNBALANCE

**Fig. 33.7** Residual specific unbalance. Divide specified unbalance
tolerance by number of correction planes.

**Description.** Figure 33.7 gives five classes of residual unbalance. Unbalance is
expressed in terms of millimeter-gram per kilogram of rotor weight as a function of frequency. Frequency means maximum rotor speed when in use. The
mmg/Kg units are the same numerically as the eccentricity expressed in microns.
Figure 33.7 may be summarized as follows.

1. The criterion of unbalance is eccentricity, the distance of center of gravity from the center of rotation.

2. The velocity in each class is constant.

3. The classes are arranged in steps of 3.16 to 1 in terms of eccentricity. Thus, two classes show a step of 10 to 1 or 20 dB; and one class shows a step of 10 dB.

4. mmg/Kg and inch-ounces per pound are the usual units. The former is likely to become the international unit.

5. One inch-ounce per pound is equivalent to 1600 mmg/Kg.

6. Displacement is given in mmg/Kg, inch-ounces per pound, microns peak and inches peak.

7. The figure makes no comment on the severity of the classes beyond indicating that B is less severe than A and so on and that constant severity is indicated by constant velocity.

8. The classes are bounded by acceleration limits of 0.005-gP and 4-gP. Velocities range from 1 in/sec (peak) to 0.00316 in/sec (peak).

9. Shaft eccentricity may be expressed as:
—millimeter gram per kilogram of rotor weight. This is numerically the same as microns, where one micron is one millionth of a meter or one-hundredth of a millimeter, or 39 microinches;
—inch-ounces per pound, which is numerically the same as 16 times the eccentricity. To obtain eccentricity in inches, divide inch-ounces per pound by 16.

**Discussion.** Standards in this section are to be considered as applying to rigid rotors balanced in the shop. Some further important points arise and are summarized as follows.

1. For symmetrical rotors divide the given overall tolerance (inch-ounce/pound residual unbalance) by two or by the number of correction planes. For very asymmetrical rotors the tolerance is often divided in relation to distribution of the mass.

2. Bearing housing vibration is about 1/3 to 1/2 that of the shaft so that housing vibration PP is likely to be in the range of 2/3 to 1 times eccentricity.

3. If acceleration (Fig. 33.7) exceeds 1-gP, stiff shafts may be expected to jump in their bearings. This becomes less noticeable at very high speeds and with very small eccentricities.

4. Balancing speed is usually far lower than operating speed, and there is no particular relation between the two. Measured eccentricity in a balancing machine is more or less independent of the speed of rotation so long as this is more than about four times the natural frequency of the suspended system.

5. Figure 33.7 has been specially prepared by the writer to give a convenient summary of data, reflecting as well as possible current standards concepts.

6. Balance specifications on drawings should show: planes of correction, methods of correction such as drilling, locations of journals or bearing to which balance is referred, and if not obvious, the radius at which correction is to be made.

Most current specifications are based on the simple constant velocity-type equation

$$\frac{U}{W} = \frac{K}{N}$$

where U is the residual moment (e.g. inch-ounces), W is the rotor weight, K is a constant and N is rpm.

This may be rewritten as

$$e = \frac{K}{N}$$

where e denotes eccentricity. It is evident that K/N always denotes a constant velocity line in Fig. 33.7. For example, the upper limit of class A is about 1000 microns at 450 rpm so that K is 450,000. Using inch-ounce units for the same line, 0.04 equals K/450 so that K is 18. A common value of K is 4, in the in-oz/lb system.

The user of standard specifications either receives them in a drawing or devises them from his own experience and that of accepted specifications. Table 44 gives a very rough indication of the practical meanings of the classes shown in Fig. 33.7.

TABLE 44
AN INTERPRETATION OF VIBRATION
SEVERITY CLASSES

| CLASS | GRADE | TYPE DEVICE |
|-------|-------|-------------|
| E | Ultra precision | Some gyroscopes |
| D | Precision | Special small motors |
| C | Commercial | Many motors and turbines |
| B | Commercial | Farm machinery |
| A | Commercial | Automotive wheels and tires |

For a better understanding the reader is referred to the bibliography.

## 9. VIBRATION OF MACHINE TOOLS

The situation may be envisaged as follows. Vibration is generated by the machine itself and by the machining process. These, in turn, affect both the machine and the quality of the work. It is important in a given application to understand and decide what are the important operative factors. For example, in a centerless grinder the balance of the stone is the most important as regards vibration, and the critical effect is that of work-piece quality. The vibration generated by grinding is secondary as a rule, and so also is the effect of any of the several other types of vibrations on the wear or degradation of the machine. The result that is the controlling factor in a lathe is normally the work-piece quality. The source of the vibration may be an unbalance of the work-piece and chuck combination or the vibration generated by the cutting process.

The great variation in local conditions and local objectives makes it quite impossible to offer any generalized standards that are of much value. Some surface grinders, for example, do an excellent job if the stone unbalance, measured horizontally on the bearing housing at the stone, is kept below about 0.75 mil PP. In such cases there is no need to balance to less than 0.25 mil PP. Other surface grinders doing a different job and having a different construction may have to be balanced when vibration exceeds 0.1 mil PP. Some lathes with an unbalance up to 2 mil PP at 1000 rpm measured at the chuck bearing when the work-piece is chucked do an acceptable job, whereas the writer has found it necessary to balance one turret lathe with about a 12-in swing and with a very unbalanced work-piece geometry to less than 0.1 mil PP for combined chuck and work-piece.

Often there is a limit to what can be achieved by balancing, either because of the variability of successive work pieces of the same nominal geometry or because of some inherent difficulty. For example, magnetic clutches used for rapid acceleration and braking of some lathes, permit balancing only on one element of the clutch or sheave. When the clutch is disengaged and reengaged after balancing, the elements take up a different phase configuration and much of the excellence of balance is lost and cannot be restored, except for a one-phase configuration, which is of little help.

As always, the best approach to setting standards for machine tool vibration is to measure, if possible, the unfiltered and filtered amplitudes on the given machine when operating satisfactorily. Confidence is greatly enhanced if several similar machines are available. Then, a higher vibration on a single machine of the family clearly indicates that trouble is present. When general levels for adequate performance have been outlined, they may be related to standards graphs, such as Fig. 33.1, by using service factors experimentally discovered. Then the graph enables the observer to make intelligent predictions of classification levels for different speeds and varying conditions of severity.

A few values are listed in Table 45 as a mere hint; they are not intended for unqualified use.

TABLE 45
SUGGESTED VALUES OF UNBALANCE DISPLACEMENT
FOR SOME MACHINE TOOLS

| ITEM | RANGE FOR NORMAL OPERATION MILS, PP |
|---|---|
| Thread and contour grinder | 0.01–0.1 |
| Centerless and cylindrical grinder | 0.05–1.0 |
| Lathe | 0.2 –1.5 |

| 40-SERIES | 20-SERIES | 10-SERIES | 5-SERIES | EXACT VALUE | MANTISSA |
|---|---|---|---|---|---|
| 1 | 1 | 1 | 1 | 10000 | 000 |
| 1.06 | | | | 10593 | 025 |
| 1.12 | 1.12 | | | 11220 | 050 |
| 1.18 | | | | 11885 | 075 |
| 1.25 | 1.25 | 1.25 | | 12589 | 100 |
| 1.32 | | | | 13335 | 125 |
| 1.4 | 1.4 | | | 14125 | 150 |
| 1.5 | | | | 14962 | 175 |
| 1.6 | 1.6 | 1.6 | 1.6 | 15849 | 200 |
| 1.7 | | | | 16788 | 225 |
| 1.8 | 1.8 | | | 17783 | 250 |
| 1.9 | | | | 18836 | 275 |
| 2 | 2 | 2 | | 19953 | 300 |
| 2.12 | | | | 21135 | 325 |
| 2.24 | 2.24 | | | 22387 | 350 |
| 2.36 | | | | 23714 | 375 |
| 2.5 | 2.5 | 2.5 | 2.5 | 25119 | 400 |
| 2.65 | | | | 26607 | 425 |
| 2.8 | 2.8 | | | 28184 | 450 |
| 3 | | | | 29854 | 475 |
| 3.15 | 3.15 | 3.15 | | 31623 | 500 |
| 3.35 | | | | 33497 | 525 |
| 3.55 | 3.55 | | | 35481 | 550 |
| 3.75 | | | | 37584 | 575 |
| 4 | 4 | 4 | 4 | 39811 | 600 |
| 4.25 | | | | 42170 | 625 |
| 4.5 | 4.5 | | | 44668 | 650 |
| 4.75 | | | | 47315 | 675 |
| 5 | 5 | 5 | | 50119 | 700 |
| 5.3 | | | | 53088 | 725 |
| 5.6 | 5.6 | | | 56234 | 750 |
| 6 | | | | 59566 | 775 |
| 6.3 | 6.3 | 6.3 | 6.3 | 63096 | 800 |
| 6.7 | | | | 66834 | 825 |
| 7.1 | 7.1 | | | 70795 | 850 |
| 7.5 | | | | 74989 | 875 |
| 8 | 8 | 8 | | 79433 | 900 |
| 8.5 | | | | 84140 | 925 |
| 9 | 9 | | | 89125 | 950 |
| 9.5 | | | | 94406 | 975 |

Fig. 33.8   Internationally preferred numbers

| MULTIPLE | PREFIX | SYMBOL |
|----------|--------|--------|
| $10^{12}$ | tera | T |
| $10^9$ | giga | G |
| $10^6$ | mega | M |
| $10^3$ | kilo | k |
| 100 | hecto | h |
| 10 | deka | da |
| $10^{-1}$ | deci | d |
| $10^{-2}$ | centi | c |
| $10^{-3}$ | milli | m |
| $10^{-6}$ | micro | $\mu$ |
| $10^{-9}$ | nano | n |
| $10^{-12}$ | pico | p |
| $10^{-15}$ | femto | f |
| $10^{-18}$ | atto | a |

Fig. 33.9   Standard prefixes

| WAVEFORM | TRUE VALUE | OSCILLOSCOPE WILL READ 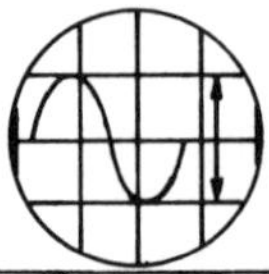 | HALF – WAVE METER WILL READ 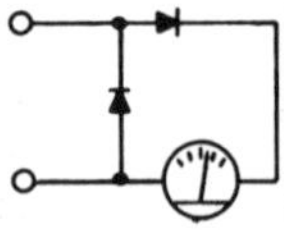 | FULL – WAVE METER WILL READ 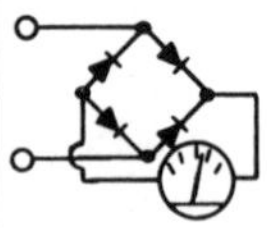 | RMS METER WILL READ  |
|---|---|---|---|---|---|
| SINE 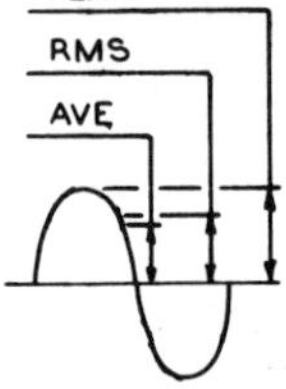 PEAK RMS AVE | I VOLT PEAK TO PEAK | 1 | .1593 | .3185 | .3535 |
| | I VOLT PEAK | 2 | .3185 | .637 | .707 |
| | I VOLT RMS | 2.828 | .45 | .9 | 1 |
| | I VOLT AVERAGE | 3.142 | .5 | 1 | 1.11 |
| TRIANGULAR 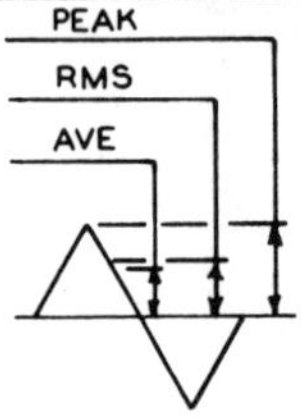 PEAK RMS AVE | I VOLT PEAK TO PEAK | 1 | .125 | .25 | .2885 |
| | I VOLT PEAK | 2 | .25 | .5 | .577 |
| | I VOLT RMS | 3.464 | .4085 | .817 | 1 |
| | I VOLT AVERAGE | 4 | .5 | 1 | 1.224 |
| SQUARE 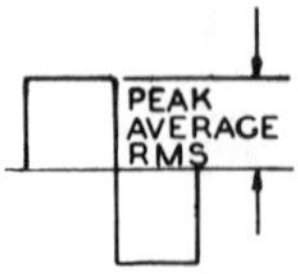 PEAK AVERAGE RMS | I VOLT PEAK TO PEAK | 1 | .25 | .5 | .5 |
| | I VOLT PEAK | 2 | .5 | 1 | 1 |
| | I VOLT RMS | 2 | .5 | 1 | 1 |
| | I VOLT AVERAGE | 2 | .5 | 1 | 1 |

SP 10557A

Fig. 33.10   The read-out of standard waveforms. After Crowhurst, N. H., *Basic Audio*, Rider Publishing Co.

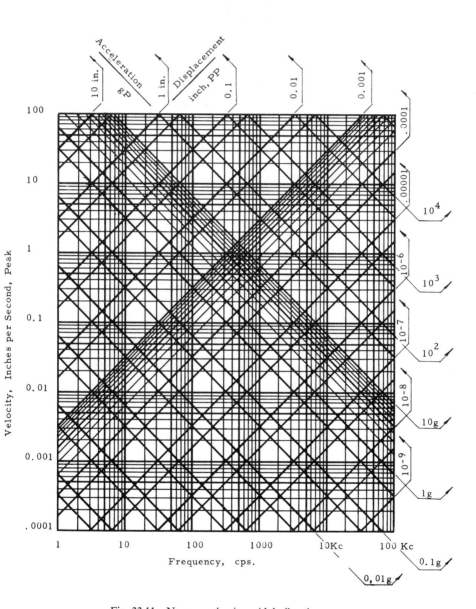

Fig. 33.11 Nomograph, sinusoidal vibration

# REFERENCES

1. Blake, Michael P. "New Vibration Standards for Maintenance." *Hydrocarbon Processing and Petroleum Refining Magazine*, January 1964.
2. Rathbone, T. C. "Vibration Tolerance." *Power Plant Engineering* 43 (1939): 271.
3. Nittinger, R. H. "Vibration Analysis Can Keep Your Plant Humming." *Chem. Engineering*, April 17, 1964, pp. 152–158.
4. Maten, S. "New Vibration Velocity Standards." *Hydrocarbon Processing* 46, no. 1 (1967): 137–141.
5. Blake, M. P. "The Vibration of Cooling Tower Fans." Presented to Cooling Tower Institute Conf., New Orleans, January 1964.
6. Blake, M. P. "New Vibration Standards for Electric Motors." *Hydrocarbon Processing and Petroleum Refining Magazine*, March 1966.
7. Goldman, D. E. Figure 44.20. In *Shock and Vibration Handbook*, vol. 3, edited by C. Harris and C. Crede. New York: McGraw-Hill Book Co., 1961.
8. Peterson, A. and Gross, E., Jr. *Handbook of Noise Measurement.* Concord, Mass: General Radio Co., 1967.
9. Harris, C., ed. *Handbook of Noise Control.* New York: McGraw-Hill Book Co., 1957.
10. Beranek, L. L., ed. *Noise Reduction.* New York: McGraw-Hill Book Co., 1960.
11. Department of Labor, Safety and Health Standards, Federal Register, vol. 34, no. 96 (May 20, 1969): 7948–7949.

# BIBLIOGRAPHY

*Acoustical Terminology, Including Vibration,* Standard S1.1. United States of America Standards Institute, New York, 1960.

*Balancing Standards,* untitled and undated. In a six-page publication of Micro Balancing, Inc., based on Standards of The Society of German Engineers' Communication from Schenck Trebel Corp., 1969.

"Rolling Bearing Vibration and Noise." In Section 13, AFBMA Standards. New York: The Anti-Friction Bearing Manufacturers Association, January 1968.

*Symposium on Dynamic Balancing.* Presented at the University of Birmingham. Birmingham, England: W. & T. Avery, March 1964.

Beranek, L. L., ed. *Noise Reduction.* New York: McGraw-Hill Book Co., 1960.

Buscarello, R. T. *Practical Solutions of Machinery and Maintenance Vibration Problems.* Denver: Update Industrial Seminars, 1968.

Filepp, L. *Proposed Dynamic Balancing Tolerances for High-Speed Gear Couplings.* Presented to meeting of American Gear Manufacturers Association, Waldron Hartig Division, 519.01., Chicago, 1967.

Harris, C. M., and Crede, C. E. *Shock and Vibration Handbook,* 1st ed. New York: McGraw-Hill Book Co., 1961.

Langlois, A. B. and Rosecky, E. J. *Field Balancing.* Allis Chalmers Manufacturing Co., 1968.

Muster, D. and Flores, B. "Balancing Criteria and Their Relation to American Practice." *Journal of Engineering for Industry,* Trans. ASME (November 1969): 1035–1046.

Richart, F. E.; Hall, J. R.; and Woods, R. D. *Vibrations of Soils and Foundations,* 1st ed. Englewood Cliffs, New Jersey: Prentice-Hall, Inc., 1970.

Tobias, S. A. *Machine-Tool Vibration.* New York: John Wiley & Sons, Inc. Authorized translation of *Schwingungen an Werkzeugmaschinen,* 1961.

*Japanese Industrial Standard for the Balance Qualtiy of Rotating Machinery,* ISO/TC 108/WG 6, British Standards Institution, 2 Park Street, London W1, England.

*Terminology for Balancing Rotating Machinery,* American National Standards Institute, Inc, 1430 Broadway, New York, N. Y.

# INDEX

# ERRATA

Page 320, Fig. 12.6: Add to caption
How the amplitude-frequency spectra reflect the vibration severity of two motors, one designated good, and the other bad. In this example the vibration signal is proportional to acceleration. (Adapted from *Instructions and Applications for Accelerometers* 4308/09/10/11, October 1960, Bruel and Kjaer, Denmark.)

Page 436, line 9 up: *For* amplitude fundamental *read* amplitude of the fundamental

Page 493, Fig. 24.5b: Add to caption
(Impact-o-Graph Corporation, Cleveland)

Page 637, line 9 up: *For* Axis, principle *read* Axis, principal

The publisher regrets that the contributors' affiliations were not shown in the text. They are:
Steve Maten, Shawinigan Chemicals Company, Quebec, Canada
Lionel J. Lortie, Vibration Specialist, Montreal, Canada
D. G. Stadelbauer, Schenk-Trebel Corporation, New York
M. G. Sharma, Pennsylvania State University, Pennsylvania